Multi-Robot Systems: Challenges, Trends and Applications

Multi-Robot Systems: Challenges, Trends and Applications

Editors

Juan Jesús Roldán Gómez
Antonio Barrientos Cruz

MDPI • Basel • Beijing • Wuhan • Barcelona • Belgrade • Manchester • Tokyo • Cluj • Tianjin

Editors
Juan Jesús Roldán Gómez
Universidad Autónoma de Madrid
Spain

Antonio Barrientos Cruz
Universidad Politécnica de Madrid
Spain

Editorial Office
MDPI
St. Alban-Anlage 66
4052 Basel, Switzerland

This is a reprint of articles from the Special Issue published online in the open access journal *Applied Sciences* (ISSN 2076-3417) (available at: https://www.mdpi.com/journal/applsci/special_issues/Multi-Robot_Systems).

For citation purposes, cite each article independently as indicated on the article page online and as indicated below:

LastName, A.A.; LastName, B.B.; LastName, C.C. Article Title. *Journal Name* **Year**, *Volume Number*, Page Range.

ISBN 978-3-0365-2846-5 (Hbk)
ISBN 978-3-0365-2847-2 (PDF)

Cover image courtesy of Juan Jesús Roldán Gómez

Contents

About the Editors

Juan Jesús Roldán Gómez is an Assistant Professor at the Department of Computer Science and Engineering of the Autonomous University of Madrid. He received a Ph.D. in Automation and Robotics (2018), an MSc in Automation and Robotics (2014), and an MSc in Industrial Engineering (2012) from the Technical University of Madrid. His research focuses on robotics, including topics such as multi-robot systems, robot swarms, artificial intelligence applied to robots, adaptive and immersive interfaces, and robotics in agriculture.

Antonio Barrientos Cruz is a Full Professor at the Technical University of Madrid. He received a Ph.D. in Robotics (1986) and an MSc in Industrial Engineering (1982) from the Technical University of Madrid, and an MSc in Biomedical Engineering (2002) from the National Distance Education University. He is the head of the Robotics and Cybernetics research group at the Centre for Automation and Robotics a (joint center of the Technical University of Madrid and the Spanish National Research Council). He has worked in robotics for more than 30 years, developing industrial and service robots for different areas. His main research interests are ground and aerial field robotics.

Editorial

Special Issue on Multi-Robot Systems: Challenges, Trends, and Applications

Juan Jesús Roldán-Gómez [1,*] and Antonio Barrientos [2]

1 Departamento de Ingeniería Informática, Escuela Politécnica Superior, Universidad Autónoma de Madrid, Francisco Tomás y Valiente, 11, 28049 Madrid, Spain
2 Centro de Automática y Robótica (UPM-CSIC), Universidad Politécnica de Madrid, José Gutiérrez Abascal, 2, 28006 Madrid, Spain; antonio.barrientos@upm.es
* Correspondence: juan.roldan@uam.es

Citation: Roldán-Gómez, J.J.; Barrientos, A. Special Issue on Multi-Robot Systems: Challenges, Trends, and Applications. *Appl. Sci.* **2021**, *11*, 11861. https://doi.org/10.3390/app112411861

Received: 20 October 2021
Accepted: 30 November 2021
Published: 14 December 2021

1. Introduction

Multi-Robot Systems (MRSs) have emerged as a suitable alternative to single robots to improve current and enable new missions. These systems offer the following advantages over single robots:

- Effectiveness: In most scenarios, using a fleet of homogeneous robots improves the performance obtained by a single one. For instance, multiple robots can cover a larger area or spend less time than one robot in an exploration task.
- Efficiency: In a wide range of missions, using a heterogeneous fleet leads to more efficient management of resources than using a single robot. That is the case of fleets with aerial and ground robots applied to search tasks, where the aerial units can cover more terrain and make faster detections, but the ground ones can provide more accurate information of these detections.
- Flexibility: A multi-robot system can adapt to changes in the mission better than a single robot. These changes can be related to the scenario (e.g., scalability of mission area in search), tasks (e.g., fire control in environmental monitoring), or fleet (e.g., robots with difficulties during missions). The availability of resources allows replanning the missions to address these situations.
- Fault tolerance: This is a particular but relevant case where a multi-robot system is more flexible than a single robot. When a robot experiences problems, such as getting trapped in an obstacle or consuming all its battery, the rest can assume its functions.

However, multi-robot systems still face challenges related to robot autonomy and human factors. The deployment, operation, and collection of these systems in real-world scenarios need autonomy in the broad sense: robots with more capabilities and intelligence to operate longer in adverse conditions. In addition, the complexity of these systems poses some challenges to operators in terms of workload, situational awareness, and stress.

The recent literature on MRSs considers these challenges and proposes new strategies to face them. That is the case of artificial intelligence, which has given rise to new algorithms that allow managing the complexity and uncertainty of real scenarios, and immersive technologies (virtual and augmented reality), which are applied to facilitate the work of operators. These technologies are opening up a wide variety of missions, such as search and rescue, environmental monitoring, and many more.

In this "Special Issue on Multi-Robot Systems: Challenges, Trends, and Applications", we have collected a set of high-quality works that discuss the main challenges of MRSs, present the trends to address these issues, and report various relevant applications.

The remainder of this editorial is organized as follows: Section 2 discusses the challenges of MRSs, Section 3 addresses the proposals to solve them, and Section 4 describes the real-world applications presented in the different articles of the Special Issue.

2. Challenges

The contributions to the Special Issue reveal a good amount of challenges for the operation of MRSs. Some are related to multi-agent mission planning, intervention in complex scenarios with uncertainty, and operation under restricted communications. Robot cooperation often helps deal with these issues through sharing information and coordinating actions. Kurt Geihs [1] reviews the state-of-the-art on teamwork in the context of multi-robot systems and dynamic environments. His paper identifies and analyzes multiple engineering challenges: dynamic coalitions, platform harmonization and configuration, knowledge base, methodology and tools, edge and cloud integration, and human in the loop and other sociotechnical concerns, among other concepts.

Rongye Shi, Peter Steenkiste, and Manuela M. Veloso [2] address the challenge of Multi-Agent Path Planning (MAPP), which is a resource allocation problem complicated by the highly dynamic and distributed environments. In contrast to most MAPP approaches, they assume soft collisions instead of hard collisions, which means that the agents can share resources or concur at the same location at the expense of reducing the quality of solutions. They propose the Soft-Collision M* (SC-M*) algorithm to solve these constrain satisfaction problems and compare its results with the most common algorithms in terms of path cost, success rate, and run time. Meanwhile, Yang Lyu, Quan Pan, and Jian Lv [3] address the problem of multi-robot collaborative self-localization in the context of target tracking missions. They propose an unscented transformation-based collaborative self-localization algorithm to deal with inter-robot and robot–target correlations during the missions.

Martin Juhás and Bohuslava Juhásová [4] address other relevant challenges for MRSs, which are communications and coordination. Their paper presents a time-synchronization solution for operations performed by a heterogeneous set of robotic manipulators grouped into a production cell. They develop a master–slave architecture without an external control element, whose communications are implemented via TCP/IP sockets. Similarly, Facundo Benavides, Caroline Ponzoni Carvalho Chanel, Pablo Monzón, and Eduardo Grampín [5] consider the multi-robot exploration problem under restricted communications. They propose a novel auto-adaptive multi-objective strategy to support the selection of tasks regarding both exploration performance and connectivity level. Compared with other algorithms, it shows effectiveness and flexibility to tackle the multi-robot exploration problem, decreasing the effects of disconnection periods without noticeable degradation of the exploration time.

Cooperation can be carried out in homogeneous fleets but also between robots with different morphology. Pablo R. Palafox, Mario Garzón, João Valente, Juan Jesús Roldán, and Antonio Barrientos [6] analyze air–ground robot cooperation in their paper. The combination of aerial and ground robots is useful in search and rescue tasks, given that aerial robots can provide valuable insight to support the navigation of ground robots in complex scenarios affected by disasters. The article proposes a state machine with algorithms that allow an aerial robot to take off, track, and land on a mobile ground platform.

Abhijeet Ravanka, Ankit A. Ravankar, Yohei Hoshino, and Yukinori Kobayashi [7] focus on another relevant challenge in multi-robot missions: uncertainty. They find information sharing as a powerful tool to deal with uncertainty in mission planning. In this way, when a robot finds a new obstacle or blocked path, it can share this information with the rest of the fleet, allowing other robots to plan better paths. The paper proposes a novel method for information sharing that works when robots have different sensors, there is positional uncertainty, and obstacles are dynamic. Aliakbar Akbari, Mohammed Diab, and Jan Rosell [8] also focus on uncertainty but in the context of mobile manipulation. In these applications, humans can collaborate with robots to execute complex actions, sharing their knowledge about the task and scenario. They propose a contingent-based task and motion planning method that generates trees of feasible plans considering robot uncertainty and human–robot interactions. This algorithm is validated in grasp tasks with occluding objects.

3. Trends

Currently, one of the most relevant multi-robot systems is swarms. David Garzón Ramos and Mauro Birattari [9] present an automatic method to design robot swarms. This method can generate control software by assembling preexisting software modules via optimization. They validate these developments with a swarm of e-pucks, which can use color-based information for handling events, communicating, and navigating. James Wilson, Jon Timmis, and Andy Tyrrell [10] address another relevant aspect of swarms: hormone systems for collective behaviors. They use a collection of virtual hormones to control the selection of behaviors that produce an effective foraging swarm.

Immersive technologies such as virtual, augmented, and mixed realities are usually proposed to improve operator workload, situational awareness, and performance. Ashish Kumar, Sugjoon Yoon, and V.R. Sanal Kumar [11] present a mixed reality simulation for Unmanned Aerial Vehicles in high-endurance missions of Earth exploration. This environment combines real and virtual quadcopters to monitor missions and find paths, among other things. Luis Pérez, Silvia Rodríguez-Jiménez, Nuria Rodríguez, Rubén Usamentiaga, and Daniel F. García [12] propose the creation of digital twins of manufacturing processes in Industry 4.0. In this way, a virtual reality testbed can be used to design, implement, and monitor the process in real-time before its physical development.

Finally, machine learning and especially reinforcement learning are increasingly popular for MRSs, especially for their operation in complex unstructured scenarios. Wenzhou Chen, Shizheng Zhou, Zaisheng Pan, Huixian Zheng, and Yong Liu [13] apply deep reinforcement learning for the collaborative formation and navigation of a robot fleet.

4. Applications

Robots are often applied in emergency scenarios because they can obtain information and even intervene, preventing dangers for human teams. Multi-robot systems are being introduced in these missions, such as outdoor and indoor fires. Juan Jesús Roldán-Gómez, Eduardo González-Gironda, and Antonio Barrientos [14] propose the use of drone swarms for the prevention, surveillance, and extinguishing of forest fires. This system consists of quadcopters that individually can visit waypoints and use payloads but, collectively, can perform complex tasks. The authors propose the use of immersive interfaces to allow operators to control multiple drones simultaneously.

However, emergency scenarios are not the only ones in which these systems can add value. Ángel Montes-Romero, Arturo Torres-González, Jesús Capitán, Maurizio Montagnuolo, Sabino Metta, Fulvio Negro, Alberto Messina, and Aníbal Ollero [15] propose a set of director tools for autonomous media production with a team of drones. They focus on a language for cinematography mission description and a procedure to translate missions into plans, so a media director that is not necessarily familiar with robots can manage the system.

Finally, multi-robot systems have also reached social robotics, although the need was not as clear as in other fields. Toshiaki Nishio, Yuichiro Yoshikawa, Kohei Ogawa, and Hiroshi Ishiguro [16] study multi-party conversations with two human-like robots. They focus on conveying information to the viewers through a natural conversation between the robots. Takamasa Iio, Yuichiro Yoshikawa, Mariko Chiba, Taichi Asami, Yoshinori Isoda, and Hiroshi Ishiguro [17] try a question-answer-response dialogue model with two humanoid robots to involve elderly users in the conversation. The results suggest that the presence of two robots might likely encourage elderly people to sustain longer talks.

Author Contributions: Conceptualization, J.J.R.-G. and A.B.; methodology, J.J.R.-G. and A.B.; writing—original draft preparation, J.J.R.-G.; writing—review and editing, J.J.R.-G.; visualization, J.J.R.-G.; supervision, A.B.; project administration, J.J.R.-G. and A.B.; funding acquisition, J.J.R.-G. and A.B. All authors have read and agreed to the published version of the manuscript.

Funding: Work produced with the support of a 2020 Leonardo Grant for Researchers and Cultural Creators, BBVA Foundation. The Foundation takes no responsibility for the opinions, statements, and contents of this project, which are entirely the responsibility of its authors.

Acknowledgments: This Special Issue has been possible thanks to the hard work of authors, reviewers, and editors. We would like to thank and congratulate all the authors for their valuable contributions to our Special Issue. In addition, we would like to express our gratitude to Daria Shi, Managing Editor of Applied Sciences (MDPI), for her outstanding work during these years. Juan Jesús Roldán-Gómez wants to dedicate this book to the memory of their grandmother Gloria Montoya Guerrero, who would have been very proud to hear about it on one of their Sunday evening calls.

Conflicts of Interest: The authors declare no conflict of interest.

References

1. Geihs, K. Engineering Challenges Ahead for Robot Teamwork in Dynamic Environments. *Appl. Sci.* **2020**, *10*, 1368. [CrossRef]
2. Shi, R.; Steenkiste, P.; Veloso, M. SC-M*: A Multi-Agent Path Planning Algorithm with Soft-Collision Constraint on Allocation of Common Resources. *Appl. Sci.* **2019**, *9*, 4037. [CrossRef]
3. Lyu, Y.; Pan, Q.; Lv, J. Unscented Transformation-Based Multi-Robot Collaborative Self-Localization and Distributed Target Tracking. *Appl. Sci.* **2019**, *9*, 903. [CrossRef]
4. Juhás, M.; Juhásová, B. Synchronization of Heterogeneous Multi-Robotic Cell with Emphasis on Low Computing Power. *Appl. Sci.* **2020**, *10*, 5165. [CrossRef]
5. Benavides, F.; Ponzoni Carvalho Chanel, C.; Monzón, P.; Grampín, E. An Auto-Adaptive Multi-Objective Strategy for Multi-Robot Exploration of Constrained-Communication Environments. *Appl. Sci.* **2019**, *9*, 573. [CrossRef]
6. Palafox, P.; Garzón, M.; Valente, J.; Roldán, J.; Barrientos, A. Robust Visual-Aided Autonomous Takeoff, Tracking, and Landing of a Small UAV on a Moving Landing Platform for Life-Long Operation. *Appl. Sci.* **2019**, *9*, 2661. [CrossRef]
7. Ravankar, A.; Ravankar, A.; Hoshino, Y.; Kobayashi, Y. On Sharing Spatial Data with Uncertainty Integration Amongst Multiple Robots Having Different Maps. *Appl. Sci.* **2019**, *9*, 2753. [CrossRef]
8. Akbari, A.; Diab, M.; Rosell, J. Contingent Task and Motion Planning under Uncertainty for Human–Robot Interactions. *Appl. Sci.* **2020**, *10*, 1665. [CrossRef]
9. Garzón Ramos, D.; Birattari, M. Automatic Design of Collective Behaviors for Robots that Can Display and Perceive Colors. *Appl. Sci.* **2020**, *10*, 4654. [CrossRef]
10. Wilson, J.; Timmis, J.; Tyrrell, A. An Amalgamation of Hormone Inspired Arbitration Systems for Application in Robot Swarms. *Appl. Sci.* **2019**, *9*, 3524. [CrossRef]
11. Kumar, A.; Yoon, S.; Kumar, V. Mixed Reality Simulation of High-Endurance Unmanned Aerial Vehicle with Dual-Head Electromagnetic Propulsion Devices for Earth and Other Planetary Explorations. *Appl. Sci.* **2020**, *10*, 3736. [CrossRef]
12. Pérez, L.; Rodríguez-Jiménez, S.; Rodríguez, N.; Usamentiaga, R.; García, D. Digital Twin and Virtual Reality Based Methodology for Multi-Robot Manufacturing Cell Commissioning. *Appl. Sci.* **2020**, *10*, 3633. [CrossRef]
13. Chen, W.; Zhou, S.; Pan, Z.; Zheng, H.; Liu, Y. Mapless Collaborative Navigation for a Multi-Robot System Based on the Deep Reinforcement Learning. *Appl. Sci.* **2019**, *9*, 4198. [CrossRef]
14. Roldán-Gómez, J.; González-Gironda, E.; Barrientos, A. A Survey on Robotic Technologies for Forest Firefighting: Applying Drone Swarms to Improve Firefighters' Efficiency and Safety. *Appl. Sci.* **2021**, *11*, 363. [CrossRef]
15. Montes-Romero, Á.; Torres-González, A.; Capitán, J.; Montagnuolo, M.; Metta, S.; Negro, F.; Messina, A.; Ollero, A. Director Tools for Autonomous Media Production with a Team of Drones. *Appl. Sci.* **2020**, *10*, 1494. [CrossRef]
16. Nishio, T.; Yoshikawa, Y.; Ogawa, K.; Ishiguro, H. Development of an Effective Information Media Using Two Android Robots. *Appl. Sci.* **2019**, *9*, 3442. [CrossRef]
17. Iio, T.; Yoshikawa, Y.; Chiba, M.; Asami, T.; Isoda, Y.; Ishiguro, H. Twin-Robot Dialogue System with Robustness against Speech Recognition Failure in Human-Robot Dialogue with Elderly People. *Appl. Sci.* **2020**, *10*, 1522. [CrossRef]

Review

Engineering Challenges Ahead for Robot Teamwork in Dynamic Environments

Kurt Geihs [1,2]

1 University of Kassel, Electrical Engineering & Computer Science, 34121 Kassel, Germany; geihs@uni-kassel.de
2 Universidad Carlos III de Madrid, Ingenieria Telematica, 28911 Leganes (Madrid), Spain

Received: 29 November 2019; Accepted: 12 February 2020; Published: 18 February 2020

Abstract: The increasing number of robots around us creates a demand for connecting these robots in order to achieve goal-driven teamwork in heterogeneous multi-robot systems. In this paper, we focus on robot teamwork specifically in dynamic environments. While the conceptual modeling of multi-agent teamwork was studied extensively during the last two decades and commercial multi-agent applications were built based on the theoretical foundations, the steadily increasing use of autonomous robots in many application domains gave the topic new significance and shifted the focus more toward engineering concerns for multi-robot systems. From a distributed systems perspective, we discuss general engineering challenges that apply to robot teamwork in dynamic application domains and review state-of-the-art solution approaches for these challenges. This leads us to open research questions that need to be tackled in future work.

Keywords: autonomous robots; multi-robot systems; teamwork; coordination; dynamic environments

---

1. Introduction and Motivation

Autonomous robots pervade our daily lives. Single autonomous robots for particular tasks are already accepted and used in private, business, and public environments, for example, in application domains such as warehouse and transportation logistics, search and rescue, smart factories, space exploration, healthcare, smart public transportation, precision farming, and domestic services. Clearly, autonomous robots will play a crucial role in more and more application domains. It is an obvious thought that these robots around us will have to talk to each other and work collaboratively as a team—a trend that one can compare with the evolution of distributed computing by connecting stand-alone computers. Teams can be more than the sum of their parts. A multi-robot system is able to perform tasks that exceed the capabilities of a single robot, not only due to workload sharing but also in terms of functionality. Just like a team of human beings can achieve more than a single individual, the teamwork of autonomous robots provides opportunities for robots to accomplish tasks that a single robot cannot do alone.

Conceptual teamwork models for multi-agent systems were a subject of intense research approximately 20 years ago. Among the seminal papers at this time were [1–5], to name just a few out of many (see the survey in [6] for more information on early research). However, the commercial adoption of agent-based solutions was low during the subsequent decade. According to [7], among the main hindrances for agent–based applications were limited awareness about the potentials, limited publicity of successful industrial projects, misunderstandings and over-expectations, and lack of mature enough design and development tools. While the subject since then never vanished from the research agenda, one may notice a recent increased interest in robot teamwork. This is due to substantial progress in robotic hardware and software. In the realm of software, which is our focus in this paper, advances in artificial intelligence techniques make autonomous robots fit for applications in private, industrial, and public environments. Thus, we are observing a shift of research focus from theoretical multi-agent models to practical software and hardware engineering issues. Moreover, the general acceptance of robotic helpers is increasing steadily in society, as demonstrated, for example, in the private sector by the increasing number of autonomous robots for lawn mowing, vacuum cleaning, window cleaning, and even for caretaking and medical applications.

Our emphasis in this paper is on the engineering challenges that arise when autonomous robots collaborate as a team to solve a joint task. In particular, we view these systems from a distributed systems perspective. In general, teamwork in multi-robot systems exhibits potential benefits and complexities as any distributed computing system. However, robot teamwork introduces a number of additional new challenges that we discuss in this paper. For each challenge, we present state-of-the-art solution approaches. This leads us to research questions that future research needs to tackle.

Our own experience with multi-robot systems originated from our participation in the Middle-Size League of international RoboCup tournaments where teams of custom-build soccer robots compete against each other. Our research focus was mainly on a software framework for the development of teamwork applications in adverse and dynamic environments, as they prevail in Robocup tournaments. Later, we evolved the framework and showed successfully that it was also fit for other robotic application domains, such as collaborative exploration, autonomous driving, and service robotics.

In particular, in this paper we are interested in robotic teamwork in dynamic and adverse application environments where adaptation and reconfiguration may be necessary due to a continuously changing runtime context. This is a far-reaching assumption that includes fewer demanding requirements. Our focus is on the software for robot teamwork. Neither the variety of theoretical models for multi-agent systems nor the mechatronic and hardware design issues are subjects of this paper. For these issues, we refer the reader to surveys such as [8,9].

In Section 2, we start with a brief clarification of terminology for collaborative robots. This is necessary because there is no general agreement on the terminology in the wider robotic community. In Section 3, we discuss engineering challenges for robot teamwork, and we point to existing solution approaches in order to explicate dimensions of the design space. Section 4 presents a summary of open research questions. Section 5 concludes the paper.

2. Terminology

Let us first briefly define the basic terminology used in this paper. A *robot* is a programmable machine capable of carrying out a complex series of actions automatically (The Oxford English Dictionary, Oxford University Press). An autonomous robot is capable of perceiving its environment through sensors, reasoning about the gained information, making decisions accordingly, and acting upon its environment through actuators, all without human intervention. These capabilities are also commonly associated with the term *agent*, whereby agent is considered a more general term, i.e., a robot is a special kind of agent that (mostly) is realized as a mechatronic construct. A robot may adopt a certain *role* based on its capabilities. It executes *tasks* that are described in a *task plan*. For example, in an autonomous driving traffic scenario, an emergency vehicle has a role that is different from regular vehicles. It has specific capabilities and rights and executes different tasks than regular traffic participants.

According to Farinelli et al. [6], a *multi-robot system* is a group of robots operating in the same environment. The authors point out that there are many different kinds of multi-robot systems. Their taxonomy is based on the two general dimensions *coordination* and *system*. The coordination dimension is subdivided into *cooperation* (do the robots cooperate to solve a problem?), *knowledge* (how much knowledge do the robots have about each other?), *coordination* (how much coordination is enforced?), and *organization* (what kind of decision structure does the multi-robot system employ?). The system dimension consists of *communication* (what kind of communication mechanisms and protocols do the robots use?), *team composition* (are the robots homogeneous or heterogeneous?), *system architecture* (does the collective as a whole deliberately cope with an unanticipated problem or just the directly affected robots), and *team size* (how scalable is the system in terms of number of robots?). For a more detailed discussion, we refer the reader to the original publication [6].

Specifically, our emphasis in this paper is on multi-robot systems consisting of autonomous robots that *collaborate* in order to achieve a common *global goal*, perhaps in addition to their own local goals. We call such a collective of collaborating robots a *multi-robot team (MRT)* or multi-robot coalition. In dynamic and unpredictable environments, roles and tasks are allocated dynamically to the members of an MRT according to their capabilities and current situation [10]. This allocation was formally modeled and analyzed as an optimization problem using various optimization techniques [11].

From a conceptual modeling perspective, an MRT is a multi-agent system that has a physical representation with specific properties determined by the mechatronic nature of the agents. The main focus of our own research is achieving adaptive goal-driven *teamwork* in a group of autonomous robots. Thus, according to the taxonomies in [6,12], we are concerned only with *cooperative* MRTs, consisting of robots that are aware of their teammates. How cooperation and awareness are achieved may differ.

Since teamwork is the main subject of this paper, we should briefly discuss the related terminology for characterizing the type of interaction that the robots employ to achieve teamwork. Here, we have to point out that, even with existing standards such as the Foundation for Intelligent Physical Agents (FIPA, http://www.fipa.org/), there is no common agreement on the definitions of these terms, i.e., different authors use different connotations. We refer to Parker [13] who differentiates between four types of interaction styles as follows:

Collective	Entities are not aware of other entities on the team, yet they do share goals, and their actions are beneficial to their teammates.
Cooperative	Entities are aware of other entities, they share goals, and their actions are beneficial to their teammates.
Collaborative	Robots have individual goals, they are aware of their teammates, and their actions do help advance the goals of others.
Coordinative	Entities are aware of each other, but they do not share a common goal, and their actions are not helpful to other team members.

We refer the reader to [13] for more information and examples. Here, it should suffice to note that our focus with respect to goal-driven teamwork is on the two interaction styles *cooperative* and *collaborative*. We rule out the other two because they lack properties that we consider essential for teamwork in an MRT: *collective* lacks awareness for other teammates, and *coordinative* lacks a common team goal and robots do not act together as a team.

Clearly, as stated by Parker, there is no sharp boundary between the two interaction styles *cooperative* and *collaborative*. For the sake of clarity and simplicity, we hereafter view the two terms as synonyms (as a side remark: The Merriam-Webster dictionary lists both words as synonyms; see https://www.merriam-webster.com) and do not differentiate between the two styles, but combine and denote them as *collaborative* interaction. It is worthwhile to note here that a collaborative interaction style does not imply a particular choice of system architecture, teamwork programming paradigm, communication protocol, decision-making technique, agreement protocol, etc.

3. Teamwork Challenges

In this section, we discuss key engineering challenges that apply to multi-robot teams in dynamic application scenarios from a software developer's point of view. It is not our intention to present a complete review of the broad spectrum of design aspects for multi-robot teamwork. Instead, we focus on those engineering challenges that are related specifically to dynamic environments. For each challenge, we present a brief look at initial approaches as examples for possible solutions. The order of the sections below does not imply any kind of priority.

3.1. Dynamic Coalitions

In open robot teams where the team members are not known a priori at design time, we need support for establishing a temporary team membership. Participants form temporary coalitions in order to solve a problem and achieve a common goal. A traffic intersection in autonomous driving scenarios is an example of a short-lived coalition with continuous team reconfiguration, while Industry 4.0 scenarios would likely imply a longer-lasting coalition. Only members of a coalition should be involved in the teamwork interactions, and agents outside of the coalition should not disturb it. This requires that all agents know about their membership. Moreover, it may require security measures to protect the interactions of a team. Key challenges are as follows:

- How are team members discovered and identified?
- Who manages team membership?
- How do team members learn about the team composition?
- How does the team protect itself against malicious intruders?

Many communication paradigms achieve interaction among distributed components based on the identities of the components. Examples are the classical client/server model, the actor model [14], or named channels in channel-based binary communication [15]. On the other hand, broadcast communication [16] may not require identities depending on the capabilities of the underlying communication system, but loses the ability to address a selection of individual agents. However, in open teams in dynamic environments, the identity of robots may not be known at design time, if robots may join and leave a team at run-time. Thus, the concept of identity is not easy to establish and may even be irrelevant [17].

In such environments, we need different ways to determine team membership and to address team members. Note that a single central team manager that monitors and controls team membership is out of the question here because we need to avoid a single point of failure in environments where robots may move out of reach temporarily or break down completely.

One solution is based on attribute-based interaction. It is a variant of publish/subscribe communication, and it was proposed in [17,18] as a paradigm to address collectives of possibly anonymous agents. In attribute-based interaction, robots of a multi-robot system explicitly expose a set of attributes that are relevant for the application at hand. Interaction between robots is based on groupcast communication, whereby sending and receiving messages is determined by predicates over the specified attributes. A *send* command expresses the intention to deliver a message to all robots satisfying the *send predicate*. Likewise, a *receive* command signals willingness to receive messages from team members according to the specified *receive predicate*.

For example, in an Industry 4.0 scenario, one might ask for "components that need to be delivered within the next 15 min" or, in autonomous driving, one might want to address "vehicles that are capable of autonomous driving and are closer than 50 m to the intersection". Thus, attribute-based interaction is a more fine-grained content-based selection of possible receivers and senders. Potentially, it allows a more efficient filtering of messages by the distribution infrastructure and reduces the communication overhead. The drawback is the need for a powerful, rather heavyweight distribution infrastructure. Attribute-based interaction is integrated into the syntax of several programming languages, such as Erlang [19] and Google Go [20].

Other well-known protocols for open coalitions include the JXTA peer-to-peer protocols that target overlay-based communication of peers across public networks, where security issues such as firewall traversal are of utmost importance [21]. An application of JXTA for the control of robots was reported in [22]. However, the JXTA project is no longer officially supported. The FIPA recommendations (http://www.fipa.org/) offer a conceptual framework for communication and management in multi-agent teams. Their focus is on intelligent (software) agents, not on mechatronic robots. The FIPA standardization is currently not active.

3.2. Heterogeneity

Heterogeneity in an MRT may refer to the hardware and software features of the individual team members. Different application domains require different robot functionalities. Thus, capabilities related to robot mobility (e.g., static, wheels, legs, aerial), sensors (e.g., optical, acoustical, temperature, air quality parameters, laser, lidar, infrared, etc.) and actuators (e.g., arm, drill, kicker, extension rails, etc.), compute power, storage capacity, operating system software, communication type and range (e.g., WiFi, Bluetooth, LoRaWAN), access to cloud computing resources, and many more will be different for different robot types. In a smart factory, the degree of heterogeneity will probably be limited and known in advance at design time, while, in autonomous driving scenarios, we cannot anticipate completely what kind of traffic participants appear and what their specific properties and capabilities are.

From the perspective of teamwork, robots in a team need to be capable of interacting with each other. Thus, not only must a common communication architecture be in place, but team-wide understood application level protocols for information exchange, coordination, and decision-making are also needed. On the one hand, the heterogeneity of robots certainly contributes to the complexity of the teamwork application design, particularly since, so far, there are no dominating standards for masking the heterogeneity through transparency solutions. On the other hand, from an application perspective, heterogeneity may be beneficial if a team is able to make use of specific capabilities of team members. Then, the whole can truly be more than its parts.

The key challenge is as follows:

- Will it be possible to agree on a small set of standardized hardware and software interfaces for the wide range of diverse robot hardware components and software services?

The situation is similar to the evolution of computing hardware and software where commonly agreed interface standards—de facto or official ones—made it possible to mix and match components from different manufacturers to prevent vendor lock-in and to respond to the variety of user requirements. If the emergence of multi-robot applications continues at the speed that we are witnessing today, more standardization is needed in order to enable a flexible combination of heterogeneous robots for application specific robot teamwork. Consequently, more practical experience and applied research are needed to fuel such a standardization.

3.3. Middleware Support

A multi-robot system, as any other distributed system, benefits from middleware that hides the complexities of distributed computing in heterogeneous environments and, thus, eases the job of the developer of a distributed application. In general, middleware for robotic applications needs to satisfy the same basic requirements as any middleware in a distributed computing system, i.e., to simplify the application design by making transparent the low-level details of the underlying hardware, software, communication, sensing, and actuating activities. Moreover, middleware facilitates the integration of new technologies, and it improves software maintenance and evolution, as well as software re-use across multiple development efforts, thus reducing application development costs.

Taking up experiences with agent reference models (e.g., FIPA and its offspring NAWG (FIPA P2P Nomadic Agents WG (P2PNA WG6), http://www.fipa.org/subgroups/P2PNA-WG.html)) and agent platform implementations (see the FIPA web page at http://www.fipa.org/resources/livesystems.html for an (outdated) list of publicly available FIPA compliant implementations of agent platforms), as well as with popular general-purpose middleware systems, many middleware architectures specifically for multi-robot systems were proposed in the literature. We point the reader to surveys such as [23,24].

Key challenges are as follows:

- Do we need specialized middleware for robot teamwork, or can we build on existing standards for general middleware architectures?
- Are the robot devices capable of running a heavyweight middleware in terms of processing and storage capacity?

- How important are quality of service guarantees for the application at hand?

There are a variety of models underlying middleware for multi-agent coordination. Middleware frameworks such as Orocos [25], CLARAty [26], and MIRO [27] use event-based behavior coordination. The events are triggered by either communication or timer events that are mostly realized as remote procedure calls. This results in an insufficient decoupling between the initiator and receiver of an event. Orocos and MIRO rely on heavyweight architectures, i.e., CORBA [28] and Ice [29], respectively. In contrast, CLARAty which was developed for communication of NASA rovers, explicitly handles unreliable communication and can operate in either centralized or decentralized mode.

The most common communication concept of robot middleware is publish–subscribe due to its higher degree of decoupling. Examples are RoboFrame [30], Spica [31], and ROS [32]. RoboFrame and Spica were designed explicitly for distributed computing in multi-robot systems. They are capable of dealing with unreliable communication, as for example encountered in RoboCup soccer tournaments where standard WiFi communication channels often suffer from bandwidth limitations and packet losses due to interferences among the many WLANs at the competition site. While the robot software framework ROS 1 had limited support for distributed multi-robot applications, the new ROS 2 includes a middleware based on the popular data distribution service (DDS) (Data Distribution Service, OMG, https://www.omg.org/spec/DDS/).

Nevertheless, there are several open issues related to middleware support for robot teamwork that need more research. A truly flexible middleware toolbox would be needed that enables the designer of a teamwork application to configure the middleware by selecting the specific set of components that matches best the application requirements. Operating systems research faced basically the same problem several decades ago when more and more electronic devices became available with very different hardware architectures and application requirements.

Another open issue is the concern for security in distributed collaboration scenarios; most existing middleware solutions for multi-robot systems assume that applications, if needed, can make use of transport-level security mechanisms. Not all applications would need such security. For example, in RoboCup tournaments, there is no real need to secure the communications between the robots (and no time to do so anyway). Likewise, in the collaborative exploration of unknown territories, e.g., on another planet as part of a space mission, secure communication between the explorer robots is not required. However, when it comes to collaborative autonomous vehicles on public roads, potential vulnerabilities are a crucial concern. Moreover, in Section 3.1, we already mentioned the need for security in managing dynamic team membership. More research specifically on adaptive security for teams of autonomous robots is needed. This seems to be a research area on its own.

3.4. Organizational Structure and Decision-Making

Teams need to take team decisions. For example, vehicles need to agree on the speed and direction of a particular vehicle or soccer robots on the location of the ball on the field. Obviously, application requirements are very different. In an autonomous driving scenario, decisions on the locations and intentions of unequipped traffic participants need to be taken very fast in very dynamic short-lived coalitions. Since safety concerns are paramount, consensus is required for most decisions. On the other hand, in robot soccer, for most decisions, we tolerate a relaxed consensus level but demand swift reactions due to the high dynamics of the game situation.

Decision-making (note that our notion of decision-making is different from Reference [9] where the term is viewed as a synonym for planning and control of a multi-agent system) in a team can be organized according to three basic structural principles, i.e., centralized, hierarchical, and distributed [33]. In a *centralized* structure, decisions are made by a central leader or controller. This structure suffers from the vulnerability of a central point of failure and the potential performance bottleneck. In a *hierarchical* structure, decisions are made at different levels by a hierarchy of leaders that have decision authority according to their rank. Such a structure is more robust than a centralized one because it can potentially react faster to "lower-level" events and tolerate partial failures. Its drawback is the incurred high organizational overhead. In a *decentralized* structure, all team members autonomously perceive their own situation and the state of the surrounding execution environment. Team members decide about their actions by themselves according to a given team plan. Team decisions that tolerate a relaxed consensus level can be taken using different kinds of voting schemes, auctions, games, and more.

Decisions are made based on the given team plan and observations about the current context. The application developer will need to evaluate and judge the required level of agreement for team actions, as well as the affordable coordination overhead. A decentralized decision structure is an obvious choice if we are concerned about the reliability of the individual robots and the communication network. Likewise, if we deal with temporary coalitions in highly dynamic environments where swift decisions are required, such as in robot soccer, there is no time for the execution of a time-consuming leader election algorithm or any other costly algorithm for establishing an organizational structure. The reader is referred to [33–35] for detailed discussions of organizational structures and decision-making in multi-agent systems.

The key questions are as follows:

- What kind of organizational structure follows from the application requirements?
- What kind of consensus level is required for team-wide decisions?
- Which decision-making protocols are appropriate considering the trade-off between consensus level and protocol overhead?

Let us look in more detail at a decentralized team organization. Generally, decision-making happens in five steps:

1. Agents collect relevant data by observing the environment and their own status;
2. Agents form their own opinion based on the outcome of step 1;
3. Agents propose their own opinion by replicating it to all team members;
4. The team discusses and resolves conflicting opinions;
5. The team takes a joint decision.

For replication and conflict resolution, there is a choice of well-known protocols depending on the application needs. Hence, a teamwork middleware should offer flexible support for decision-making that is tunable to different application requirements. The core functionality of such a middleware function is to support the team decision-making process with respect to the current values of specified decision variables. Below, we present one concrete example for such a middleware.

The middleware PROViDE [36] is part of a multi-agent framework called ALICA [37]; ALICA teams have a decentralized team structure. PROViDE offers a choice of replication and agreement protocols for common decision variables. If a team decision about the value of an environment variable needs to be made, all team members broadcast their opinion to their teammates. The developer may choose the level of replica consistency depending on the specific application requirements in the face of unreliable communications, temporarily disconnected robots, and diverging sensor readings by the robots. The replicated values of a decision variable can lead to a situation where a robot receives several divergent observations from its teammates in addition to its own observed value. Thus, after the replication phase, a robot needs to decide which value from the set of available opinions it will accept locally as its own value. This may lead to a situation where the individual team members accept different values of the decision variable as their own individual "view of the world". Now, we need a third coordination phase where the robots agree on a single joint value. Such a decision could be based on majority voting, priorities, timestamps, or other criteria.

Thus, there are three distinct phases in team decision-making that resemble the typical process of decision-making in human teams (added in parentheses):

1. Replication of individually perceived values of the decision variable to teammates (team members learn about diverse opinions in the team).

2. Team members locally commit to a value (team members determine their own opinion).
3. If needed, conflicting choices are resolved by a specified conflict resolution protocol (the team consolidates diverging opinions and arrives at a joint decision).

In summary, based on the PROViDE middleware, the application developer can tune the middleware by choosing from a set of provided protocols and, thus, can adapt the quality and overhead of decision-making to diverse application requirements. However, this raises another question. One might argue that such an abundant choice of strategies shifts the complexity onto choosing the right combination of strategies. This argument cannot be ignored. One solution might be to identify reusable typical patterns of strategy combinations for specific application scenarios. This can be addressed in future work.

3.5. Programming

The complexity of teamwork in multi-robot systems in dynamic and adverse environments requires software architectures and integrated toolchains that ease the development process. Model-driven engineering (MDE) allows developers to shift their focus from implementation to modeling in the domain knowledge space. MDE is expected to promote separation of concerns, efficiency, flexibility, and evolution in application development. From a practical engineering point of view, MDE demands a toolchain that not only automates the required model transformations, but also includes tools for examining the models through simulation (e.g., using Gazebo (http://gazebosim.org/)) or model checking (e.g., using UPPAAL [38]).

In order to ease the modeling and implementation of executable plans for robot applications, domain-specific languages (DSL) were proposed. A DSL is a *computer programming language of limited expressiveness focused on a particular domain* [39]. The "limited" in this definition should not be seen as a negative point; instead, it signals that a DSL is targeted at a specific application domain. Typically, a DSL for developing plans for robots consists of two parts, i.e., a modeling language and an associated execution engine. While there are a number of DSLs available for programming single robots (e.g., [3,40–45]), only a few DSLs explicitly address teamwork for multi-robot systems (e.g., [46,47]) (see [48] for a detailed review of robot DSLs). We claim that the complexity of teamwork in dynamic environments makes such a high-level abstraction a necessity, i.e., a DSL that enables the developer to concentrate on the teamwork behavior of the distributed robot system.

Dynamic environments typically imply a dynamic allocation of tasks to individual team members. A good example is robot soccer. A soccer team continuously needs to be aware of the game situation, which may change instantaneously. Thus, tasks such as defending, attacking the ball, dribbling, blocking an opponent, etc. need to be assigned dynamically based on conditions such as whether the team possesses the ball, proximity of robots to the goal, position of the ball, distance to opponents, etc. Clearly, dynamic task allocation in a decentralized formation is a team decision where all team members should agree on their current duties. In contrast, in an Industry 4.0 scenario, allocation of tasks to robots will typically be static.

General research questions are as follows:

- Do we need different teamwork DSLs for different application domains? Ideally, a single DSL would be suitable for programming a wide spectrum of teamwork scenarios in order to enable reuse of models and development know-how.
- Does the modeling and execution environment support a dynamic task allocation to team members instead of fixed allocations?
- How can we efficiently integrate simulation and automated verification into the application development environment in order to examine the models for desired MRT properties, such as safety, fairness, freedom from deadlocks and livelocks, no starvation, etc.?

Let us look at three examples for high-level modeling languages for robot teamwork, i.e., STEAM, BITE, and ALICA.

Shell for Teamwork (STEAM) [5] is a modeling approach for teamwork. STEAM builds on two well-known teamwork theories, i.e., joint intentions theory [3] and shared plans theory [1]. It tries to combine their benefits in order to achieve a coordinated behavior of the team members. In particular, STEAM assigns sub-teams of agents to a hierarchical shared plan structure. Agents need to establish a joint intention before acting together. This makes the teamwork susceptible to degraded or failed communication links.

The *Bar Ilan Teamwork Engine* (BITE) by Kaminka and Frenkel [49] divides the team modeling into three structures. A tree-like structure, similar to hierarchical task networks [50], represents the global execution plan of the team. Another structure describes the organizational hierarchy of sub-team memberships. This results in a hierarchical task structure that provides a team-wide allocation of robots and sub-teams to behaviors. The third structure describes the social interaction behaviors, i.e., explicit communication and coordination activities between agents. A major drawback of BITE is the fact that it requires a successful negotiation before any physical action can take place. As a result, BITE is not appropriate for domains that require swift reactive behavior.

Let us look at one framework in more detail to make the descriptions more concrete. ALICA (A Language for Interactive Collaborative Agents) is a language and execution environment for developing teamwork applications. ALICA provides a formally defined modeling language, tool support for development, and an execution engine for highly adaptive multi-agent team behavior [37,46]. The design of ALICA targets dynamic environments with fast changing situations, imperfect network communication, and possibly diverging sensor data from team members. It supports the known design blocks of multi-robot systems [51], i.e., task decomposition, team formation, and task allocation, as well as task execution and control. ALICA was developed and used originally for robot soccer and then evolved and applied to other application domains such as collaborative exploration of unknown territories, autonomous driving, and service robotics [52].

The team behavior is specified from a global perspective in a single ALICA program which is deployed to all team members and executed without central control. ALICA uses hierarchically arranged annotated state machines to model robot tasks. Figure 1 shows an example where agents collaborate to explore a territory, collect objects, and assemble some structure. Note that this plan is not complete; the figure only shows the highest specification level. A characteristic feature of ALICA is that task allocation to the individual robots is not static but adaptive to the current context and capabilities of the involved robots. State transitions depend on the situation at hand as perceived by a robot. For further information on the syntax and semantics of ALICA, the reader is referred to [37].

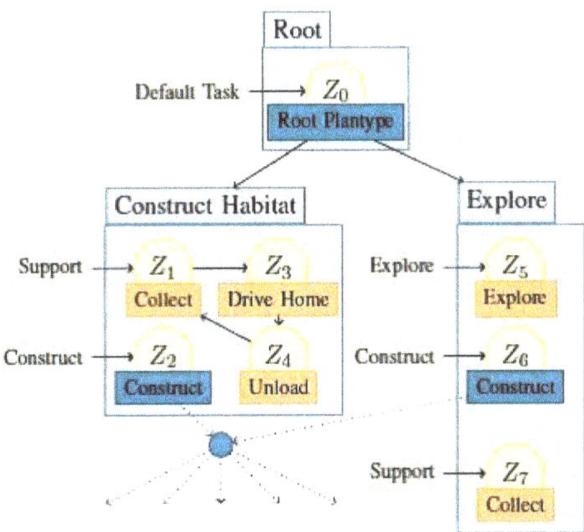

Figure 1. Example of ALICA (A Language for Interactive Collaborative Agents) program for an exploration.

ALICA features a strictly decentralized team organization. Team decisions, e.g., about task allocation, result from individual decisions of the team members according to the given team plan, their observations of the environment, and the information exchange with their teammates. Nevertheless, there may be application situations where the team has to agree unanimously on the value of a certain decision variable. This might lead to decision conflicts that have to be resolved. For example, to execute a ball passing plan in robot soccer, at least the pass executor and pass receiver have to agree on their own positions, the position of the ball, and the opponents' positions. Thus, these positions represent joint decision variables. Note that "agree" in this context may mean different levels of agreement on a spectrum from simple broadcasts of opinions to strict consensus [53]. The developer of the respective ALICA plans must decide what kind of agreement is appropriate for an application in a certain situation. To facilitate this choice, ALICA contains a specific decision-support middleware (see Section 3.4 above).

Other examples of languages suitable for the specification of MRT behavior are Buzz [54], ISPL [55], and SCEL [56]. These languages differ in many properties according to their specific application focus and design paradigms. SCEL is the only language that supports open teams using attribute-based interaction (see [19] for a detailed comparison of the three languages).

3.6. Shared Knowledge

We already mentioned that our viewpoint of robot teamwork is closely linked to mutual awareness in the robots, i.e., teammates have—in addition to their local knowledge—some knowledge about their colleagues in the team. This awareness often, but not necessarily, is based on the provision of a shared global knowledge base for the entire team. The knowledge base, which typically would be implemented as a distributed replicated knowledge store, contains the concepts, objects, and relations known to the robots, as well as the fused perceptions of the state of the execution environment. In many works, the individual local view of a robot is called the *local world model*, while the shared team knowledge base is called the *shared world model* [57].

Potentially, such a shared knowledge base with frequently consulted thousands of objects and relations is a very resource-consuming and a performance-critical element of the teamwork. This creates several challenges, as listed below.

- How do we formally define the shared knowledge such that reasoning at run time satisfies performance requirements?
- How do we equip robots with commonsense knowledge that human beings would have implicitly about the environment?
- How can we integrate individual, heterogeneous knowledge representations of heterogeneous robots into a shared world model?
- What kind of consistency guarantees are necessary and feasible for the distributed, replicated storage of the (dynamic) contents of the knowledge base?
- How robust and scalable is the common knowledge base in view of, e.g., unreliable communication connections, imprecise sensor data, and predefined time barriers?
- Can the knowledge base structure be adapted and extended at run time?

While there is a large set of publications focusing on knowledge representation techniques for robotic applications (e.g., [58–60]), little was published specifically on distributed knowledge bases for multi-robot systems. One thread of research—in particular for service robots—looked at offloading the knowledge base to the cloud [61]. The viability of such an approach clearly depends on the application requirements in terms of access performance and availability. Other approaches exploit a decentralized storage of knowledge [57] leading to well-known questions of consistency in data replication.

The spectrum of diverse requirements in multi-robot applications seems to be so large, that a harmonization of techniques for knowledge representation and storage is out of reach. On the other hand, a vast body of know-how is available for knowledge representation, reasoning, distributed data storage systems, and replication and consensus protocols. In this situation, it is worth considering the question of whether the variety of existing solutions could be narrowed down to a few solution patterns, which would satisfy the majority of application scenarios.

3.7. Robustness and Dynamic Adaptation

Robustness is the ability of a system to cope with errors during execution and with erroneous input (https://en.wikipedia.org/wiki/Robustness_(computer_science)). A lack of robustness in dynamic team coordination may be due to various technical causes. Communication links may be unreliable, i.e., different communication technologies and conditions in the runtime environment may lead to message loss and network partitions such that standard communication protocols cannot provide a guaranteed error-free service. Individual robots may move out of communication range and be temporarily disconnected from the team. This has implications for the design of the application-level protocols. Centralized configurations are not appropriate in this case, since a single point of failure and performance bottleneck can severely hinder the teamwork and make the whole MRT useless. Sensors of team members may deliver different values for the same environment variable. This raises the question of what level of agreement the application requires for collective perceptions and whether the fusion of different types of sensor information can help in such a situation to improve the quality of the information in the shared world model. The amount of overhead for achieving reliable sensor information and consensus building may be prohibitive in very dynamic environments where swift decisions are more important than lengthy computation and communication activities. Moreover, run time execution errors may be caused by situations that were not foreseen at design time. Robustness in this case would mean that the system is able to perform an unanticipated adaptation. Thus, the team as a whole should be able to evolve its team plan, as well as the plans of the individual team members based on, e.g., input from other agents or machine learning techniques. In general, unanticipated software adaptations, which were not planned by the developer at design time, are a challenging problem. Only a few attempts on a general solution for unanticipated on-the-fly adaptation appeared in the literature [62,63]. Most adaptive systems assume that the adaptation state space is known completely at design time [64].

Related research questions are as follows:

- How does the MRT cope with diverging sensor readings?
- Can the MRT tolerate temporarily impaired communications?
- Is the MRT capable of evolving its plans on-the-fly in order to integrate a learned or otherwise derived behavior?
- What are appropriate strategies for unanticipated adaptation?

In teamwork scenarios, the arrival of new team members with new capabilities or the departure of team members with individual capabilities might require changes in the team plans. Likewise, the evolution of global team goals and/or individual robot goals might demand a re-planning. Note that we are not concerned about the generation of the new plans. This may be done manually by a human developer or automatically by machine learning techniques and planning algorithms. Our emphasis is on the implications of openness of teams and, thus, on the capability for dynamic evolution and interchange of team plans.

A possible approach to unanticipated adaptation in an MRT is based on semantic annotations of team plans using a declarative logic programming language such as answer set programming (ASP) [65,66]. ASP adheres to a similar programming model as Prolog [67]. A number of projects showed that ASP meets the requirements for semantic specifications in a wide range of application areas in terms of expressiveness, efficiency, dynamic extensibility, and scalability. Examples are semantic service adaptation [68], dynamic information stream configuration in crisis management scenarios [57], and service robotics [59]. Thus, by adding semantic annotations to team plans, the developer lays the foundation for re-planning at runtime based on the specified properties and constraints for the robots and their relationships. The semantic compatibility of annotated team plans can be checked using established techniques for semantic matching and adaptation [10,69]. Nevertheless, this is still largely unchartered territory in respect to the applicability to different robot scenarios. More research and practical experience are needed on the scope, expressiveness, and performance implications of different paradigms for unanticipated adaptation.

3.8. Socio-Technical Concerns

Even if a team of robots is able to operate autonomously and perform application tasks without human intervention, experience with self-adaptive applications shows that the human user does not always appreciate being out of the loop [70]. Self-adaptive systems may fail to meet user expectations, and autonomous actions may be inappropriate in certain user situations. In other words, the user wants to stay in control in certain situations, or, even more importantly, in safety critical application domains such as autonomous driving, the user must be able to override automatic decisions.

This automation paradox, also called the irony of automation [70], is known since automated control systems took over tasks that were previously carried out by human operators. Psychologists identified human contribution in automated systems as not less but more important. A more advanced automated system denotes a more demanding interaction with the human user. In cases of failures or irregular conditions, humans should still have a chance to intervene. At all times, humans need to be protected against harm caused by the robot behavior. Clearly, this general insight related to automation applies also to the engineering of an MRT, especially if the MRT may self-adapt its plans to situations that the designer did not anticipate.

In addition to the *human in the loop* aspect, concerns about the social embedding of a robotic application solution arise when a team of robots operates in a dynamic environment where users and robots interact. Most of the concerns are of a general nature for adaptive systems, such as *transparency of decisions, trust in technology, fairness, privacy of context information, liability*, and more. Surely, these concerns play a crucial role for the user acceptance of any technical system and particularly in safety-critical applications. They apply to single robots, as well as multi-robot systems. However, one question remains unanswered so far in the literature:

- Will the envisaged teamwork of robots, in comparison to a single robot application, create more complex or even additional challenges in respect to socio-technical design concerns?

4. Summary

The wide spectrum of applications that require teamwork of robots poses the following question: can we discuss engineering concerns at all from a general, all-encompassing point of view? Application domains such as autonomous driving, Industry 4.0, and search and rescue clearly have very different requirements. Nevertheless, our answer is positive, looking at a comparison of robot teamwork with the evolution of distributed systems technologies where models, architectures, and techniques emerged that provide a strong foundation for practical implementations.

In contrast to classical distributed systems technologies, we assume that robot teamwork happens in dynamic environments; robots are mobile, robots use unreliable wireless communications, robots move out of communication range, new team members appear, robots sense the state of the runtime environment and reason about appropriate reactions, specific components of robots fail without rendering the whole robot useless, the team encounters unforeseen situations, and more. Below, we summarize from a general, systems-oriented perspective the discussions in the previous chapter about engineering challenges for robot teamwork in dynamic environments. Thus, we point to research areas that need to be tackled in future work.

4.1. Dynamic Coalitions

The dynamic environment, as described above, implies a need for a highly flexible team organization and collaboration infrastructure. Team membership must be managed to cope with varying team membership. The capabilities of the team as a whole may change if robots leave, join, or experience a breakdown of components or the whole robot. We need a semantic description of the capabilities that are currently available in the team; consequently, we need reasoning about the appropriate dynamic allocation of tasks to team members and the modification of execution plans. All of this should be supported by the teamwork collaboration infrastructure such that the developer is relieved as much as possible from the nitty-gritty details. The individual building blocks for such an infrastructure are known. However, one difficult engineering question remains: how can the different independently developed pieces be put together?

4.2. Platform Harmonization and Configurability

In order to facilitate the reuse of software components, portability of applications, and exchange of know-how, a harmonization—if not standardization—of robot platforms and their application programming interfaces is desirable. The diversity of robot application domains creates a need for a flexibly configurable and customizable collaboration platform architecture that reflects the different computational capacities of robots. Thus, in analogy to operating systems for embedded systems, we need highly configurable, component-based teamwork collaboration platforms (i.e., middleware) that can be tailored to specific application needs and properties of the involved robots. Such a toolbox is missing.

4.3. Knowledge Base

The ability to share knowledge is a crucial prerequisite for teamwork. As discussed in Section 3.6, there are various approaches for building a shared knowledge base in multi-robot systems. Some standardization would also be helpful here. Important research questions to ask in this realm concern the integration of heterogeneous knowledge representations, the implementation of the knowledge base in a distributed system with largely diverging agent capabilities, the satisfaction of stringent application performance requirements, the extensibility of the knowledge base structure at run-time, i.e., adding new concepts and relations, as well as removing invalid facts, and the inclusion of "common-sense knowledge" that humans would have implicitly, but which needs to be provided explicitly to a robot team.

For all of these aspects, the state of the art provides individual solutions. However, how to forge these solutions into a shared knowledge base that satisfies the specific requirements of an application domain is an open problem. Moreover, we need to explore whether it is feasible to reduce the large number of approaches to a few consolidated ones.

4.4. Methodology and Tools

Like other software, the development of robot teamwork applications should be supported by an effective development methodology as well as corresponding development tools. Many proposals for domain-specific languages for robotic applications exist, as mentioned above in Section 3.5. How to filter out a few approaches that would serve a larger number of application domains is an open question. The formal verification of teamwork plans for dynamic environments with respect to correctness and properties such as liveness and freedom from deadlocks is not well developed so far and requires more research. Moreover, facing the large variety of protocols for agreement, data consistency, synchronization, etc., the identification of agreed-upon reusable best-practice design patterns for teamwork applications would greatly facilitate the software development process. Ideally, all of this should be integrated into a robot teamwork development environment built around a powerful DSL.

4.5. Edge and Cloud Integration

Offloading computation-intensive tasks or large data quantities from robots to edge or cloud computing resources is an attractive option for resource-scarce robots. However, the challenges and open questions that arise in such a scenario are manifold. Which part of an application can/should be offloaded to improve the performance or to save battery capacity taking into account the communication overhead and latencies? When and under what conditions should it be done? How can the state of the robot's execution context be provided to the offloaded computation? These questions were solved already for mobile cloud computing scenarios on mobile devices, primarily looking at offloading computation and data from a single device [68]. For teamwork scenarios in multi-robot systems with a high degree of agent cooperation and coupling, the viability and effectiveness of these solutions have to be re-examined.

4.6. Human in the Loop and Other Sociotechnical Concerns

In many application environments, robots interact with humans. For example, humans may use the services of a robot team. Alternatively, humans may augment the capabilities of the team, effectively making them a member of the team, or humans may give instructions to a robot team to control the execution. Research in social robotics delivered various means of interacting with robots, e.g., based on voice or gestures. However, interaction with and control of a whole team of robots received little attention so far. Some general questions remain. How would the team and the individual team members be addressed? How would an "emergency button" be implemented that immediately stops the execution of the whole MRT? How would the possible actions of a human be modeled in the team plan? How would the MRT react to unanticipated actions of the human?

Sociotechnical concerns for technical innovations, as presented in Section 3.8 above, received increasing attention in society recently. Clearly, these concerns also apply to robot teamwork. As in other technical domains, the big question is as follows: how do we translate abstract sociotechnical requirements that are mostly specified in natural language into concrete engineering artefacts for multi-robot applications? For example, liability issues for a team of heterogeneous robots from different manufacturers could be difficult to decide. Likewise, explanations for team decisions and actions may be even more difficult to understand for a user who demands transparency for MRT activities; this will undermine the user´s trust in the technology and may lead to a lack of acceptance. Experts from different disciplines must work together to solve these interdisciplinary puzzles.

4.7. General Remarks

As for any distributed system, the engineering of robot teamwork raises questions about concerns such as scalability, fault tolerance, performance, software evolution, and the like. It is important that developers of teamwork applications respond to these concerns. We do not discuss these concerns here because they do not create specific questions for the engineering of robot teamwork.

5. Conclusions

The proliferation of robotics is likened often to the evolution of the personal computer (PC). Many expect that—like the PC—autonomous robots, in whatever form, will become everyday assistants that will surround and support us in all kinds of application domains. Naturally, over the years, the increasing number of robots will lead to "distributed robot systems" where autonomous robots form (temporary) teams and collaborate to achieve a common goal. Due to the manifold technical, contextual, and situational dependencies, these teams will often act under dynamically changing conditions, and not all teamwork can be planned and implemented at design time. Hence, dynamic team building and adaptive team behavior will become important concerns.

In this paper, we focused particularly on the engineering of teamwork for multi-robot systems that operate in dynamically changing environments. We tried to raise the awareness for crucial issues in the realization of such teamwork, and we pointed out exemplary solution approaches. Clearly, the diversity of application requirements is huge, and the design space is vast. This will keep the research and development community busy in the coming years.

Funding: This research received no external funding.

Acknowledgments: Parts of this paper were written while the author was a guest scientist at IMT Lucca (Italy). The author sincerely thanks Rocco de Nicola and the members of his group for insightful discussions and contributions. The author gratefully acknowledges the support from the Banco Santander Chairs of Excellence program and the insightful discussions with researchers from Universidad Carlos III de Madrid (UC3M) and IMDEA Networks, as well as the constructive comments by the anonymous reviewers.

Conflicts of Interest: The authors declare no conflict of interest.

References

1. Grosz, B.; Kraus, S. Collaborative plans for complex group action. *Artif. Intell.* **1996**, *86*, 269–357. [CrossRef]
2. Jennings, N. Commitments and conventions: The foundation of coordination in multi-agent systems. *Knowl. Eng. Rev.* **1993**, *8*, 223–250. [CrossRef]
3. Levesque Hector, J.; Cohen Philip, R.; Nunes José, H.T. On Acting Together. In Proceedings of the Eighth National Conference on Artificial Intelligence (AAAI-90), Boston, MA, USA, 29 July–3 August 1990; pp. 94–99.
4. Parker, L.E. Current state of the art in multi-robot teams. In *Distributed Autonomous Robotic Systems*; Springer: Berlin/Heidelberg, Germany, 2000; pp. 3–12.
5. Tambe, M. Towards flexible teamwork. *J. Artif. Intell. Res.* **1997**, *7*, 83–124. [CrossRef]
6. Farinelli, A.; Iocchi, L.; Nardi, D. Multi-Robot Systems: A classification focused on coordination. *IEEE Trans. Syst. Man Cybern. B* **2004**, *34*, 2015–2028. [CrossRef] [PubMed]
7. Pěchouček, M.; Mařik, V. Industrial deployment of multi-agent technologies: review and selected case studies. *J. Auton. Agents Multi Agent Syst.* **2008**, *17*, 397–431. [CrossRef]
8. Chennareddy, S.; Agrawal, A.; Anupama, K.R. Modular Self-Reconfigurable Robotic Systems: A Survey on Hardware Architectures. *J. Robot.* **2017**, *2017*, 1–19.
9. Rizk, Y.; Awad, M.; Tunstel, E.W. Cooperative Heterogeneous Multi-Robot Systems: A Survey. *ACM Comput. Surv.* **2019**, *52*, 29–31. [CrossRef]
10. Gerkey, B.P.; Mataric, M.J. A formal analysis and taxonomy of task allocation in multi-robot systems. *Int. J. Robot. Res.* **2004**, *23*, 939–954. [CrossRef]
11. Mosteo, A.R.; Montano, L. *A Survey of Multi-Robot Task Allocation*; Technical Report AMI-009-10-TEC; University of Zaragoza: Zaragoza, Spain, 2010.
12. Doran, J.; Franklin, S.; Jennings, N.; Norman, T. On cooperation in multi-agent systems. *Knowl. Eng. Rev.* **1997**, *12*, 309–314. [CrossRef]
13. Parker, L.E. Distributed intelligence: Overview of the field and its application in multi-robot systems. *J. Phys. Agents* **2008**, *2*, 5–14. [CrossRef]
14. Agha, G. *Actors: A Model of Concurrent Computation in Distributed Systems*; MIT Press: Cambridge, MA, USA, 1986.
15. Sangiorgi, D.; Walker, D. *The Pi-Calculus: A Theory of Mobile Processes*; Cambridge University Press: Cambridge, UK, 2003.
16. Prasad, K. A calculus of broadcasting systems. In *TAPSOFT'91*; Springer: Berlin/Heidelberg, Germany, 1991; pp. 338–358.
17. De Nicola, R.; Di Stefano, L.; Inverso, O. Towards formal models and languages for verifiable Multi-Robot Systems. *Front. Comput. Sci.* **2018**, *5*. [CrossRef]
18. Alrahman, Y.A.; De Nicola, R.; Loreti, M. On the Power of Attribute-Based Communication. In Proceedings of the 36th IFIP WG 6.1 International Conference, FORTE 2016, Heraklion, Greece, 6–9 June 2016; pp. 1–18.
19. De Nicola, R.; Duong, T.; Inverso, O.; Trubiani, C. *AErlang: Empowering Erlang with Attribute-Based Communication*; Jacquet, J.-M., Massink, M., Eds.; Springer: Berlin/Heidelberg, Germany, 2017; pp. 21–39.
20. Alrahman, Y.A.; De Nicola, R.; Garbi, G. GoAt: Attribute-based Interaction in Google Go. *Comput. Sci.* **2018**. [CrossRef]

21. Sun Microsystems, JXTA Java Standard Edition V2.5: Programmers Guide. 2007. Available online: https://www.tamps.cinvestav.mx/~{}vjsosa/clases/redes/JXTA_SE_ProgGuide_v2.5.pdf (accessed on 18 February 2020).
22. Ogata, Y.; Spaho, E.; Matsuo, K.; Barolli, L.; Moreno, J.; Xhafa, F. JXTA-Overlay P2P Platform and Its Application for Robot Control. In Proceedings of the 13th International Conference on Network-Based Information Systems (NBiS 2010), Takayama, Gifu, Japan, 14–16 September 2010; pp. 133–138.
23. Elkady, A.; Sobh, T. Robotics Middleware: A Comprehensive Literature Survey and Attribute-Based Bibliography. *J. Robot.* **2012**, *2012*, 1–15. [CrossRef]
24. Mohamed, N.; Al-Jaroodi, J.; Jawhar, I. Middleware for Robotics: A Survey. In Proceedings of the 2008 IEEE Conference on Robotics, Automation and Mechatronics, Chengdu, China, 21–24 September 2008; pp. 736–742.
25. Bruyninckx, H.; Soetens, P.; Koninckx, B. The Real-Time Motion Control Core of the Orocos Project. In Proceedings of the 2003 IEEE International Conference on Robotics and Automation, Taipei, Taiwan, 14–19 September 2003; pp. 2766–2771.
26. Volpe, R.; Nesnas, I.A.D.; Estlin, T.; Mutz, D.; Petras, R.; Das, H. CLARAty: Coupled Layer Architecture for Robotic Autonomy. Tech. rep. NASA Jet Propulsion Laboratory: Pasadena, CA, USA, 2000.
27. Utz, H.; Sablatnog, S.; Enderle, S.; Kraetzschmar, G. Miro—Middleware for mobile robot applications. *IEEE Trans. Robot. Autom.* **2002**, *18*, 493–497. [CrossRef]
28. Object Management Group (OMG). *The Common Object Request Broker: Architecture and Specification (CORBA 3.3)*; Object Management Group: Needham, MA, USA, 2012.
29. Shumko, S. Ice Middleware in the New Solar Telescope's Telescope Control System. In *Astronomical Data Analysis Software and Systems XVIII*; Astronomical Society of the Pacific: Orem, UT, USA, 2009; Volume 411.
30. Petters, S.; Thomas, D.; von Stryk, O. RoboFrame—A Modular Software Framework for Lightweight Autonomous Robots. In Proceedings of the Workshop on Measures and Procedures for the Evaluation of Robot Architectures and Middleware, IEEE/RSJ IROS, San Diego, CA, USA, 29 October – 2 November 2007.
31. Baer, P.A. Platform-Independent Development of Robot Communication Software. Ph.D. Thesis, University of Kassel, Kassel, Germany, 2008.
32. Robot Operating System. Available online: https://index.ros.org/ (accessed on 10 February 2020).
33. Abbas, H.A.; Shaheen, S.I.; Amin, M.H. Organization of Multi-Agent Systems: An Overview. *Int. J. Intell. Inf. Syst.* **2015**, *4*, 46–57.
34. Bulling, N. *A Survey of Multi-Agent Decision-Making, KI - Künstliche Intelligenz*; Springer: Berlin/Heidelberg, Germany, 2014; pp. 147–158.
35. Grossi, D.; Dignum, F.; Dastani, D.; Royakkers, L. Foundations of Organizational Structures in Multiagent Systems. In Proceedings of the 4th International Joint Conference on Autonomous Agents and Multiagent Systems (AAMAS 2005), Utrecht, The Netherlands, 25–29 July 2005; pp. 690–697.
36. Geihs, K.; Witsch, A. Decentralized decision-making in adaptive multi-robot teams. *Inf. Technol.* **2018**, *60*, 239–248. [CrossRef]
37. Skubch, H. Modelling and Controlling Behaviour of Cooperative Autonomous Mobile Robots. Ph.D. Thesis, Universität Kassel, Kassel, Germany, 2012.
38. Nguyen Van, T.; Fredivianus, N.; Tran Huu, T.; Geihs, K.; Binh Huynh, T. Formal Verification of ALICA Multi-agent Plans Using Model Checking. In Proceedings of the 9th Int. Symposium on Information and Communication Technology, Danang, Vietnam, 6–7 December 2018.
39. Fowler, M. *Domain-Specific Languages*; Addison-Wesley: Boston, MA, USA, 2010.
40. Verma, V.; Estlin, T.; Jonsson, A.; Pasareanu, C.; Simmons, R.; Sing Tso, K. Plan Execution Interchange Language (PLEXIL) for Executable Plans and Command Sequences. In Proceedings of the Internatational Symposium on Artificial Intelligence, Robotics and Automation in Space (iSAIRAS), Munich, Germany, 5–8 September 2005.
41. Urmson, C.; Anhalt, J.; Bagnell, D.; Baker, C.; Bittner, R.; Dolan, J.; Duggins, D.; Ferguson, D.; Galatali, T.; Geyer, C.; et al. *Tartan Racing: A Multi-Modal Approach to the DARPA Urban Challenge*; CMU TR Urmson-2007-9708; Carnegie Mellon University: Pittsburgh, PA, USA, 2007.
42. Kammel, S.; Ziegler, J.; Pitzer, B.; Werling, M.; Gindele, T.; Jagszent, D.; Schroder, J.; Thuy, M.; Goebl, M.; Hundelshausen, F.; et al. Team AnnieWAY's Autonomous System for the 2007 DARPA Urban Challenge. *J. Field Robot.* **2008**, *25*, 615–639. [CrossRef]

43. Dhouib, S.; Kchir, S.; Stinckwich, S.; Ziadi, T.; Ziane, M. *Robotml, A Domain-Specific Language to Design, Simulate and Deploy Robotic Applications*; Noda, I., Ando, N., Brugali, D., Kuffner, J.J., Eds.; Springer: Berlin/Heidelberg, Germany, 2012; pp. 149–160.
44. Arda, K.; Ümit, O. Hierarchical Finite State Machines for Autonomous Mobile Systems. *Control Eng. Pract.* **2013**, *21*, 184–194.
45. The Eclipse Foundation: Papyrus. Available online: https://www.eclipse.org/papyrus-rt (accessed on 2 April 2019).
46. Skubch, H.; Wagner, M.; Reichle, R.; Geihs, K. A Modelling Language for Cooperative Plans in Highly Dynamic Domains. *Mechatronics* **2011**, *21*, 423–433. [CrossRef]
47. Zweigle, O.; Lafrenz, R.; Buchheim, T.; Käppeler, U.P.; Rajaie, H.; Schreiber, F.; Levi, P. Cooperative Agent Behavior Based on Special Interaction Nets. In Proceedings of the 9th International Conference on Intelligent Autonomous Systems—IAS, Tokyo, Japan, 7–9 March 2006; pp. 651–659.
48. Nordmann, A.; Hochgeschwender, N.; Wigand, D.; Wrede, S. A Survey on Domain-Specific Modeling and Languages in Robotics. *J. Softw. Eng. Robot.* **2016**, *7*, 75–99.
49. Kaminka, G.A.; Frenkel, I. *Flexible Teamwork in Behavior-Based Robots*; Veloso, M.M., Kambhampati, S., Eds.; The MIT Press: Cambridge, MA, USA, 2005; pp. 108–113.
50. Tate, A. Generating Project Networks. In Proceedings of the 5th International Joint Conference on Artificial Intelligence IJCAI'77, Cambridge, MA, USA, 22–25 August 1977; Morgan Kaufmann Publishers Inc.: Burlington, MA, USA, 1977; Volume 2, pp. 888–893.
51. Kiener, J.; Von Stryk, O. Towards cooperation of heterogeneous, autonomous robots: A case study of humanoid and wheeled robots. *Robot. Auton. Syst.* **2010**, *58*, 921–929. [CrossRef]
52. Opfer, S.; Jakob, S.; Jahl, A.; Geihs, K. ALICA 2.0—Domain-Independent Teamwork. In Proceedings of the 42nd German Conference on Artificial Intelligence (KI2019), Kassel, Germany, 23–26 September 2019.
53. Lamport, L. The Part-time Parliament. *ACM Trans. Comput. Syst.* **1998**, *16*, 133–169. [CrossRef]
54. Pinciroli, C.; Beltrame, G. Buzz: An Extensible Programming Language for Heterogeneous Swarm Robotics. In Proceedings of the 2016 IEEE/RSJ International Conference on Intelligent Robots and Systems (IROS), Daejeon, Korea, 9–14 October 2016; pp. 3794–3800.
55. Lomuscio, A.; Qu, H.; Raimondi, F. MCMAS: An open-source model checker for the verification of multi-agent systems. *Int. J. Softw. Tools Technol. Transf.* **2017**, *19*, 9–30. [CrossRef]
56. De Nicola, R.; Loreti, M.; Pugliese, R.; Tiezzi, F. A Formal Approach to Autonomic Systems Programming. *ACM Trans. Auton. Adapt. Syst.* **2014**, *9*, 1–29. [CrossRef]
57. Niemczyk, S.; Opfer, S.; Fredivianus, N.; Geihs, K. ICE: Self-Configuration of Information Processing in Heterogeneous Agent Teams. In Proceedings of the Symposium on Applied Computing 2017, Marakesh, Marocco, 4–6 April 2017; pp. 417–423.
58. Opfer, S.; Jakob, S.; Geihs, K. Reasoning for Autonomous Agents in Dynamic Domains: Towards Automatic Satisfaction of the Module Property. In *Agents and Artificial Intelligence*, 1st ed.; Springer International Publishing: Berlin/Heidelberg, Germany, 2018; pp. 22–47. ISBN 978-3-319-93581-2.
59. Opfer, S.; Jakob, S.; Geihs, K. Teaching Commonsense and Dynamic Knowledge to Service Robots. In Proceedings of the 11th Conference on Social Robotics (ICSR2019), Madrid, Spain, 26–29 November 2019.
60. Paulius, D.; Sun, Y. A Survey of Knowledge Representation in Service Robotics. *Robot. Auton. Syst.* **2019**, *118*, 13–30. [CrossRef]
61. Riazuelo, L.; Tenorth, M.; Di Marco, D.; Salas, M.; Gálvez-López, D.; Mösenlechner, L.; Kunze, L.; Beetz, M.; Tardos, J.D.; Montano, L.; et al. Roboearth semantic mapping: A cloud enabled knowledge-based approach. *IEEE Trans. Autom. Sci. Eng.* **2015**, *12*, 432–443. [CrossRef]
62. Keeney, J. Completely Unanticipated Dynamic Adaptation of Software. Ph.D. Thesis, The University of Dublin, Dublin, Ireland, 2004.
63. Khan, M.U. Unanticipated Dynamic Adaptation of Mobile Applications. Ph.D. Thesis, University of Kassel, Kassel, Germany, 2010.
64. Floch, J.; Fra, C.; Fricke, R.; Geihs, K.; Wagner, M.; Lorenzo, J.; Cantero, E.S.; Mehlhase, S.; Paspallis, N.; Rahnama, H.; et al. Playing MUSIC—Building context-aware and self-adaptive mobile applications. *Softw. Pract. Exp.* **2013**, *43*, 359–388. [CrossRef]
65. Gebser, M.; Kaminski, R.; Kaufmann, B.; Schaub, T. *Answer Set Solving in Practice*; Morgan & Claypool Publishers: San Rafael, CA, USA, 2012; Volume 6.

66. Gelfond, M.; Kahl, Y. *Knowledge Representation, Reasoning, and the Design of Intelligent Agents: The Answer-Set Programming Approach*; Cambridge University Press: Cambridge, UK, 2014.
67. Clocksin, W.F.; Mellish, C.S. *Programming in PROLOG*; Springer Science & Business Media: Berlin/Heidelberg, Germany, 2003.
68. Baraki, H.; Schwarzbach, C.; Jakob, S.; Jahl, A.; Geihs, K. SAM: A Semantic-Aware Middleware for Mobile Cloud Computing. In Proceedings of the 11th IEEE International Conference On Cloud Computing (IEEE CLOUD 2018), San Francisco, CA, USA, 2–7 July 2018.
69. Scioni, E. Online Coordination and Composition of Robotic Skills: Formal Models for Context-aware Task Scheduling. Ph.D. Thesis, KU Leuven, Leuven, Belgium, 2016.
70. Geihs, K.; Evers, C. User Intervention in Self-Adaptive Context-Aware Applications. In Proceedings of the 17th Australasian User Interface Conference (AUIC), Canberra, Australia, 2–5 February 2016.

Article

SC-M*: A Multi-Agent Path Planning Algorithm with Soft-Collision Constraint on Allocation of Common Resources

Rongye Shi [1,*], Peter Steenkiste [1,2,*] and Manuela M. Veloso [3,*]

[1] Department of Electrical and Computer Engineering, Carnegie Mellon University, Pittsburgh, PA 15213, USA
[2] Computer Science Department, Carnegie Mellon University, Pittsburgh, PA 15213, USA
[3] Machine Learning Department, Carnegie Mellon University, Pittsburgh, PA 15213, USA
* Correspondence: rongyeshi@cmu.edu (R.S.); prs@cs.cmu.edu (P.S.); mmv@cs.cmu.edu (M.M.V.)

Received: 30 July 2019; Accepted: 21 September 2019; Published: 26 September 2019

Featured Application: SC-M* generalizes the M* algorithm to address real-world multi-agent path planning problems in the soft-collision context, which considers the allocation of common resources requested by agents. Application examples include but are not limited to city-scale passenger routing in mass transit systems, network traffic engineering and planning for large-scale autonomous vehicles.

Abstract: Multi-agent path planning (MAPP) is increasingly being used to address resource allocation problems in highly dynamic, distributed environments that involve autonomous agents. Example domains include surveillance automation, traffic control and others. Most MAPP approaches assume hard collisions, e.g., agents cannot share resources, or co-exist at the same node or edge. This assumption unnecessarily restricts the solution space and does not apply to many real-world scenarios. To mitigate this limitation, this paper introduces a more general class of MAPP problems—MAPP in a soft-collision context. In soft-collision MAPP problems, agents can share resources or co-exist in the same location at the expense of reducing the quality of the solution. Hard constraints can still be modeled by imposing a very high cost for sharing. This paper motivates and defines the soft-collision MAPP problem, and generalizes the widely-used M* MAPP algorithm to support the concept of soft-collisions. Soft-collision M* (SC-M*) extends M* by changing the definition of a collision, so paths with collisions that have a quality penalty below a given threshold are acceptable. For each candidate path, SC-M* keeps track of the reduction in satisfaction level of each agent using a collision score, and it places agents whose collision scores exceed its threshold into a soft-collision set for reducing the score. Our evaluation shows that SC-M* is more flexible and more scalable than M*. It can also handle complex environments that include agents requesting different types of resources. Furthermore, we show the benefits of SC-M* compared with several baseline algorithms in terms of path cost, success rate and run time.

Keywords: multi-agent systems; planning; M* algorithm; shortest path finding; collision-free constraint; optimality and completeness

---

1. Introduction

Multi-agent path planning (MAPP) involves finding the set of least-cost paths for a set of agents co-existing in a given *graph* such that each of the agents is free from collision, where a collision is defined as at least two agents moving to the same location at the same time. MAPP attracts increasing attention due to its practical applications in multi-robot systems for surveillance automation, video gaming, traffic control, and many other domains [1–4]. This problem is, however, difficult to solve because the configuration space grows exponentially with the number of agents in the system, incurring extremely heavy computational efforts. It is an NP-hard problem to find optimal solutions for MAPP in its general form [5].

Approaches to solving MAPP problems fold into three main categories: *coupled, decoupled* and *intermediate* [6]. *Coupled* approaches search the *joint configuration space* of the multi-agent system, which is the *Tensor product* of the free configuration spaces of all the individual agents. A popular coupled planner is the A* algorithm [7] that directly searches the whole joint configuration space, making such an approach computationally infeasible when the number of agents is large. Enhanced variants of A*, such as operator decomposition (OD), enhanced partial expansion A* (EPEA*), and iterative deepening A* (IDA*), can—to some extent—mitigate the exponential growth in the number of neighbors by improving the admissible heuristics [8–11]. Coupled approaches are optimal and complete, but usually at high computational cost. *Decoupled* approaches plan for each agent separately and then adjust the path to avoid collisions. Algorithms in this category are generally faster because they perform a graph search and collision-avoidance adjustment in low-dimensional spaces. However, optimality and completeness are not guaranteed [3,12].

Intermediate approaches lie between coupled and decoupled ones because they dynamically couple agents and grow the search space during the planning. In this way, the search space is initially small and grows when necessary. A few intermediate MAPP algorithms can guarantee optimality and completeness. State-of-the-art examples include Conflict-Based Search (CBS) [6,13]. CBS is a two-level algorithm. At the high level, conflicts are added into a *conflict tree* (CT). At the low level, solutions consistent with the constraints given by the CT are found and updated to agents. CBS behaves poorly when a set of agents is strongly coupled. Meta-agent CBS (MA-CBS) is then proposed by merging strongly coupled agents into a meta-agent to handle the strongly coupled scenarios.

The M* algorithm is a state-of-the-art coupled approach. It starts with decoupled planning and applies a strategy called *sub-dimensional expansion* to dynamically increase the dimensionality of the search space in regions in which agent collisions occur. In this way, an efficient graph search with a strict collision-free constraint can be achieved, while minimizing the explored portion of the joint configuration space. M* identifies which subsets of agents can be safely decoupled and hence plans for multi-agents in a lower-dimensional space. Compared to CBS and its variant MA-CBS, M* and its variants, e.g., recursive M* (rM*), have much more fine-grained control over some technical details, such as the management of conflict sets for better scalability. The fine-grained nature of M* allows it to be integrated into MA-CBS to take advantage of both [14]. Recent work extended both M* and CBS algorithms to handle the imperfect path execution due to unmodeled environments and delays [15,16].

Most fundamental MAPP approaches assume *hard collisions*, which means that solutions in which agents share resources (nodes or edges) are rejected. In many real world scenarios, some degree of resource sharing between agents is acceptable, so the hard-collision constraint needlessly over-constrains the solution space. This paper relaxes the hard collisions constraint by allowing some sharing of resources, including space and various services on edges/nodes, by agents. Such sharing reduces the quality of the path, i.e., the satisfaction level of the agent using it, but as long as the quality reduction for each path is below a settable threshold, the solution is acceptable. We call this concept *soft collisions*. Hard collisions are still supported by having a very strict threshold, i.e., a penalty for sharing is very high. The reduction in satisfaction level experienced by an agent caused by soft collisions on resources in its path is quantified using a *collision score*. In this paper, we develop a generalized version of the M* algorithm, called *soft-collision M** (SC-M*), for solving the MAPP problem in the soft-collision context. Note that we that we are not simply replacing hard with soft collisions, but instead introducing soft collisions as a generalization that allows modeling different types of collisions.

SC-M* extends M* by taking the perspective of soft collision on common resources. Specifically, SC-M* tracks the collision score of each agent and places agents whose collision scores exceed certain thresholds into a *soft-collision set* for *sub-dimensional expansion*, a technique that limits the search space while maintaining the optimality of the algorithm with respect to the objective. In this way, SC-M* achieves improved scalability to handle a larger number of agents while limiting the probability of collisions on resources to a bound.

In this paper, we show that SC-M* has advanced flexibility and scalability for efficiently solving the MAPP problem in the soft-collision context where common resources are considered, and can handle complex environments (e.g., with multiple types of agents requesting multiple types of resources). We theoretically prove that SC-M* is *complete* and *suboptimal* under the soft-collision constraints on resources. Experimental results demonstrate the advantages/trade-offs of SC-M* in terms of path cost, success rate and run time against baseline SC-based MAPP planners, such as SC-A* and SC-CBS.

The rest of the paper is organized as follows. Section 2 discusses the motivation of soft collisions. Section 3 gives technical briefing of the M* algorithm. Section 4 presents our proposed SC-M* approach. Section 5 evaluates SC-M* in a grid public transit network. Finally, Section 6 concludes our work.

2. Motivation

In some planning problems, solutions in which agents share resources, i.e., they collide using the traditional MAPP problem definition, may be acceptable, at the cost of having a reduced level of agent satisfaction. Problems of this type have two properties in common: (1) Agents' satisfaction conditions are reduced when meeting at the same place; and (2) the extent of reduction in satisfaction depends on how long the dissatisfying situation lasts in terms of distance or time.

One motivating example of this type of problems involves mass transit systems, in which passengers have various preferences, even necessities, in terms of *common resources*, such as seat availability (necessary for seniors) and on-vehicle Wi-Fi supply (preferred by video viewers and game players during the trip). Passengers may interfere with one another on common resources in crowded situations. Individually optimal paths can cause serious interference, leading to low-quality experiences. Interference between passengers is soft because it is possible that they do not call for the same resource when they are on the same public vehicle. In addition, even when they call for the same resource and interfere, they are able to tolerate each other over a short time and distance. Intuitively, how likely a collision (intolerable interference) actually happens depends on: (1) whether the resource supply is less than the demands; and (2) how long the lack-of-supply condition lasts in terms of the time and distance that the passengers stay together. Passengers can be viewed as agents, moving through the transportation network. When planners plan for all the agents, sticking to eliminating any hard collision is neither necessary nor feasible. Thus, people are more interested in another problem: How can the resource received by all agents be maximized such that the probability of collision of each agent is less than a bound? This is an important topic of passenger-centered research [17–19].

In addition to public transit scenarios, other examples include: network traffic engineering, where multiple data streams can route through a router. Long streams will have a higher chance of being blocked when unexpected traffic spikes pop up, exceeding the link capacity [20]. How to maximize the throughput with a bounded chance of blocking is of great interest to researchers in the field of communications and computer networks.

Another example is planning for large-scale self-driving cars, where multiple cars can share the same lane, and the number of cars on a road will influence the chance of crashes among autonomous vehicles [21,22]. Scholars and engineers dealing with the fundamentals of autonomous vehicles in unstructured and dynamic environments aim to increase the road traffic while bounding the crash risk.

Military transportation also has the soft-collision property, in which transport aircrafts or vehicles are subject to higher risks to be detected and attacked by enemy troops when many of they move together due to path overlap for a long distance. Formally, as the transportation volume on a road increases, the degree of concealment decreases [23]. The dispatcher must bound the security risk when attempting to maximize the military transportation efficiency.

To support these application classes, we introduce the soft-collision property (related to common resources) to MAPP. SC-M*, introduced in this paper, is the first attempt to generalize M* to handle real MAPP problems in a soft-collision context, considering various common resources requested by agents. Specifically, SC-M* changes M*'s definition of a collision so it can represent soft collisions on resources and their impact on an agent's dissatisfaction level. We show the advantages of the SC-M* against other SC-based MAPP solvers.

3. Technical Briefing of M*

Before introducing the SC-M* method, this section reviews the traditional MAPP problem and the M* algorithm [6].

3.1. MAPP Problem Definition

In this problem, we have m agents indexed by the set $I = \{1, \ldots, m\}$. Let the free configuration space of agent j be represented by the directed graph $G^j = \{V^j, E^j\}$. For any agent j, graph G^j is the same. The joint configuration space, which describes the set of all possible states of the multi-agent system, is defined as the *tensor product* of the graphs of all individual agents: $G = G^1 \times \cdots \times G^m$. G consists of a *joint vertex* set V and a *joint edge* set E. As an example, in a 2-D joint configuration space given by the agents j and k, the two 2-D joint vertexes $v_p = (v_p^j, v_p^k)$ and $v_q = (v_q^j, v_q^k)$ is connected by the joint edge (e_{pq}^j, e_{pq}^k). Note that $v_p^j \in V^j$ and $e_{pq}^j \in E^j$. Let $\pi^j(v_p^j, v_q^j)$ denote a sequence of joint vertexes, called a *path* in G^j from v_p^j to v_q^j. The cost of a path $\pi(v_p, v_q)$ in G is defined as

$$g(\pi(v_p, v_q)) = \sum_{j=1}^{m} g(\pi^j(v_p^j, v_q^j)), \tag{1}$$

where $g(\pi)$ is the sum of all edge costs involved in the joint path π.

The goal of MAPP is to find a collision-free path, which is optimal with respect to minimal cost, from the source configuration $v_s = v_s^1 \times \cdots \times v_s^m$ to the destination configuration $v_d = v_d^1 \times \cdots \times v_d^m$. To determine the collision between agents, a collision function $\psi(v_p)$ is defined to return the set of conflicting agents at v_p.

Most fundamental MAPP approaches use hard collisions, where no intersection is allowed between any two agents in terms of the occupation of any *resource*, such as a workspace. This implies that the capacity of each resource can support only one agent at a time (i.e., a collision happens immediately once agents intersect at any resource). Suppose we have a set of resources $A = \{A_1, \ldots, A_L\}$ requested by each agent in the multi-agent system, where A_i is defined as the set of resource of type i on all edges and vertexes in G. A_i is a continuous set because only continuous resources are considered in the paper. A traditional hard-collision constrained MAPP problem is formulated as follows:

$$\begin{aligned} & \min_{\pi} g\left(\pi\left(v_s, v_d\right)\right) \\ s.t. \quad & \bigcup_{\forall i \neq j \in I} \left(A_k(v_p^i) \cap A_k(v_p^j)\right) = \emptyset, \ \forall A_k \in A, \ \forall v_p \in \pi, \end{aligned} \tag{2}$$

where $A_k(v_p^j)$ denotes the subset of resource A_k occupied by the agent j at the joint vertex v_p. One state-of-the-art solver to this problem is M*, which uses the sub-dimensional expansion strategy to dynamically increase the dimensionality of the search space in regions featuring some agent collisions. M* enables a relatively cheaper graph search under the strict hard-collision constraint.

3.2. Graphic-Centric Description of M*

This section uses the graphic-centric description introduced by wanger [6] to illustrate M*. M* is a complete and optimal MAPP algorithm. The main idea of M* is to iteratively construct/update a so-called *search graph* G^{sch} (i.e., to iteratively remove the collision configuration vertexes and expand necessary neighbors) and apply the A* algorithm on the new G^{sch} until the optimal collision-free path to v_d exists in the G^{sch} and is found by the A* search. Specifically, G^{sch} is a sub-graph of G and consists of three other sub-graphs: the *expanded graph* G^{exp}, *neighbor graph* G^{nbh}, and *policy graph* G^{ϕ}. The expanded graph G^{exp} is the sub-graph of G that has been explored by M*. G^{nbh} contains the *limited neighbors* of all the joint vertexes in G^{exp}. The definition of limited neighbors is given below. G^{ϕ} consists of the paths induced by the *individually optimal policy* ϕ that connects each joint vertex in $G^{nbh} \cup G^{exp}$ to v_d without the collision-free constraint. Specifically, ϕ^j is the individually optimal policy for the agent j that leads any v^j in $G^{nbh} \cup G^{exp}$ to v_d^j without considering collisions. Examples of policy ϕ include the standard *Dijkstra's* algorithm [24] and A* [5]. Using the above graphic concepts, we can define the *collision set* C_p as

$$C_p = \begin{cases} \psi\left(v_p\right) \cup \left\{\cup_{v_q \in \boldsymbol{V_p}} \psi\left(v_q\right)\right\}, & for\ v_p \in G^{exp} \\ \varnothing & , for\ v_p \notin G^{exp} \end{cases}, \tag{3}$$

where $\boldsymbol{V_p} = \{v_q | \exists \pi(v_p, v_q) \subseteq G^{exp}\}$ is the set of the joint vertexes to which there exists a path to from v_p in G^{exp}. Let $\phi^j(v^j)$ be the immediate *successor vertex* of v^j in the policy path, then the set of *limited neighbors* V_p^{nbh} for the joint vertex v_p in G^{nbh} is defined as

$$V_p^{nbh} = \left\{ v_q \middle| \begin{cases} e_{pq}^j \in E^j, & for\ j \in C_p \\ v_q^j = \phi^j\left(v_p^j\right), & for\ j \notin C_p \end{cases} \right\}, \tag{4}$$

where $e_{pq}^j = edge\left(v_p^j, v_q^j\right)$. The definition of the limited neighbors implies the sub-dimensional expansion strategy: We only expand the search space at the dimensions where the collision occurs ($j \in C_p$), otherwise for collision-free dimensions ($j \notin C_p$), M* will not expand, limiting the unexpanded search space to the graph that only consists of individually optimal path induced by the policy ϕ.

3.3. Algorithm Description of M*

The high-level description of M* is as follows [6]: Initially, M* computes the individually optimal policy ϕ for each agent from source v_s to destination v_d. The initial search graph G^{sch} only consists of an individually optimal path: Initial G^{exp} contains v_s only; initial G^{nbh} contains $\phi(v_s)$ only, which is the successor of v_s along the individually optimal policy; and initial G^{ϕ} contains the optimal policy path from the vertex in G^{nbh} and G^{exp} all the way to v_d. $C_p = \varnothing$ for all v_p in initial G^{sch}. Given the initial G^{sch}, the A* algorithm is applied using the following *admissible heuristic*

$$h\left(v_p\right) = g(\pi_{\phi}(v_p, v_d)\,) \leq g(\pi_*(v_p, v_d)), \tag{5}$$

where π_{ϕ} is the individually optimal path induced by policy ϕ, and π_* is the ground-truth optimal multi-agent path we want to find. The initial *open list* (i.e., *priority queue*) contains v_s only, with zero cost. The open list is sorted according to $v_p.cost + h\left(v_p\right)$, where $v_p.cost$ is the current cost of v_p from the source.

In each iteration, M* expands the first-ranked vertex v_p from the open list to G^{exp} and investigates each joint vertex v_q in the limited neighbors of v_p (i.e., $v_q \in V_p^{nbh}$) if no collision occurs at v_p; otherwise, it jumps to the next iteration. If there exists a collision (i.e., $\psi\left(v_q\right) \neq \varnothing$), M* will update the collision set C_q with $C_q \cup \psi\left(v_q\right)$, and this update will back-propagate from v_q to: (1) its immediate predecessor v_p; and (2) all the way back to any ancestors that have at least one path inside of G^{exp} leading to v_q (see Equation (3) for details). After this pre-processing, the algorithm:

- investigates and updates the cost of the vertex v_q and records its corresponding predecessor; and
- adds v_q and all its predecessors/ancestors, of which the collision sets are changed, to the open list.

This process is repeated until v_d is expanded or open list is empty.

The critical point is that: Only when a collision set C_p is changed will the search graph G^{sch} change. It is the operation of updating the collision set in a back-propagation way that makes the story different: By including $\psi(v_p)$ to C_p, M* can tell which agents are *immediately* collided at the current v_p; by including all $\psi(v_q)$ for $v_q \in V_p$ to C_p (i.e., the collision information of all the expanded downstream successors from v_p), M* can preview which agents will collide in the future, making it possible to pre-plan to avoid that. Therefore, using the limited neighbor set in Equation (4) makes sense: It advises M* to only expand the dimensions where there exists an immediate collision at v_p or there will be collisions in the future, starting from v_p, in the current expanded graph G^{exp}. Figure 1 shows an example of how M* solves the optimal collision-free path planning for the two agents.

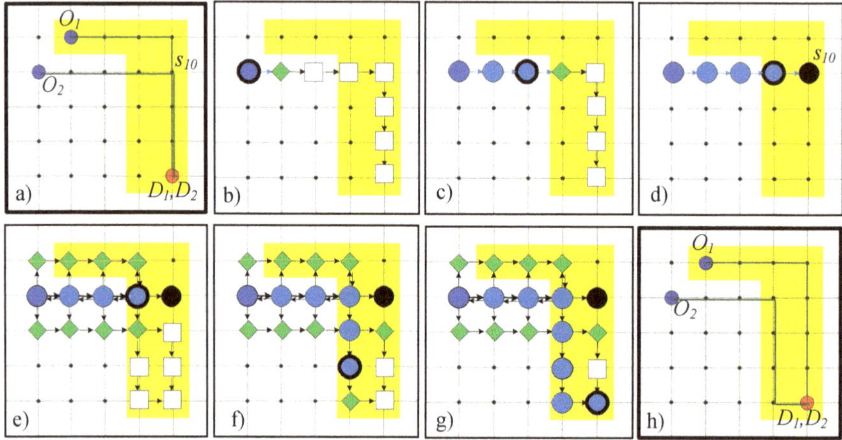

Figure 1. Illustration of traditional M* for two agents, where we show the evolution of the expanded graph G^{exp} (circle), neighbor graph G^{nbh} (diamond), and policy graph G^{ϕ} (square) for Agent 2 as the M* algorithm proceeds. (**a**) Individually optimal paths; (**b**) the first expanded vertex; (**c**) the third expanded vertex; (**d**) collision occurs at vertex s_{10}; (**e**) sub-dimensional expansion; (**f**) search in the expanded space; (**g**) the destination of Agent 2 founded; (**h**) collision-free optimal paths for both agents founded by M*.

In Figure 1, we can visualize the evolution of the search graph G^{sch} of Agent 2. G^{sch} consists of an expanded graph G^{exp} (circle), a neighbor graph G^{nbh} (diamond), and a policy graph G^{ϕ} (square). Edge cost and direction-changing cost are considered during planning. Yellow zones are preferred areas with lower edge cost. In M*, individually optimal paths are induced by ϕ for each individual agent (Figure 1a). We can observe that there will be a collision at vertex s_{10}, which is ignored by ϕ. For Agent 2, M* searches in the subspace, and the most promising vertex is expanded at each iteration (Figure 1b,c). Then, a collision occurs at vertex s_{10} and triggers the removal of the rest of G^{sch} (Figure 1d), which is equivalent to jumping to the next iteration. Following the sub-dimensional expansion strategy, M* extends the search space to include the limited neighbors, and a new G^{sch} is obtained (Figure 1e). By searching in the new G^{sch}, M* finds the optimal collision-free path for Agent 2 (Figure 1f,g). On the other hand, the planning for Agent 1 is conducted simultaneously, and, finally, the collision-free optimal paths for both agents are found by M* (Figure 1h).

4. Soft-Collision M* (SC-M*)

M* assumes hard-collision constraint which does not apply to many real-world applications. Our contribution in this paper is to generalize M* to soft-collision context where common resources are considered, and to introduce soft collisions as a generalized concept allows us to model different types of collisions. In addition, we show the advantages and trade-offs of the proposed algorithm in this new scenario. The proposed SC-M* extends M* by changing the definition of a collision, so paths with hard collisions but with a level of dissatisfaction on resources below a given threshold are acceptable. In this section, we formulate the concept of soft collision on common resources, describe the generalized M* (i.e., soft-collision M*) for planning in the soft-collision context, and extend our approach to a more complex environment with multiple types of agents requesting multiple types of resources.

4.1. Soft-Collision Constraint on Common Resources

Inspired by real-world scenarios, we introduce the recourse-related soft-collision property to the model of an agent. We define that all the agents have the following properties: (1) a collision among agents is *soft*, quantified using some *collision scores*; and (2) different agents have different collision scores, according to their individual experiences through the paths. We suppose that each agent cares about a set of resources $A = \{A_1, \ldots, A_L\}$. To obtain the properties, we introduce to each agent an additional attribute called *resource experience* (for each resource) and use the resource experience to calculate the *collision score*.

In doing so, this section first uses the *resource experience* (as defined in Section 4.1.1 Definition 1) to quantify how dissatisfying the agent is about the resource allocated to it. Then, we combine this information of all the resources into a *collision score* (as defined in Section 4.1.2 Definition 2) that indicates the probability of the agent announcing a collision given its resource experience. *Threshold of collision* is used to limit the collision score, implying to what degree of unpleasantness we want to pursue the solution. The agent, of which the collision score exceeds the threshold, will be placed into a soft-collision set via the *soft-collision function* for sub-dimensional expansion (as defined in Section 4.1.3 Definition 3).

4.1.1. Definition 1 (Resource Experience)

We define *resource experience* to quantify the dissatisfying experience per resource about which an agent cares.

Let

- $\pi = \pi\left(v_s, v_b\right)$ be a path from the source v_s to some v_b;
- $v_q = \pi\left(v_p\right)$ be the immediate successor of v_p along the path π;
- $A_k(e^j_{pq})$ be the capacity (amount) of the subset of the resource A_k on the edge e^j_{pq}, given by the graph model; and
- $A^j_k(e^j_{pq})$ be the amount of the subset of the resource A_k *actually allocated* to the agent j on the edge e^j_{pq}, called the *allocated resource value*.

The *resource experience* is then defined as the *dissatisfying* experience of agent j on resource A_k along the path π^j:

$$D\left(\pi^j, A_k\right) = \sum_{v_p | v_p \in \pi / v_b} \mathbf{1}\left(A_k(e^j_{pq}) \geq \varepsilon_k \wedge A^j_k(e^j_{pq}) < \varepsilon_k\right) \cdot g(e^j_{pq}), \tag{6}$$

where $\mathbf{1}(\cdot)$ is the indicator function, whose value is one if the logical condition is true, else zero; $\varepsilon_k \in \varepsilon = \{\varepsilon_1, \ldots, \varepsilon_L\}$ is the *satisfying value* regarding the resource A_k, which is a positive real value; $g(e^j_{pq})$ is the edge cost regarding travel time/distance given by the graph model; and $A^j_k(e^j_{pq})$ is formulated as:

$$A^j_k(e^j_{pq}) = \frac{A_k(e^j_{pq})}{\sum_{k \in I} \mathbf{1}\left(e^k_{pq} = e^j_{pq}\right)}. \tag{7}$$

Obviously, $A^j_k(e^j_{pq}) = A_k(e^j_{pq})$ if and only if no other agents are physically moving along with agent j on the edge e^j_{pq}. The allocated resource value $A^j_k(e^j_{pq})$ quantifies the level of interference incurred by other agents when they physically move together. In contrast, the traditional hard-collision setting will always label a collision to the agent j and all other involved agents whenever $A_k^j(e_{pq}^j)$ is (even slightly) smaller than $A_k(e_{pq}^j)$. The resource experience is implemented as an attribute of the vertex class and can be calculated incrementally using Algorithm 1.

Algorithm 1 Function: experience(v_k, v_l, A).

Input: v_k: base vertex; v_l: immediate successor of the base vertex; A: list of resources
Output: v_l with updated experience

```
for A_p in A do
    for j in I do
        v_l.exp[A_p][j] ← v_k.exp[A_p][j] + D(π(v_k, v_l)^j, A_p)
    end for
end for
return v_l      //the successor with updated experience
```

Combined with the allocated resource value, which serves as a proxy of the interference level, the definition of resource experience in Equation (6) actually defines a property of an agent: Only the situation in which the resource allocated to an agent is dissatisfying because of the co-existence of other agents (i.e., $A_k(e_{pq}^j) \geq \epsilon_k$ should hold), will contribute to the dissatisfying experience of that agent. Furthermore, each dissatisfying condition is weighted by the edge cost $g(e_{pq}^j)$. In this way, we can quantify the resource experience in terms of how long such a dissatisfying condition lasts in travel time or distance, which is quantified by $g(e_{pq}^j)$. As discussed below, the resource experience of an agent will determine its collision score, which is defined from a probabilistic point of view.

4.1.2. Definition 2 (Collision Score)

We use the resource experience results from Definition 1 to calculate the *collision scores*. This is defined from the view point of collision probability, that must be constrained under some threshold.

Let

- Col_j be the event that agent j announces a collision (i.e., when agent j calls for one of the resources, the allocated resource is less than satisfying);
- $D^j = \{D_1^j, \ldots, D_L^j\}$, where $D_k^j = D(\pi^j, A_k)$, be the set of dissatisfying experiences of agent j along path π^j on the resource A_k; and
- $f_k \in f = \{f_1, \ldots, f_L\}$ be a customized cumulative distribution function (CDF) defined on $[0, +\infty)$, mapping the resource experience D to a probability of collision on the resource A_k.

The *collision score* of the agent j is defined as the probability of how likely a collision occurs to the agent j on *at least* one of the resources given its resource experience D^j:

$$P\left(Col_j \middle| D^j\right) = 1 - \prod_{k \in \{1,\ldots,L\}} \left(1 - f_k(D_k^j)\right). \tag{8}$$

Note that $P\left(Col_j | D^j\right)$ calculates the *complement* of the success probability—the joint probability of being tolerable at all resources.

Figure 2 shows two example designs of f: $f_1(D) = sigmoid(D - \delta)$, with a discontinuity point $f_1(0) = 0$, is a sigmoid-based CDF function, featuring a surge in the collision score (the derivative is bump-shaped) at the experience value around δ. This function is suitable to important resources that are sensitive to the agent; $f_2(D) = min(1, D/(4\delta))$ is a linear CDF with a shallow slope (the derivative is flat). This function can apply to trivial resources that are not very sensitive to the agent but still accumulate to contribute to the collision score. We use the offset parameter δ to adjust the *tolerance level* of the dissatisfying experience. With larger δ, the agent will tolerate a longer unpleasant experience before announcing a collision.

Although the definition of the collision score can be customized according to different practices, the probabilistic definition of collision score introduced here is a general one: Different types of resources may have different value ranges, and Equation (8) standardizes the resource ranges, mapping them to a value within $[0, 1]$ and enabling an efficient integration of different types of resources to the framework.

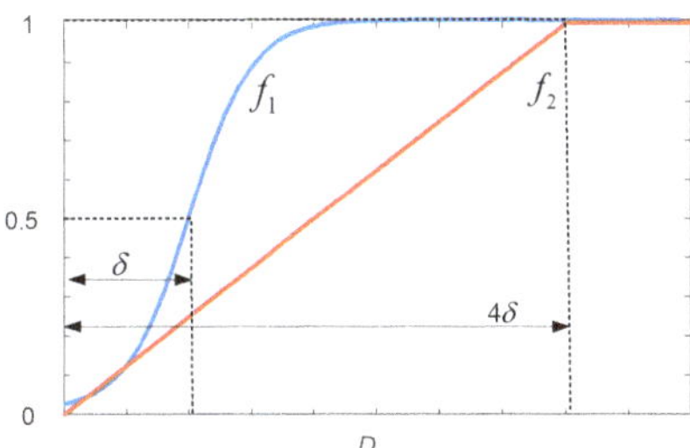

Figure 2. Example designs of cumulative distribution functions (CDFs), mapping the resource experience D of an agent to a collision probability on certain resource. f_1: sigmoid-based CDF for important (sensitive) resources. f_2: linear CDF for trivial (insensitive) resource. δ: offset parameter adjusting the *tolerance level*.

4.1.3. Definition 3 (Soft-Collision Function)

Now, according to the collision scores from Definition 2, we want to pick out the above-threshold agents and place them into the soft-collision set via the *soft-collision function* for the purpose of applying the sub-dimensional expansion.

Given a path $\pi = \pi(v_s, v_b)$ and corresponding resource experience D^j for the agent j, the *soft-collision function* of the agent j is

$$\widetilde{\psi}^j(v_b) = \begin{cases} \{j\}, & for\ P\left(Col_j \middle| D^j\right) \geq T \\ \varnothing, & otherwise \end{cases}, \tag{9}$$

where T is the *threshold of collision*. The definition of the *global soft-collision function* is then defined as

$$\widetilde{\psi}(v_b) = \bigcup_{j \in I} \widetilde{\psi}^j(v_b). \tag{10}$$

Based on Definition 3, we can formally construct the soft-collision constraint on common resources and obtain the soft-collision constrained MAPP problem:

$$\begin{aligned} & \min_{\pi} g\left(\pi\left(v_s, v_d\right)\right) \\ s.t.& \\ & \widetilde{\psi}\left(v_p\right) = \varnothing, \quad \forall v_p \in \pi. \end{aligned} \tag{11}$$

This problem setting is general and can be utilized to express the hard collision setting in Equation (2) by setting $T = 0$ or changing the condition inside the indicator function of Equation (6) to $A_k(e^j_{pq}) \neq A^j_k(e^j_{pq})$ with infinite cost.

4.2. SC-M* Description

SC-M* is a general solver to the MAPP problem in Equation (11) by adjusting M* to the soft-collision constraints on common resources. The pseudocode for SC-M* is presented in Algorithm 2, where critical commands relative to the soft-collision constraint are underscored. In this algorithm, Lines 1–7 initialize each vertex v in the vertex set V with infinite cost from the source v_s (the cost of v_s itself is zero), set dissatisfying experience to zero and make collision set C_k empty. The initial open list contains v_s only (Line 8). In each iteration, SC-M* expands the first-ranked vertex v_k in the open list ordered by the total cost $v_k.cost + heuristic[v_k]$ (Lines 10 and 11). The algorithm terminates and returns the result if the expanded v_k is the destination v_d (Lines 12–14) or jumps to the next iteration if immediate collision occurs at v_k, i.e., $\widetilde{\psi}(v_k) \neq \emptyset$ (Lines 15–17). Line 18 constructs the *limit neighbors* V_k^{nbh} of v_k using Equation (4). For each vertex v_l in V_k^{nbh} (Line 19), it adds v_l to the descendant set V_k of v_k (Line 20), updates the dissatisfying experience of v_l using Algorithm 1 (Line 21), and merges the immediate collision at v_l to its soft-collision set C_l (Line 22). On top of the new collision set of v_l, SC-M* backpropagates to update all the affected ancestor vertexes from v_l (see Equation (3)) and adds them back to the open list for re-expanding (Line 23). After this collision set updating operation, if v_l is free from collision and has improved cost, the algorithm accepts the new cost by save the trace-back information and adding v_l to the open list for expansion (Lines 24–28). This process repeats until the open list is empty (Line 9) when no solution exists or the optimal solution is found (Lines 12–14).

Algorithm 2 Soft-collision M*.

Input: v_s: source joint vertex; v_d: destination joint vertex; $\{V, E\}$: joint configuration graph; A: list of resources
Output: Path finding results

1: **for** all v_k in V **do**
2:   v_k.cost $\leftarrow +\infty$
3:   v_k.exp $\leftarrow$ all zero experience
4:   $C_k \leftarrow \emptyset$
5:   v_k.traceBack $\leftarrow \emptyset$
6: **end for**
7: v_s.cost $\leftarrow 0$
8: open $\leftarrow \{v_s\}$
9: **while** open$\neq \emptyset$ **do**
10:   open.sort() by v.cost+heuristic[v]   //i.e., sort the open list from small to large
11:   $v_k \leftarrow$ open.pop()
12:   **if** $v_k = v_d$ **then**
13:     **return** back_track_path[v_k]   //optimal path found
14:   **end if**
15:   **if** $\widetilde{\psi}(v_k) \neq \emptyset$ **then**
16:     **continue**   //skip the vertex in collisions
17:   **end if**
18:   conduct the construction of V_k^{nbh} using Equation (4)
19:   **for** v_l in V_k^{nbh} **do**
20:     add v_l to V_k //note $V_k = \{v_q | \exists \pi(v_k, v_q) \subseteq G^{exp}\}$
21:     $v_l \leftarrow$ experience(v_k, v_l, A)   //update experience using Algorithm 1
22:     $C_l \leftarrow C_l \cup \widetilde{\psi}(v_l)$
23:     backpro_update(v_k, C_l, open)   // 1) update all the affected soft-collision sets using Eq.(3)
        //2) add all affected vertexes back to open list (see reference [6] for details)
24:     **if** $\widetilde{\psi}(v_l) = \emptyset$ **and** v_k.cost+e_{kl}.cost $< v_l$.cost **then**
25:       v_l.cost $\leftarrow v_k$.cost+e_{kl}.cost
26:       v_l.traceBack$\leftarrow v_k$
27:       open.add(v_l)
28:     **end if**
29:   **end for**
30: **end while**
31: **return** no path exists

SC-M* can make a transition from a decoupled individual A* ($T = 1$) to a standard hard-collision constrained M* ($T = 0$), providing more flexibility to the performance of the algorithm with bounded soft-collision scores.

4.3. Completeness and Cost-Suboptimality

A MAPP algorithm is complete if it guarantees that it will either return a path, or determine that no path exists in finite time. An algorithm is optimal if it guarantees returning an optimal path if such a solution exists. SC-M* is complete and suboptimal conditioned on the soft-collision constraint (i.e., $P\left(Col_j \middle| D^j\right) < T$, for a given collision threshold T).

4.3.1. Completeness

Theorem 1. *SC-M* is a complete algorithm.*

Proof of Theorem 1. SC-M* inherits the sub-dimensional expansion from M* (i.e., it changes the G^{sch} only when one of the soft-collision sets C_p changes). The algorithm applies A* in the updated search graph. Due to the merging operation applied to collision set C_p, as shown in Equation (3), C_p for each vertex will change finite times (at most m times, which is the number of agents). Because A* is complete, applying A* to a given G^{sch} takes finite time to return a result. Therefore, SC-M* is complete. □

4.3.2. Cost-Suboptimality

Different from M*, which is optimal, SC-M* is suboptimal because Equations (9) and (10) only include the immediate conflicting agents to the soft-collision set; the agents that softly interfere with the conflicting agents in the upstream path are excluded. Those excluded agents also contribute to the announced collision (i.e., making the collision score above the threshold). Because of this, SC-M* cannot guarantee the *inclusion property*, which is the basis to ensure the optimality in M* [6]: *The optimal path for some subset of agents costs no more than the optimal joint path for the entire set of agents.* Without the inclusion property, SC-M* may not guarantee cost optimality.

Figure 3 provides a counterexample of the inclusion property of SC-M* in the soft-collision MAPP context defined in this paper. Let $\pi'_{\Omega}(v_k, v_f)$ be the joint path constructed by combining the optimal path for a subset $\Omega \in I$ of agents with the individually optimal paths for the agents in $I \backslash \Omega$. The inclusion property is defined as follows: If the configuration graph contains an optimal path $\pi_*(v_k, v_f)$, then $\forall \Omega \subset I$, $g(\pi'_{\Omega}(v_k, v_f)) \leq g(\pi_*(v_k, v_f))$. See *Lemma* 6 in [6].

In the soft-collision context, this inclusion property does not always hold. In Figure 3, we have a three-agent MAPP problem ($I = \{r1, r2, r3\}$) in the soft-collision context. Agents $r1$, $r2$, and $r3$ attempt to move from the vertexes a, f, and h to the vertexes e, g, and i, respectively. The individually optimal paths (shortest distance) are $a \rightarrow b \rightarrow c \rightarrow d \rightarrow e$ with distance 4 for $r1$, $f \rightarrow c \rightarrow d \rightarrow g$ with distance 3 for $r2$, and $h \rightarrow b \rightarrow c \rightarrow i$ with distance 3 for $r3$. The total cost of the joint individually optimal path is 10. However, assuming that the agents can only tolerate a dissatisfying experience with distance 1, $r1$ will announce a collision at vertex d because of the interference on the edge $b \rightarrow c$ and $c \rightarrow d$ from agents $r3$ and $r2$, respectively.

If we choose $\Omega = \{r1, r2\} \in I$, as can be seen in Figure 3, the only solution would be that $r1$ takes a detour through the vertex x to avoid the collision on the edge $c \rightarrow d$, resulting in a cost of 5 for $r1$, and the total $g(\pi'_{\Omega}(v_k, v_f))$ is 11 (3 for $r2$, 5 for $r1$ and 3 for $r3$). On the other hand, by searching through all three dimensions, a better solution would be that $r3$ detours through the vertex y, and $r1$ is free from collision because the interference on the edge $b \rightarrow c$ disappears. The total cost of this joint path is 10.5, and we have $g(\pi'_{\Omega}(v_k, v_f)) = 11 > g(\pi_*(v_k, v_f))$ =10.5, which is contradictory to the inclusion property.

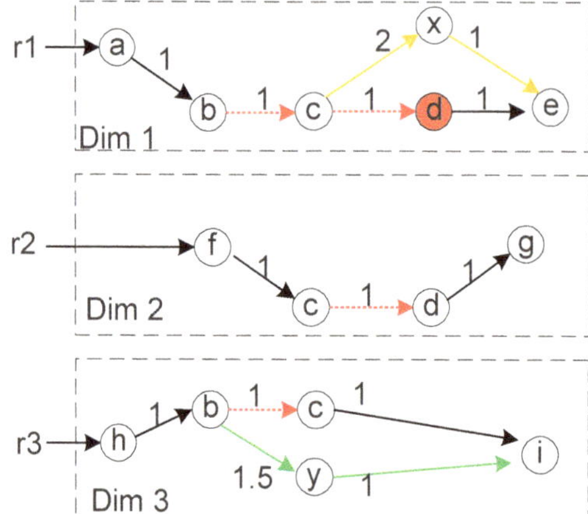

Figure 3. Counterexample of the inclusion property of soft-collision M* (SC-M*) in the soft-collision context. Agents $r1, r2$, and $r3$ have the planning O-D demands $(a, e), (f, g)$, and (h, i), respectively. Vertexes in the system are labeled as a, b, c, etc.

The reason for this phenomenon is that, in the hard-collision context, only the immediate conflicting agent $r2$ contributes to the collision of $r1$ at vertex d. However, in the soft-collision context, both $r2$ and $r3$ contribute to the collision of $r1$ at vertex d, and thus, the inclusion property does not apply. Without this inclusion property, which is the basis of the optimality of M*, the optimality of SC-M* cannot be guaranteed.

However, we notice that suboptimal methods have long been used successfully to solve many interesting MAPP problems [15,25,26]. Given the fact that we show in the next section that SC-M* is superior to other alternative SC-based MAPP solvers (e.g., SC-A* and SC-CBS) in terms of scalability, run time, and path cost, we demonstrate that the proposed method, which is adjusted to MAPP in the soft-collision context, is a powerful tool in practice.

5. Experiments and Results

We evaluated SC-M* in simulation on a grid public mass transit network with an Intel Core i7-6700 CPU at 3.4 GHz with 16 GB RAM. As shown in Figure 4, the grid transit environment has 20×20 stops. There are 20 bidirectional horizontal lines. Likewise, 20 bidirectional vertical lines are deployed in the environment. At each stop, agents can switch lines. The yellow areas are covered by some resources, such as the on-vehicle free Wi-Fi in our experiments. Agents traversing those areas can enjoy high-quality on-vehicle Wi-Fi connections. A fully covered edge has a Wi-Fi resource value of 100, and the Wi-Fi value of an edge is proportional to the length of coverage. Each agent wants to move from its source (square) to its destination (circle) with the lowest cost (i.e., a linear combination of distance cost and Wi-Fi cost) as well as bounded collision score. The second resource is the space on the edge, which is fixed at 5. The satisfying values are $\varepsilon_1 = 20$ and $\varepsilon_2 = 1$ for Wi-Fi and space resources, respectively.

We randomly generated a source–destination pair for each agent. Each trial was given a 1000-s run-time limitation to find a solution. For each configuration (including the number of agents, collision threshold T, and offset parameter δ), we ran 20 random trials to calculate the average metrics (i.e., the success rate and run time). The success rate is the number of trials ending with a solution divided by the number of trials. The run time is the average over trials ending with a solution or a no-solution declaration. If all trials under a certain configuration exceeded 1000 s, we used ">1000" to represent the run time of the corresponding configuration. We used the standard A* as the coupled planner and policy generator in the SC-M* framework and compared our results to the baselines.

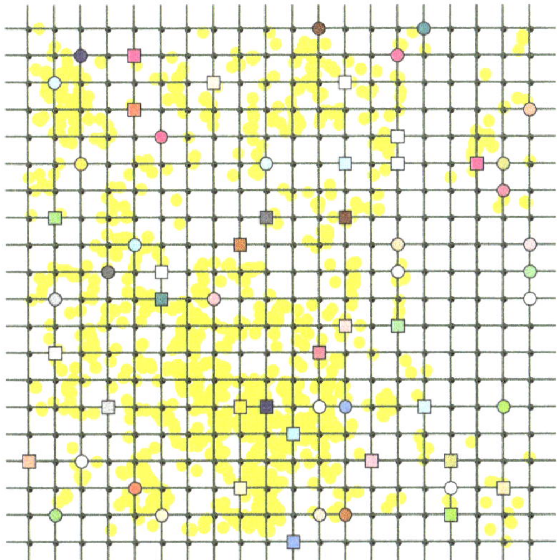

Figure 4. Grid system with 20 × 20 stops and 40 bidirectional lines. Square and circle of the same color correspond to the source and destination of an agent, respectively.

5.1. Planning for the One-Resource-One-Agent-Type

The first experiment considered Wi-Fi as the only resource requested by agents (i.e., $A = \{A_1 : \text{"}WiFi\text{"}\}$). Only one agent type exists, and all agents use sigmoid-based function f_1 as the collision CDF.

We first studied the influence of the collision threshold T and the offset parameter δ on performance. Figure 5a shows the success rate of the one-resource-one-type SC-M* with different thresholds T =0 (equivalent to the basic M*), 0.2, 0.4, and 0.45, while the offset parameter is fixed to $\delta = 6.0$. Table 1 (left) shows the run time in seconds for the experiment. The results clearly show that larger thresholds bring improvement in performance with a higher success rate and lower run time for a large system size ($m > 50$). The improvement in performance results from the property of SC-M* that larger thresholds render more relaxed constraints, and thus, agents are less likely to collide on resources.

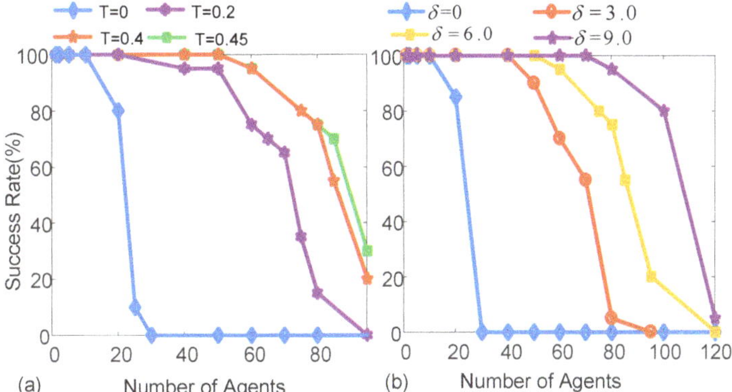

Figure 5. Impact of the collision threshold T (given $\delta = 6.0$) and offset parameter δ (given $T = 0.35$) on one-resource-one-type SC-M*.

Figure 5b shows the success rate of the SC-M* with different offset parameters $\delta = 0$, 3.0, 6.0, and 9.0, with fixed $T = 0.35$. Table 1 (right) shows the run time for the experiment. The results illustrate that SC-M* is sensitive to δ and can efficiently handle up to 100 agents with $\delta = 9.0$. These results are reasonable because the sigmoid-based CDF is used in the experiments, featuring a surge in the collision probability at the experience value around the offset, and the offset parameter poses a cutoff value on the resource experience, with collision always announced once the resource experience is larger than the offset. The standard M* ($T = 0$) can only scale to fewer than 30 agents. Taking advantage of this property, one can tune the parameters to trade off the scalability against the tightness of constraints.

Table 1. Run time of one-resource-one-type SC-M* under different parameters.

m	$T = 0$	$T = 0.2$	$T = 0.4$	$T = 0.45$	m	$\delta = 3.0$	$\delta = 6.0$	$\delta = 9.0$
5	0.556389	0.52489	0.5472	0.3616	5	0.506	0.3616	0.575
10	1.25143	1.18687	0.7057	0.7965	10	1.0765	0.7965	1.0427
20	403.3011	2.72513	1.4871	1.4488	20	2.2578	1.4488	2.1034
40	>1000	56.1898	4.2336	4.4318	40	17.201	4.4318	4.525
70	>1000	370.059	257.59	255.78	80	477.96	292.17	59.31
95	>1000	>1000	951.34	774.40	120	>1000	>1000	857.0
Left: $\delta = 6.0$					Right: $T = 0.35$			

5.2. Planning for the Two-Resource-Two-Agent-Type

We also evaluated SC-M* in more complex environments: two agent types requesting two resources. This experiment considered both Wi-Fi and space capacity (i.e., $A = \{A_1 : \text{"}WiFi\text{"}, A_2 : \text{"}Space\text{"}\}$). Type I agents use f_1 in Figure 2 as the collision CDF for the Wi-Fi resource, and the linear CDF f_2 for the space resource, implying that they treat Wi-Fi and space as important and trivial, respectively. On the other hand, Type II agents use f_1 for space and f_2 for Wi-Fi. Each agent has a 50% chance of being Type I. Both CDFs are adjusted using the same δ at each trial, as illustrated in Figure 2.

Figure 6a shows the success rate of the two-resource-two-type SC-M* with different thresholds $T = 0$ (equivalent to the basic M*), 0.2, 0.35, and 0.45, and with a fixed offset parameter $\delta = 9.0$. Table 2 (left) shows the run time for the experiments. As can be seen from the results, in general, SC-M* can handle the two-resource-two-type systems and plan for more than 80 agents. Because more resources contribute more factors to increasing the collision score, a relatively large offset ($\delta = 9.0$) is needed to achieve comparable performance to the one-resource-one-type SC-M*.

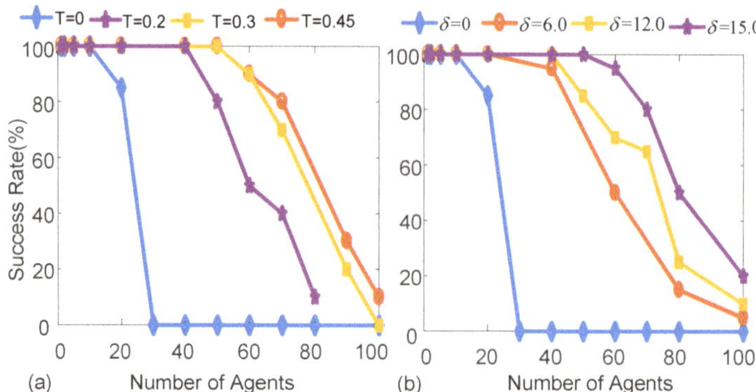

Figure 6. Impact of the collision threshold T (given $\delta = 9.0$) and offset parameter δ (given $T = 0.35$) on two-resource-two-type SC-M*.

Table 2. Run time of two-resource-two-type SC-M* under different parameters.

m	$T = 0.2$	$T = 0.35$	$T = 0.45$	m	$\delta = 6.0$	$\delta = 12.0$	$\delta = 15.0$
5	0.3438	0.3493	0.363948	5	0.333498	0.3677	0.527464
20	1.2485	1.8549	1.807993	20	1.479236	1.4032	2.369483
40	10.102	3.2387	4.415021	40	61.49792	4.5024	4.04331
60	503.94	106.02	104.6499	60	521.2721	306.46	60.30944
90	901.91	801.47	702.4526	70	627.0925	347.14	209.5725
100	>1000	909.0	901.1799	100	901.91	751.78	606.6522
Left: $\delta = 9.0$				Right: $T = 0.35$			

Figure 6b and Table 2 (right) present the impact of the offset parameter δ on performance. Different from the first experiment, SC-M* with the above configurations is less sensitive to δ, when compared to Figure 5. The reason is that 50% of the agents are insensitive to one of the resources because of the linear CDF f_2, thus increasing δ does not contribute to a significant reduction in collisions. This property implies that we can control the importance levels of resources efficiently through the design of collision CDFs. This experiment demonstrates that, with the proper parameter settings, SC-M* can feasibly handle a complex environment with multiple resources and multiple agent types.

5.3. Comparison of SC-M* to Baselines

We next compared the SC-M* to other SC-based MAPP algorithms, including SC-A* (optimal) and SC-CBS (suboptimal), in the one-resource-one-type environment.

5.3.1. Path Cost

Firstly, we compared the path cost of the three algorithms. We designed 60 planning tasks for environments with 4–6 agents (20 tasks for each), in which agents will encounter at least one collision along the individually optimal paths under the $T = 0.05$, $\delta = 1$ setting. We start with small agent numbers because SC-A* cannot handle a large number of agents.

Figure 7 shows the average difference of the three SC-based solvers relative to the individually optimal cost (i.e., the sum of the optimal cost of each agent when the agent is the only one in the system). In other words, the Y-axis represents the cost of collisions. We observe that SC-A* and SC-CBS have the lowest and highest additional cost, respectively. SC-M* solutions cost more than SC-A* but noticeably less than SC-CBS.

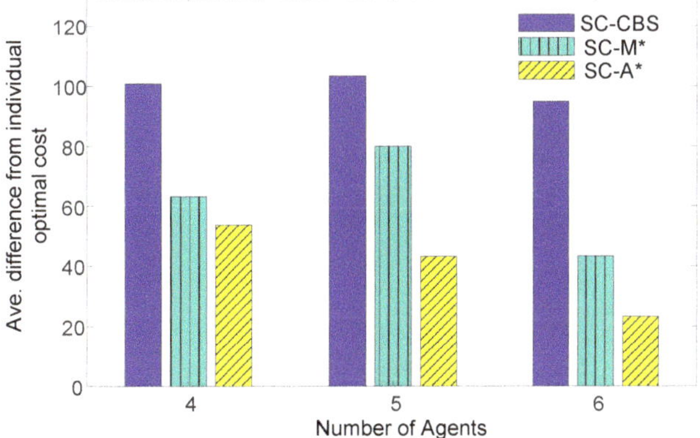

Figure 7. Average cost difference of soft-collision-based multi-agent path planning (SC-based MAPP) solvers from the individually optimal cost in the one-resource-one-type context.

To be more detailed, in the experiments, we designed MAPP tasks for environments containing 4–6 agents with 20 tasks for each. All tasks were designed to encounter at least one collision along the individually optimal paths under the above-mentioned configuration. Thus, additional costs relative to the individually optimal path are expected for each of the three SC-based MAPP solvers. Table 3 compares the results of SC-M* and SC-CBS to the optimal solutions obtained by SC-A*. The top half of the table shows the increase in cost relative to the cost for SC-A*; the costs for SC-A* for all scenarios vary within a small range so the results are in absolute numbers. The bottom half shows the ratio in run time with respect to SC-A*; the run time for SC-A* varies greatly across the experiments so we show the cost reduction as a percentage. In the table, we observe that the additional cost of SC-M* from the SC-A* is consistently lower than that of SC-CBS. We also observe that SC-M* is significantly faster than SC-A* and competitive relative to the run time of SC-CBS. The standard deviations show the fluctuations of the solutions for SC-M* and SC-CBS around the optimal solutions for SC-A*.

Table 3. Results of the path cost experiments.

		$m = 4$		$m = 5$		$m = 6$	
	idx	SC-CBS	SC-M*	SC-CBS	SC-M*	SC-CBS	SC-M*
Cost difference from SC-A*	1	50.60	**0.00**	287.00	**216.60**	22.00	**0.00**
	2	11.00	**0.00**	5.50	**0.00**	50.60	**50.60**
	3	1.10	**1.10**	45.10	**7.70**	5.50	**0.00**
	4	136.50	**0.00**	6.60	**0.00**	11.00	**0.00**
	5	62.70	**31.80**	81.40	**81.40**	177.10	**0.00**
	6	9.90	**9.90**	38.93	**34.31**	0.00	**0.00**
	7	22.00	**22.00**	270.50	**204.50**	33.14	**26.65**
	8	16.50	**16.50**	3.30	**3.30**	182.18	**0.00**
	9	13.20	**0.00**	27.50	**0.00**	211.10	**169.30**
	10	104.50	**0.00**	78.30	**67.08**	32.33	**20.89**
	11	58.08	**24.08**	22.00	**0.00**	79.20	**0.00**
	12	20.90	**0.00**	36.84	**5.94**	52.80	**15.40**
	13	11.00	**0.00**	115.40	**56.00**	72.16	**59.73**
	14	28.60	**28.60**	17.15	**9.34**	63.70	**26.84**
	15	13.20	**0.00**	66.00	**0.00**	19.11	**14.05**
	16	94.60	**0.00**	35.53	**23.65**	18.70	**1.10**
	17	16.83	**2.86**	14.90	**10.71**	318.90	**0.00**
	18	205.70	**22.00**	1.10	**1.10**	48.86	**8.94**
	19	53.24	**27.71**	14.90	**10.71**	1.10	**1.10**
	20	12.10	**3.30**	34.75	**2.53**	32.61	**8.62**
Std. dev		52.44	**12.00**	80.49	**64.14**	84.64	**39.15**
Run time percentage to SC-A*	1	38.39%	**52.84%**	24.54%	**18.24%**	0.33%	**0.58%**
	2	10.41%	**20.12%**	4.37%	**3.42%**	0.23%	**0.26%**
	3	14.53%	**24.58%**	0.09%	**0.30%**	2.32%	**4.60%**
	4	25.12%	**15.42%**	0.29%	**0.71%**	0.92%	**1.35%**
	5	0.81%	**0.54%**	21.62%	**56.30%**	0.55%	**0.82%**
	6	19.17%	**15.67%**	0.52%	**0.59%**	3.48%	**3.07%**
	7	14.02%	**23.87%**	16.39%	**17.05%**	0.43%	**0.44%**
	8	29.82%	**19.11%**	1.40%	**1.89%**	0.07%	**0.18%**
	9	64.67%	**14.87%**	6.16%	**8.93%**	0.13%	**0.14%**
	10	46.67%	**37.19%**	0.45%	**0.73%**	0.10%	**0.15%**
	11	5.94%	**16.49%**	1.96%	**5.56%**	0.15%	**0.23%**
	12	5.16%	**36.22%**	2.36%	**3.29%**	0.46%	**1.19%**
	13	11.40%	**19.94%**	0.44%	**1.12%**	0.47%	**0.61%**
	14	6.18%	**17.72%**	0.76%	**2.11%**	0.78%	**0.58%**
	15	33.07%	**44.81%**	25.38%	**34.96%**	0.48%	**0.56%**
	16	86.83%	**37.86%**	1.92%	**3.10%**	0.26%	**0.20%**
	17	16.10%	**29.91%**	0.92%	**2.10%**	4.01%	**3.95%**
	18	11.68%	**31.59%**	10.04%	**9.57%**	0.28%	**0.67%**
	19	3.97%	**14.12%**	0.92%	**2.10%**	1.67%	**1.80%**
	20	4.37%	**13.83%**	1.01%	**1.54%**	0.60%	**0.52%**
Std. dev.		22.34%	**12.57%**	8.65%	**14.08%**	1.12%	**1.30%**

The reason for the results is that SC-A* is an optimal solver for this type of MAPP problem because it always explores cheaper paths in the entire multi-agent joint space before considering the paths that cost more [7]. SC-M* is suboptimal because of the process discussed in Section 4.3.2. Compared to SC-M*, SC-CBS suffers from more path cost due to the way it collects a collision: CBS collects collisions into a *conflict tree* and arranges the collision into the form [agent j, vertex v, step s], indicating that agent j collides at vertex v at step s. In each iteration, CBS conducts decoupled planning to avoid agent j reaching vertex v at step s. This might lose some information in the soft-collision context because there might exist another path that leads j to vertex v at step s without announcing a collision, by avoiding one of the upstream vertexes involved in soft interference. In contrast, SC-M* can explore those paths excluded by SC-CBS because it searches the entire space of the immediate colliding agents. Figure 8 provides an example to visualize the difference in planning among the three SC-based MAPP solvers.

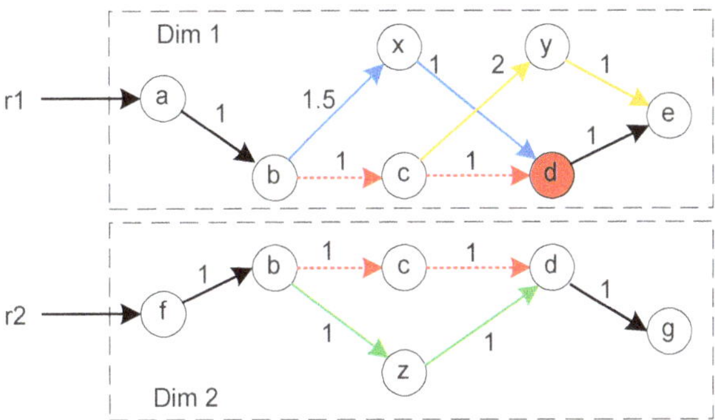

Figure 8. Illustration of the difference in planning among soft-collision A* (SC-A*), soft-collision M* (SC-M*), and soft-collision conflict-based search (SC-CBS).

Figure 8 shows a two-agent MAPP problem in the soft-collision context. Agents $r1$ and $r2$ attempt to move from vertexes a and f to vertexes e and g, respectively. The individually optimal paths (shortest distance) for both agents are $a \rightarrow b \rightarrow c \rightarrow d \rightarrow e$ with distance 4 for $r1$ and $f \rightarrow b \rightarrow c \rightarrow d \rightarrow g$ with distance 4 for $r2$, respectively. The total cost of the joint individually optimal path is 8. $r1$ and $r2$ softly collide on the edge $b \rightarrow c$ and $c \rightarrow d$, where $r2$ can tolerate the dissatisfying experience with distance 2. However, $r1$ can only tolerate the dissatisfying experience with distance 1 and announces a collision at the vertex d.

When using SC-CBS, we record the collision that occurred to $r1$ as [$r1$, d, 3], indicating that agent $r1$ will collide at vertex d at the third step. Then, SC-CSB will avoid any paths leading $r1$ to d at Step 3 (including $a \rightarrow b \rightarrow x \rightarrow d \rightarrow e$ and $a \rightarrow b \rightarrow c \rightarrow d \rightarrow e$) and will end up with a longer detour through vertex y. The SC-CBS solution has a cost of 5 for $r1$ and 9 in total.

When using SC-M*, the collision at d triggers the sub-dimensional expansion of the search graph in dimension 1, which includes both x and y. Thus, it can find a cheaper collision-free path through x and end up with a path $a \rightarrow b \rightarrow x \rightarrow d \rightarrow e$ with a dissatisfying experience of distance 1 and a cost of 4.5 for $r1$ (8.5 in total). However, SC-M* does not expand dimension 2 because no collision has been announced by $r2$.

When using SC-A*, the joint search space of both dimension 1 and dimension 2 is expanded and searched. Instead of vertexes x and y, SC-A* will first investigate vertex z in dimension 2 according to some heuristics. This process leads to another cheaper path $f \rightarrow b \rightarrow z \rightarrow d \rightarrow g$ with distance 4 for $r2$ (8 in total, which is the same as the individually optimal cost) and avoids all interference by moving through this path. As a result, SC-A* returns an optimal solution that satisfies the soft-collision constraint at the expense of search space.

The example in Figure 8 illustrates the optimality of SC-A* and the advantage of SC-M* in path cost over SC-CBS. To be specific, SC-M* provides a better solution than SC-CBS by searching thoroughly through the expanded dimensions, whereas the way SC-CBS identifies collisions is inappropriate in the soft-collision context. To the best of our knowledge, no other methodology capable of dealing with the soft-collision path planning defined in Equation (11) has been developed. It is expected that, in the future, more high-performance algorithms will be developed for solving the problem.

5.3.2. Run Time

Table 4 shows the average run time of the three SC-based MAPP solvers and we observe that both SC-M* and SC-CBS are significantly faster than SC-A* in terms of run time. This is reasonable because SC-A* always searches the global high-dimensional joint space, which is expensive. SC-CBS is faster than SC-M* because it always searches in one individual dimension at a time, whereas the SC-M* needs to occasionally deal with high-dimensional space when collisions occur.

Table 4. Average run time of SC-based MAPP solvers in the one-resource-one-type context.

m	SC-CBS	SC-M*	SC-A*
4	1.971	2.002	47.35
5	1.798	3.312	473.7
6	1.942	2.969	390.0

5.3.3. Scalability

We compared the scalability of the three SC-based MAPP solvers in terms of planning for a large system size ($m > 50$). Figure 9 presents the success rate, average additional cost (i.e., how much more cost than the individually optimal path), and run-time ratio over SC-CBS under different thresholds T, where the run-time ratio of SC-CBS is compared to itself and thus is constant. SC-A* has the slackest constraint ($T = 0.35, \delta = 9.0$) but poorest performance because of the prohibitively large search space. SC-CBS has the best success rate because of the property of the decoupled searching. However, this is at the expense of path cost. SC-M* performs decently in terms of both the success rate (significantly superior to SC-A*) and cost (noticeably lower than SC-CBS) as the number of agents increases.

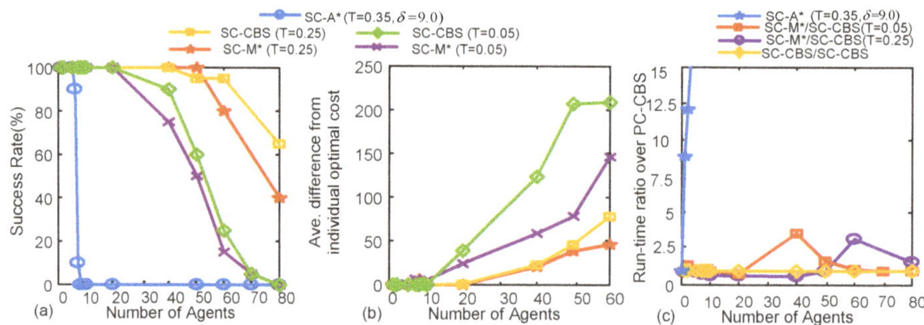

Figure 9. Success rate, cost, and run time ratio of the three SC-based MAPP solvers under different T.

The run time of the SC-M* is generally longer than that of SC-CBS. In another experiment, we observe that the run-time ratio of SC-M* over that of the SC-CBS starts to decrease after a peak. This is because we force all algorithms to terminate after 1000 s, and both curves will converge to value one when their success rates decline to zero. We conducted another scalability experiment with different offsets δ (given $T = 0.25$) and observe the same results in terms of scalability. Figure 10 shows the experimental results.

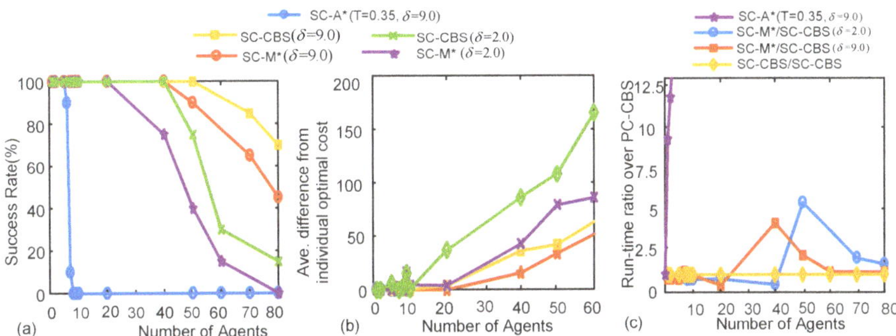

Figure 10. Success rate, cost, and run-time ratio of the three SC-based MAPP solvers under different δ.

Considering the scalability and path cost altogether, SC-M* demonstrates its overall advantages over alternative SC-based solvers.

6. Conclusions

This paper proposes SC-M*, a generalized version of M* with soft-collision constraints on common resources, which can scale to solving the multi-agent path planning problem in the soft-collision context. The SC-M* tracks the collision score of each agent and place agents, whose collision scores exceed some thresholds into a soft-collision set for sub-dimensional expansion. We show that the SC-M* has advanced flexibility and scalability for efficiently solving MAPP problems in the soft-collision context and can handle complex environments (e.g., with multiple types of agents requesting multiple types of resources). We compare the SC-M* to other SC-based MAPP solvers and show the advantages and trade-offs of the SC-M* against baselines in terms of path cost, success rate, and run time.

Future work will focus on leveraging advanced variants of M*, such as EPErM*, ODrM*, etc., to remove the basic A* component in our planner. We believe that better performance can be obtained this way because these variants improve the coupled planner and policy generator (two important components in the basic M*), which are directly related to the M* bottlenecks that limit the planning scalability. We are also interested in applying SC-M* to real-world applications for case studies. One promising research direction is to use the proposed algorithm to serve the passengers in public transits. It is expected that SC-M* will handle large-scale mobility demands in cities

Author Contributions: Conceptualization, R.S., P.S. and M.M.V.; Methodology, R.S. and M.M.V.; experimental design, R.S. and P.S.; software, R.S.; analysis, R.S.; writing, original draft preparation, R.S.; writing, review and editing, R.S., P.S. and M.M.V.; supervision, M.M.V. and P.S.; and funding acquisition, M.M.V. and P.S.

Funding: This research was funded by the Fundação para a Ciência e a Tecnologia (FCT), the Portuguese national funding agency, under the Sensing and Serving a Moving City (S2MovingCity) project (Grant CMUP-ERI/TIC/0010/2014).

Acknowledgments: The authors would like to thank Stephen F. Smith and Carlee Joe-Wong for helpful discussions.

Conflicts of Interest: The authors declare no conflict of interest.

Abbreviations

The following abbreviations are used in this manuscript:

MAPP	Multi-Agent Path Planning
SC-M*	Soft-Collision M*
OD	Operator Decomposition
EPEA*	Enhanced Partial Expansion A*
IDA*	Iterative Deepening A*
CBS	Conflict-Based Search
MA-CBS	Meta-Agent Conflict-Based Search

References

1. Dresner, K.; Stone, P. A multiagent approach to autonomous intersection management. *J. Artif. Intell. Res.* **2008**, *31*, 591–656. [CrossRef]
2. Pallottino, L.; Scordio, V.G.; Bicchi, A.; Frazzoli, E. Decentralized cooperative policy for conflict resolution in multivehicle systems. *IEEE Trans. Robot.* **2007**, *23*, 1170–1183. [CrossRef]
3. Silver, D. Cooperative Pathfinding. In Proceedings of the 1st Conference on Artificial Intelligence and Interactive Digital Entertainment (AIIDE), Marina del Rey, CA, USA, 1–3 June 2005; pp. 117–122.
4. Wagner, G.; Choset, H. M*: A complete multirobot path planning algorithm with performance bounds. In Proceedings of the 2011 IEEE/RSJ International Conference on Intelligent Robots and Systems (IROS), San Francisco, CA, USA, 25–30 September 2011; pp. 3260–3267.
5. Ratner, D.; Warmuth, M.K. Finding a shortest solution for the NxN extension of the 15-PUZZLE is intractable. In Proceedings of the 5th AAAI Conference on Artificial Intelligence (AAAI), Philadelphia, PA, USA, 11–15 August 1986; pp. 168–172.
6. Wagner, G.; Choset, H. Subdimensional expansion for multirobot path planning. *Artif. Intell.* **2015**, *219*, 1–24. [CrossRef]
7. Hart, P.E.; Nilsson, N.J.; Raphael, B. A formal basis for the heuristic determination of minimum cost paths. *IEEE Trans. Syst. Sci. Cybern.* **1968**, *4*, 100–107. [CrossRef]
8. Standley T.S. Finding optimal solutions to cooperative pathfinding problems. In Proceedings of the 24th AAAI Conference on Artificial Intelligence (AAAI), Atlanta, GA, USA, 11–15 July 2010; pp. 173–178.
9. Felner, A.; Goldenberg, M.; Sharon, G.; Stern, R.; Beja, T.; Sturtevant, N.; Schaeffer, J.; Holte, R. Partial-expansion A* with selective node generation. In Proceedings of the 26th AAAI Conference on Artificial Intelligence (AAAI), Toronto, ON, Canada, 22–26 July 2012; pp. 471–477.
10. Goldenberg, M.; Felner, A.; Stern, R.; Sharon, G.; Sturtevant, N.; Holte, R.C.; Schaeffer, J. Enhanced partial expansion A*. *J. Artif. Intell. Res.* **2014**, *50*, 141–187. [CrossRef]
11. Korf, R.E. Depth-first iterative-deepening: An optimal admissible tree search. *Artif. Intell.* **1985**, *27*, 97–109. [CrossRef]
12. Sanchez, G.; Latombe, J.C. Using a PRM planner to compare centralized and decoupled planning for multi-robot systems. In Proceedings of the 2002 IEEE International Conference on Robotics and Automation (ICRA), Washington, DC, USA, 11–15 May 2002; pp. 2112–2119.
13. Sharon, G.; Stern, R.; Felner, A.; Sturtevant, N.R. Conflict-based search for optimal multi-agent pathfinding. *Artif. Intell.* **2015**, *219*, 40–66. [CrossRef]
14. Ferner, C.; Wagner, G.; Choset, H. ODrM* optimal multirobot path planning in low dimensional search spaces. In Proceedings of the 2013 IEEE International Conference on Robotics and Automation (ICRA), Karlsruhe, Germany, 6–10 May 2013; pp. 3854–3859.
15. Wagner, G.; Choset, H. Path planning for multiple agents under uncertainty. In Proceedings of the 27th International Conference on Automated Planning and Scheduling (ICAPS), Pittsburgh, PA, USA, 18–23 July 2017; pp. 577–585.
16. Ma, H.; Kumar, T.S.; Koenig, S. Multi-agent path finding with delay probabilities. In Proceedings of the 31st AAAI Conference on Artificial Intelligence (AAAI), San Francisco, CA, USA, 4–9 Feburary 2017; pp. 3605–3612.
17. Shi, R.; Steenkiste, P.; Veloso, M.M. Second-order destination inference using semi-supervised self-training for entry-only passenger data. In Proceedings of the 4th IEEE/ACM International Conference on Big Data Computing, Applications and Technologies (BDCAT), Austin, TX, USA, 5–8 December 2017; pp. 255–264.
18. Shi, R.; Steenkiste, P.; Veloso, M.M. Generating synthetic passenger data through joint traffic-passenger modeling and simulation. In Proceedings of the 21st IEEE International Conference on Intelligent Transportation Systems (ITSC), Maui, HI, USA, 4–7 November 2018; pp. 3397–3402.
19. Shi, R. Optimizing Passenger on-Vehicle Experience Through Simulation and Multi-Agent Multi-Criteria Mobility Planning. Ph.D. Dissertation, Carnegie Mellon University, Pittsburgh, PA, USA, May 2019.
20. Wang, H.; Xie, H.; Qiu, L.; Yang, Y.R.; Zhang, Y.; Greenberg, A. COPE: Traffic engineering in dynamic networks. In Proceedings of the 2006 Conference of the ACM Special Interest Group on Data Communication (SIGCOMM), Pisa, Italy, 11–15 September 2006; pp. 99–110.

21. Fletcher, L.; Teller, S.; Olson, E.; Moore, D.; Kuwata, Y.; How, J.; Leonard, J.; Miller, I.; Campbell, M.; Huttenlocher, D.; et al. The MIT—Cornell Collision and Why It Happened. *J. Field Robot.* **2008**, *25*, 775–807. [CrossRef]
22. Leonard, J.; How, J.; Teller, S.; Berger, M.; Campbell, S.; Fiore, G.; Fletcher, L.; Frazzoli, E.; Huang, A.; Karaman, S.; et al. A perception-driven autonomous urban vehicle. *J. Field Robot.* **2008**, *25*, 727–774. [CrossRef]
23. Zhou, W.; Zhang, C.; Wang, Q. Optimal flow distribution of military supply transportation based on network analysis and entropy measurement. *Eur. J. Oper. Res.* **2018**, *264*, 570–581. [CrossRef]
24. Dijkstra, E.W. A note on two problems in connexion with graphs. *Numer. Math.* **1959**, *1*, 269–271. [CrossRef]
25. Jansen, M.R.; Sturtevant, N.R. Direction Maps for Cooperative Pathfinding. In Proceedings of the 4th Conference on Artificial Intelligence and Interactive Digital Entertainment (AIIDE), Stanford, CA, USA, 22–24 October 2008; pp. 185–190.
26. Van Den Berg, J.; Abbeel, P.; Goldberg, K. LQG-MP: Optimized path planning for robots with motion uncertainty and imperfect state information. *Int. J. Robot. Res.* **2011**, *30*, 895–913. [CrossRef]

Article

Unscented Transformation-Based Multi-Robot Collaborative Self-Localization and Distributed Target Tracking

Yang Lyu [1,*], Quan Pan [1] and Jian Lv [2,*]

1 School of Automation, Northwestern Polytechnical University, Xi'an 710072, China; quanpan@nwpu.edu.cn
2 Key Laboratory of Advance Manufacturing Technology, Ministry of Education, Guizhou University, Guiyang 550025, China
* Correspondence: lincoln1587@mail.nwpu.edu.cn (Y.L.); jlv@gzu.edu.cn (J.L.)

Received: 26 December 2018; Accepted: 27 February 2019; Published: 3 March 2019

Abstract: The problem of multi-robot collaborative self-localization and distributed target tracking in practical scenarios is studied in this work. The major challenge in solving the problem in a distributed fashion is properly dealing with inter-robot and robot–target correlations in order to realize consistent state estimates of the local robots and the target simultaneously. In this paper, an unscented transformation-based collaborative self-localization and target tracking algorithm is proposed. Inter-robot correlations are approximated in a distributed fashion, and robot–target correlations are safely discarded with a conservative covariance intersection method. Furthermore, the state update is realized in an asynchronous manner with different kinds of measurements while accounting for measurement and communication limitations. Finally, to deal with nonlinearity in the processes and measurement models, the unscented transformation approach is adopted. Unscented transformation is better able to characterize nonlinearity than the extended Kalman filter-based method and does not require computation of the Jacobian matrix. Simulations are extensively studied to show that the proposed method can realize stable state estimates of both local robots and targets, and results show that it outperforms the EKF-based method. Moreover, the effectiveness of the proposed method is verified on experimental quadrotor platforms carrying off-the-shelf onboard sensors.

Keywords: collaborative localization; distributed tracking; nonlinear model; unscented transformation

1. Introduction

Multi-robot systems (MRSs) have garnered tremendous research interest in recent years [1]. Compared with a single robot, an MRS usually has greater efficiency and operational capability in accomplishing complex tasks, such as transportation [2], search and rescue [3], and mapping [4]. Among these MRS applications are the fundamental tasks of obtaining reliable localization information for the local robot and the uncooperative target using various measurements; these two processes are often referred to as collaborative self-localization (CL) [5–9] and distributed target tracking (DT) [10–13], respectively. In the CL process, each robot measures the relative quantities with regard to neighboring cooperative robots. By cooperating with other robots, each robot is able to refine its own positioning information. In the DT process, each robot performs a measurement function on the uncooperative targets to be tracked. Then, the states of the target can be estimated cooperatively through interactions with other robots. Although the problems of CL and DT are usually solved by two separate techniques, such as in [5–11,13–17], they are correlated in most practical scenarios. In the DT process, the target tracking accuracy is dependent on the localization information of the corresponding robots, as well as the relative measurements between the robot and target. The target tracking results obtained by each local robot, in turn, can be implemented to improve the localization performance of the robots. To realize MRS self-localization and target tracking simultaneously, a combined collaborative self-localization and distributed target tracking (CLAT) framework is studied in this paper.

The problem of multi-robot collaborative localization has drawn significant attention in recent years. In [18], the state of the art in collaborative localization is surveyed, and the theoretical limits, algorithms, and practical challenges are discussed. As one of the fundamental challenges in CL, the application of a proper data fusion strategy to deal with the correlations between robots was studied in [5–9,14,15]. A direct approach involves the local robot treating the states of neighboring robots as fully confident variables that will lead to zero correlations between robots [5]. However, this impractical assumption of neighboring positions can lead to overconfident estimates. A more practical method to fuse the relative measurements when the correlation is unknown is the implementation of conservative correlation approximation methods, such as covariance intersection (CI) [6] or split covariance intersection (SCI) [7]. A CL approach using CI was proposed in [14]. This method is provably consistent and can handle asynchronous communication and measurement. The SCI-based approach, as studied in [7,8], further separates the covariance into correlated and uncorrelated parts, and the latter is fused using the CI method. Despite the ability of CI-based methods to preserve the consistency of the estimates, they often have overly conservative results. Making a trade-off between estimation accuracy and the corresponding cost during the CI-based collaborative localization process was investigated using the optimal scheduling problem in [19]. Another popular method to deal with the CL problem is based on factor graphs, which are formulated on the basis of entire trajectories, such as in [15]. The correlation can be explicitly tracked in the factor-graph-based method. However, storing all of the measurements resulting from this method requires significantly more storage space than the recursive method. To address the above drawbacks, a recursive extended Kalman filter (EKF)-based CL method was proposed in [20], in which the correlation was accurately tracked in a decentralized manner. In [21], the processing and storage costs were further reduced by introducing a server that broadcasts an update message when an inter-robot relative measurement is taken. However, in this method, when a relative measurement is taken between two robots, communication involves all robots rather than just the two in the relative measurement, and this significantly increases the communication burden. Another recursive EKF-based CL method was proposed in [9]. This method efficiently approximates the correlation and only stores the current measurement. When the relative measurement is taken, only the communication between the two robots is required.

The distributed tracking problem has also been extensively studied [22]. Early-stage algorithms that have been proposed to solve this problem can be roughly split into two categories: consensus-based algorithms [16] and diffusion-based algorithms [10]. The former category, in general, requires multiple communication iterations during each sampling time interval and hence could lead to a heavy communication burden. To reduce the communication bandwidth, a distributed Kalman filter with event-triggered communication was proposed in [23], and the stability is guaranteed. The latter category does not have such drawbacks, but it may require local joint detectabilities at every single agent, and such a requirement might not be satisfied in a general multi-robot target tracking scenario. A more practical DT approach called distributed hybrid information fusion (DHIF) [11] is able to guarantee stability and is asymptotically unbiased with very mild sufficient conditions. To further solve the distributed tracking problem with a nonlinear process and sensing models, an EKF-based paradigm was proposed in [24], and the stability was analyzed in [25]. Also, the unscented transformation-based approach, which has been regarded as a superior alternative to the EKF when the systems are highly nonlinear, was integrated with the DT process in [26]. However, both the EKF and unscented Kalman filter (UKF) mentioned above are consensus-based and hence may generally result in a heavy communication burden. Recently, the aforementioned DHIF was extended to a nonlinear scenario using the DT approach in [12], and the stochastic stability was analytically studied. Besides the methodology research and theoretical analyses above, the performance of different fusion strategies in terms of their communication rate, information type, and memory size were compared in [27].

The CLAT framework has gained attention in recent years. There are two main kinds of methods: batch and recursive methods. The batch method estimates the entire state trajectory on the basis of all measurements and motion information up to the present, and the recursive method uses only the current measurement and control information. The batch-based method is supposed to outperform the recursive method but at the cost of significantly larger computation and storage requirements and, if in a distributed fashion, communication requirements. One batch-based method was proposed in [28]. By introducing a factor graph that contained robot and target nodes and relative measurements, the problem was formulated as a least-square minimization problem and was solved with sparse optimization methods. Another batch-based method was presented in [29], where the CLAT problem was formulated as a maximum a posteriori estimation problem, and the unscented transformation (UT) technique was implemented to better characterize the nonlinear process. Furthermore, the observability condition was extensively studied. For an MRS with limited computation and storage capacity, the recursive method is often preferred. A recursive-filter-based CLAT was studied in [30], and the error bounds are theoretically provided. Nevertheless, the results in [30] are based on a specially designed measurement graph so that the correlation can be tracked properly. A recursive Bayesian method was proposed in [31] to perform the CLAT in a distributed sequential fashion; however, this method needs synchronous communication at each time instance and will therefore add a significant communication burden. Further, an error propagation analysis was carried out, and the convergence conditions are given in [32], which showed that the localization and tracking accuracy only depends on the expectation of the measurement precision.

In this paper, the multi-robot localization and target tracking problem with a general nonlinear process and various measurement models is studied, and a UT-based CLAT scheme is proposed with consideration of the communication and memory limitations. The main contributions of this paper are summarized as follows: First, the proposed UT-based CLAT is recursive, and it does not store measurements; each robot only keeps the latest estimates of its own and the target, so the storage requirement is significantly reduced. Furthermore, communication is limited to the two robots to obtain a cooperative relative measurement, and no communication with other robots is needed. Meanwhile, to guarantee estimation consistency, inter-robot correlations are approximated in a distributed fashion on the basis of the covariance split method, and the robot–target correlation is discarded using the conservative CI method. Finally, the overall system is modeled on the basis of general nonlinear models and is characterized on the basis of the UT approach rather than the EKF method. Thus, the computation of a Jacobian is avoided. Simulations were carried out, and they indicate that the proposed UT-CLAT method is able to realize stable state estimates of both local robots and targets. More importantly, a hardware platform containing three quadrotors was implemented to verify the effectiveness of the proposed UT-CLAT method. Specifically, three types of measurements (absolute measurement, relative cooperative, and uncooperative measurement) from, respectively, the navigation system, ultra-wide bandwidth (UWB) transmitters, and onboard cameras were utilized to effectively estimate the states of the local robots and targets.

The rest of this paper is organized as follows: Section 2 formulates the CLAT problem, and Section 3 describes the proposed CLAT method. Sections 4 and 5, respectively, provide the simulation results, which are based on synthetic data, and experimental results, which are based on hardware platforms. Section 6 concludes the paper.

2. Problem Formulation

2.1. Models

Consider N homogeneous cooperative robots, denoted as $i \in \mathcal{V}$, performing collaborative self-localization and distributed tracking of a target of interest, denoted as t. The dynamics of the cooperative robots and the target are expressed respectively with the following nonlinear process models:

$$\mathbf{x}_{i,k+1} = f_v(\mathbf{x}_{i,k}, \mathbf{u}_{i,k}), \tag{1}$$

$$\mathbf{x}_{t,k+1} = f_t(\mathbf{x}_{t,k}, \mathbf{w}_{t,k}), \tag{2}$$

where $\mathbf{x}_{i,k} \in \mathbb{R}^{n_v}$ and $\mathbf{x}_{t,k} \in \mathbb{R}^{n_t}$ respectively denote the state of robot i and target t at time k. $\mathbf{u}_{i,k} \in \mathbb{R}^{n_u}$ is the control input of robot i, which is assumed to be subject to a Gaussian distribution $\mathbf{u}_{i,k} \sim \mathcal{N}(\bar{\mathbf{u}}_{i,k}, \mathbf{Q}_i)$. $\bar{\mathbf{u}}_i$ denotes the control command, and $\mathbf{Q}_i$ is the control input covariance. $\mathbf{w}_{t,k} \in \mathbb{R}^{n_w}$ is the process noise of the target and is assumed to be drawn from a zero-mean Gaussian distribution $\mathbf{w}_{t,k} \sim \mathcal{N}(\mathbf{0}, \mathbf{Q}_t)$. It is assumed that all robots $i \in \mathcal{V}$ share the same nonlinear transformation $f_v : \mathbb{R}^{n_v} \times \mathbb{R}^{n_u}$, and the moving target nonlinear transformation $f_t : \mathbb{R}^{n_t} \times \mathbb{R}^{n_w}$ is known to all robots.

In the cooperative localization and target tracking scenario, each robot i is able to measure three pieces of information: its absolute state and the relative pairwise measurements to neighboring robots and to the target. The measurements at time instance k are denoted respectively as $\mathbf{z}^a_{i,k} \in \mathbb{R}^{n_{za}}$, $\mathbf{z}^c_{ij,k} \in \mathbb{R}^{n_{zc}}, j \in \mathcal{V}\backslash\{i\}$, and $\mathbf{z}^t_{it,k} \in \mathbb{R}^{n_{zt}}$. The corresponding measurement functions are listed below:

$$\mathbf{z}^a_{i,k} = h^a(\mathbf{x}_{i,k}, \mathbf{v}^a_{i,k}), \tag{3}$$

$$\mathbf{z}^c_{ij,k} = h^c(\mathbf{x}_{i,k}, \mathbf{x}_{j,k}, \mathbf{v}^c_{i,k}), \tag{4}$$

$$\mathbf{z}^t_{it,k} = h^t(\mathbf{x}_{i,k}, \mathbf{x}_{t,k}, \mathbf{v}^t_{i,k}), \tag{5}$$

where $\mathbf{v}^a_{i,k}, \mathbf{v}^c_{i,k}$ and $\mathbf{v}^t_{i,k}$ are the measurement noise of the above three measurement processes and assumed to be drawn from zero-mean Gaussian distributions, i.e., $\mathbf{v}^a_{i,k} \sim \mathcal{N}(\mathbf{0}, \mathbf{R}^a_i)$, $\mathbf{v}^c_{i,k} \sim \mathcal{N}(\mathbf{0}, \mathbf{R}^c_i)$, and $\mathbf{v}^t_{i,k} \sim \mathcal{N}(\mathbf{0}, \mathbf{R}^t_i)$.

Defining the CLAT as a graph $G(V, E)$, the node set $V = \mathcal{V} \cup \{t\} \cup \{0\}$, and the special node 0 denotes the absolute position origin. The edge set is denoted as $E \subseteq \mathcal{V} \times V$. For a robot $i \in \mathcal{V}$, an edge $(i, 0)$ denotes a robot that can access its own absolute position. In this paper, it is assumed that a subset of the robots $\mathcal{V}$ can obtain the absolute measurement $\mathbf{z}^a_k$. An edge (i, t) indicates that a robot i is able to detect a target t. Since the sensing range of an uncooperative sensor is limited, the availability of an uncooperative target measurement depends on the relative positions of the robot and the target. An edge $(i, j), j \in \mathcal{V}\backslash i$ denotes that a pairwise measurement between robots i and j is obtained. Similarly, a cooperative relative measurement is available when two robots are within the cooperative sensing range. Moreover, in this paper, it is assumed that whenever a relative measurement $\mathbf{z}^c_{ij,k}$ is taken, a communication link is established simultaneously between robots i and j so that they can share information.

2.2. Motivation and Objective

Although the CL and DT problems have been extensively studied, their combination still draws limited attention, especially when considering practical multi-robot operation conditions such as nonlinear models or limited sensing and communication capabilities. One answer to the above challenge was provided in [9] by implementing the EKF scheme and asynchronized measurement update. However, this only covered the CL task, and the uncooperative target was not considered. On the other hand, it is also well known that the computation of Jacobian matrices is required by EKF-based algorithms. This may cause difficulties during implementation. Moreover, the estimation performance may deteriorate if the assumption of local linearity is not valid (e.g., bearing sensors). An alternative approach to extending the algorithm while avoiding the aforementioned potential drawbacks is to use the UT.

Motivated by the above observations, in this paper, a UT-based CLAT scheme (UT-CLAT) is proposed that can realize self-localization and target tracking simultaneously in practical multi-robot operation scenarios. The correlations are properly addressed by implementing split covariance methods, similar to the method in [9], and the covariance intersection method. The UT approach was adopted to approximate the statistics of random variables in nonlinear models. In the end, the effectiveness of the proposed UT-CLAT algorithm is illustrated using not only simulations with synthetic data but also experiments with a networked quadrotor system and off-the-shelf sensors (cameras and UWB transmitters).

3. UT-Based CLAT

In this section, the proposed UT-CLAT is described. The states of local robots and targets are estimated using a recursive UT-based Kalman filter, with the aforementioned three types of measurements updated in an asynchronous fashion. For each robot i, the local states, covariance, and the correlation between it and other robots $j \in \mathcal{V}$ are tracked. Specifically, the correlation term is approximately tracked in a distributed fashion, similar to [9]. As a matter of fact, the target may be detected by different robots at different times. It is difficult to track the robot–target correlation in a local robot when there are inter-robot correlations. To realize consistent state estimation under unknown robot–target correlations, a conservative CI method is introduced to safely remove the robot–target correlation terms and the correlation between target estimates from different robots. The above algorithm consists of state propagation (Section 3.1) and three types of measurement update processes (Section 3.2). In particular, the communication link is supposed to be established only during the cooperative relative measurement update process and the data from different robots are fused.

Suppose that at time k, each robot i has a posterior estimated state and its error covariance at a previous time instance, denoted as $\hat{\mathbf{x}}_{i,k-1}$ and $\mathbf{P}_{i,k-1}$, respectively. If a relative measurement between robots i and j is taken before time instance k, then the correlated term $\mathbf{P}_{ij,k-1}$ is arbitrarily decomposed as

$$\mathbf{P}_{ij,k-1} = \sigma_{ij,k-1}\sigma_{ji,k-1}^{\top} \tag{6}$$

and respectively stored in robots i and j. Robot i also holds an estimation of the target t locally, denoted as $\hat{\mathbf{x}}_{t_i,k-1}$ and $\mathbf{P}_{t_i,k-1}$.

3.1. Propagation

The propagation process involves the local robots as well as the target. According to the dynamics of Equations (1) and (2), each robot propagates its own state estimates and the local estimate of the target.

Let the augmented state vector and the corresponding augmented covariance matrix for each robot's local state at time $k-1$ be denoted respectively as $\hat{\mathbf{x}}^{a}_{i,k-1} \in \mathbb{R}^{n_a}$ and $\mathbf{P}^{a}_{i,k-1} \in \mathbb{R}^{n_a \times n_a}$, where $n_a = n_v + n_u$,

$$\hat{\mathbf{x}}^{a}_{i,k-1} \triangleq \begin{bmatrix} \hat{\mathbf{x}}_{i,k-1} \\ \bar{\mathbf{u}}_{i,k-1} \end{bmatrix}, \text{ and } \mathbf{P}^{a}_{i,k-1} \triangleq \begin{bmatrix} \mathbf{P}_{i,k-1} & \mathbf{0} \\ \mathbf{0} & \mathbf{Q}_i \end{bmatrix}. \tag{7}$$

A set of $2n_a + 1$ sigma points, denoted as $\mathcal{X}^a$, is selected as follows:

$$\begin{aligned}
\mathcal{X}^{a,0}_{i,k-1} &= \hat{\mathbf{x}}^a_{i,k-1}, \\
\mathcal{X}^{a,r}_{i,k-1} &= \hat{\mathbf{x}}^a_{i,k-1} + \left\{\sqrt{(n_a+\gamma)\,\mathbf{P}^a_{i,k-1}}\right\}_{(:,r)}, \text{if } r \in \{1,\cdots,n_a\}, \\
\mathcal{X}^{a,r}_{i,k-1} &= \hat{\mathbf{x}}^a_{i,k-1} - \left\{\sqrt{(n_a+\gamma)\,\mathbf{P}^a_{i,k-1}}\right\}_{(:,r-n_a)}, \text{otherwise.}
\end{aligned}$$

Here, $\gamma = \alpha^2(n_a+\kappa) - n_a$ is a scaling parameter, with $0 < \alpha \le 1$ and $\kappa \in \mathbb{R}$ as tuning parameters to control the spread of the sigma points. The weights for propagating the mean and covariances, denoted respectively as W^r_m and W^r_c, are computed as

$$\begin{aligned}
W^0_m &= \gamma/(n_a+\gamma), \\
W^0_c &= \gamma/(n_a+\gamma) + \left(1-\alpha^2+\beta\right), \\
W^r_m &= W^r_c = 1/2\,(n_a+\gamma), \quad r = 1,\ldots,2n_a,
\end{aligned}$$

where β is used to incorporate extra higher-order effects. Note that the definition of the sigma points directly implies that

$$\sum_{r=0}^{2n_a}(W^r_m)\mathcal{X}^{a,r}_{i,k-1} = \mathcal{X}^{a,0}_{i,k-1} = \hat{\mathbf{x}}^a_{i,k-1},$$

or equivalently,

$$\sum_{r=0}^{2n_a} W^r_m \mathcal{X}^r_{i,k-1} = \hat{\mathbf{x}}_{i,k-1}, \quad \sum_{r=0}^{2n_a} W^r_m \mathcal{U}^r_{k-1} = \bar{\mathbf{u}}_{i,k-1},$$

where $\mathcal{X}^r_{i,k-1}$ and $\mathcal{U}^r_{i,k-1}$ collect the components of $\mathcal{X}^{a,r}_{i,k-1}$ corresponding to, respectively, $\mathbf{x}_{i,k-1}$ and $\mathbf{u}_{i,k-1}$.

The above unscented transform is summarized below:

$$\mathcal{X}^{a,r}_{i,k-1} = UT(\hat{\mathbf{x}}^a_{i,k-1}, \mathbf{P}^a_{i,k-1}), \qquad r = 0,\cdots,2n_a.$$

By defining the augmented state vector and the covariance with regard to the local estimates of the target t within robot i similar to Equation (7), the UT of the target t can be summarized as

$$\mathcal{X}^{a,r}_{t_i,k-1} = UT(\hat{\mathbf{x}}^a_{t_i,k-1}, \mathbf{P}^a_{t_i,k-1}), \qquad r = 0,\cdots,2n_a.$$

Then, the prior local estimates and corresponding error covariance of the current state and target are computed respectively as

$$\hat{\mathbf{x}}_{i,k} = \sum_{r=0}^{2n_a} W^r_m \mathcal{X}^r_{i,k}, \tag{8}$$

$$\mathbf{P}_{i,k} = \sum_{r=0}^{2n_a} W^r_c \left(\mathcal{X}^r_{i,k} - \hat{\mathbf{x}}_{i,k}\right)\left(\mathcal{X}^r_{i,k} - \hat{\mathbf{x}}_{i,k}\right)^\top, \tag{9}$$

and

$$\hat{\mathbf{x}}_{t_i,k} = \sum_{r=0}^{2n_a} W^r_m \mathcal{X}^r_{t_i,k}, \tag{10}$$

$$\mathbf{P}_{t_i,k} = \sum_{r=0}^{2n_a} W^r_c \left(\mathcal{X}^r_{t_i,k} - \hat{\mathbf{x}}_{t_i,k}\right)\left(\mathcal{X}^r_{t_i,k} - \hat{\mathbf{x}}_{t_i,k}\right)^\top, \tag{11}$$

where

$$\mathcal{X}^r_{i,k} = f_v\left(\mathcal{X}^r_{i,k-1}, \mathcal{U}^r_{i,k-1}\right), \quad r = 0,\cdots,2n_a,$$

and

$$\mathcal{X}^r_{t_i,k} = f_t\left(\mathcal{X}^r_{t_i,k-1}, \mathcal{W}^r_{t_i,k-1}\right), \quad r = 0, \cdots, 2n_a.$$

$\mathcal{X}^r_{t_i,k-1}$ and $\mathcal{W}^r_{t_i,k-1}$ are the sigma points corresponding to $\mathbf{x}_{t_i,k-1}$ and $\mathbf{w}_{t,k-1}$, and

$$\sum_{r=0}^{2n_a} W^r_m \mathcal{X}^r_{t_i,k-1} = \hat{\mathbf{x}}_{t_i,k-1}, \sum_{r=0}^{2n_a} W^r_m \mathcal{W}^r_{k-1} = \mathbf{0}.$$

The propagation of the correlation term $\mathbf{P}_{ij}$ involves the pose and control inputs of both i and j, and therefore cannot be propagated locally by robot i. To avoid communication, the local correlation term σ_{ij} is instead propagated as

$$\sigma_{ij,k} = \boldsymbol{\mathcal{F}}_v \sigma_{ij,k-1}, \tag{12}$$

where $\boldsymbol{\mathcal{F}}_v$ is the inferred Jacobian matrix with regard to the dynamic function in Equation (1) and, according to [33], is defined as

$$\boldsymbol{\mathcal{F}}_v = \mathbf{P}^{xx}_{i,k|k-1}\left(\mathbf{P}_{i,k-1}\right)^{-1},$$

where $\mathbf{P}^{xx}_{i,k|k-1} \approx \sum_{r=0}^{2n_a}(\mathcal{X}^r_{i,k} - \hat{\mathbf{x}}_{i,k})(\mathcal{X}^r_{i,k-1} - \hat{\mathbf{x}}_{i,k-1})^\top$.

3.2. Update

In the update stage, three types of measurement (Equations (3)–(5)) are considered. When a private measurement or a target measurement is taken by robot i, the information is updated locally to avoid communication. When two robots i and j are within the relative range, a relative measurement is taken, and local beliefs of both robot and target and the inter-robot correlation term are exchanged to update the estimates of the local robots and target. For clarity, $\hat{\mathbf{x}}^-$, $\mathbf{P}^-$ and $\hat{\mathbf{x}}$, $\mathbf{P}$ are used respectively to denote the state estimate and covariance prior to and after a certain measurement update process.

3.2.1. Private Update

During each private update process, the local robot measures its local pose through, for example, a GPS receiver and magnetometer, to refine its local estimation. Only the local pose participates in the private update process.

First, the inferred Jacobian $\boldsymbol{\mathcal{H}}_{i,k}$ corresponding to the measurement function in Equation (3) is obtained as

$$\boldsymbol{\mathcal{H}}_{i,k} = \mathbf{P}^{xzr}_{i,k}\left(\mathbf{P}^-_{i,k}\right)^{-1},$$

where

$$\mathbf{P}^{xzr}_{i,k} = \sum_{r=0}^{2n_a} W^r_c\left(\mathcal{X}^r_{i,k} - \hat{\mathbf{x}}^-_{i,k}\right)\left(h^a(\mathcal{X}^r_{i,k}) - \mathbf{z}^a_{i,k}\right).$$

Then, the state and covariance can be updated as

$$\hat{\mathbf{x}}_{i,k} = \hat{\mathbf{x}}^-_{i,k} + \mathbf{K}_{i,k}\left(\mathbf{z}^a_{i,k} - h^a_i(\hat{\mathbf{x}}^-_{i,k})\right), \tag{13}$$

$$\mathbf{P}_{i,k} = (\mathbf{I} - \mathbf{K}_{i,k}\boldsymbol{\mathcal{H}}_{i,k})\mathbf{P}^-_{i,k}, \tag{14}$$

where

$$\mathbf{S}_{i,k} = (\boldsymbol{\mathcal{H}}_{i,k})^\top \mathbf{P}^-_{i,k}\boldsymbol{\mathcal{H}}_{i,k} + \mathbf{R}^a_i,$$
$$\mathbf{K}_{i,k} = \mathbf{P}^-_{i,k}\boldsymbol{\mathcal{H}}^\top_{i,k}\mathbf{S}^{-1}_{i,k}.$$

The correlation term within local robot i is updated as

$$\sigma_{ij,k} = (\mathbf{I} - \mathbf{K}_{i,k}\boldsymbol{\mathcal{H}}_{i,k})\sigma^{-}_{ij,k}, j \in \mathcal{V}\backslash i. \tag{15}$$

3.2.2. Target Measurement Update

When a target is detected by robot i, a relative measurement related to the pose of both robot i and target t, denoted as $\mathbf{z}^t_{it}$, is obtained. The measurement update involves the estimates of the robot and the target, as well as their correlation term. As a matter of fact, the correlation term, denoted as $\mathbf{P}_{it_i}$, is difficult to track in a distributed fashion owing to the existence of the inter-robot correlation term. Therefore, in this part, a conservative CI-based method [34] is used to remove the robot–target correlations and guarantee consistency at the same time.

$$\begin{bmatrix} \frac{1}{w_1}\mathbf{P}^{-}_{i,k} & \mathbf{0} \\ \mathbf{0} & \frac{1}{1-w_1}\mathbf{P}^{-}_{t_i,k} \end{bmatrix} \succcurlyeq \begin{bmatrix} \mathbf{P}^{-}_{i,k} & \mathbf{P}^{-}_{it_i,k} \\ \mathbf{P}_{t_i i,k} & \mathbf{P}^{-}_{t_i,k} \end{bmatrix}. \tag{16}$$

The weight w is determined according to [34]. Let $\bar{\mathbf{P}}^{-}_{i,k} \triangleq \frac{1}{w}\mathbf{P}^{-}_{i,k}$ and $\bar{\mathbf{P}}^{-}_{t_i,k} \triangleq \frac{1}{1-w}\mathbf{P}^{-}_{t_i,k}$. The augmented state can be defined as

$$\hat{\mathbf{x}}^{b-}_{i,k} = \begin{bmatrix} \mathbf{x}^{-}_{i,k} \\ \mathbf{x}^{-}_{t_i,k} \\ \mathbf{0} \end{bmatrix} \text{ and } \mathbf{P}^{b}_{i,k} = \begin{bmatrix} \bar{\mathbf{P}}^{-}_{i,k} & \mathbf{0} & \mathbf{0} \\ \mathbf{0} & \bar{\mathbf{P}}^{-}_{t,k} & \mathbf{0} \\ \mathbf{0} & \mathbf{0} & \mathbf{R}^{t}_{i} \end{bmatrix}.$$

Then, the augmented sigma points are obtained as

$$\mathcal{X}^{b}_{i,k} = UT\left(\hat{\mathbf{x}}^{b-}_{i,k}, \mathbf{P}^{b-}_{i,k}\right), \quad r = 0, \cdots, 2n_a. \tag{17}$$

The inferred measurement Jacobian is

$$\begin{bmatrix} \boldsymbol{\mathcal{H}}_{i,k} & \boldsymbol{\mathcal{H}}_{t_i,k} & \boldsymbol{\mathcal{H}}_{v,k} \end{bmatrix} = \mathbf{P}^{xzb}_{i,k}(\mathbf{P}^{b-}_{i,k})^{-1}, \tag{18}$$

where

$$\mathbf{P}^{xzb}_{i,k} = \sum_{r=0}^{2n_a} W^{r}_{c}\left(\mathcal{X}^{b}_{i,k} - \mathbf{x}^{b-}_{i,k}\right)\left(h^{t}(\mathcal{X}^{b}_{i,k}) - \mathbf{z}^{t}_{it,k}\right).$$

The target measurement update process is finally summarized as Equations (19)–(22):

$$\hat{\mathbf{x}}_{i,k} = \hat{\mathbf{x}}^{-}_{i,k} + \mathbf{K}_{i,k}\left(\mathbf{z}^{t}_{it,k} - h^{t}_{i}(\mathbf{x}^{b-}_{i,k})\right), \tag{19}$$

$$\hat{\mathbf{x}}_{t_i,k} = \hat{\mathbf{x}}^{-}_{t,k} + \mathbf{K}_{t_i,k}\left(\mathbf{z}^{t}_{it,k} - h^{t}_{i}(\mathbf{x}^{b-}_{i,k})\right), \tag{20}$$

$$\mathbf{P}_{i,k} = (\mathbf{I} - \mathbf{K}_{i,k}\boldsymbol{\mathcal{H}}_{i,k})\bar{\mathbf{P}}^{-}_{i,k}, \tag{21}$$

$$\mathbf{P}_{t_i,k} = (\mathbf{I} - \mathbf{K}_{t_i,k}\boldsymbol{\mathcal{H}}_{i,k})\bar{\mathbf{P}}^{-}_{t_i,k}, \tag{22}$$

where the innovation covariance and gain are calculated as

$$\mathbf{S}_{i,k} = \begin{bmatrix} \mathcal{H}_{i,k} & \mathcal{H}_{i_t,k} \end{bmatrix} \begin{bmatrix} \bar{\mathbf{P}}_{i,k}^- & \mathbf{0} \\ \mathbf{0} & \bar{\mathbf{P}}_{t,k}^- \end{bmatrix} \begin{bmatrix} \mathcal{H}_{i,k} & \mathcal{H}_{i_t,k} \end{bmatrix}^\top + \mathbf{R}_i^u,$$
$$\mathbf{K}_{i,k} = \bar{\mathbf{P}}_{i,k}^- \mathcal{H}_{i,k} \mathbf{S}_{i,k}^{-1},$$
$$\mathbf{K}_{t_i,k} = \bar{\mathbf{P}}_{t,k}^- \mathcal{H}_{t_i,k} \mathbf{S}_{i,k}^{-1}.$$

Formally, the correlation between robots i and j should be updated as

$$\mathbf{P}_{ij,k} = (\mathbf{I} - \mathbf{K}_{i,k}\mathcal{H}_{i,k})\mathbf{P}_{ij,k}^-.$$

On the basis of the decomposition in Equation (6), the correlation term σ_{ij} can be calculated as below without communication:

$$\sigma_{ij,k} = (\mathbf{I} - \mathbf{K}_{i,k}\mathcal{H}_i)\sigma_{ij,k}^-, j \in \mathcal{V}\backslash i.$$

3.3. Neighbor Measurement Update and Target Information Fusion

When two robots i and j are within a given range, a relative measurement is taken, denoted as $\mathbf{z}_{ij,k}$, and a communication link between the two robots is established. The target update process is as follows. First, the covariance between two robots $\mathbf{P}_{ij}^-$ is recovered according to Equation (6). Similar to the target measurement update process, we define the augmented state prior to the measurement update as

$$\hat{\mathbf{x}}_k^{c-} = \begin{bmatrix} \hat{\mathbf{x}}_{i,k}^- \\ \hat{\mathbf{x}}_{j,k}^- \\ \mathbf{0} \end{bmatrix} \text{ and } \mathbf{P}_k^{c-} = \begin{bmatrix} \mathbf{P}_i^- & \mathbf{P}_{ij}^- & \mathbf{0} \\ \mathbf{P}_{ij}^{-\top} & \mathbf{P}_j^- & \mathbf{0} \\ \mathbf{0} & \mathbf{0} & \mathbf{R}_i^c \end{bmatrix}.$$

Then, the augmented sigma points are obtained as

$$\mathcal{X}_k^c = UT\left(\hat{\mathbf{x}}_k^{c-}, \mathbf{P}_{i,k}^{c-}\right), \quad r = 0, \cdots, 2n_a.$$

The inferred measurement Jacobian is

$$\begin{bmatrix} \mathcal{H}_{i,k} & \mathcal{H}_{j,k} & \mathcal{H}_{v,k} \end{bmatrix} = \mathbf{P}^{xzc}(\mathbf{P}_{i,k}^{c-})^{-1}$$

where

$$\mathbf{P}_{i,k}^{xzc} = \sum_{r=0}^{2n_a} W_c^r \left(\mathcal{X}_{i,k}^c - \hat{\mathbf{x}}_{,k}^{c-}\right)\left(h^c(\mathcal{X}_{i,k}^c) - \mathbf{z}_{ij,k}^c\right).$$

Consequently, the update process for the relative measurement between robots i and j is as below:

$$\hat{\mathbf{x}}_{i,k} = \hat{\mathbf{x}}_{i,k}^- + \mathbf{K}_{i,k}\left(\mathbf{z}_{ij,k} - h_i^c(\hat{\mathbf{x}}_{i,k}^{c-})\right), \tag{23}$$

$$\hat{\mathbf{x}}_{j,k} = \hat{\mathbf{x}}_{t,k}^- + \mathbf{K}_{t,k}\left(\mathbf{z}_{it,k} - h_i^c(\hat{\mathbf{x}}_{i,k}^{c-})\right), \tag{24}$$

$$\mathbf{P}_{i,k} = (\mathbf{I} - \mathbf{K}_{i,k}\mathcal{H}_{i,k})\mathbf{P}_{i,k}^- - \mathbf{K}_{i,k}\mathcal{H}_{j,k}\mathbf{P}_{ji}^-, \tag{25}$$

$$\mathbf{P}_{j,k} = (\mathbf{I} - \mathbf{K}_{j,k}\mathcal{H}_{j,k})\mathbf{P}_{j,k}^- - \mathbf{K}_{j,k}\mathcal{H}_{i,k}\mathbf{P}_{ij}^-, \tag{26}$$

$$\mathbf{P}_{ij,k} = (\mathbf{I} - \mathbf{K}_{i,k}\mathcal{H}_{i,k})\mathbf{P}_{ij,k}^- - \mathbf{K}_{i,k}\mathcal{H}_{j,k}\mathbf{P}_j^-, \tag{27}$$

where

$$\mathbf{S}_{i,k} = \begin{bmatrix}\mathcal{H}_{i,k} & \mathcal{H}_{j,k}\end{bmatrix}\begin{bmatrix}\mathbf{P}_{i,k} & \mathbf{P}_{ij,k}, \\ \mathbf{P}_{ij,k}^{\top} & \mathbf{P}_{j,k}\end{bmatrix}\begin{bmatrix}\mathcal{H}_{i,k} & \mathcal{H}_{j,k}\end{bmatrix}^{\top} + \mathbf{R}^{c},$$
$$\mathbf{K}_{i,k} = (\mathbf{P}_{i,k}\mathcal{H}_{i,k} + \mathbf{P}_{ij,k}\mathcal{H}_{j,k})\mathbf{S}_{k}^{-1},$$
$$\mathbf{K}_{j,k} = (\mathbf{P}_{ji,k}\mathcal{H}_{i,k} + \mathbf{P}_{j,k}\mathcal{H}_{j,k})\mathbf{S}_{k}^{-1}.$$

After the relative measurement update, the correlation $\mathbf{P}_{ij,k}$ is decomposed again as two multiplicative parts $\sigma_{ij,k}$ and $\sigma_{ji,k}$ according to Equation (22). Then, $\sigma_{ij,k}$ and $\sigma_{ji,k}$ are stored in i and j, respectively.

The relative measurement update process also involves the correlation term $\mathbf{P}_{il,k}, l \in \mathcal{V}, l \neq i, j$. Formally, the $\mathbf{P}_{il,k}$ should be updated as

$$\mathbf{P}_{il,k} = (\mathbf{I} - \mathbf{K}_{i,k}\mathcal{H}_{i,k})\mathbf{P}_{il,k} - \mathbf{K}_{i,k}\mathcal{H}_{j,k}\mathbf{P}_{jl,k}.$$

The correlation term $\mathbf{P}_{jl,k}$ is not available to robot i. To reduce the overall communication and avoid communication with l, the split term σ_{il}^{-} in robot i is instead updated in an approximate form, similar to the process in [9], as below:

$$\sigma_{il,k} = \mathbf{P}_{i,k}(\mathbf{P}_{i,k}^{-})^{-1}\sigma_{il,k}^{-}. \tag{28}$$

In addition to the measurement update, the target beliefs $\{\hat{\mathbf{x}}_{t_i,k}, \mathbf{P}_{t_i,k}\}$ and $\{\hat{\mathbf{x}}_{t_j,k}, \mathbf{P}_{t_j,k}\}$ are fused simultaneously. As a matter of fact, the correlation between the two estimates is unknown owing to the unknown target–robot correlation. Again, the conservative CI algorithm can be used as below:

$$\mathbf{P}_{t,k} = \left(w_2(\mathbf{P}_{t_i,k})^{-1} + (1 - w_2)(\mathbf{P}_{t_j,k})^{-1}\right)^{-1}, \tag{29}$$
$$\hat{\mathbf{x}}_{t,k} = \mathbf{P}_{t,k}\left(w_2\mathbf{P}_{t_i,k}^{-1}\hat{\mathbf{x}}_{t_i,k} + (1 - w_2)\mathbf{P}_{t_j,k}^{-1}\hat{\mathbf{x}}_{t_j,k}\right). \tag{30}$$

The weight w_2 can be determined according to [11]. The fused results are then stored in both robots i and j.

4. Simulation

In this section, the proposed UT-CLAT method is validated using synthetic data. Without loss of generality, the scenario contains four cooperative robots, labeled 1–4, tracking an uncooperative target in 2D space (as shown in Figure 1). The robots and the target are assumed to be subject to similar nonlinear unicycle models, as below:

$$\mathbf{x}_{k+1} \triangleq \begin{bmatrix}x_{k+1} \\ y_{k+1} \\ \theta_{k+1}\end{bmatrix} = \begin{bmatrix}x_k + \Delta_T(v_c + w_k^{v})\cos(\theta_k) \\ y_k + \Delta_T(v_c + w_k^{v})\sin(\theta_k) \\ \theta_k + \Delta_T(\omega_c + w_k^{\omega})\end{bmatrix},$$

A subscriber i or t is used to distinguish the robots or the target. The state vector $\mathbf{x}_k$ to be estimated contains three entries—x_k, y_k, and θ_k—which represent the 2D position and the orientation of the robots and the target with respect to the global frame. It is assumed that at the initial time, the robots are randomly placed on different circles centered at $[-10, 10], [10, 10], [-10, -10], [10, -10]$. The same control command $\mathbf{u} = [v_c, \omega_c]^{\top} = [0.3, 0.0375]^{\top}$ is applied to each robot to form four approximated circles with radii 8. The velocity and angular velocity noise are assumed to be subject to Gaussian distributions with the covariance $\mathbf{Q}_i = \mathrm{diag}([0.1^2, (0.5\pi/180)^2])$. The target is initialized at $[-15, -15]^{\top}$ in the global frame, and the control input is set as $\mathbf{u}_t = [0.05, 0]^\top$. Similarly, the target control is subject to zero-mean Gaussian noise with $\mathbf{Q}_t = \text{diag}([0.02^2, (2\pi/180)^2])$.

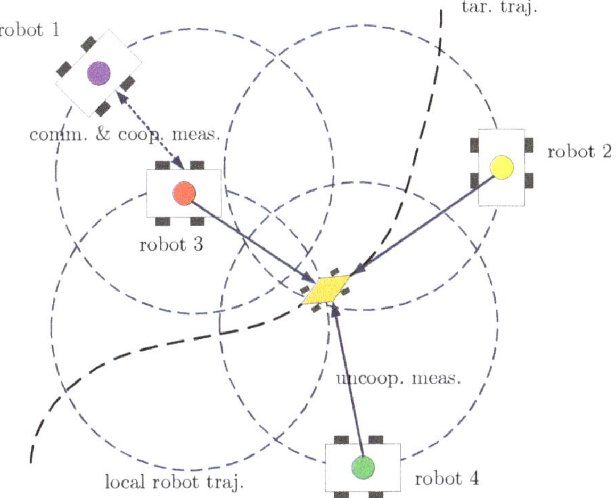

Figure 1. Simulation with four local robots and one moving target.

In the simulation, robot 1 is assumed to be accessible to the global position and orientation in the global frame with the following measurement model:

$$\mathbf{z}^a_{i,k} = \mathbf{x}_{i,k} + \mathbf{v}^a_{i,k},$$

where $\mathbf{v}^a_{i,k} \sim \mathcal{N}(\mathbf{0}, \text{diag}\left[(0.5^2, 0.5^2, 0.5\pi/180)^2\right])$ is the control noise.

Both the cooperative robots and uncooperative measurement are subject to a relative range measurement model as follows:

$$z^c_{ij,k} = \|x_j - x_i\|_d + v^c_{i,k}, \tag{31}$$

$$z^t_{it,k} = \|x_t - x_i\|_d + v^t_{i,k}, \tag{32}$$

$\|\cdot\|_d$ is the operator that calculates the relative range between two robots or a robot and the target. The sensing range for the target is set as $r_t = 20$, and the sensing range for cooperative robots, as well as the communication range, is set as $r_c = 10$. The measurement noises are $v^c_i \sim \mathcal{N}(0, 0.05^2)$ and $v^t_i \sim \mathcal{N}(0, 0.05^2)$, respectively.

4.1. Scenario 1

One trial of the simulation described above was carried out. In this scenario, the target is jointly observed by the four robots intermittently. The observation measurement availability for both cooperative measurement and target measurement is based on the sensing ranges r_c and r_t, respectively, and is shown in Figures 2 and 3.

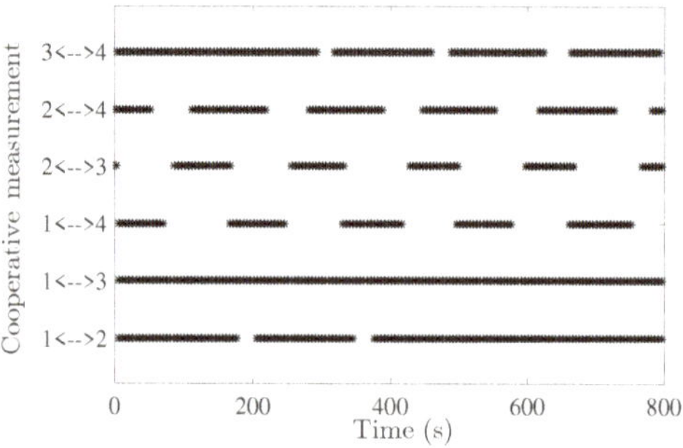

Figure 2. Measurement and communication link availability between two robots with $r_c = 10$.

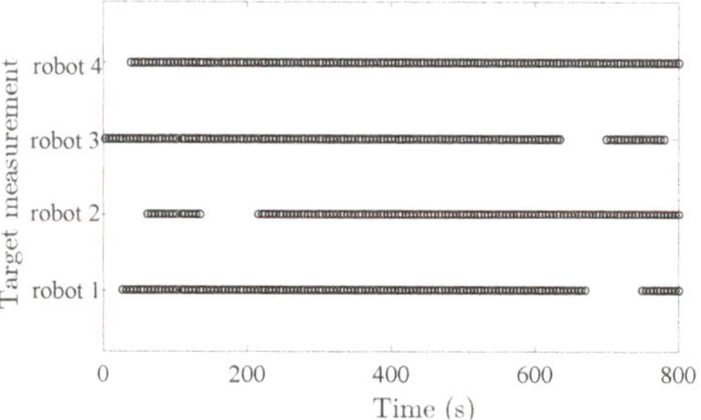

Figure 3. Measurement availability of target to four robots with $r_t = 20$.

Although, for each robot, the observability of the local state and the target's state cannot be guaranteed owing to the discontinuous range-only measurement, the joint observability for the entire system over a period of time can still be guaranteed through communication with neighbors according to [35].

The estimated trajectories of both robots and the target are plotted in Figure 4 in different colors. Each robot's self-localization result and local target tracking result are drawn with solid lines of the same color. As observed in Figure 4, the estimated trajectories indicate that each robot is able to localize its true pose and track the true trajectory of the target. The four robots' self-localization errors and covariances ($\pm 3\sigma$ bounds) are plotted in Figure 5, with solid lines in color and dashed lines in the same color, respectively. It shows that the self-localization errors by each robot are bounded by the $\pm 3\sigma$ envelopes in the steady state. Robot 1 has the lowest tracking error as it can access its own absolute measurement. The target tracking results from the four robots are plotted in Figure 6, where, for each robot, the target tracking errors (solid line in colors) are bounded by the corresponding $\pm 3\sigma$ envelopes (dashed line in the same colors) in the steady state. On the basis of Figures 5 and 6, the min/max self-localization and target tracking errors for time instance $k \in [200, 800]$ are listed in Table 1.

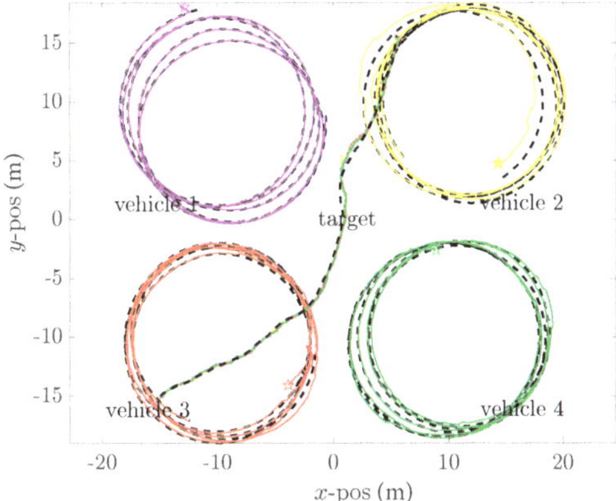

Figure 4. Estimated trajectories of robots and the target in different colors (black dashed lines indicate the ground truth).

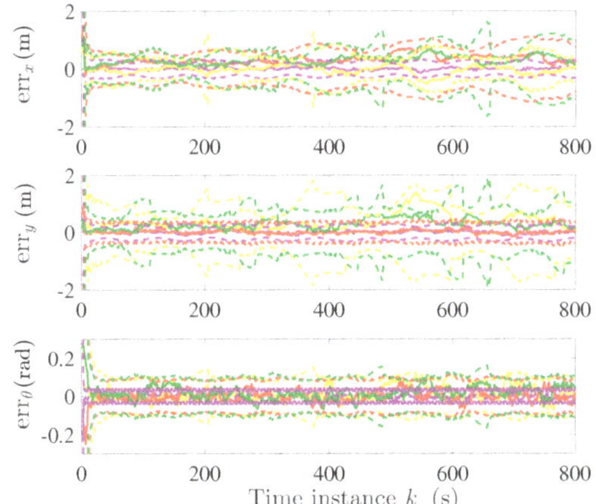

Figure 5. Self-localization errors and covariances of robots 1–4.

Table 1. Self-localization and target tracking errors ($k \in [200, 800]$).

	Absolute Self-Localization Error (min/max)			Absolute Target Tracking Error (min/max)		
	x(m)	**y(m)**	**θ(rad)**	**x(m)**	**y(m)**	**θ(rad)**
robot 1	0.05/0.35	0.03/0.41	0/0.08	0.03/0.5	0.03/0.66	0.01/0.65
robot 2	0.04/0.94	0.1/1.13	0/0.11	0.04/0.54	0.05/0.54	0.01/0.71
robot 3	0.01/1.31	0.12/0.96	0/0.07	0.02/0.61	0.05/0.47	0/0.72
robot 4	0.02/1.08	0.04/1.24	0/0.12	0.04/0.44	0.03/0.55	0/0.75

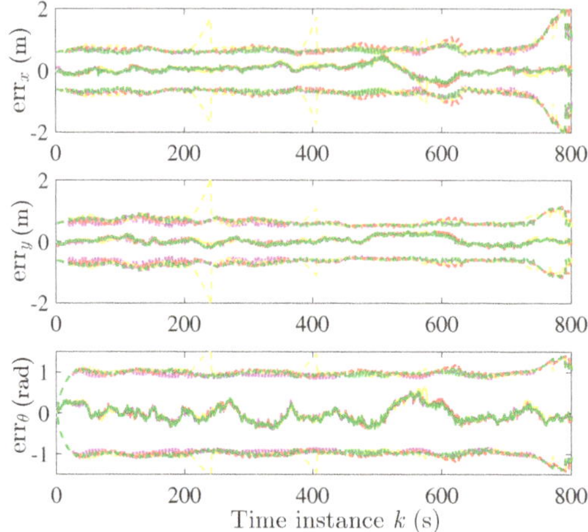

Figure 6. Target tracking errors and covariance of robots 1–4.

4.2. Scenario 2

In this part, the performance results of the proposed UT-CLAT method are presented on the basis of 1000 Monte Carlo simulations. Specifically, the simulation in Scenario 1 was repeated 1000 times with $1 \leq k \leq 1200$. For each robot, the position root-mean-square errors (PRMSEs) of the local posterior estimates and target posterior estimates were computed for all trails. Moreover, to demonstrate the effectiveness in a nonlinear scenario, the proposed UT-CLAT method is compared to the EKF-CLAT method by extending the CL algorithm in [9] to the CLAT scenario. In Figure 7, the averaged PRMSEs of the collaborative localization results of 1000 Monte Carlo simulations are plotted using both the UT-CLAT and EKF-CLAT methods.

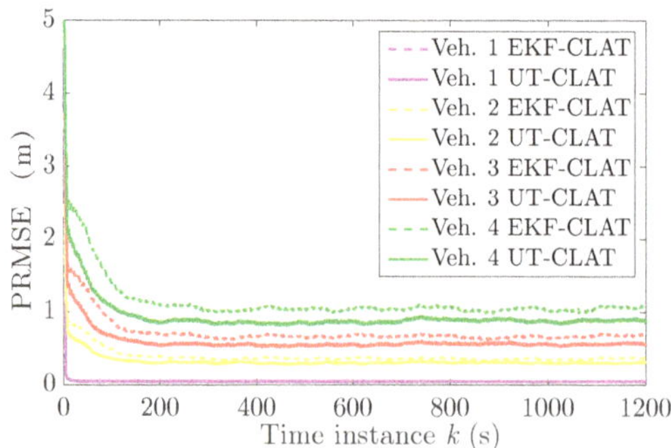

Figure 7. Self-localization estimation position root-mean-square errors (PRMSEs) of the unscented transformation-based collaborative self-localization and target tracking scheme (UT-CLAT) vs. the extended Kalman filter-based CLAT (EKF-CLAT) for robots 1–4.

As observed in Figure 7, both methods can realize stable self-localization in around 200 time instances. In general, the UT-CLAT method is able to achieve more accurate self-localization results. In Figure 8, the averaged PRMSEs of the target tracking results of different robots are plotted. Similar to the CL result in Figure 7, the UT-CLAT is able to realize stable target tracking, and it outperforms the EKF-CLAT method for each robot.

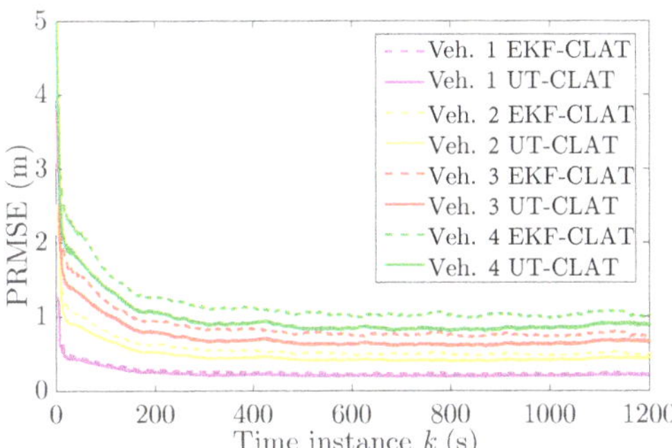

Figure 8. Target distributed estimation PRMSEs of the UT-CLAT vs. the EKF-CLAT for robots 1–4.

5. Experimental Validation of Quadrotors

In this section, the validation results of the proposed UT-CLAT method are presented. Validation was performed using hardware platforms that included three quadrotors tracking a ground robot. As shown in Figure 9, the system consists of three Intel Aero RTF quadrotors, referred to as quad1–3, and one TurtleBot ground robot. Each quadrotor is equipped with UWB transmitters and a downward monochrome camera. The UWB sensors measure the relative distance and transmit information when two quadrotors are within the functional range of the UWB sensors. A camera is rigidly connected to the body frame of each quadrotor. A target is detected by the camera when the target is within the field-of-view (FOV). Furthermore, quad1 is assumed to have access to its position through the onboard navigation system. In addition to the above onboard devices, UWB ground anchors are used to record the ground truth states of both robots and the target.

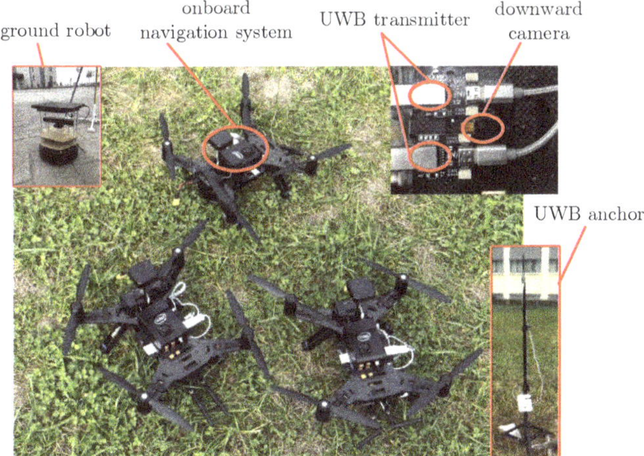

Figure 9. Intel Aero RTF quadrotors (equipped with an onboard navigation system, ultra-wide bandwidth (UWB) transmitters, and downward cameras) and the TurtleBot ground robot.

5.1. Robot and Target Dynamics Model

For target monitoring, each quadrotor is controlled to follow a plenary circular trajectory in 3D space. The state $\mathbf{x}_i$ includes the position in 3D space and the heading angle, namely, $\mathbf{x}_t = [x_i, y_i, z_i, \theta_i]^\top$. $\mathbf{u}_i \triangleq [v_i, \omega_i]^\top$ denotes the control input command, namely, the linear plenary velocity and angular velocity. The actual velocity and angular velocity are contaminated by zero-mean Gaussian noises, $w^v_{i,k} \sim \mathcal{N}(0, Q^v_i)$, $w^\omega_{i,k} \sim \mathcal{N}(0, Q^\omega_i)$. An extra altitude noise $w^z_{i,k} \sim \mathcal{N}(0, Q^z_i)$ is added to the process noise. The overall process noise covariance is denoted as $\mathbf{Q}_i = \text{blkdiag}(Q^v_i, Q^\omega_i, Q^z_i)$. On the basis of the above definition, the process model for each robot f_i is defined as follows:

$$\mathbf{x}_{i,k+1} \triangleq \begin{bmatrix} x_{i,k+1} \\ y_{i,k+1} \\ z_{i,k+1} \\ \theta_{i,k+1} \end{bmatrix} = \begin{bmatrix} x_{i,k} + \Delta_T(v_i + w^v_{i,k})\cos(\theta_{i,k}) \\ y_{i,k} + \Delta_T(v_i + w^v_{i,k})\sin(\theta_{i,k}) \\ z_c + w^z_{i,k} \\ \theta_{i,k} + \Delta_T(\omega_i + w^\omega_{i,k}) \end{bmatrix} \tag{33}$$

The target ground robot is modeled with a unicycle model f_t similar to Equation (33). Correspondingly, the state, control input, and process noise are denoted as $\mathbf{x}_t$, $\mathbf{u}_t$, and $\mathbf{Q}_t = \text{blkdiag}(Q^v_t, Q^\omega_t, Q^z_t)$, respectively.

5.2. Measurement Model

The three types of measurement are utilized to realize the CLAT purpose in this system setup: private absolute measurement from the onboard navigation system, cooperative relative range measurement from the UWB sensors, and angle measurement relative to the target from the downward cameras.

Although the position information from the onboard navigation system is a fusion result from multiple sensors, in this part, it is treated as a private absolute measurement and is modeled as below:

$$\mathbf{z}^a_{i,k} = H^a_i \mathbf{x}_{i,k} + \mathbf{v}^a_{i,k} \tag{34}$$

The measurement noise is assumed to be subject to a zero-mean Gaussian distribution, $\mathbf{v}^a_{i,k} \sim \mathcal{N}(\mathbf{0}, \mathbf{R}^a)$. Obviously, the absolute measurement model is a linear model. Therefore, we can substitute the inferred Jacobian matrix (defined in Section 3.2.1) with H^a_i. In this paper, we assume that only a subset of all robots has access to the onboard navigation signal.

The cooperative relative measurement between robots i and j from UWB is a scalar distance modeled as the following nonlinear model:

$$z^c_{ij,k} = \|\mathbf{x}_{i,k} - \mathbf{x}_{j,k}\|_d + v^c_{i,k}, \tag{35}$$

The measurement noise $v^c_{ij,k}$ is assumed to be subject to the zero-mean Gaussian noise $v^c_{ij,k} \sim \mathcal{N}(0, \mathbf{R}^c_i)$.

The target detection measurement is the position of the target on the captured image plane, and the measurement function is defined as

$$\mathbf{z}^t_{it,k} = d_f \frac{\bar{\mathbf{x}}^{(1:2)}_{it,k}}{\bar{\mathbf{x}}^{(3)}_{it,k}} + \mathbf{v}^t_{i,k} \tag{36}$$

where d_f denotes the focal length in pixels, $\bar{\mathbf{x}}_{it,k}$ denotes the position of the target in the camera coordinate, i.e., $\bar{\mathbf{x}}_{it,k} = \mathcal{R}(\theta_{i,k})\mathcal{R}_{c,k}(\mathbf{x}_{t,k} - \mathbf{x}_{i,k})$. $\mathcal{R}(\theta_{i,k})$ denotes the yaw angle rotation matrix, and

$$\mathcal{R}(\theta_{i,k}) = \begin{bmatrix} \cos(\theta_{i,k}) & \sin(\theta_{i,k}) & 0 \\ -\sin(\theta_{i,k}) & \cos(\theta_{i,k}) & 0 \\ 0 & 0 & 1 \end{bmatrix}.$$

$\mathcal{R}_{c,k}$ denotes the roll and pitch rotation matrix and is assumed to be retrieved from the quadrotor navigation system, and therefore, it is treated as a known variable. The measurement noise is $v^t_{i,k} \sim \mathcal{N}(\mathbf{0}, \mathbf{R}^t_i)$. In this paper, target detection is carried out using kernelized correlation filters (KCFs) [36] on the image plane.

5.3. Experiment Results and Analysis

According to the above model description, an experiment with three quadrotors tracking one ground robot on the basis of the UT-CLAT algorithm was carried out at 10 Hz. The experimental setup is shown in Table 2.

One experimental snapshot of the three quadrotors with the corresponding captured images is shown in Figure 10 when the ground robot is within the FOV of all three quadrotor cameras.

According to the CLAT algorithm, the trajectories of the three quadrotors' self-localization results and target tracking results are plotted in Figure 11. As observed in Figure 11, the three quadrotors are able to localize themselves while stably tracking the target. It is obvious that the self-localization result from quad1 is better than that of the other two quadrotors as it can obtain the navigation signal from the onboard navigation system. The errors of self-localization and target tracking are plotted in Figures 12 and 13, respectively. The local estimation errors (solid line in colors) and the corresponding approximated $\pm 3\sigma$ envelopes (dashed lines in the same color) of the aforementioned three quadrotors are plotted. As observed in Figures 12 and 13, the estimation errors by each quadrotor are bounded by the $\pm 3\sigma$ envelopes in the steady state.

Table 2. Experiment setup.

Item	Quantity
robot setup	circular center (in meters): $c_1 = [-5, -5, 15]^\top$, $c_2 = [5, -5, 16.5]^\top$ $c_3 = [-5, 5, 18]^\top$ control input: $v_i = 0.35$ m/s, $\omega_i = 0.035$ rad/s $Q^v_i = 0.05^2$, $Q^w_i = (\pi/180)^2$, $Q^z_i = 0.1^2$ initial state: $\mathbf{P}_{i,0} = \text{diag}([1^2, 1^2, 0.5^2, (10\pi/180)^2])$ $\mathbf{x}_{1,0} = [-8.5, 11.5, 15, 2.35]^\top$, $\mathbf{x}_{2,0} = [-10, -11, 16.5, -0.53]^\top$, $\mathbf{x}_{3,0} = [-1, 1, 18, 0.78]^\top$
target setup	control input: $v_i = 0.2$ m/s, $\omega_i = 0$ rad/s $Q^v_t = 0.02^2$, $Q^w_t = (0.5\pi/180)^2$, $Q^z_t = 0.05^2$ initial state: $\mathbf{P}_{t_i,0} = \text{diag}([0.1^2, 0.1^2, 0.5^2, (\pi/180)^2])$, $\mathbf{x}_{t_i,0} = [-15, -15, 0, 0.785]^\top$
measurement Setup	$\mathbf{R}^a = \text{diag}([0.2^2, 0.2^2, 0.1^2])$, $\mathbf{R}^c = 0.1^2$, $\mathbf{R}^t = \text{diag}([5^2, 5^2])$

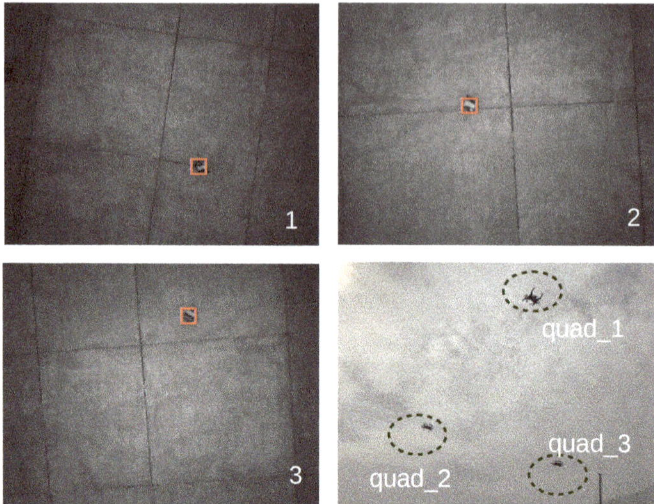

Figure 10. Snapshot of the CLAT experiment setup (three quadrotors and corresponding captured image).

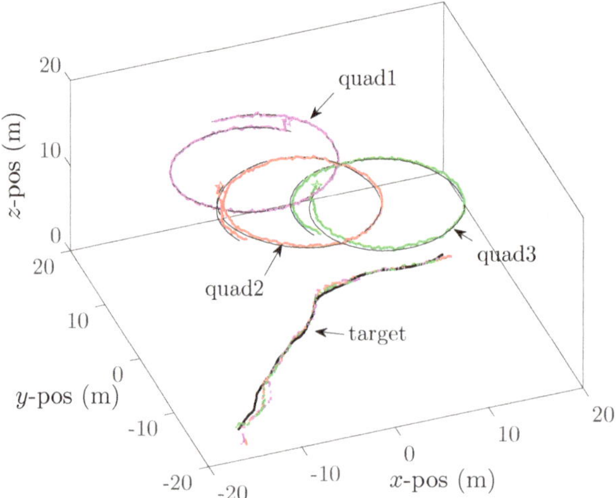

Figure 11. Trajectories of three quadrotors' self-localization results and target tracking results with different colors. Ground truth is indicated by black lines.

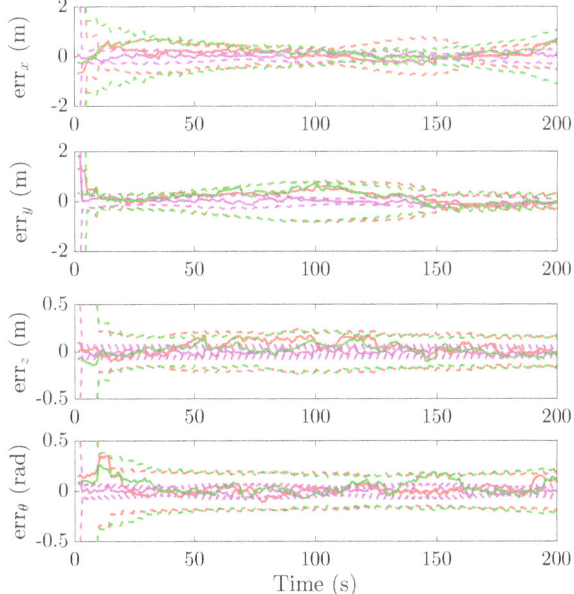

Figure 12. The self-localization errors and covariance of three quadrotors.

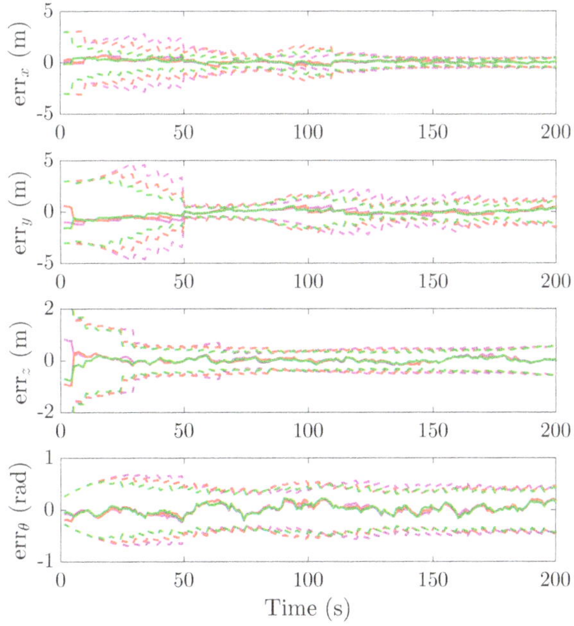

Figure 13. Object tracking errors and covariance of three quadrotors.

6. Conclusions

A UT-based CLAT method is proposed to realize multi-robot self-localization and target tracking in a distributed fashion. The proposed method is recursive, and only the most recent estimation is stored within each local robot. The communication is limited to the two robots within the relative measurement, and estimation consistency is guaranteed with the covariance split and covariance intersection method. To deal with the nonlinearity in the dynamics models and measurement models, a UT was integrated into the CLAT framework. Both simulation and experimental results show that the proposed method can fulfill the self-localization and target tracking task in practical multi-robot operation scenarios. Future works will focus on the theoretical analysis of the error bounds of both self-localization and target tracking on the basis of different measurement setups.

Author Contributions: Investigation, Y.L.; Methodology, Y.L.; Project administration, Q.P.; Software, J.L.; Supervision, Q.P.

Funding: This work is supported by the National Natural Science Foundation of China under Grant 61603303, 61473230, the Natural Science Foundation of Shaanxi Province under Grant 2017JQ6005, 2017JM6027, the China Postdoctoral Science Foundation under Grant 2017M610650 and the Fundamental Research Funds for the Central Universities under Grant 3102017JQ02011.

Conflicts of Interest: The authors declare no conflict of interest.

References

1. Ren, W.; Beard, R.W.; Atkins, E.M. Information consensus in multivehicle cooperative control. *IEEE Control Syst.* **2007**, *27*, 71–82.
2. Alonso-Mora, J.; Baker, S.; Rus, D. Multi-robot formation control and object transport in dynamic environments via constrained optimization. *Int. J. Robot. Res.* **2017**, *36*, 1000–1021. [CrossRef]
3. Liu, Y.; Nejat, G. Multirobot cooperative learning for semiautonomous control in urban search and rescue applications. *J. Field Robot.* **2016**, *33*, 512–536. [CrossRef]
4. La, H.M.; Sheng, W. Distributed sensor fusion for scalar field mapping using mobile sensor networks. *IEEE Trans. Cybern.* **2013**, *43*, 766–778. [PubMed]
5. Panzieri, S.; Pascucci, F.; Setola, R. Multirobot localisation using interlaced extended Kalman filter. In Proceedings of the 2006 IEEE/RSJ International Conference on Intelligent Robots and Systems, Beijing, China, 9–15 October 2006; pp. 2816–2821.
6. Chen, L.; Arambel, P.O.; Mehra, R.K. Fusion under unknown correlation-covariance intersection as a special case. In Proceedings of the Fifth International Conference on Information Fusion, Annapolis, MD, USA, 8–11 July 2002; Volume 2, pp. 905–912.
7. Li, H.; Nashashibi, F. Cooperative multi-vehicle localization using split covariance intersection filter. *IEEE Intell. Transp. Syst. Mag.* **2013**, *5*, 33–44. [CrossRef]
8. Li, H.; Nashashibi, F.; Yang, M. Split covariance intersection filter: Theory and its application to vehicle localization. *IEEE Trans. Intell. Transp. Syst.* **2013**, *14*, 1860–1871. [CrossRef]
9. Luft, L.; Schubert, T.; Roumeliotis, S.I.; Burgard, W. Recursive decentralized localization for multi-robot systems with asynchronous pairwise communication. *Int. J. Robot. Res.* **2018**, *37*. [CrossRef]
10. Hu, J.; Xie, L.; Zhang, C. Diffusion Kalman filtering based on covariance intersection. *IEEE Trans. Signal Process.* **2012**, *60*, 891–902. [CrossRef]
11. Wang, S.; Ren, W. On the convergence conditions of distributed dynamic state estimation using sensor networks: A unified framework. *IEEE Trans. Control Syst. Technol.* **2018**, *26*, 1300–1316. [CrossRef]
12. Wang, S.; Ren, W.; Chen, J. Fully distributed state estimation with multiple model approach. In Proceedings of the 2016 IEEE 55th Conference on Decision and Control (CDC), Las Vegas, NV, USA, 12–14 December 2016; pp. 2920–2925.

13. Wang, S.; Lyu, Y.; Ren, W. Unscented-Transformation-Based Distributed Nonlinear State Estimation: Algorithm, Analysis, and Experiments. *IEEE Trans. Control Syst. Technol.* **2018**, *99*, 1–14. [CrossRef]
14. Carrillo-Arce, L.C.; Nerurkar, E.D.; Gordillo, J.L.; Roumeliotis, S.I. Decentralized multi-robot cooperative localization using covariance intersection. In Proceedings of the 2013 IEEE/RSJ International Conference on Intelligent Robots and Systems (IROS), Tokyo, Japan, 3–7 November 2013; pp. 1412–1417.
15. Cunningham, A.; Paluri, M.; Dellaert, F. DDF-SAM: Fully distributed slam using constrained factor graphs. In Proceedings of the 2010 IEEE/RSJ International Conference on Intelligent Robots and Systems (IROS), Taipei, Taiwan, 18–22 October 2010; pp. 3025–3030.
16. Ren, W.; Beard, R.W.; Kingston, D.B. Multi-agent Kalman consensus with relative uncertainty. In Proceedings of the American Control Conference, Portland, OR, USA, 8–10 June 2005; pp. 1865–1870.
17. Farahmand, S.; Roumeliotis, S.I.; Giannakis, G.B. Set-membership constrained particle filter: Distributed adaptation for sensor networks. *IEEE Trans. Signal Process.* **2011**, *59*, 4122–4138. [CrossRef]
18. Vaghefi, R.M.; Buehrer, R.M.; Wymeersch, H. Collaborative Sensor Network Localization: Algorithms and Practical Issues. *Proc. IEEE* **2018**, *PP*, 1–26.
19. Chang, T.K.; Mehta, A. Optimal Scheduling for Resource-Constrained Multirobot Cooperative Localization. *IEEE Robot. Autom. Lett.* **2018**, *3*, 1552–1559. [CrossRef]
20. Kia, S.S.; Rounds, S.; Martinez, S. Cooperative Localization for Mobile Agents: A Recursive Decentralized Algorithm Based on Kalman-Filter Decoupling. *IEEE Control Syst.* **2016**, *36*, 86–101.
21. Kia, S.S.; Hechtbauer, J.; Gogokhiya, D.; Martínez, S. Server-Assisted Distributed Cooperative Localization Over Unreliable Communication Links. *IEEE Trans. Robot.* **2017**, *99*, 1–8. [CrossRef]
22. Chong, C.Y.; Chang, K.C.; Mori, S. A Review of Forty Years of Distributed Estimation. In Proceedings of the 2018 21st International Conference on Information Fusion (FUSION), Bonn, Germany, 10–12 October 2018; pp. 1–8.
23. Battistelli, G.; Chisci, L.; Selvi, D. A distributed Kalman filter with event-triggered communication and guaranteed stability. *Automatica* **2018**, *93*, 75–82. [CrossRef]
24. Battistelli, G.; Chisci, L.; Mugnai, G.; Farina, A.; Graziano, A. Consensus-based linear and nonlinear filtering. *IEEE Trans. Autom. Control* **2015**, *60*, 1410–1415. [CrossRef]
25. Battistelli, G.; Chisci, L. Stability of consensus extended Kalman filter for distributed state estimation. *Automatica* **2016**, *68*, 169–178. [CrossRef]
26. Li, W.; Wei, G.; Han, F.; Liu, Y. Weighted average consensus-based unscented Kalman filtering. *IEEE Trans. Cybern.* **2016**, *46*, 558–567. [CrossRef] [PubMed]
27. Ajgl, J.; Straka, O. Covariance Intersection in Track-to-Track Fusion: Comparison of Fusion Configurations. *IEEE Trans. Ind. Inform.* **2018**, *14*, 1127–1136. [CrossRef]
28. Ahmad, A.; Tipaldi, G.D.; Lima, P.; Burgard, W. Cooperative robot localization and target tracking based on least squares minimization. In Proceedings of the 2013 IEEE International Conference on Robotics and Automation (ICRA), Karlsruhe, Germany, 6–10 May 2013; pp. 5696–5701.
29. Huang, G.; Truax, R.; Kaess, M.; Leonard, J.J. Unscented iSAM: A consistent incremental solution to cooperative localization and target tracking. In Proceedings of the 2013 European Conference on Mobile Robots (ECMR), Barcelona, Spain, 25–27 September 2013; pp. 248–254.
30. Morbidi, F.; Mariottini, G.L. Active target tracking and cooperative localization for teams of aerial vehicles. *IEEE Trans. Control Syst. Technol.* **2013**, *21*, 1694–1707. [CrossRef]
31. Meyer, F.; Hlinka, O.; Wymeersch, H.; Riegler, E.; Hlawatsch, F. Distributed Localization and Tracking of Mobile Networks Including Noncooperative Objects. *IEEE Trans. Signal Inf. Process. Netw.* **2016**, *2*, 57–71. [CrossRef]
32. Zhou, B.; Chen, Q.; Xiao, P. The error propagation analysis of the received signal strength-based simultaneous localization and tracking in wireless sensor networks. *IEEE Trans. Inf. Theory* **2017**, *63*, 3983–4007. [CrossRef]
33. Huang, G.P.; Roumeliotis, S.I. *An Observability Constrained UKF for Improving SLAM Consistency*; Tech. Rep.; University of Minnesota: Minneapolis, MN, USA, 2008.
34. Zhu, J.; Kia, S.S. Consistent loosely coupled decentralized cooperative navigation for team of mobile agents. In Proceedings of the ION's International Technical Meeting, Monterey, CA, USA, 30 January–2 February 2017.

35. Chakraborty, A.; Misra, S.; Sharma, R.; Taylor, C.N. Observability Conditions for Switching Sensing Topology for Cooperative Localization. *Unmanned Syst.* **2017**, *5*, 141–157. [CrossRef]
36. Henriques, J.F.; Caseiro, R.; Martins, P.; Batista, J. High-speed tracking with kernelized correlation filters. *IEEE Trans. Pattern Anal. Mach. Intell.* **2015**, *37*, 583–596. [CrossRef] [PubMed]

Article

Synchronization of Heterogeneous Multi-Robotic Cell with Emphasis on Low Computing Power

Martin Juhás * and Bohuslava Juhásová

Faculty of Materials Science and Technology in Trnava, Slovak University of Technology in Bratislava, Trnava 917 24, Slovakia; bohuslava.juhasova@stuba.sk

* Correspondence: martin_juhas@stuba.sk

Received: 26 June 2020; Accepted: 23 July 2020; Published: 27 July 2020

Abstract: This paper presents a time-synchronization solution for operations performed by a heterogeneous set of robotic manipulators grouped into a production cell. The cell control is realized using master–slave architecture without an external control element. Information transmission in a cell is provided by a TCP/IP channel in which communication is ensured via sockets. The proposed problem solution includes an algorithm, which is verified and validated by simulation and tested in real environment. This algorithm requires minimal computational power thanks to an empirically oriented approach, which enables its processing directly by the control unit of each participating element of the robotic cell. The algorithm works on the basis of monitoring and evaluating time differences among sub-operations of master and slave devices. This ensures defined production cycle milestones of each robotic manipulator in the cell at the same time are attained. Dynamic speed adaptation of slave manipulators utilizing standard instructions of their native language is used. The proposed algorithm also includes a feedforward form of operations synchronization which responds to changes in the operating cycle of the master manipulator. The application of the solution proposal is supplemented with a visualization part. This part represents a complementary form of designed solution implementation.

Keywords: time-based synchronization; heterogeneous multi-robotic cell; socket communication; low computational power; native language application

---

1. Introduction

A current significant challenge for industrial production is vision substantiation of a future factory within the concept of Industry 4.0, with the aid of the Internet of Things (IoT) concept. The implementation of this concept into production requires high flexibility and adaptability of production lines and their smaller units. The deployment of this concept in practice is widely supported in Germany, and the first solutions in isolated production systems were developed there [1,2]. Qin et al. [3] note that the Internet of Things is a well-known concept that represents the next generation of products and communication among them. It has a direct correlation with the Industry 4.0 standard, where the existence of smart products is one of the prerequisites for intelligent manufacturing implementation [4].

In a smart factory, individual customer orders determine manufacturing processes and the associated supply chains. This results in the need for high production flexibility with shorter production times, which require the implementation of measures to improve production efficiency, often at a low cost associated with solving these problems. The new term "smart factory" is introduced here to refer to the current trend of automation and data exchange in manufacturing technologies. It includes cyber-physical systems, the Internet of Things, and cloud computing, as declared by Hermann et al. [5], Jasperneite [6], and Kagermann et al. [7]. One of the important indicators for such production is decentralized decision making, that is, the ability of equipment to make its own decisions, which are the most independent inside closed production complexes. In the case of essential decisions and conflicts, it is clearly necessary to assign tasks to the next level of production control [5]. Closed production complexes require continuity of individual manufacturing operations, synchronization of individual production facilities, and the possibility of rapid adaptation and production time changes of individual production facilities, depending on the control structure of such complexes.

An alternative deployment area of robotic manipulators with a potential requirement for synchronization of their activities is healthcare, particularly that related to the rehabilitation of patients. Techniques dealing with robot-assisted therapy are included, for example, in the works of authors Chang and Kim or Yoon et al. [8,9].

Various methods of synchronization and control of individual production operations and processes are currently used. It is necessary to differentiate whether these are autonomous mobile systems or manipulators and what usage is expected of them. Many published works deal with the synchronization of mobile platform activities. Rubenstein et al. [10] offer a solution for the synchronization of a large group of mobile robots as an open-source, low cost robot, designed to test collective algorithms on hundreds or thousands of robots accessible to robotics researchers. This solution allows for easier testing of algorithms designed to control robot groups because these control algorithms, due to their cost, time, and complexity, are confirmed only through simulations. Popular synchronization methods are also based on the observation of nature, for example, fireflies as presented by Werner-Allen et al. [11]. A modification of this approach is intended to operate on systems that use a communication channel where contention and delays are possible. In addition, the coordination mechanisms that enable the execution of cooperative tasks with modular robotic systems are presented in the contribution of Baca et al. [12]. They describe the implementation of a tight cooperation strategy through Intra M-Robot communication based on a closed-loop discrete time method and the remote clock across the robot configuration enables proper coordination inside the colony.

The work of Chung and Slotine [13] presents a new synchronization tracking control law that can be directly applied to the cooperative control of multi-robot systems and oscillation synchronization in robotic manipulation and locomotion, where a common desired trajectory can be explicitly given.

Rodriguez-Angeles et al. [14] describe a controller utilization that solves the problem of position synchronization of two (or more) robotic systems, under a cooperative scheme, in the case when only position measurements are available. In the work of Yasuda [15], a Petri-net-based prototyping tool is presented to implement the control flow of parallel processes in multiple robot systems. The next variant of synchronization is presented by Markus et al. [16], where the coordination control of two flexible joint robotic manipulators using flat outputs is implemented by means of simulations. The differential flatness technique of trajectory generation enables easy estimation of synchronization parameters and trivializes stabilization of these trajectories around predefined points. Bouteraa et al. [17] describe a new adaptive algorithm, which was proposed for synchronization and trajectory tracking of multiple robot manipulators. The same authors also discuss other techniques in this problem area. They describe the possibility of designing decentralized control laws to cooperatively command a team of general actuated manipulators in the article "Distributed synchronization control to trajectory tracking of multiple robot manipulators" [18], or an approach to position synchronization of multiple robot manipulators based on emergent consensus algorithms [19]. Synchronization of activities in task-oriented robotic rehabilitation training using iterative learning synchronization (ILS) and immediate error correction (IEC) techniques is addressed by Duschau-Wicke et al. [20].

In the above-mentioned cases, the authors based their solutions on complex mathematical algorithms and derivation of complex relations, or by exploring new approaches using specialized hardware. These methods require investment in hardware infrastructure, which is their major disadvantage.

In some cases in homogenous production complexes, it is possible to use tools that are directly implemented in control systems, such as the MultiMove option of ABB robots [21,22] or the RoboTeam software of KUKA [22,23].

However, if limited possibilities exist to upgrade infrastructure and it is desired to use available hardware that may not be from the latest production series, there could be a serious problem with the use of the previously described solutions. We used an alternative approach in a production or technical process of a simple solution implementation even in production facilities without the possibility of using sophisticated methods of robotic set or robotic cell synchronization. Our aim is the simplest solution possible in terms of computing power demands. The basis is to design the least complex algorithms that can be easily implemented in existing controllers in their native programming languages. Therefore, our goal is to develop a solution in which synchronization algorithms can be performed directly on the control units of robotic manipulators. This is based on the utilization of our rich empirical knowledge and experiences in algorithms, in addition to the implementation of various tasks in the field of robotics or modeling and visualization of processes.

The main idea does not lead to a specific use or solution to a precisely specified problem, for example in a production process. The aim of the proposed solution can have a wide range. From the analyzed areas it is possible to use this solution either in production or in healthcare in rehabilitation. Another area of application could be the control of collaboration-oriented workplaces with a master manipulator connected to movements of a human as a cell control element. Finally, it could be used in presentation events oriented to Industry 4.0 or the latest trend, Internet-of-Robotic-Things (IoRT).

2. Problem Area Definition

The multi-robotic cell consists of a set of n robotic manipulators $M = \{M_1, \ldots, M_i, \ldots, M_n\}$. In the organizational structure of cell control, manipulator M_1 represents the master element and a set of manipulators $MS \subset M$, $MS = \{M_2, \ldots, M_i, \ldots, M_n\}$ represents the slave elements. Each manipulator M_i performs a set of m operations $O = \{O_1, \ldots, O_j, \ldots, O_m\}$ cyclically. Each operation O_j is in one cycle executed by manipulator M_i. The duration time of this operation is $T_{i,j} \in T = \left\{ \begin{matrix} T_{1,1} & \ldots & T_{1,m} \\ \ldots & \ldots & \ldots \\ T_{n,1} & \ldots & T_{n,m} \end{matrix} \right\}$.

It is clear that duration times $T_{x,j}$ for $x = (1, \ldots, n)$ of operation O_j for every manipulator M_i are different without using a synchronization algorithm. Duration time depends on a movement speed $V_{i,j} \in V = \left\{ \begin{matrix} V_{1,1} & \ldots & V_{1,m} \\ \ldots & \ldots & \ldots \\ V_{n,1} & \ldots & V_{n,m} \end{matrix} \right\}$ of manipulator M_i endpoint, defined as a parameter of a movement instruction.

The goal is to modify the movement speed of manipulators MS separately for every operation O_j in such a way that the duration times of the same operation for every slave manipulator MS_i will be equal or very similar.

The main requirements for the synchronization algorithm design are listed in Figure 1.

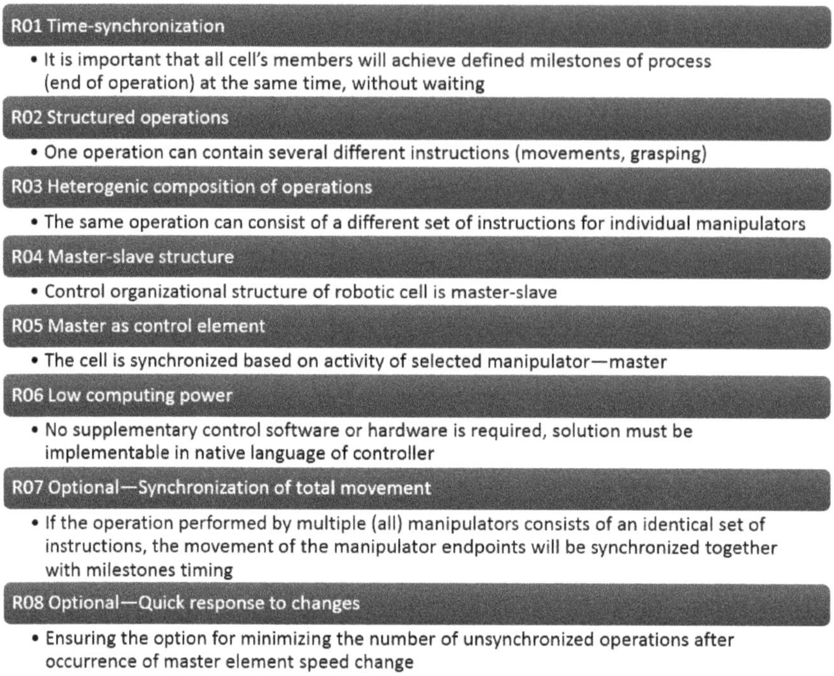

R01 Time-synchronization

- It is important that all cell's members will achieve defined milestones of process (end of operation) at the same time, without waiting

R02 Structured operations

- One operation can contain several different instructions (movements, grasping)

R03 Heterogenic composition of operations

- The same operation can consist of a different set of instructions for individual manipulators

R04 Master-slave structure

- Control organizational structure of robotic cell is master-slave

R05 Master as control element

- The cell is synchronized based on activity of selected manipulator—master

R06 Low computing power

- No supplementary control software or hardware is required, solution must be implementable in native language of controller

R07 Optional—Synchronization of total movement

- If the operation performed by multiple (all) manipulators consists of an identical set of instructions, the movement of the manipulator endpoints will be synchronized together with milestones timing

R08 Optional—Quick response to changes

- Ensuring the option for minimizing the number of unsynchronized operations after occurrence of master element speed change

Figure 1. Requirements for algorithm design.

3. Design of Multi-Robotic Cell Synchronization Algorithm

The principle of the synchronization algorithm design is to dynamically adjust the speed of operation performing $V_{i,j}$ on the slave side. The speed adjustment means the change of manipulator endpoint movement speed, based on the synchronization coefficient $K_{i,j} \in K = \{K_{1,1}, \ldots, K_{n,m}\}$. This coefficient $K_{i,j}$ is evaluated for each operation O_j for each slave manipulator MS_i based on Equation (1).

$$K_{i,j} = T_{i,j}/T_{1,j} \tag{1}$$

The synchronization coefficient $K_{i,j}$ therefore represents the ratio of the operation duration $T_{1,j}$ of the master element and the operation duration of the slave element $T_{i,j}$ for $i = (2, \ldots, n)$.

It is necessary to emphasize that although the speed $V_{i,j}$ is based on coefficient $K_{i,j}$ recalculated immediately after operation O_j execution, the new speed value is used only in the next cycle.

The ability of the master manipulator M_1 to distribute operation O_i duration time $T_{1,j}$ after performing for each slave manipulator MS is the obligatory condition for feasibility of this proposal.

3.1. Basic Synchronization Algorithm

The elemental analysis results in the basic synchronization algorithm, which is the same for each slave manipulator MS_i; its structure is depicted in Figure 2.

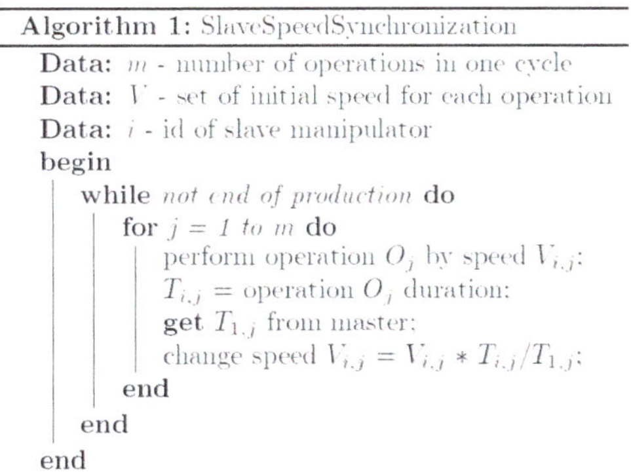
Algorithm 1: SlaveSpeedSynchronization
Data: m - number of operations in one cycle
Data: V - set of initial speed for each operation
Data: i - id of slave manipulator
begin
  while *not end of production* **do**
    for $j = 1$ *to* m **do**
      perform operation O_j by speed $V_{i,j}$;
      $T_{i,j}$ = operation O_j duration;
      get $T_{1,j}$ from master;
      change speed $V_{i,j} = V_{i,j} * T_{i,j}/T_{1,j}$;
    end
  end
end

Figure 2. Basic algorithm for synchronization of robotic cell operations.

As is later mentioned in Section 4.1., the main disadvantage of the proposed algorithm is low reaction speed to changes in master or slave behavior. This is contrary to the requirement R08 from Figure 1. Every change of master manipulator M_1 (or slave manipulator MS_i) activity is captured immediately so the synchronization coefficient $K_{i,j}$ is evaluated. This leads to the adjustment of endpoint movement speed $V_{i,j}$ of each slave manipulator MS_i. However, the adjusted endpoint movement speed $V_{i,j}$ is actually used during movement of the slave manipulator MS_i in the next production cycle.

3.2. Advanced Synchronization Algorithm

If the master manipulator M_1 (or slave manipulator MS_i) activity change has a global character (endpoint movement speed increasing/decreasing), then it is possible to indicate this change using variable D. This variable represents the percentage proportion deviations among operation durations of master and slave manipulators, where $D_{i,j} \in D = \begin{Bmatrix} D_{2,1} & \ldots & D_{2,m} \\ \ldots & \ldots & \ldots \\ D_{n,1} & \ldots & D_{n,m} \end{Bmatrix}$ is calculated based on Equation (2).

$$D_{i,j} = (1 - (T_{i,j}/T_{1,j})) \times 100 \tag{2}$$

The variable D is calculated in each step of the algorithm, but is only usable from the second production cycle. The first production cycle has diverse values of deviations $D_{i,j}$, from first step of second cycle, $D_{i,j}$ within defined tolerance represent synchronized state. Every change of $D_{i,j}$ values exceeding the limit can be considered as a desynchronization indicator.

If two (or another specified number d) differences $D_{i,j\text{-}(d-1)} \ldots D_{i,j}$ of consecutive operations $O_{j\text{-}(d-1)} \ldots O_j$ are sufficiently similar based on the similarity threshold D_{ratio_limit} (Equation (3))

$$(|(1 - (D_{i,j-1}/D_{i,j}))| \times 100 < D_{ratio\_limit}) \wedge \ldots \wedge (|(1 - (D_{i,j\text{-}(d-1)}/D_{i,j\text{-}(d-2)}))| \times 100 < D_{ratio\_limit}) \tag{3}$$

and this difference exceeds the given threshold of change D_{limit} (Equation (4)),

$$|D_{i,j}| > D_{limit} \tag{4}$$

then advanced synchronization coefficient K_i (Equation (5)) is applied to all other operation execution speeds $V_{i,j}$, except the actual operation (which is already adjusted in this step) and previous $d-1$ values (which were adjusted in previous $d-1$ steps).

$$K_i = 1 - \left(\left(\frac{1}{d} \sum_{z=0}^{d-1} D_{i,j-z} \right) / 100 \right) \tag{5}$$

Both algorithms were designed for (theoretically) endless repetition of production cycle. The above-mentioned previous values of the current $D_{i,j}$ in the initial steps of every cycle are based on an endless loop of variable j. For example, in the case of 10 operations in one production cycle, after the initialization cycle, the previous two elements $V_{2,j}$ (endpoint speed of slave manipulator M_2) to element $V_{2,1}$ are $V_{2,10}$ and $V_{2,9}$.

The synchronization algorithm with a feedforward reaction to the general change of the master manipulator M_1 (or slave manipulators MS_i) endpoint movement speed is identical for every slave manipulator MS_i and is depicted in Figure 3.

```
Algorithm 2: SlaveAdvancedSpeedSynchronization
Data: m - number of operations in one cycle
Data: V - set of default speed for each operation
Data: i - id of slave manipulator
Data: d - number of compared values
Data: D_off - deviation offset
begin
    while not end of production do
        for j = 1 to m do
            perform operation O_j by speed O_{i,j};
            T_{i,j} = operation O_j duration;
            get T_{1,j} from master;
            change speed V_{i,j} = V_{i,j} * T_{i,j}/T_{1,j};
            D_{i,j} = percentage proportion deviation between T_{i,j} and T_{1,j};
            if
            D_{i,j} ≥ D_off
            and
            D_{i,j} is significantly similar to previous d values of D_{i,?}
             then
                evaluate K_i based on last d values of D_{i,j};
                begin
                    change speed V_{i,x} = V_{i,x} *K_i
                    where x ∈ (1,...,m) − (last d values of j)
                end
            end
        end
    end
end
```

Figure 3. Advanced algorithm for synchronization of robotic cell operations.

4. Validation of the Proposed Solution

The proposed solution was experimentally validated in two phases. In the first phase, the algorithms were implemented as scripts in the MATLAB software environment. To evaluate the efficiency of the algorithms, monitoring of the percentage proportion deviation $D_{i,j}$ (Equation (2)) was used. The aim of the solution was to reach a $D_{i,j}$ value as close as possible to zero. The synchronization state was indicated by a value close to zero within the specified tolerance. In contrast, a value exceeding the limit can be considered as a desynchronization indicator.

In the second phase, the designed algorithms were implemented in the native language of real robotic manipulators in a specific multi-robotic cell.

4.1. Simulation Validation of Solution Functionality

The functionality of the proposed solution was validated by simulations using the MATLAB software tool (R2018a, The MathWorks, Inc., Natic, MA, USA, 2018). The designed algorithms were processed by a script and the results were represented in graphical form.

Duration times $T_{1,j}$ of master manipulator M_1 operations were simulated as generated random values of a vector in the range of 0–5000 ms. Each value $T_{1,j}$ was modified during its processing with a random element from the range (−2%, +2%), that represents a stochastic process part. Because the synchronization algorithm for all slave manipulators MS_i is identical, the case with one slave manipulator MS_i, where $i = 2$, was validated by simulation. Operation durations $T_{i,j}$ of this slave manipulator were simulated on the basis of Equation (6).

$$T_{i,j} = S_{i,j}/V_{i,j} \tag{6}$$

In Equation (6) the variable $S_{i,j} \in S = \{S_{2,1}, \ldots, S_{n,m}\}$ represents the path length of operation O_j performed by manipulator MS_i. An idealized kinematic model of a manipulator with omission of the non-linear character of robot arm movement in acceleration and deceleration was used for simulation validation purposes. Elements of the S set were generated as random values of a vector in the range of 0–1000 mm in the simulation validation process. The default endpoint movement speed $V_{i,j}$ of the slave manipulator was set to 200 mm/s. In processing of each operation duration $T_{i,j}$, an additional modification with the stochastic part from the interval (−2%, +2%) of $T_{i,j}$ was also used.

4.1.1. Experiment 1a—Master Speed Change, Basic Algorithm

Simulation Experiment 1a includes:

- eight production cycles of production cell
- five operations in every production cycle ($m = 5$)
- change of master manipulator M_1 operation speed
    - in the 3rd production cycle by +20% of current speed value
    - in the 4th production cycle by +30% of current speed value
    - in the 7th production cycle by −50% of current speed value
- master manipulator M_1 speed change in different time of production cycle

4.1.2. Experiment 2a—Slave Speed Change, Basic Algorithm

Simulation Experiment 2a includes:

- six production cycles of production cell
- 10 operations in every production cycle ($m = 10$)
- change of slave manipulator MS_2 operation speed
    - in the 2nd production cycle by −30% of current speed value
    - in the 4th production cycle by +30% of current speed value
- slave manipulator MS_2 speed change in different time of production cycle

4.1.3. Summary of Experiment 1a and Experiment 2a

The obtained simulation results of the experiment with master speed change are depicted in Figures 4 and 5. Results of both experiments, experiment with master speed change and the experiment with slave speed change, is aggregated in Figure 6, and the results of the experiment with slave speed change in Figures 7 and 8. Figure 9 also presents results of both experiments, experiment with master speed change and the experiment with slave speed change.

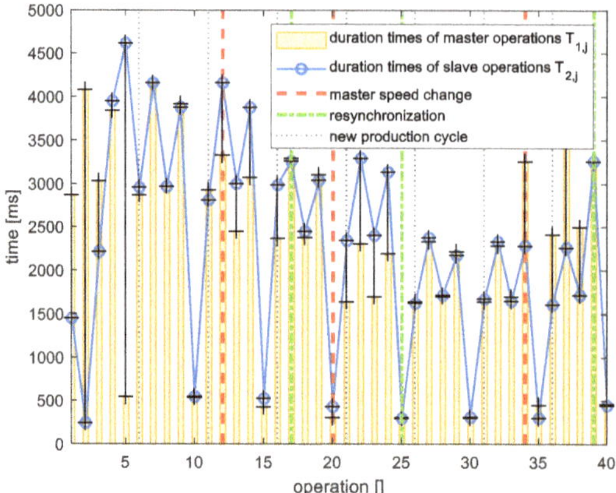

Figure 4. Operation duration times of master and slave manipulators without feedforward synchronization—Experiment 1a.

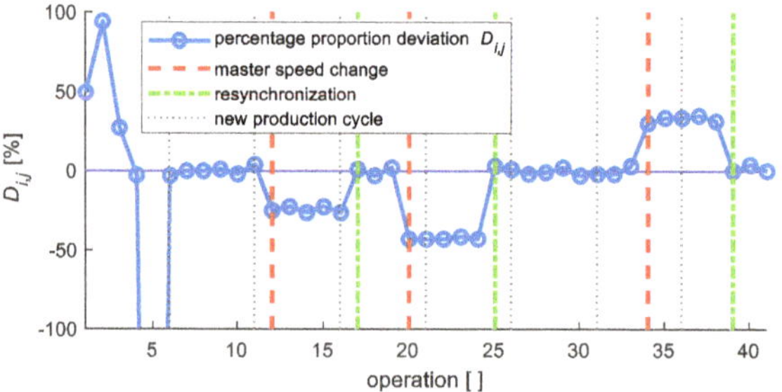

Figure 5. Percentage proportion deviation without feedforward synchronization—Experiment 1a.

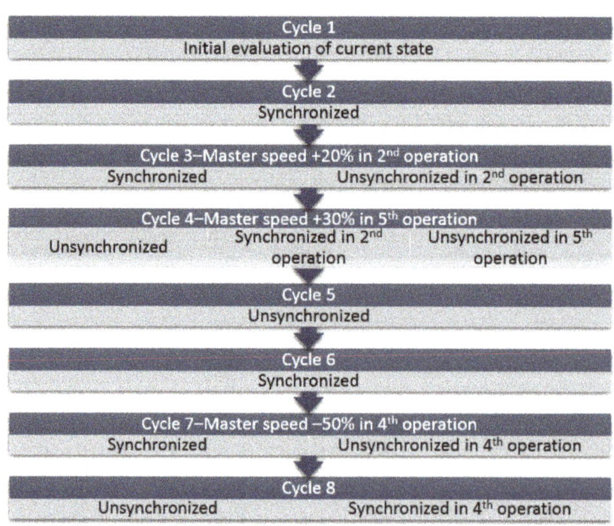

Figure 6. Brief summary of Experiment 1a results.

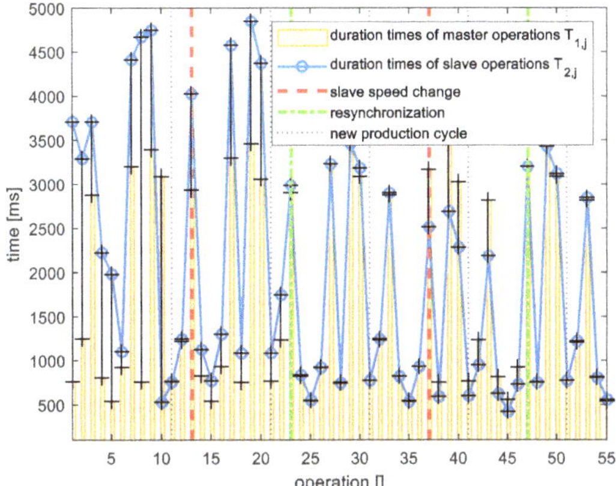

Figure 7. Operation duration times of master and slave manipulators without feedforward synchronization—Experiment 2a.

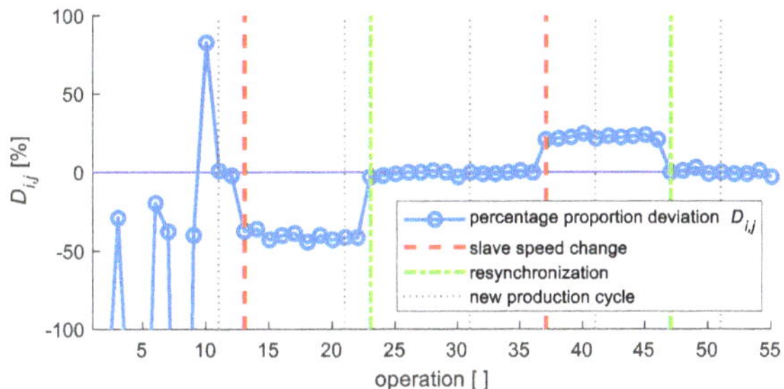

Figure 8. Percentage proportion deviation without feedforward synchronization—Experiment 2a.

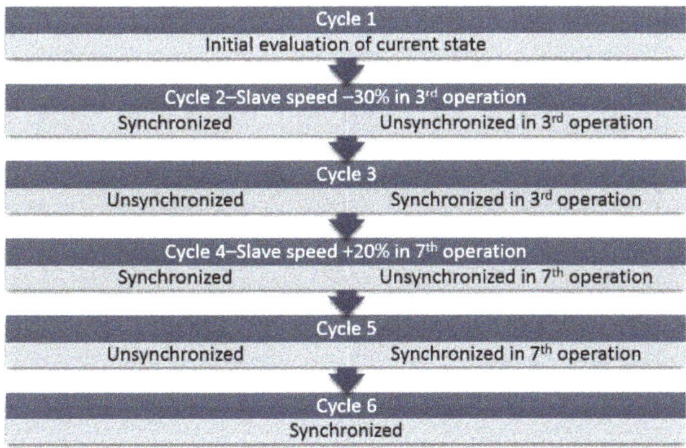

Figure 9. Brief summary of Experiment 2a results.

4.1.4. Experiment 1b—Master Speed Change, Advanced Algorithm

The case with a 10% limit for the error of the difference between operation duration performed by master and slave manipulators and a 15% significance limit for the similarity of two successive deviations $D_{i,j}$ and $D_{i,j+1}$ was validated via simulation.

In simulation Experiment 1b, equal parameters of the whole simulated system were used as in simulation Experiment 1a, with the addition of the feedforward synchronization feature to the synchronization algorithm.

4.1.5. Experiment 2b—Slave Speed Change, Advanced Algorithm

Identical parameters of the whole system as in simulation Experiment 2a were used in simulation Experiment 2b. The synchronization algorithm was supplemented by a feedforward synchronization function.

4.1.6. Summary of Experiment 1b and Experiment 2b

The simulation results of the experiment with a change of master manipulator speed are depicted in Figures 10 and 11. Results of the experiment with a change of slave manipulator speed are shown in Figures 12–14. Both experiments results, where an advanced synchronization algorithm was used, are aggregated in Figures 12 and 15. These results prove that the proposed algorithm can identify a global change of speed of the master element (Experiment 2a) and also of the slave element (Experiment 2b). Most importantly, the algorithm is able to modify the speed of the slave element (elements) feedforward unlike in the case of the standard algorithm. Based on monitoring the operation duration of master and slave manipulators (Figures 10 and 13), and according to evaluation of percentage proportion deviations (Figures 11 and 14), it is clear that the algorithm ensured the time-synchronized movement in the 3rd step of the production cycle from such a change. This resynchronized activity is indicated by a minimized absolute value of percentage proportion deviation $D_{i,j}$ in the two operations after this change.

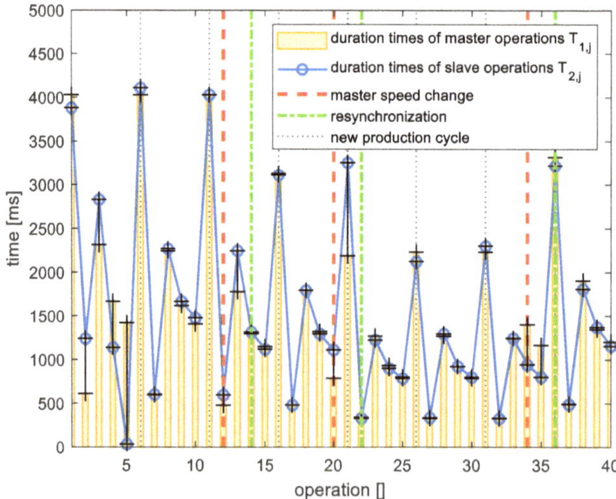

Figure 10. Operation duration times of master and slave manipulators with feedforward synchronization—Experiment 1b.

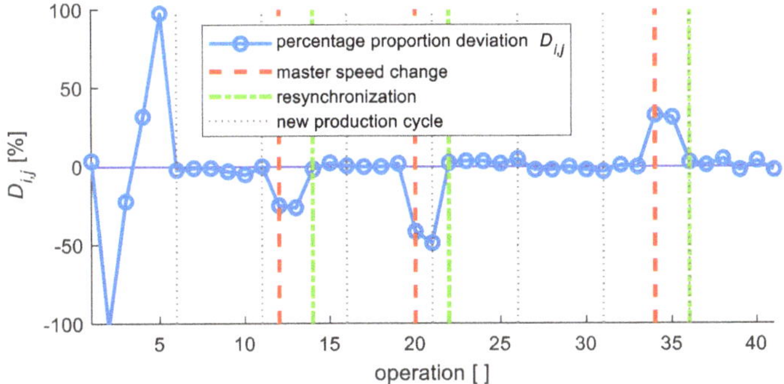

Figure 11. Percentage proportion deviation with feedforward synchronization—Experiment 1b.

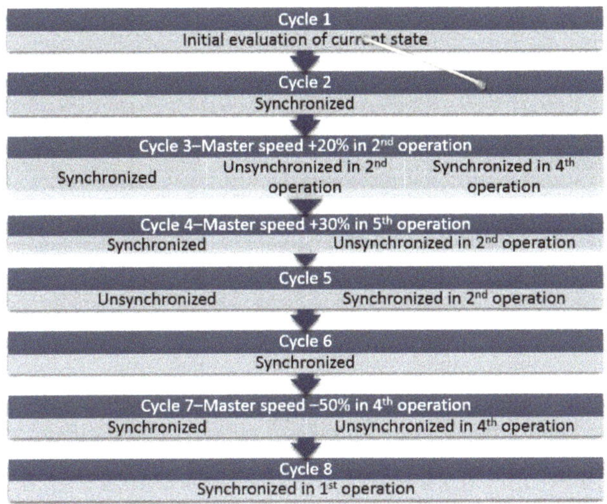

Figure 12. Brief summary of Experiment 1b results.

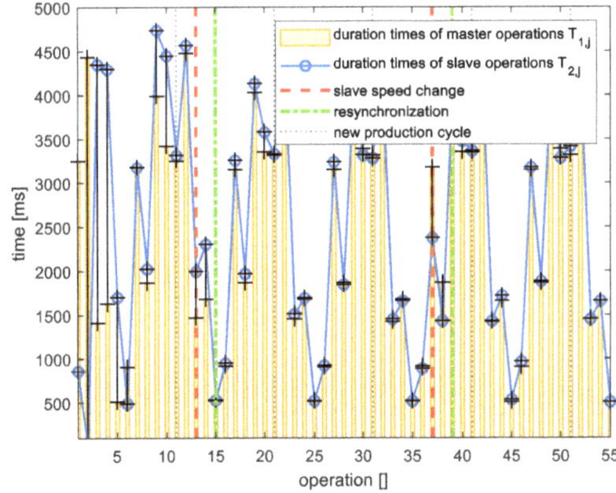

Figure 13. Operation duration times of master and slave manipulators with feedforward synchronization—Experiment 2b.

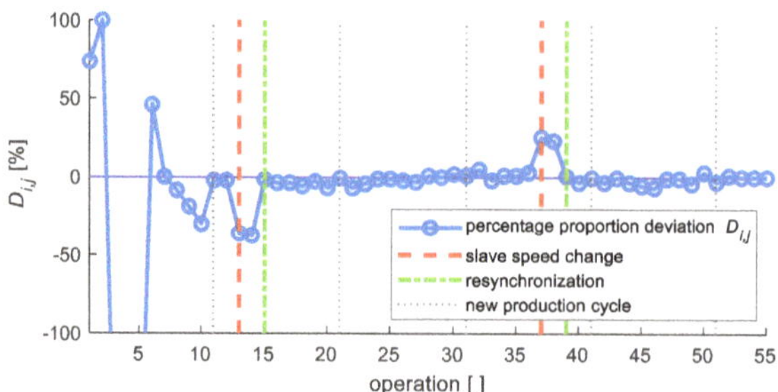

Figure 14. Percentage proportion deviation with feedforward synchronization—Experiment 2b.

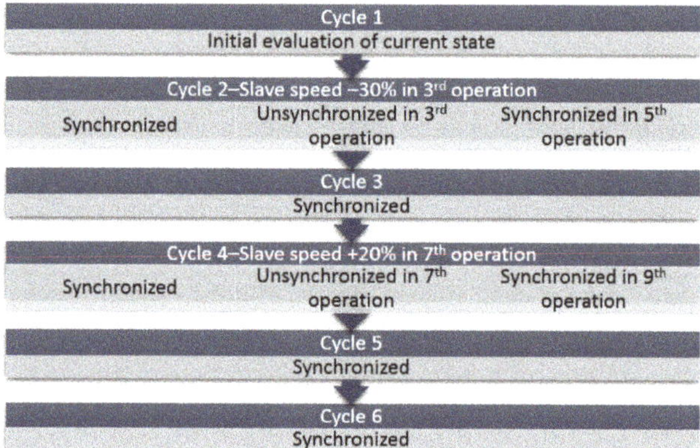

Figure 15. Brief summary of Experiment 2b results.

4.2. Implementation of the Proposed Solution and Discussion

The functionality of the designed algorithms was tested in the final phase on real robot manipulators in a multi-robotic cell. The algorithms were processed in the native language of each robotic manipulator controller using its built-in standard functions.

The model robotic cell, used for the purposes of this paper, contains a heterogeneous triplet of robotic manipulators as shown in Figure 16.

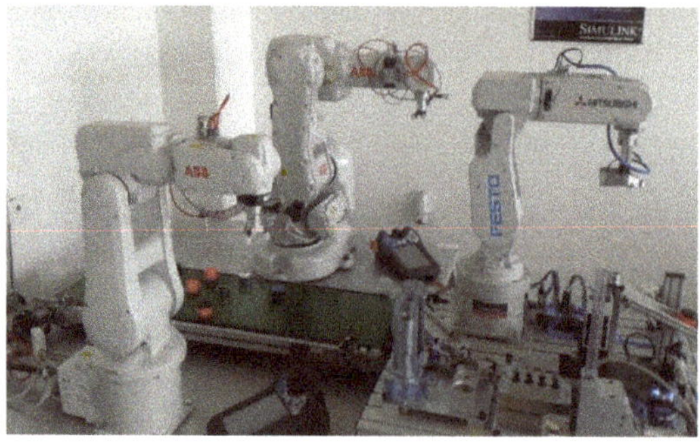

Figure 16. Heterogeneous multi-robotic cell.

One production cycle consists of five operations ($m = 5$), and mainly contains instructions for circular interpolation movement in combination with linear interpolation movement. The whole production cycle is schematically shown in Figure 17.

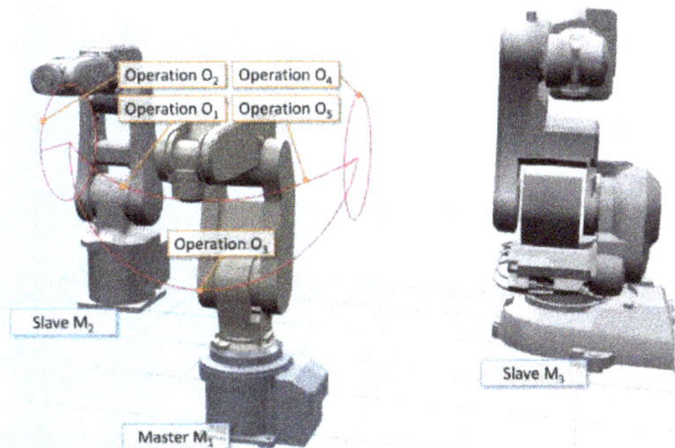

Figure 17. Desired set of operations in one production cycle.

The organizational structure of this robotic cell control was realized as master–slave [24,25] without an external control element. The master object (master manipulator) chosen was a Mitsubishi Melfa RV-2FB-D robot with the Mitsubishi CR750-D control system [26]. Robots ABB IRB 120 and ABB IRB 140, with IRC5 Compact control systems, were the child objects-slaves [27]. A personal computer was considered an element of the slave group and provided visualization of the cell activity synchronization process. Configurations of every cell element are listed in Tables 1 and 2. The TCP/IP channel provided communication among cell elements on the basis of socket exchange [25,28].

Table 1. Slave Configuration.

Object	IP Address
Robot ABB IRB 120 Controller ABB IRC5 COMPACT	192.168.1.120
Robot ABB IRB 140 Controller ABB IRC5 COMPACT	192.168.1.140
PC	192.168.1.111

Table 2. Master Configuration.

Parameter	Value
Object	Robot Mitsubishi Melfa RV-2FB-D Controller Mitsubishi CR750-D
NETIP	192.168.1.20
COMDEV	5, 6, 7 (OPT16, OPT17, OPT18)
NETMODE	1 (SERVER)
NETPORT	10006, 10007, 10008
CRPCE	2 (DATALINK)
PORT	COM6, COM7, COM8

Confirmed coordination of operations control among robotic manipulators and unconfirmed communication between master and monitoring computer were used in the cell. Confirmed coordination means that after every operation is executed by the master element, a terminating message is sent to the slave side and the master element waits for a confirmation message. The next operation can only be initiated after receiving a confirmation message from the slave object. Unconfirmed communication in this case involves simply receiving messages about the synchronization process by the slave object, without sending any confirmation message to the master object [25].

4.2.1. Master—Mitsubishi Melfa RV-2FB-D

In the case of the Mitsubishi CR750-D controller of the robotic manipulator Mitsubishi Melfa RV-2FB-D, the algorithm application was implemented in MELFA BASIC V language [29]. Communication with the slave object was realized through the initiated TCP/IP channel (Box 1).

Box 1. Communication initiation in MELFA BASIC V.

```
Open "COM7:" As#1 'IRB120 IP:192.168.1.120 port 10007
Open "COM8:" As#2 'IRB140 IP:192.168.1.140 port 10008
Open "COM6:" As#3 'PC IP:192.168.1.111 port 10006
```

A duration time of each operation was evaluated and this information was, after the message was received on the completion of each operation by all slave objects, sent in a defined format to these slave objects. This message was also a confirmation message that enabled the beginning of the next operation execution. The operation duration time evaluated by the master object (*Time*) was measured in milliseconds. All of the obtained information was also sent to the visualization part of the application for further processing (Box 2).

Box 2. Socket communication in MELFA BASIC V.

```
M_TIMER(1) = 0
{Operation instructions}
Time = M_TIMER(1)
Input #1,msg1$
Input #2,msg2$
Print #1,STR$(Operation_ID)+":"+STR$(Time)
Print #3,"1-"+STR$(ID)+":"+STR$(Time)
Print #3,"2-"+msg1$
Print #3,"3-"+msg2$
```

4.2.2. Slaves—ABB IRB 120/IRB 140

In the case of controllers ABB IRC5 Compact of robotic manipulators ABB IRB 120 and ABB IRB 140, the designed algorithms were implemented in RAPID language, which is the native language for ABB robot programming [30]. Communication with master object was realized through the activated TCP/IP channel (Box 3).

Box 3. Communication initiation in RAPID.

```
VAR socketdev client_socket;
SocketCreate client_socket;
SocketConnect client_socket,"192.168.1.20",10007\Time:=5;
```

Each operation time duration was measured in each production cycle and, on the basis of the decoded information from the received socket sent by the master object (custom function *DecodeSocket*), an actual speed for that operation was modified (custom function *ChangeSpeed*). This modified speed was used for the presently evaluated operation in the next production cycle. If there is a need for feedforward form of reaction on speed change then a custom function *FeedforwardChange* can be used (Box 4).

Box 4. Socket communication in RAPID.

```
ClkStart clock1;
{Operation instructions}
ClkStop clock1;
n_time: = ClkRead(clock1);
SocketSend client_socket\Str: = ValToStr(op_id)+":"+ValToStr(n_time&#13; × 1000);
SocketReceive client_socket\Str: = s_receive_string\Time:= 10;
DecodeSocket;
ChangeSpeed;
[FeedforwardChange];
```

The function *DecodeSocket* provided information extraction from received socket (*s_receive_string*). This information consisted of operation identifier (*op_id*) and time of its duration (*t_in*) for the master object. Built-in functions of RAPID language for string processing *StrPart*, *StrMatch*, *StrLen* and *StrToVal* were used.

The custom function *ChangeSpeed* provided the modification of speed used for the current operation execution in the array of all operations speeds. These speed values were used in the next production cycle. A numerical array was also used for storing percentage proportion deviations between durations of the operation performed by master and slave objects. This array can be optionally used for feedforward synchronization. The operation duration time evaluated by a slave object (*n_time*) was measured in seconds (Box 5).

Box 5. The function ChangeSpeed in RAPID.

```
speed{op_id}:= speed{op_id} × (n_time × 1000/t_in);
diff{op_id}:= (1-(n_time × 1000/t_in)) × 100;
```

The optional function *FeedforwardChange* provides an evaluation of whether the indicator of current operation change is supraliminal and, at the same time, is sufficiently similar to the indicator of the previous operation change. In the case of positive evaluation, all records in the operation speed array, except those of the current and previous operation, are modified utilizing a dynamically modified array of operation indexes (Box 6).

Box 6. The feedforward function in RAPID.

```
op_indx:= [1, ... , n]; diff_ratio:=Abs(1-(diff{op_indx{n}}/diff{op_indx{1}}))&#13; × 100;
IF (Abs(diff{op_indx{1}})>diff_limit) AND&#13; (diff_ratio<diff_ratio_limit) THEN
k:=1-(((diff{op_indx{1}}+diff{op_indx{n})/2)/100);
FOR i FROM 2 TO n-1 DO
speed{op_indx{i}}:=speed{op_indx{i}} × k;
ENDFOR
ENDIF
Rotate_op_indx_left;
```

4.2.3. Visualization—MATLAB Application

For the visualization of the robotic cell behavior, any development tool supporting functionalities for sockets processing can be used. In this case, the visualization part of application was implemented using the MATLAB software tool since it was already used for simulation validation of the proposed solution. The capability of MATLAB to access existing system Java classes to be used in the MATLAB workspace [31–34] was utilized in this application part. The communication with the master object was realized through the activated TCP/IP channel (Box 7).

Box 7. Communication initiation in MATLAB.

```
import java.net.Socket
import java.io.×
input_socket = java.net.Socket();
input_socket.connect(java.net.InetSocketAddress(server,port));
```

The received message from the master object consisted of cell object identifier (*station*), operation identifier (*index*), and operation duration (*value*). After decoding, the message was used as a data source for graphical representation (3D bar graph) of the production process in the robotic cell, as shown in Figures 18–20 (Box 8).

Box 8. Socket communication in MATLAB.

```
input =&#13; BufferedReader(InputStreamReader(input_socket.getInputStream));
message = char(input.readLine());
{graph(station)}.ZData(index) = value;
```

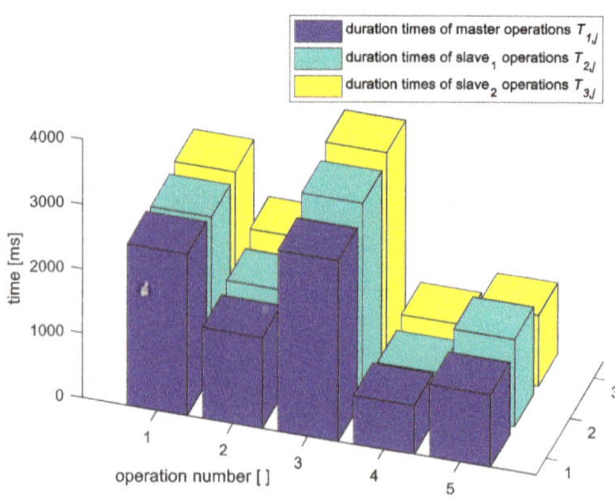

Figure 18. Visualization of operation duration in the production process.

The graph in Figure 18 was updated after each operation. Information on the operation duration of all participating executive elements was collected and subsequently distributed by the master manipulator.

4.2.4. Implementation Results

Implementation results are represented based on the visualization part of the application. The operation durations of all cell elements during the production process with the implemented basic algorithm without feedforward synchronization are depicted in Figure 19.

The change of endpoint movement speed of the master manipulator occurred in this case during operation number two in production cycle five (Figure 19e). The production cell activities were resynchronized after the 2nd operation in the next production cycle (Figure 19f).

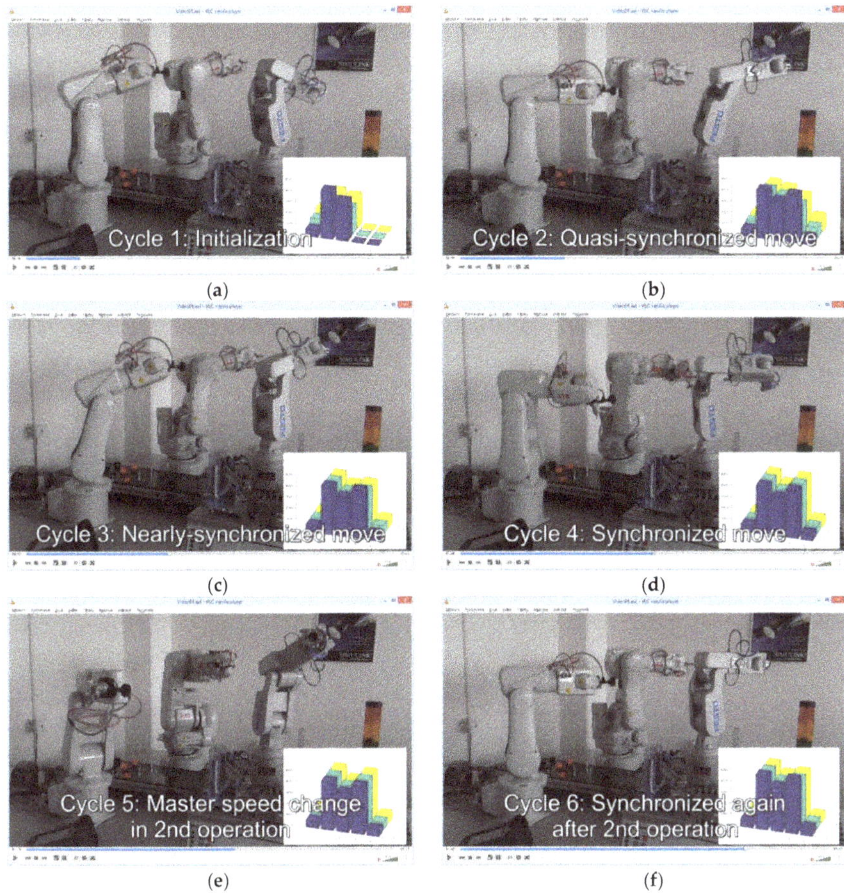

(a) (b) (c) (d) (e) (f)

Figure 19. Implementation and visualization–synchronization without feedforward effect. (**a**) Initial cycle; (**b**) Quasi-synchronized state; (**c**) Nearly-synchronized state; (**d**) Synchronized state; (**e**) Master speed change cycle; (**f**) Resynchronized state.

In the case of the advanced algorithm with feedforward synchronization implementation (Figure 20), the movement speed change of the master manipulator endpoint occurred in operation one during production cycle five (Figure 20d). The feedforward effect ensured production cell activities were resynchronized in the next two operations in the same production cycle (Figure 20e).

On the basis of the obtained results depicted in Figures 19 and 20, it is clear that the proposed and simulation-validated algorithm for the synchronization of the movement speeds of manipulators (representing endpoint movement speeds) grouped in heterogeneous robotic work cells is functional and applicable. The deviations in achieved operation duration times of master and slave robotic manipulators in the initial production cycles (Figure 19b,c and Figure 20b) are noticeable. These are caused by non-linear characteristics of the movement of robot arms (acceleration, deceleration), and because operations two and four are composed of several movements of different types (linear, circular). However, this disproportion is fully eliminated by dynamic speed correction of the slave manipulators during subsequent cycles.

In the case of basic non-feedforward synchronization, the system reaction to the endpoint movement speed change of the master manipulator was as expected. The process was resynchronized in the next production cycle in the same operation, where the change occurred.

The results also indicate that the reaction of slave manipulators to the global change of the master manipulator speed with the use of feedforward synchronization persists in real conditions for an additional cycle step (Figure 20e) than is presented in Section 3.1 for the simulated system with idealized conditions. The reason for this difference is that the master manipulator speed change occurred during operation execution. The percentage proportion deviation among these operation durations (between master and each slave) is thus different from the percentage proportion deviation among durations of the next operation that is completely performed at the changed speed.

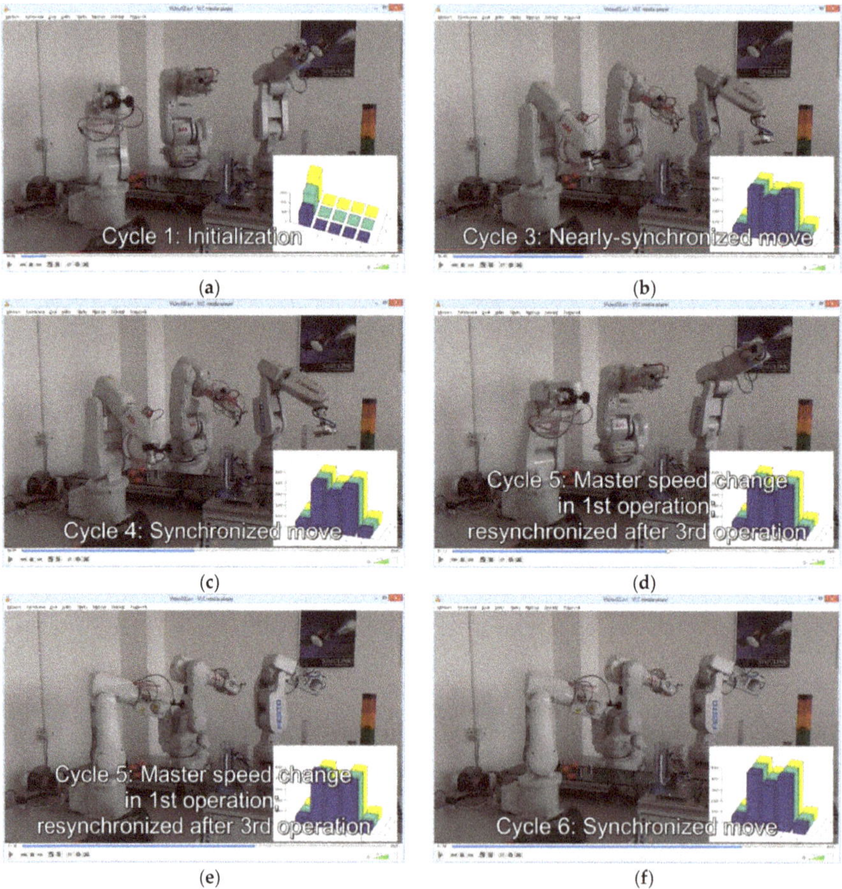

Figure 20. Implementation and visualization–synchronization with feedforward effect.(**a**) Initial cycle; (**b**) Nearly-synchronized state; (**c**) Synchronized state; (**d**) Master speed change cycle; (**e**) Resynchronized state in the same cycle; (**f**) Synchronized state.

5. Conclusions

In this study, an algorithms for the time synchronization of operations performed by a heterogeneous set of robotic manipulators grouped into a production cell was proposed, validated, and implemented. The organizational structure of this robotic cell control was realized as master–slave without an external control element. Communication in the cell was provided by a TCP/IP channel via sockets. The proposed problem solution requires minimal computational power due to an empirically oriented approach. We relied on our wide empirical knowledge and experiences in process algorithmizing, as well as on various previously implemented tasks in the field of robotics or the modeling and visualization of processes. This approach enabled the solution to be processed directly by the control unit of each participating element of the robotic cell with utilization of standard instructions in their native language. The main aim was to dynamically adapt the movement speed of slave manipulator endpoints to the master manipulator activity. Therefore, the algorithms ensure the defined milestones of the production cycle of each robotic manipulator in the cell are attained at the same time, while all operations may include various sets of different motion or manipulation instructions.

The proposed solution also includes an advanced feedforward form of operation synchronization which responds to changes in the operating cycle of the master manipulator or slave manipulators more effectively. The main difference between the two proposed algorithms is the number of unsynchronized operations performed after the change of the master or the slave behavior. In the basic algorithm case, after desynchronization, the operations of one cycle are performed unsynchronized. In contrast, the advanced algorithm ensures resynchronization after a defined number (in our case two) of asynchronously performed operations.

The application of the solution proposal is supplemented with a visualization part created using MATLAB software for technical computing. This application illustrates each intervention of the synchronization algorithms, and enables more efficient monitoring and evaluation of the multi-robotic cell activity with a focus on the synchronization process. This application part complements the validation of the functionality of the designed solution.

Finally, it can be stated that all requirements were successfully met and our solution for synchronization of the heterogeneous multi-robotic cell with emphasis on low computing power is functional and feasible.

Our goal in the future is to continue to develop this idea based on current trends in industrial automation [35]. There is a possibility in master–slave architecture to distribute more process or control information among elements, e.g., target position or movement type together with operation duration as used in our solution. Visualization, as an important aspect of the production of tomorrow, can be realized using virtual or augmented reality [35].

Author Contributions: Conceptualization, M.J.; formal analysis, B.J.; project administration, M.J.; software, M.J. and B.J.; supervision, M.J.; validation, M.J.; visualization, M.J.; writing—original draft, M.J. and B.J.; writing—review & editing, M.J. and B.J. All authors have read and agreed to the published version of the manuscript.

Funding: This research was funded by VEGA agency, grant number 1/0232/18—"Using the methods of multi-objective optimization in production processes control".

Conflicts of Interest: The authors declare no conflict of interest. The funders had no role in the design of the study; in the collection, analyses, or interpretation of data; in the writing of the manuscript, or in the decision to publish the results.

References

1. Weyer, S.; Schmitt, M.; Ohmer, M.; Gorecky, D. Towards Industry 4.0—Standardization as the crucial challenge for highly modular, multi-vendor production systems. In Proceedings of the IFAC Symposium on Information Control in Manufacturing (INCOM), Ottawa, ON, Canada, 11–13 May 2015. [CrossRef]
2. Halenar, I.; Juhasova, B.; Juhas, M. Proposal of communication standardization of industrial networks in Industry 4.0. In Proceedings of the IEEE 20th Jubilee International Conference on Intelligent Engineering Systems (INES), Budapest, Hungary, 30 June–2 July 2016. [CrossRef]
3. Qin, Y.; Sheng, Q.Z.; Falkner, N.J.G.; Dustdar, S.; Wang, H.; Vasilakos, A.V. When things matter: A survey on data-centric internet of things. *J. Netw. Comput. Appl.* **2016**, *64*, 137–153. [CrossRef]

4. Leitao, P.; Rodrigues, N.; Barbosa, J.; Turrin, C.; Pagani, A. Intelligent products: The grace experience. *Control. Eng. Pract.* **2015**, *42*, 95–105. [CrossRef]
5. Hermann, M.; Pentek, T.; Otto, B. Design Principles for Industrie 4.0 Scenarios. In Proceedings of the 49th Hawaii International Conference on System Sciences (HICSS), Kauai, Hawaii, 5–8 January 2016.
6. Jasperneite, J. Was hinter Begriffen wie Industrie 4.0 steckt. *Comput. Autom.* **2014**, *12*, 24–28.
7. Kagermann, H.; Wahlster, W.; Helbig, J. Recommendations for implementing the strategic initiative Industrie 4.0. In *Final Report of the Industrie 4.0 Working Group*; National Academy of Science Engineering: Munich, Germany, 2013.
8. Chang, W.H.; Kim, Y.H. Robot-assisted Therapy in Stroke Rehabilitation. *J. Stroke* **2013**, *15*, 174–181. [CrossRef] [PubMed]
9. Yoon, J.; Novandy, B.; Yoon, C.; Park, K. A 6-DOF Gait Rehabilitation Robot With Upper and Lower Limb Connections That Allows Walking Velocity Updates on Various Terrains. *IEEE/ASME Trans. Mechatron.* **2010**, *15*, 201–215. [CrossRef]
10. Rubenstein, M.; Ahler, C.; Hoff, N.; Cabrera, A.; Nagpal, R. Kilobot: A low cost robot with scalable operations designed for collective behaviors. *Robot. Auton. Syst.* **2014**, *62*, 966–975. [CrossRef]
11. Werner-Allen, G.; Tewari, G.; Patel, A.; Welsh, M.; Nagpal, R. Firefly-inspired sensor network synchronicity with realistic radio effects. In Proceedings of the 3rd International Conference on Embedded Networked Sensor Systems, San Diego, CA, USA, 2–4 November 2005; ACM: New York, NY, USA. [CrossRef]
12. Baca, J.; Pagala, P.; Rossi, C.; Ferre, M. Modular robot systems towards the execution of cooperative tasks in large facilities. *Robot. Auton. Syst.* **2015**, *66*, 159–174. [CrossRef]
13. Chung, S.; Slotine, J.E. Cooperative robot control and concurrent synchronization of Lagrangian systems. *IEEE Trans. Robot.* **2009**, *25*, 686–700. [CrossRef]
14. Rodriguez-Angeles, A.; Nijmeijer, H. Mutual synchronization of robots via estimated state feedback: A cooperative approach. *IEEE Trans. Control. Syst. Technol.* **2004**, *12*, 542–554. [CrossRef]
15. Yasuda, G. A distributed autonomous control architecture for synchronization and coordination of multiple robot systems. In Proceedings of the SICE Annual Conference 2012 (SICE), Akita, Japan, 20–23 August 2012.
16. Markus, E.D.; Yskander, H.; Agee, J.T.; Jimoh, A.A. Coordination control of robot manipulators using flat outputs. *Robot. Auton. Syst.* **2016**, *83*, 169–176. [CrossRef]
17. Bouteraa, Y.; Poisson, G.; Ghommam, J.; Derbel, N. Adaptive multi-robots synchronization. In Proceedings of the IEEE International Symposium on Industrial Electronics, Bari, Italy, 4–7 July 2010. [CrossRef]
18. Bouteraa, Y.; Ghommam, J.; Poisson, G.; Derbel, N. Distributed synchronization control to trajectory tracking of multiple robot manipulators. *J. Robot.* **2011**, *9*, 1–10. [CrossRef]
19. Bouteraa, Y.; Ghommam, J. Synchronization control of multiple robots manipulators. In Proceedings of the 6th International Multi-Conference on Systems, Signals and Devices, Djerba, Tunisia, 23–26 March 2009. [CrossRef]
20. Duschau-Wicke, A.; von Zitzewitz, J.; Banz, R.; Riener, R. Iterative Learning Synchronization of Robotic Rehabilitation Tasks. In Proceedings of the 10th International Conference on Rehabilitation Robotics, Noordwijk, The Netherlands, 13–15 June 2007. [CrossRef]
21. ABB MultiMove Functionality Heralds a New Era in Robot Applications. Available online: https://searchext.abb.com/library/Download.aspx?DocumentID=9AKK105152A2837&Action=Launch (accessed on 1 December 2019).
22. Gan, Y.; Dai, X.; Li, D. Off-Line Programming Techniques for Multirobot Cooperation System. *Int. J. Adv. Robot. Syst.* **2013**, *10*, 1–17. [CrossRef]
23. KUKA. KUKA. RoboTeam. Available online: https://www.kuka.com/en-gb/products/robotics-systems/software/hub-technologies/kuka_roboteam (accessed on 1 December 2019). [CrossRef]
24. Juhasova, B.; Juhas, M.; Halenar, I. TCP/IP Protocol Utilisation in Process of Dynamic Control of Robotic Cell According Industry 4.0 Concept. In Proceedings of the 15th IEEE International Symposium on Applied Machine Intelligence and Informatics (SAMI), Herl'any, Slovakia, 26–28 January 2017.
25. Mitsubishi Electric. *MELFA Industrial Robots Instruction Manual (Ethernet Interface CRn-500 Series)*; Mitsubishi Electric: Ratingen, Germany, 2002.
26. ABB Automation Technologies AB Robotics. *Product Specification Controller Software IRC5 Manual*; ABB: Zurich, Sweden, 2004.

27. Youm, B.; Park, J. TCP/IP protocol over ieee-1394 network for real-time control applications. In Proceedings of the 16th IFAC World Congress, Prague, Czech Republic, 3–8 July 2005; Volume 38, pp. 37–42.
28. Mitsubishi Electric. *MELFA Industrial Robots Instruction manual (Detailed Explanations of Functions and Operations)*; Mitsubishi Electric: Ratingen, Germany, 2005.
29. ABB AB Robotics Products. *Technical Reference Manual—RAPID Instructions, Functions and Data Types*; ABB: Zurich, Sweden, 2010.
30. MathWorks Matlab, Call Java Libraries. Available online: https://www.mathworks.com/help/matlab/using-java-libraries-in-matlab.html (accessed on 1 December 2019).
31. Oracle, Socket (Java Platform SE 7). Available online: https://docs.oracle.com/javase/7/docs/api/java/net/Socket.html (accessed on 1 December 2019).
32. Oracle, Buffered Reader (Java Platform SE 7). Available online: https://docs.oracle.com/javase/7/docs/api/java/io/BufferedReader.html (accessed on 1 December 2019).
33. MathWorks Matlab, Plot 3-D Bar Graph. Available online: https://www.mathworks.com/help/matlab/ref/bar3.html (accessed on 1 December 2019).
34. Pérez, L.; Rodríguez-Jiménez, S.; Rodríguez, N.; Usamentiaga, R.; García, D.F. Digital Twin and Virtual Reality Based Methodology for Multi-Robot Manufacturing Cell Commissioning. *Appl. Sci.* **2020**, *10*, 3633. [CrossRef]
35. Juhás, M.; Juhásová, B.; Halenár, I. Augmented reality in education 4.0. In Proceedings of the 2018 IEEE 13th International Scientific and Technical Conference on Computer Sciences and Information Technologies (CSIT 2018), Lviv, Ukraine, 11–14 September 2018. [CrossRef]

Article

An Auto-Adaptive Multi-Objective Strategy for Multi-Robot Exploration of Constrained-Communication Environments

Facundo Benavides [1,2,†,*], Caroline Ponzoni Carvalho Chanel [2,‡], Pablo Monzón [3,‡] and Eduardo Grampín [1,‡]

1 Computer Science Institute, Faculty of Engineering, Universidad de la República, 11300 Montevideo, Uruguay; grampin@fing.edu.uy
2 Aerospace Vehicles Design and Control Department (DCAS), ISAE-SUPAERO, Université de Toulouse, 31055 Toulouse, France; caroline.chanel@isae-supaero.fr
3 Electrical Engineering Institute, Faculty of Engineering, Universidad de la República, 11300 Montevideo, Uruguay; monzon@fing.edu.uy
* Correspondence: fbenavid@fing.edu.uy; Tel.: +598-2714-2714(12)
† Current address: J. H. y Reissig 565, Facultad de Ingeniería (InCo), 11300 Montevideo, Uruguay.
‡ These authors contributed equally to this work.

Received: 21 December 2018; Accepted: 2 February 2019; Published: 9 February 2019

Featured Application: In application fields where strong communication requirements do not condition the mission, the present approach represents a proper option for coping with real communication constraints, being more fault tolerant and still having good performance simultaneously.

Abstract: The exploration problem is a fundamental subject in autonomous mobile robotics that deals with achieving the complete coverage of a previously unknown environment. There are several scenarios where completing exploration of a zone is a main part of the mission. Due to the efficiency and robustness brought by multi-robot systems, exploration is usually done cooperatively. Wireless communication plays an important role in collaborative multi-robot strategies. Unfortunately, the assumption of stable communication and end-to-end connectivity may be easily compromised in real scenarios. In this paper, a novel auto-adaptive multi-objective strategy is followed to support the selection of tasks regarding both exploration performance and connectivity level. Compared with others, the proposed approach shows effectiveness and flexibility to tackle the multi-robot exploration problem, being capable of decreasing the last of disconnection periods without noticeable degradation of the completion exploration time.

Keywords: exploration missions; cooperative systems; multi-robot coordinated systems; constrained-communication environments

1. Introduction

The exploration problem is a fundamental subject in autonomous mobile robotics that deals with achieving the complete coverage of a previously unknown environment. There are several scenarios where completing exploration of a zone is a central part of the mission, e.g., planetary exploration, reconnaissance, search and rescue, agriculture, cleaning, or dangerous places as mined lands and radioactive zones. Additionally, due to the inner qualities—mainly efficiency and robustness—of multi-robot systems, exploration is usually done cooperatively [1].

Schematically, the exploration of an environment can be seen as the composition of *Mapping* and *Motion Planning* tasks. A map is needed in order to plan new motions. Moreover, choosing a correct motion sequence based on this map is also needed to expand the knowledge about the environment optimally. Consequently, *Mapping* is regularly interleaved with *Motion Planning*, and vice versa, during the whole process [1–3].

Given that the lack of knowledge is essentially inherent to exploration missions, the best choice for the robots is to visit the places where the gain of information can be potentially higher. The *Task Identification* problem concerns the identification of the points of interest that should be visited next. It strongly depends on both the sensory robot capabilities and the underlying environment representation. The most widely used representation for this purpose is the well-known *Occupancy Grid* structure [4]. Based on it, a method to identify points of interest was proposed by [5]. The strategy assumes that the closer to the frontier between known and unknown regions the tasks are defined, the more information the team can gather. Since then, the majority of exploration proposals has adopted this scheme known as *Frontier Points* or *Frontier Regions* [6–8].

When multiple robots are involved, it is advisable to avoid several of them moving to the same place. The *Task Allocation* problem concerns the search for a distribution of tasks to robots that maximises the overall system utility and minimises the amount of overlapped information obtained by the robots [1,9,10].

There exist a wide variety of proposed solutions to this problem where a family of methods based on market economies are probably the most popular ones. These methods are based on the notion of *Auctions* from which the robots can bid for the tasks to decide who goes to where at each moment. The market may be managed centrally either by a virtual agent at the base station as in [3] where the bids are processed centralised by a greedy algorithm or by a robotic agent as in [1]. Conversely, the fleet can manage to exchange the bids among all the members in order to take decentralised decisions [11,12], avoiding, in turn, the single point of failure. All these methods owe their popularity to their simplicity and ease of implementation, but they suffer from a significant shortcoming: falling in local minima [13].

Far from economy inspired approaches, a scheduling based approach is presented in [2]. This method combines an environment segmentation technique with the centralised task allocation method proposed by [14]. The exploration is performed after dividing the environment into disjoint segments. Thus, the expected sensory overlap between agents is decreased as much as possible.

In [15] the authors address coordination implicitly through localisation data exchanging. Robots are forced to wait for others before making a decision. Task selection is made iteratively—one robot after another—employing an objective function which rewards the right choices. In [16] a centralised approach is used. The tasks-to-robots distribution is computed balancing information gain, localisation quality, and navigation costs. Another centralised approach computes a utility function enabling the robots to locally prioritise the tasks within its scope and, potentially, also enabling the whole team to search for the best global distribution as well [9].

On the contrary, a decentralised approach, called *minPos* [17], attempts to distribute the robots over the unexplored locations as much as possible. By doing so, it has outperformed several reference proposals decreasing the completion-exploration time for a big set of practical scenarios. The working principle is to rank robots concerning their distance to every possible task. The robots coordinate their actions implicitly and may choose to visit the tasks for which they are best ranked at each point in time.

Finally, the strategy described in [18] is mainly devoted to deal with uncertainties in sensing and motion processes of a multi-robot system. To this end, the authors model the exploration and mapping problem as a POMDP that is solved centrally. In [19] the assignment algorithm works in an asynchronous fashion assuming that not all robots must be ready for new plans at the same time.

Wireless communication plays an important role in collaborative multi-robot strategies. Unfortunately, the assumption or requirement of stable communication and end-to-end connectivity may be easily compromised in real scenarios due to interference, fading, or simply robots moving beyond the communication range. When robots are unconnected they have no possibilities to coordinate their actions and damages or inner failures can lead to information losses. Therefore, depending on the application field, the exploration strategy should take this into account to prevent isolation situations.

1.1. Communication Issues

Mobile Ad-hoc NETworks (MANETs) constitute a particular example of scenarios where the topology of the robot network varies dynamically over time. This kind of network is recommended when the fixed infrastructure is no longer available, e.g., in disasters to support the communication among rescue team members. In such cases, connectivity is of utmost importance because the loss of communication could imply human losses.

A first critical issue concerns the collective knowledge of the environment. Under communication restrictions, such knowledge cannot be assumed to be always accessible and depending on the coordination mechanism could be the cause of significant performance degradation [20]. Therefore, depending on the application, the exploration strategy should take this into account in order to prevent the robots from becoming completely unconnected, let say isolation situations. Such an isolation situation, as well as its possible effects, are illustrated in the example scene depicted in Figure 1.

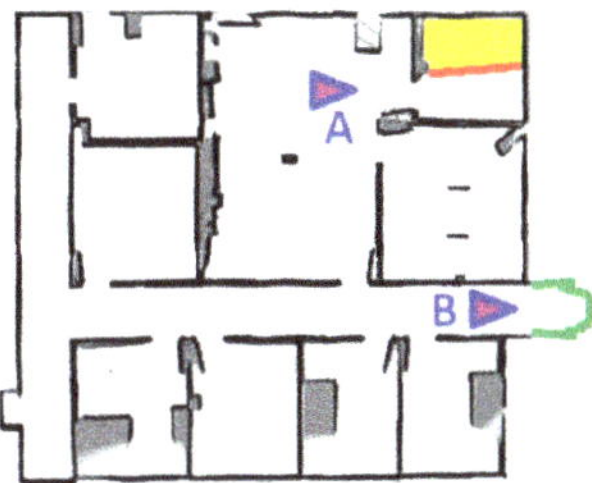

Figure 1. Re-exploration caused by restricted communication. [20]. The yellow portion of the map is only known by robot B. Thus, robot A goes to re-explore the region beyond the red frontier.

1.1.1. Connection Requirements

Three categories are mainly identified [20]:

- *None.* Robots are not required to communicate.
- *Event-based connectivity.* The need for regaining connectivity is triggered by particular events such as the discovery of new information or just periodically.
- *Continuous connectivity.* Every robot must be connected at all times to any other fleet member either directly or in a multi-hop manner.

Please note that these requirements could have an impact on the fleet mobility and, in turn, on the availability of exploration strategies to be adopted. For instance, under a continuous connectivity scheme, the fleet is more restricted to move around than in other cases.

1.1.2. Communication Models

Communication model refers to the prior knowledge about communication capabilities that support the decision making of the robots along the exploration. Nevertheless, sometimes no communication model is assumed and, consequently, robots do not depend on communicating to decide where to go next. In such cases, explicit coordination only occurs opportunistically due to random encounters [20].

The communication models typically adopted are [20]:

- *None.* Robots do not make any assumption on the communication possibilities between any pair of arbitrary locations.

- *Line-of-sight (LoS)*. Two robots can communicate if and only if their positions belong to a free-of-obstacle line segment. Usually, the distance is also restricted to a maximum value that is often related to the scope of the communication device.
- *Disc or Circle*. Communication with any other robot is permitted when its location is within a fixed maximum distance (communication radius) regardless of the presence of obstacles.
- *Signal*. Communication is available with a certain probability that depends on the estimated signal power between the robot positions. The higher the signal power, the higher the probability.
- *Traces*. Robots can communicate with each other by dropping messages in the environment.

Additionally, to these five categories that cover an essential aspect of the communications, say connectivity, there exist other formulations aimed at cover bandwidth or throughput as well. Clear examples of its use are the applications with a strong dependence on video streaming like search and rescue applications.

1.2. Connectivity-Based Proposals

Despite its well-known inefficiencies, there exist some few approaches without any connection requirements where robots meet each other by chance. Nevertheless, this section only surveys the proposals that depend on connectivity in one way or another.

In [21] a behaviour-based approach is presented. The architecture is designed to guide the exploration constraining the fleet to keep within the communication range, establishing a mobile network. The well-known disk model and a graph structure are used to model the network connectivity and identify possible disconnections. Frontier cells are evaluated regarding costs (computed utilising a flooding algorithm) and information utility (based on the ideas proposed in [3]). Behaviours are selected according to the network topology conditions.

In [22] a centralised communicative exploration algorithm is proposed. Communicative exploration implies that the team of robots have to maintain connections between each other at all times. The target selection is based on a utility function that weights the benefits of exploring new regions versus the goal of keeping connected. While connectivity is valued using the classic *disc* model, the costs of the shortest paths are computed from the Manhattan distance notion. Due to spatial and movement restrictions, specific behaviours are defined to deal with deadlocks. Also following a centralised approach, [23] presents four fully reactive exploration strategies. They consist in translating the distance to tasks and disconnection situations into artificial forces that pull and push the robot to reach new positions smoothly, avoiding them to lose connectivity. The radio signal quality is modelled considering both the communication range and the distance attenuation effect. Deadlocks are avoided by assigning tasks to a cluster of robots. This allocation guarantees that robots belonging to the same cluster do not exert conflicting forces upon each other towards different directions.

In [24], the authors propose a decentralised version of the strategy proposed in [22] based on message exchanging and a graph structure where the group always tries to keep a biconnected network efficiently. Communication model is based on the classic *disc* model. In consequence, robot mobility is restricted by the communication range. Using the same graph theory, in [25] the experimental validation of a distributed algorithm that preserves connectivity is also discussed. Nevertheless, a different coordination mechanism—supported by a market-based negotiation algorithm—is adopted. Unfortunately, only results on connectivity maintenance are shown, lacking exploration metrics reports.

The proposal of [26] aims to maintain and repair the underlying wireless mesh network while the coverage task is being performed, all at once. The system works in a fully asynchronous and distributed way. Differently from previous works, the authors propose a network disconnection detection by checking the real state of connections without assumptions on communication range or propagation model. On the contrary, all nodes require knowledge about the area to be covered and on global positions.

In [15] the robots can disconnect as long as they regain connectivity periodically following a distributed but synchronous strategy. Authors address coordination implicitly through localisation data exchanging. Robots are forced to wait for others before making a decision. The system works as an optimisation method where each variable is optimised at a time in a round-robin while the others remain unchangeable. In [27] the authors describe a heterogeneous multi-robot system for exploration tasks. They consider several explorer robots and conceive a particular robot playing the role of relay dispenser. This agent is in charge of place relays when and where it is necessary to support the video/audio streaming generated by explorers.

In [28] the problem of exploration and mapping is addressed by using a *Decentralised POMDP*. This technique takes advantage of local interaction and coordination from the interaction-oriented resolution of decentralised decision makers. *Distributed value functions (DVS)* are used by decoupling the multi-agent problem into a set of individual agent problems. In order to address full local observability, limited information sharing, and communication breaks, an extension of the *DVS* methodology is proposed and applied in multi-robot exploration so that each robot computes locally a strategy that minimises the interaction between fleet members and maximises the coverage achieved by the team, even in communication constrained environments. A decision step consists in building the model, computing the policy from the *DVS* and producing a trajectory.

Rendezvous-based techniques have also been used to deal with limited communication ranges. In [16] robots are enabled to move out of the communication range but forced to rejoin the group frequently. After moving out the communication range robots have to return to a pre-arranged meeting point to exchange the information gathered during the disconnection period in order to avoid exploration overlaps.

The proposal presented in [29] describes a *Particle Swarm Optimisation* based approach to achieving fault-tolerance in preventing communication network splits. The principal objective is to keep the fleet k-connected. Considering that the application domain defines the fault-tolerance level required to the system, a MANET connectivity algorithm is extended with the concept of k-fault-tolerance.

A multi-robot system for crisis management is described in [30]. The system is composed of mobile sensors (ground robots—UGV) and mobile relays (aerial vehicles—UAV). However, some robots may change roles dynamically during the mission (e.g., UAVs equipped with both wireless routers and cameras). The problem is modelled and solved using constrained-based local search on a communication model based on graph theory.

In [31] a fully distributed approach for multi-robot sweep exploration is introduced. The proposal aims to guarantee full coverage using a minimum number of messages and to maintain connectivity at all times, even under severe restrictions on the communication type, range, and quality. The algorithm proposed uses communication not only to exchange information but to direct the robot movements. Communication intensity is used in order to disperse the fleet while beacons are used to mark locations of interest.

In [32] a multi-robot exploration algorithm based on multiple behaviours is proposed. Quad-rotors are asked to explore and map an indoor zone with unreliable communication and limited battery life. Robots are enabled to change roles both dynamically according to intrinsic and extrinsic factors (e.g., boundaries/distances and battery level) and hierarchically in order to explore and avoid collision among each other. The remaining battery level is considered in order to avoid losing gathered information. Quad-rotors are also able to leave the network, but after a fixed period they search for regaining connectivity. Relay robots are designated to forward information from/to the more distant robots improving communication between team members. Although no optimal relay placement is computed, the existence of relays is crucial in the proposed scheme.

In [33,34] the relay node dynamic re-positioning problem is tackled. The proposed solution relies on optimisation procedures and evolutionary algorithms to find the best relay locations and how the robots should move to these points. The authors follow a centralised multi-stage approach where one node is in charge of computing the best assignment regarding both connectivity and throughput.

In [35], the problem of how to connect one or more remote units to a base station investing a limited number of intermediate relay robots in constrained communication environments is investigated.

The authors study the complexity of the optimal relay placement problem and propose methodologies to create chains or trees of relays as required by different static scenarios. By contrast, in changing environments static solutions cannot be successfully applied because the location optimality does not hold over time.

In [36] the exploration problem is addressed ensuring a time-varying connected topology in *3D* cluttered environments but following a decentralised control strategy which enables simultaneous multi-task exploration.

Another centralised but asynchronous strategy is followed in [19,37] in order to address the problem of multi-robot exploration under recurrent connectivity. In these works, the authors leverage a variant of the Steiner tree problem that appears as a particular case of different known graph optimisation problems. Robot placement is treated as an optimisation problem through Integer Linear Programming. Exact and approximated algorithms are compared on particular scenarios.

1.3. Conclusions

Some conclusions arise from this brief survey of recent works. Firstly, it is remarkable that despite being the most restrictive class of exploration algorithms, the exploration strategies based on continuous connectivity are prevalent in applications where real-time image streaming are needed (e.g., search and rescue), or simply when human operators at the base station need to enforce timely information updates, or even when a high level of coordination is needed (i.e., when globally shared knowledge between robots is assumed). Additionally, robustness is also highly appreciated in hostile or inaccessible scenarios. In these missions, fault-tolerance is typically achieved adding redundancy (e.g., systems that guarantee k-connected time-varying network topologies) and employing distributed systems.

Nevertheless, when these strong requirements do not condition the mission, the event-based connectivity—that is less restrictive than the former concerning the fleet mobility—seems to be more appropriate.

Now, moving up from essential aspects as communication to the top of the software architecture stack. There exists a large set of distributed reactive and behaviour-based proposals. Compared to the centralised approaches, distributed approaches have the advantage of not presenting the single-point-of-failure weakness. However, in many cases, it suffers from deadlocks at the individual or collective level.

Market-based coordination methods represent another popular option. There exists a wide variety of implementations that mainly differ from each other in the way the bids are computed by the robots (e.g., single-item or multiple-item auctions). These difference are not insignificant and typically trade simplicity and computational efficiency off for proper coordination and local optima avoidance. Besides, since each auction involves a period of synchronicity between robots, fully asynchronous market-based systems have no place. Nevertheless, asynchronous systems may be advantageous over those that periodically ask the robots to wait for others before making a decision.

Finally, in communication-restricted environments, there seems to be a general agreement on the benefits of spreading out the fleet as long as the robots can regain connectivity in disconnection case. From this, and trying to balance these potentially opposed goals, some multi-objective utility-based approaches have been proposed. Also, defining multiple roles (including communication relays) has demonstrated to be a worthy strategy to address the multi-robot exploration problem when communication restrictions are present.

In conclusion, the survey suggests that in the context of decentralised systems there is room to try new ideas related to connectivity-regaining policies and rendezvous places. On the one hand, the event-based connectivity framework imposes the execution of connectivity-regaining actions in the presence of some events. On the other hand, rendezvous-based approaches imply the definition of particular meeting points where robots have to meet in order to regain connectivity. Leaving apart the fact that the selection of these places could be a hard issue itself, once the connectivity-regaining action is triggered and the meeting place is known by robots, they should interrupt its exploration plans deviating from its current trajectories in order to accomplish the new goal. This action probably leads to global time performance degradation and individual energy consumption increasing. However, what would happen if robots are only influenced to keep or recover connectivity at all times instead of being demanded to regain connectivity? Furthermore, what would happen if they are free to meet by chance, having been motivated to stay close but without having to meet at specific places?

1.4. Contributions

This work tries to answer these research questions from the development of a novel multi-objective approach where the robots, when selecting their targets, are always considering travelling costs and the opportunity cost of keeping connected or regaining connectivity. A simple yet useful model for the signal strength and attenuation effects provide the robots with connectivity awareness. Thereupon, connectivity level measurements and path costs are considered together into a *task utility* function for finding solutions with a right balance between the benefit of visiting the closer targets and the usefulness of keeping the team connectivity level as high as possible.

For the sake of robustness, a decentralised approach is followed. Robots make decisions asynchronously addressing coordination implicitly through localisation and mapping data exchanging. The human operator is asked to use his application-field expertise to play a part in the task assessment process.

The main contributions of this proposal can be summarised as follows.

Ease to Deploy and Flexibility

In order to establish the task selection criterion, the human operator only needs to choose the extra distance he is willing to ask the robots to travel in order to keep or enlarge the connectivity level of the fleet. From this criterion and through formal analysis, the weights of these potentially conflicting objectives are derived. This way the robots can deal with communication constraints auto-adjusting the weights of each objective in a more intuitive manner. Furthermore, by eliminating the need for training stages the system is more adaptable to different environments.

Good Performance

Asynchronism is taken as a natural way of avoiding waiting times to make a decision as well as decreasing the number of robots that are simultaneously making a decision. Since the task allocation computation strongly depends on the number of robots under consideration, asynchronism also makes optimal decisions can be linearly computable most of the time. As a consequence, robots can compute optimal tasks-to-robots distributions in a short time, achieving high levels of dispersion efficiently. Besides, regarding reconnections, the proposal consists of a rendezvous policy where the locations of the selected tasks become the meeting points themselves, avoiding deviations from the planned paths. Compared with others, the proposed approaches are capable of decreasing the last of disconnection periods without noticeable degradation of the completion exploration time.

1.5. Outline

The present document is organised as follows. Section 2 provides the exploration problem formalisation including models and goals. Next, an *Auto-Adaptive Multi-Objective (AAMO)* task selection approach, as well as the task allocation algorithm and the decentralised coordination mechanism, are thoroughly described in Sections 3–5, respectively. Experimental results related to a baseline and to the *AAMO* approach itself are discussed in Section 6. Finally, the document is concluded highlighting some future research directions in Section 7.

2. Problem Formulation

This section defines the instance of the multi-robot exploration problem, which constitutes the basis for the proposal formulated in this work. All particular assumptions are mentioned throughout the following sections. Firstly, the environment, robot and communication models are defined. Namely, some real communication constraints are taken into account and formalised into the model. A *task* definition is given as well as the task identification method. Finally, the global exploration objectives are stated.

2.1. Environment Model

The environment E is defined as a bounded planar workspace $E \subseteq \mathbb{R}^2$ previously unknown. Besides, E is represented by an occupancy grid structure [4] where each cell c can belong to three different probabilistic states $S = \{f, o, u\}$, standing for free, occupied and unknown, respectively. Typically, $\mathbb{P}(state(c) = f) = 1 - \mathbb{P}(state(c) = o)$ is assumed. When $|\mathbb{P}(state(c) = f) - 0.5| < \epsilon$ the cell c is labelled as *unknown*; otherwise it is labelled as *free* or *occupied*, accordingly. These states represent all possible theoretical situations in which a point of the environment can be classified over time. The mapping algorithm frequently updates the probability value of each cell on each robot. Despite this, only the current classification of each cell at a given decision time step is considered. Consequently, the representation of E belongs to the domain of matrices $S^{m \times n}$. Furthermore, the region already explored E_{known} and the remaining that is yet unexplored $E_{unknown}$ at time t may be defined from this representation as follows: $E_{unknown}(t) = \{c \in E \mid |\mathbb{P}(state(c,t) = f) - 0.5| < \epsilon\}$ and $E_{known}(t) = \{c \in E \setminus E_{unknown}(t)\}$.

2.2. Robot Model

Given a robot team $R = \{R_1, R_2, \ldots, R_M\}$ consisting of M homogeneous circular rigid mobile robots with wireless communication capabilities, a traditional representation defines each robot: $R_i = (x_i, y_i, \theta_i, r_i, s_i, c_i)$, where $i \in [1..M]$ and $X_i(t) = \{x_i(t), y_i(t), \theta_i(t)\}$ represents the configuration vector of the robot i at time t (position of its centre and heading with respect to the inertial frame), r_i represents the radius of the robot body, and s_i, c_i represent the sensory capabilities as maximum radius of sensing and maximum range of communication, respectively.

2.3. Communication Model

This model aims to support the connectivity awareness ability of robots needed to deal with disconnection situations during the exploration. Given the position of their teammates and obstacles, robots can estimate the connectivity degree of a specific location considering some of the communication constraints that are widely present in real scenarios, mainly indoor (e.g., office-like and buildings).

The signal strength function (Γ_i represents a slight adaptation of the signal strength function presented in [38]) $\Gamma_i : \mathbb{N} \times S^{m \times n} \times \mathbb{R} \to \mathbb{R}$ is defined as follows:

$$\begin{aligned}
\Gamma_i(j, E_{known}(t), t) &= \Gamma_i^0 - d_{Att}(i,j,t) - w_{Att}(i,j,E_{known}(t),t) \\
\Gamma_i^0 &= 10 \cdot D_{af} \cdot log_{10}(c_i / r_i) \\
d_{Att}(i,j,t) &= 10 \cdot D_{af} \cdot log_{10}(d_i(j,t)/r_i) \\
d_i(j,t) &= \left\| X_i(t), X_j(t) \right\|_2 \\
w_{Att}(i,j,E_{known}(t),t) &= \begin{cases} w_i(j, E_{known}(t), t) \cdot W_{af} & \text{if } w_i(j, E_{known}(t), t) < C \\ C \cdot W_{af} & \text{otherwise} \end{cases}
\end{aligned} \tag{1}$$

where, d_{Att} and w_{Att} stand for *distance attenuation* and *wall attenuation* terms, respectively. In addition, $d_i(j,t)$ represents the Euclidean distance between two robot locations at time t: typically the transmitter

$(X_i(t))$ and receiver $(X_j(t))$, $w_i(j, E_{known}(t), t)$ represents the number of walls (robots cannot distinguish between different kind of rigid obstacles, but the term *wall* is used for simplicity and in order to be consistent with the underlying proposal) present in the known region between transmitter and receiver locations at time t, D_{af} represents a distance attenuation factor, and W_{af} represents a wall attenuation factor. Finally, C represents the maximum number of walls up to which the W_{af} factor causes a significant effect in function Γ_i. When $w_i(j, E_{known}(t), t) \geq C$, the distance attenuation effect dominates. Finally, note that in [38] the independent term Γ_i^0 is suggested to be either derived empirically or obtained directly from the wireless network device specification. Nevertheless, in this work the model is adapted in order to become independent from specific deployments (communication devices), deriving the Γ_i^0 value so that the signal strength $\Gamma_i(j, E_{known}(t), t) = 0$ when $d_i(j, t) = c_i$.

In Figure 2 the shape of the function Γ_i, as well as the attenuation effects caused by both distances and walls, are plotted.

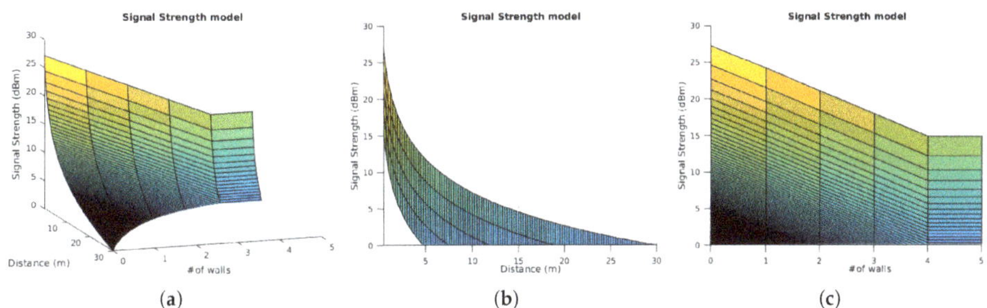

Figure 2. Behaviour of the signal strength model. (**a**) The signal strength function Γ_i (dBm) is plotted for a [0..5] range of walls and [0..30] (m) range of distances; (**b**) Attenuation caused by distance; (**c**) Attenuation caused by wall interference.

Unfortunately, due to uncertain and incomplete knowledge, the Γ_i function only can either confirm the absence of connectivity or deliver an optimistic estimation of connectivity level instead. Although this model represents a valuable improvement in relation to others (e.g., the classic disk or line of sight models [20]), for the sake of simplicity other impairments also common in communication (e.g., bandwidth, information losses, fading, and multi-path propagation phenomenon [39,40]) are not considered in this work.

2.4. Task Identification Method

The task identification problem is addressed following a frontier point approach [5] where the *free* cells (cf. Section 2.1) that belong to a frontier are over labelled as *frontier points (FP)*. Besides, the resulting set of *FP* cells is clustered (using procedures such as *K-Means* [41] or *Affinity Propagation* [42]) in order to identify the cells that better represent each frontier, defining a set of tasks (in the remainder of the document, the terms *task* and *target* are used indistinctly) $T = \{T_1, T_2, \ldots, T_N\} \mid T_j \in \mathbb{R}^2, \forall j \in \{1 \ldots N\}$. Thus, T represents, at each moment, the smallest set of promising locations that the robots could be interested in visiting to explore all frontiers. In Figure 3 these *task* cells are coloured in yellow.

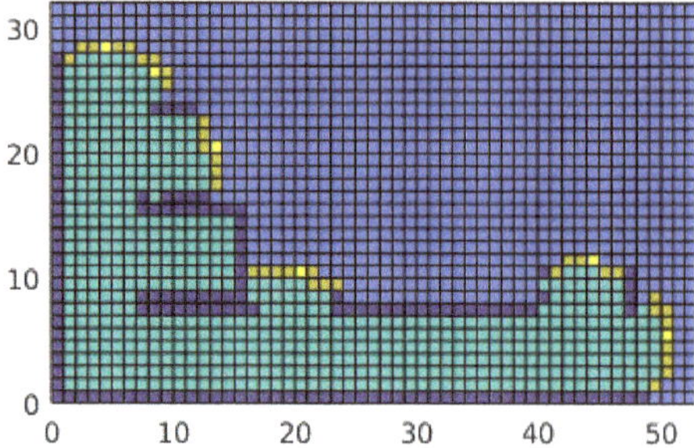

Figure 3. Frontier points. The different cell types are identified according to the following colour code: dark blue cells are **Obs**tacles, light blue cells are **Unk**nown, green cells are **Free**, orange cells are **FP** cells, and yellow cells are *tasks*.

2.5. Multi-Robot Task Allocation Problem—MRTA

Following the classification proposed in [10], the *MRTA* problem to be tackled is described as a *single-task robots (ST)*, *single-robot tasks (SR)*, and *instantaneous assignment (IA)* problem. *ST* means that each *robot* is able to visit at most one *task* at a time. *SR* means that each *task* requires only one *robot* to be explored. *IA* means that the available information about the robots, the tasks, and the environment permits only an instantaneous allocation of tasks to robots, preventing the possibility to plan future allocations. Additionally, an *ST-SR-IA* can be formulated as an instance of the well known *Optimal Assignment Problem (OAP)* as follows. Given M robots, N tasks, and utility estimates U for each MN possible robot-task pair, the goal is to assign tasks to robots so as to maximise overall expected utility. Finally, from an *Integer Linear Programming* perspective, the problem can be formalised as: Find the MN non-negative integers α_{ij} that maximise (2).

$$\sum_{i=1}^{M}\sum_{j=1}^{N}\alpha_{ij}U_{ij} \tag{2}$$

s.t.

$$\sum_{i=1}^{M}\alpha_{ij} = 1, 1 \leq j \leq N$$

$$\sum_{j=1}^{N}\alpha_{ij} = 1, 1 \leq i \leq M$$

2.6. Global Objectives

The exploration aims for the full coverage of a bounded indoor environment, a priori totally unknown, with a team of terrestrial robots, in minimal time and avoiding isolation situations as much as possible. In this context, isolation refers to the fact of being unconnected from any other fleet member. In this work, the multi-robot system is designed to address these objectives from the following definitions.

2.6.1. Full Coverage

Given the E_{known} and $E_{unknown}$ previously defined in Section 2.1, it is possible to claim that the completion condition is reached when $E = E_{known}$ or equivalently $E_{unknown} = \emptyset$. Although this condition is straightforward, it is useless in practice. Alternatively, the completion condition is conceived considering the sensing activity of the robots over time. Let $sen_i(t) = sen(X_i(t))$ be the information gathered by the robot i at time t in the configuration $X_i(t)$. From this, E_{known} at completion time $\mathcal{T}$ is defined as follows:

$$E_{known} = \bigcup_{i=1}^{M} \bigcup_{t=0}^{\mathcal{T}} sen_i(t) \tag{3}$$

Finally, the completion condition may be written as in (4) implying that there are no reachable configurations where any robot can gather new information.

$$\nexists X_i \mid sen_i \bigcap E_{unknown} \neq \emptyset \tag{4}$$

2.6.2. Completion Time Optimisation

Additionally to full coverage, the multi-robot system is asked to perform the exploration in minimal time. Therefore, from (4), the minimal completion time condition can be expressed as:

$$\min \mathcal{T} \mid \nexists X_i(\mathcal{T}), sen_i(\mathcal{T}) \bigcap E_{unknown} \neq \emptyset \tag{5}$$

2.6.3. Isolation Avoidance

In multi-robot exploration missions, the individual isolation situations (when a robot becomes unconnected from any other) are non-desirable. The key motivations to avoid them are (i) When robots are unconnected they have no possibilities to coordinate their actions, hence they could visit the same regions. Therefore, keeping the fleet connected is a way to decrease inefficiency; (ii) Damages or inner failures during isolation periods can lead to information losses. Therefore, keeping the fleet connected is also a way to decrease the risk of re-work and to prevent time performance degradation, consequently.

Thus, in addition to (5), the last of possible individual disconnections should be minimised. To this end, concepts of graph theory are borrowed in order to model a time-varying network topology of mobile robots. Such network is represented employing an undirected graph defined as $\mathcal{G}(t) = (\mathcal{V}, \mathcal{E}(t))$ where the nodes $\mathcal{V} = \{1 \ldots M\}$ represent the robots $R_i \mid i \in [1..M]$ and the edges $\mathcal{E}(t) = \{i, j \mid i, j \in \mathcal{V}, j \in \mathcal{N}_i(E_{known}(t), t)\}$ represent the operative communication links between any pair of robots (R_i, R_j).

The function $\mathcal{N}_i(E_{known}(t), t) = \{j \mid \Gamma_i(j, E_{known}(t), t) > 0\}$ computes the neighbours of a robot i at time t. From this it is possible to define the isolation situations of any robot i like the periods when the corresponding node i has no incident edges ($degree(i) = 0$). Furthermore, isolation situations may repeat several times along the exploration.

In Figure 4 an example of an exploration timeline concerning disconnections is depicted.

Figure 4. Disconnection events representation. The disconnection events dE_k can appear distributed along the exploration timeline. Its last is variable and depends on the movements realised by the fleet during the exploration. The starting and ending times of each disconnection are represented by the timestamps t^i_{sk} and t^i_{ek}, respectively.

From this model, the expression for the disconnection last optimisation may be obtained as follows:

$$\min \sum_{i \in \mathcal{V}} \sum_{k} \Delta dE_k \tag{6}$$

$$\begin{aligned} \text{where}: \quad & k \text{ indexes the disconnection events } dE \\ & \Delta dE_k = t^i_{ek} - t^i_{sk}, \text{ represents the last of the disconnection event } dE_k \\ & t^i_{sk} = \min t_k \mid \mathcal{N}_i(E_{known}(t), t_k) = \emptyset, \text{ represents the starting time of the disconnection } dE_k \\ & t^i_{ek} = \max t_k \mid \mathcal{N}_i(E_{known}(t), t_k) = \emptyset, \text{ represents the ending time of the disconnection } dE_k \end{aligned}$$

3. Auto-Adaptive Multi-Objective Task Selection Approach

In this section, a novel multi-objective based approach for multi-robot exploration missions is introduced. As was mentioned above, in exploration missions the best choice for the robots is to visit the places where the gain of information can be potentially higher. Gaining information is, actually, the only way to conclude the exploration task. Therefore, the connection between path-cost-based target selection strategies and the completion time performance obtained resides in the fact that this way the fleet expand its territorial knowledge potentially faster. Besides, when the environment presents communication restrictions, individual failures or incoordinations can lead to inefficiency more likely.

In order to make the system robust and efficient, a decentralised and asynchronous coordination mechanism is defined. An auto-adaptive multi-objective task utility function is defined in accordance with both the task identification method presented in Section 2.4 and the objectives of the exploration problem defined in Section 2.6. Its primary purpose is to integrate travelling costs and connectivity levels finding solutions with a right balance between the benefit of visiting the closer targets and the usefulness of keeping the team connectivity level as high as possible.

Furthermore, to make the system more flexible, an analytic approach through which the relative importance of each goal is set independently of the scenarios, is followed. As a result, an auto-adaptive procedure—where the human operator is asked to use his application field expertise in order to influence the robot decisions defining a criterion to balance the importance of both objectives—is developed. Several proofs of correctness on such a procedure are conducted demonstrating that the robots are always capable of auto-adapt the objectives weights to select the tasks accordingly with the human-operator criterion.

3.1. Task Utility Function

This function will guide the optimal task distribution search regarding well-balanced solutions where both the travelling cost and the team connectivity level are considered to evaluate the current targets. The objectives are implemented using utility functions such as (i) *path* utility function takes the travelling costs to deliver a notion of how beneficial—concerning distance—the tasks under consideration are; (ii) *connectivity* utility function gives the robots a connectivity awareness ability.

The *task* utility function $\Phi_i : [0,1] \times T \times R^M \times S^{m \times n} \rightarrow [0,1]$, is defined as follows:

$$\Phi_i(\alpha, T_j, R, E_{known}) = \alpha \cdot \Psi_i(T_j, E_{known}) + \beta \cdot \Omega_i(T_j, R) \tag{7}$$

$$\text{s.t.} \quad \begin{aligned} & 1 \leq i \leq M = |R|, 1 \leq j \leq N = |T| \\ & \alpha + \beta = 1 \mid \alpha, \beta \in [0,1] \end{aligned}$$

Given the current state of the fleet R and the current environment knowledge E_{known}, the function Φ_i estimates the utility obtained by a robot R_i in case of selecting the task T_j. The current fleet state refers to both the location of the assigned tasks in case of assigned robots and the robot positions otherwise. The terms Ψ and Ω represent *path* utility and *connectivity* utility functions, respectively. The weights α and β work as tuning parameters that permit to adjust the kind of solutions the system will search for. If $\alpha = 1$ during the whole exploration, then the system would only intent to spread out the fleet. On the contrary, if $\alpha = 0$ then the system would always search for potentially fully connected solutions. Otherwise, when ($0 < \alpha < 1$) the system will balance both *path* utility and *connectivity* utility. As a result, sometimes the robots could choose other tasks than the closest to favour the team connectivity level. This possibility is deeply analysed further below in Section 4.

Although in this double-objective function the symbol β could be substituted by $1 - \alpha$, it is preserved for the sake of generality: if the weighted sum had more than two terms, it would not be possible to express all weights as α functions.

3.2. Path Utility

Path utility measures the relative effort needed for a robot to reach a task from its current location. The *path* utility function $\Psi_i : T \times S^{m \times n} \rightarrow [0, 1]$ is defined as follows:

$$\Psi_i(T_j, E_{known}) = 2^{\left(\frac{\overline{\Delta} - \Delta_i(T_j)}{\overline{\Delta}}\right)^{\gamma}} - 1 \quad (8)$$

s.t.

$$1 \leq i \leq M = |R|, 1 \leq j \leq N = |T|$$

where:

$$\overline{\Delta} = \overline{d} - \underline{d}$$
$$\overline{d} = \max \|X_i, T_j\|_{sp}, \forall j$$
$$\underline{d} = \min \|X_i, T_j\|_{sp}, \forall j$$
$$\Delta_i(T_j) = \|X_i, T_j\|_{sp} - \underline{d}$$
$$\|X_i, T_j\|_{sp} = \min_{\substack{wp_k \in E_{known} \forall k \in \{1 \ldots e\} \\ wp_1 = X_i, wp_e = T_j}} \sum_{k=1}^{e-1} \|wp_k - wp_{k+1}\|_2$$

Given the current environment knowledge E_{known}, the function Ψ_i estimates the path utility obtained by a robot R_i in case of selecting the task T_j. The parameter γ works as a shaping factor that could be used to tune the relation between distance and utility. The ordered sequence of waypoints wp_k represents the shortest path between the robot configuration X_i and the target T_j. All segments (wp_k, wp_{k+1}) are safe given that they are always built regarding only the collision-free pathways present in the known region E_{known}. The wavefront propagation method proposed by [43] is employed to determine the waypoint sequence. The shape and behaviour of the Ψ function are depicted in Figure 5.

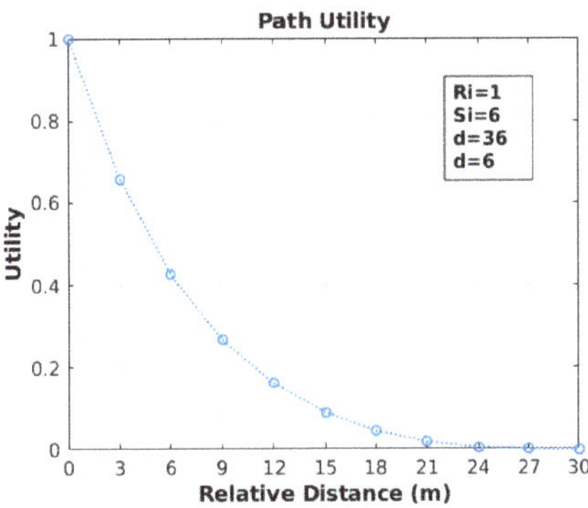

Figure 5. Path utility function behaviour. There are several tasks in the scene (blue circles). The closest is located 6 m away from the robot while the furthest is 36 m far away. The closest and furthest tasks always return 1.0 and 0.0, respectively.

3.3. Connectivity Utility

Connectivity utility computes, optimistically, the connectivity level present in a location at a certain moment. The *connectivity* utility function $\Omega_i : T \times R^M \times S^{m \times n} \rightarrow [0, 1]$ is defined as follows:

$$\Omega_i(T_j, R, E_{known}) = \frac{log_2\left((2^\rho - 1) \cdot \frac{|\mathcal{N}_i(E_{known}(t), t)|}{M-1} + 1\right)}{\rho} \quad (9)$$

s.t.

$$1 \leq i \leq M = |R|, 1 \leq j \leq N = |T|$$

Given the current state of the fleet R and the current environment knowledge E_{known}, the function Ω_i estimates the connectivity utility obtained by a robot R_i in case of selecting the task T_j. Particularly, it is interesting to do so concerning the arrival time to T_j. The current fleet state refers to both the location of the assigned tasks in case of assigned robots and the robot positions otherwise. The parameter ρ works as a shaping factor that could be used to tune the relation between connectivity level and utility. Note that the utility is decreasing in the number of robots, and may favour the adoption of MANET compliant connectivity techniques. In such networks, messages travel from source to destination members in more than one hop, where intermediate nodes forward messages until the destination is reached. The shape and usefulness of the Ω function may be appreciated in Figure 6.

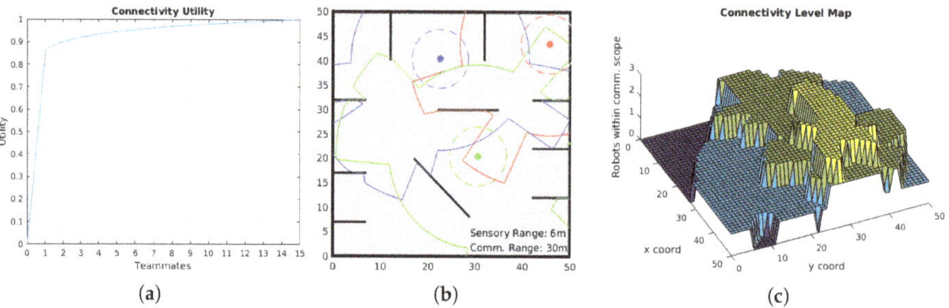

Figure 6. (**a**) Connectivity utility function shape; (**b**) A scene where the usefulness of the Ω function can be appreciated. Three robots (coloured dots) and several walls were arranged simulating an ongoing exploration process. Robots are surrounded with two lines of the same dot colour indicating sensory (dashed) and communication ranges (solid); (**c**) Shows the connectivity level map corresponding to (**b**). Therefore, as long as the will of another robot is to keep connected with the fleet, it would be able to take this perspective into account when deciding where going to.

3.4. General Considerations

The definition of multi-objective weights is usually accomplished as an empirical matter. Typically, a search process is run in order to find—after a lot of trials—values that fit some optimal criteria. This kind of methods is typically used when the parametric function is planned to be used many times. However, in the exploration context, this assumption or even the possibility of running trials are frequently out of the question. It is not possible to assume that all scenarios where the exploration will be conducted will be similar between each other and, for this reason, is neither possible to assume that the best α and β values can remain unchangeable.

Furthermore, when these procedures are followed, at the end of the training stage it is often tough to associate the resultant parameter values with real aspects of the problem (e.g., performance metrics like time, distances, energy, or even connectivity levels). This lack of understanding may, in turn, wrongly influence the fine tuning of such parameters without rerunning a portion of trials. Taking those shortcomings into account, an analytic approach—through which the α and β values might be set independently of the scenarios—is explored.

4. Adaptive α-Value Computation

When a multi-robot exploration process is going to run under communication constrained conditions, choosing between *only exploring* or *exploring preserving connectivity level* is a crucial decision. The first choice would be suitable when connectivity is out of the question, or it is impossible for a robot to keep connected and explore at once. In such a case, connectivity does not play any role in the decision-making process. On the contrary, the second choice is suitable when it is necessary to interleave high-performance exploration (minimising the total exploration time) and acceptable connectivity level (avoiding robot isolation as much as possible).

To this end, the human operator is let to use his application field expertise in order to influence the robot decision—defining a criterion to balance the importance of both objectives—by merely setting a parameter before the exploration starts.

Therefore, since α and β parameters determine the behaviour of the robots concerning target selection, two questions come up: (i) How can the value of those parameters be defined in order to ensure the applicability of the human-operator criterion along the exploration process? (ii) Should these values be adapted during the exploration process?

Henceforth, the task selection framework and the human-operator criterion are formalised. Besides, several proofs to demonstrate the existence and correctness of an adaptive α-value that makes the robots behave following the criterion mentioned above are conducted.

4.1. Task Selection Framework

This process is always made iteratively from a list, comparing the currently best task against the rest, one by one. Therefore, without loss of generality, the most relevant aspects can be studied just analysing all the possible relations between an arbitrary pair of tasks. Regarding the distance to a specific robot location and the connectivity level (number of connections with the rest of the fleet), any task can be classified according to Table 1.

Table 1. Task classification.

Connectivity \ Distance	Closest (Cl)	Furthest (F)
Connected (Co)	Cl/Co	F/Co
Non-Connected (NC)	Cl/NC	F/NC

Therefore, the meaning of these categories is straightforward: regarding the assignment of the fleet, *Co* means that the task location would offer to the robot at least the minimum level of connectivity (i.e., one connection to another fleet member); *NC* means the opposite; regarding the spatial distribution of tasks, *Cl* means that the task under consideration is closest to the robot than any other; *F* means that the task is furthest to the robot than any other task.

Moreover, let R_i a robot and T_j and T_k two tasks such that $class(T)$ can belong to any class defined in Table 1. In any scenario, these tasks can be related to each other according to Table 2. Given that T_j and T_k are arbitrary tasks, the matrix can be considered symmetric. Thus, taking one of the triangular matrices is enough to study all possible cases.

From the lower triangular, it is possible to identify some cases where one task is better (regarding both path utility and connectivity utility) than the other. Such an example is the [Cl/Co;F/NC] where T_j is closer to the robot than T_k, and it is the only one that keeps the robot connected as well. Similarly, in the [Cl/NC;F/NC] case neither task can keep the robot connected, and in consequence, the closest task T_j results more convenient than T_k. Thus, in both previous cases, the criterion to choose a task is clear: the closest task should be selected. However, in the other cases, it is not clear at all which task should be selected. In one case, [Cl/NC;F/Co], whichever selection implies either traversing longer distances or losing connectivity. In the other case, [Cl/Co;F/Co], selecting the closest task T_j ensures traversing the shortest path but could imply losing connectivity. By contrast, selecting the furthest task T_k would be acceptable only when the gain in connectivity oppose a more significant travelling effort.

Table 2. Possible cases when selecting from two tasks.

T_k \ T_j	**Cl/Co**	**F/Co**	**Cl/NC**	**F/NC**
Cl/Co		[F/Co Cl/Co]		[F/NC Cl/Co]
F/Co	[Cl/Co F/Co]		[Cl/NC F/Co]	
Cl/NC		[F/Co Cl/NC]		[F/NC Cl/NC]
F/NC	[Cl/Co F/NC]		[Cl/NC F/NC]	

Definition 1. *The human operator threshold* HO-Threshold *expresses the human operator criterion through a distance that represents the extra effort made by robots that the human operator is willing to accept in order to maintain or enlarge the size of the robot communication network.*

In other words, the human operator criterion is determined by setting the distance threshold until which the targets that preserve or enlarge connectivity are preferred over the rest, for all robots.

For instance, in the [Cl/NC;F/Co] case the selection will be conditioned as follows: T_k will be selected if and only if the length of the shortest path between T_k and the robot location is less than or equal to *HO-Threshold*. T_j will be selected otherwise.

In order to make the influence of *HO-Threshold* clearer, an example scene is depicted in Figure 7. Note that all tasks are within the *HO-Threshold*, but only T_3 can enlarge the connectivity level of the robot R_1. Thus, applying Definition 1 leads to the selection of task T_3 because it enables the robot R_1 to travel more distance to gain connectivity. On the contrary, whether the *HO-Threshold* $\leq$ 3, T_3 would be no longer preferred over the rest, and consequently the closest task T_2 would be selected instead.

Hence, in the presence of some specific conditions, it is expected that the application of the *HO* criterion can make the fleet more cohesive than following approaches that do not take communication constraints into account and less restrictive than the ones that do not permit disconnections or force re-connections as well.

Next, the proofs of correctness and existence of α (and β) values that implement the *HO* criterion are conducted regarding the cases present in the lower triangular of Table 2. The cases {[Cl/Co,F/NC];[Cl/NC,F/NC]} are considered first, while the remaining {[Cl/Co,F/Co];[F/Co,Cl/NC]} are considered afterwards.

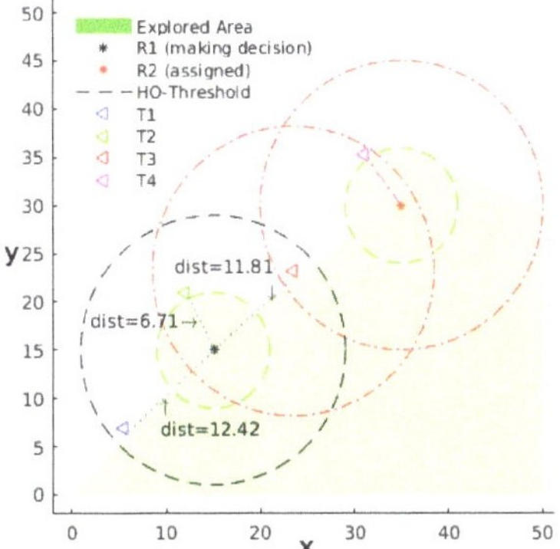

Figure 7. Two robots are carrying out an exploration mission. The communication and sensory ranges are drawn around the robots with red and green dashed lines, respectively. It is assumed that R_2 has already chosen the task T_4 whereas R_1 is still selecting from T_1, T_2, and T_3. Dotted lines are used to show the sight-line between R_2 and the tasks. The corresponding Euclidean distance is also shown. *HO-Threshold* is set to 6.

4.2. [Cl/Co,F/NC] and [Cl/NC,F/NC] Cases

In Figure 8a,b, two instances of these cases are depicted, respectively.

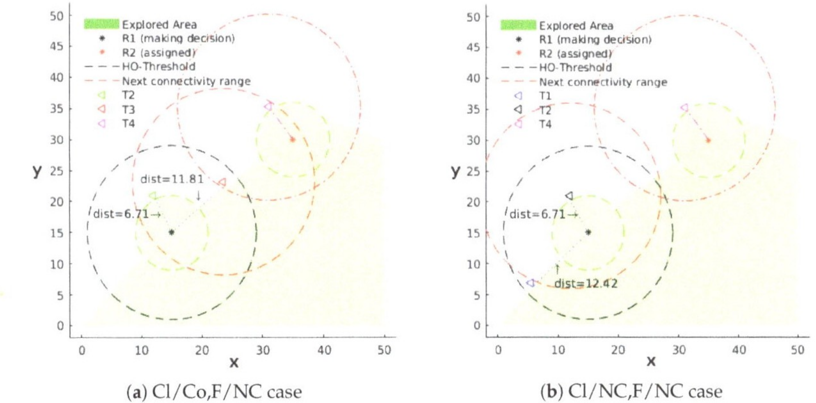

(**a**) Cl/Co,F/NC case (**b**) Cl/NC,F/NC case

Figure 8. Two robots are carrying out an exploration mission. It is assumed that R_2 has already chosen the task T_4 whereas R_1 is still making its decision. The communication and sensory ranges are drawn around the robots with red and green dashed lines, respectively. Dotted lines are used to show the sight-line between R_2 and the tasks. The corresponding Euclidean distance is also shown. (**a**) [Cl/Co,F/NC] case: robot R_1 is selecting from targets T_2 that is the closest and keeps it connected and T_3 that is the furthest and cause a disconnection; (**b**) [Cl/NC,F/NC] case: robot R_1 is selecting from targets T_1—the closest—and T_2—the furthest—given that both targets cause a disconnection.

Proposition 1. *When T_j and T_k belong to $[Cl/Co, F/NC]$ or $[Cl/NC, F/NC]$, the values of α and β do not make any difference in the selection process.*

Proof. This claim can be derived directly from the following facts:

- in the [Cl/Co,F/NC] case the furthest task T_k makes the robot disconnected, and then applying (7) to T_j and T_k leads to:

$$\Phi_i(\alpha, T_k, R) = \alpha \cdot \Psi_i(T_k) \leq \alpha \cdot \Psi_i(T_j) + \beta \cdot \Omega_i(T_j, R) = \Phi_i(\alpha, T_j, R), \ \forall \alpha \tag{10}$$

$$\text{s.t.} \quad \Omega_i(T_j, R) > 0, \Omega_i(T_k, R) = 0$$
$$\Psi_i(T_k) \leq \Psi_i(T_j)$$

- in the [Cl/NC,F/NC] case both tasks make the robot to be disconnected, and thus the Φ function value will depend only on the Ψ term:

$$\Phi_i(\alpha, T_k, R) = \alpha \cdot \Psi_i(T_k) \leq \alpha \cdot \Psi_i(T_j) = \Phi_i(\alpha, T_j, R), \quad \forall \alpha \tag{11}$$

$$\text{s.t.} \quad \Omega_i(T_j, R) = 0, \Omega_i(T_k, R) = 0$$
$$\Psi_i(T_k) \leq \Psi_i(T_j)$$

In conclusion, in any of these cases, the task selection is not affected by α. □

4.3. [Cl/Co,F/Co] and [F/Co,Cl/NC] Cases

In the [Cl/Co,F/Co] case both tasks offer the possibility to be connected. On the contrary, in the [F/Co,Cl/NC] case opposite objectives are present: one task is closer but disconnected while the other is connected but further. Thus, the latter case is taken to prove the existence of an α, that can respect any given *HO* criterion. The former case is finally used to corroborate the non-existence of any possible unwanted side effect caused by the achieved α expression.

4.3.1. [F/Co,Cl/NC] case.

Based on the human-operator criterion (set by a threshold value) we want an α-value that makes, following the scenario depicted in Figure 9, T_3 preferred over T_2 if and only if T_3 belongs to the circular area defined by the *HO-Threshold*.

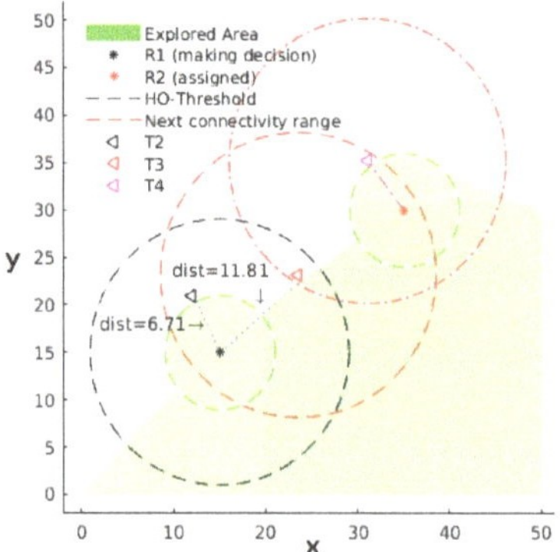

Figure 9. [F/Co,Cl/NC] case.

Next, the existence of such an α parameter will be demonstrated, and its value will be derived as well.

Proposition 2. *When T_j and T_k belong to [F/Co,Cl/NC] is always possible to find an α-value that satisfies the following inequality:*

$$\Phi_i(\alpha, T_j, R) = \alpha \cdot \Psi_i(T_j) + \beta \cdot \Omega_i(T_j, R) \geq \alpha \cdot \Psi_i(T_k) = \Phi_i(\alpha, T_k, R) \tag{12}$$
$$\begin{aligned} s.t. \quad & \Omega_i(T_j, R) > 0, \Omega_i(T_k, R) = 0 \\ & \Psi_i(T_j) \leq \Psi_i(T_k) \end{aligned}$$

Proof. Let Ω_1 the utility assigned to the fact of being connected with only one teammate. Then, from (9), it is possible to state that: if T_j belongs to any $[*/Co]$ class, $\Omega_1 \leq \Omega_i(T_j, R), \forall(i,j)$ over time. Moreover, if the number of robots does not change, it is also possible to state that Ω_1 remains invariant over time. Applying this result into (12) leads to the inequality presented next in (13):

$$\begin{aligned} \Phi_i(\alpha, T_j, R) \geq \alpha \cdot \Psi_i(T_j) + \beta \cdot \Omega_1 \geq & \alpha \cdot \Psi_i(T_k) = \Phi_i(\alpha, T_k, R) \\ \alpha \cdot \Psi_i(T_j) + (1-\alpha) \cdot \Omega_1 \geq & \alpha \cdot \Psi_i(T_k) \\ \alpha \cdot (\Psi_i(T_j) - \Omega_1) + \Omega_1 \geq & \alpha \cdot \Psi_i(T_k) \\ \frac{\Omega_1}{\Psi_i(T_k) - \Psi_i(T_j) + \Omega_1} \geq & \alpha \end{aligned} \tag{13}$$

Besides, substituting $\Omega_1 = x$ and $\Psi_i(T_k) - \Psi_i(T_j) = u$, equation (13) may be rewritten as follows:

$$\alpha \leq \frac{x}{u+x} \implies \alpha \leq inf\left(\frac{x}{u+x}\right) \tag{14}$$
$$\begin{aligned} \text{s.t. } & 0 < c \leq x \leq 1 \\ & 0 \leq u \leq 1 \\ \text{given that: } & \Omega_1 = c \\ & \Psi_i(T_k) \geq \Psi_i(T_j) \end{aligned}$$

From (14) is possible to claim the existence of an α-value that obey any *HO-Threshold* if and only if the function $\frac{x}{u+x}$ presents an absolute minimum on the domain:

$$D = \{(x,u) \mid 0 < c \leq x \leq 1, 0 \leq u \leq 1\}.$$

This fact can be stated employing *Weierstrass* theorem (a function f has an absolute extreme if it is continuous and its domain is compact). Besides, the minimum point might be calculated analysing both: (i) the relative extrema and (ii) the points lying on the border of D. Following this procedure, it is possible to find the absolute extreme of the function $\frac{x}{u+x}$ in $(x,u) = (c,1)$.

Moreover, it is remarkable that this extreme represents a place where the most demanding conditions are reached: task T_j presents the lowest positive connectivity utility, and the distance between both tasks is the largest. Hence, the existence of a positive value $\alpha \leq \frac{\Omega_1}{1+\Omega_1}$ (regardless of how demanding can be the distance relation between tasks) that might alter the task selection in favour of connectivity has been demonstrated. □

Nevertheless, in (13) α is independent of the *HO-Threshold*. Consequently, its direct application would result in a strictly connectivity-guided exploration, where tasks that offer connectivity are always preferred over the rest no matter how far they are. Therefore, to relate it with an *HO-Threshold* the value of the term $\Psi_i(T_j)$ in (13) must be substituted by the utility of being *HO-Threshold* far from the robot, say $\Psi_i(T_{\text{HO}})$. Next, the value of the term $\Psi_i(T_k)$ is substituted by 1 since $\Psi_i(T_k) = 1$ represents the necessary condition to reach the extreme coordinate $u = 1$ that arose from (14). Finally, the expression for an *HO-Threshold* dependent α, say α_{HO}, is expressed in (15) as follows:

$$\alpha_{\text{HO}} = \frac{\Omega_1}{1 - \Psi_i(T_{\text{HO}}) + \Omega_1} \tag{15}$$

Proposition 3. *The applicability of the α_{HO} referred to in (15) causes any task within the threshold scope that also offers any positive connectivity level to be favoured over the rest of the tasks that do not offer any connectivity level, regardless of how close to the robot they are.*

Proof. $\Phi_i(\alpha, T_j, R) \geq \Phi_i(\alpha, T_k, R)$ is imposed to any tasks (T_j, T_k) that respect the [F/Co,Cl/NC] conditions:

$$\begin{aligned}
\Phi_i(\alpha, T_j, R) = \alpha \cdot \Psi_i(T_j) + \beta \cdot \Omega_i(T_j, R) &\geq \alpha \cdot \Psi_i(T_k) = \Phi_i(\alpha, T_k, R) \\
\alpha \cdot \Psi_i(T_j) + (1 - \alpha) \cdot \Omega_i(T_j, R) &\geq \alpha \cdot \Psi_i(T_k) \\
\alpha \cdot (\Psi_i(T_j) - \Omega_i(T_j, R)) + \Omega_i(T_j, R) &\geq \alpha \cdot \Psi_i(T_k) \\
\frac{\Omega_i(T_j, R)}{\Psi_i(T_k) - \Psi_i(T_j) + \Omega_i(T_j, R)} &\geq \alpha
\end{aligned}$$

Then, applying (15) leads to (16):

$$\begin{aligned}
&\frac{\Omega_i(T_j, R)}{\Psi_i(T_k) - \Psi_i(T_j) + \Omega_i(T_j, R)} \geq \frac{\Omega_1}{1 + \Omega_1 - \Psi_i(T_{\text{HO}})} \\
&\Psi_i(T_j) \geq \frac{\Omega_i(T_j, R)}{\Omega_1} \cdot (\Psi_i(T_{\text{HO}}) - 1) + \Psi_i(T_k) \\
&\Psi_i(T_j) = sup\left(\frac{\Omega_i(T_j, R)}{\Omega_1} \cdot (\Psi_i(T_{\text{HO}}) - 1) + \Psi_i(T_k)\right)
\end{aligned} \tag{16}$$

Since *i*) Ω_1 is constant, *ii*) $\frac{\Omega_i(T_j,R)}{\Omega_1} \geq 1$, and *iii*) $(\Psi_i(T_{\text{HO}}) - 1) \leq 0$, it is possible to conclude that:

- $\Psi_i(T_j)$ is monotonically decreasing concerning $\Omega_i(T_j, R)$.
- the upper bound is reached when:
  - (a) $\Psi_i(T_{\text{HO}}) = 1$
  - (b) $0 \leq \Psi_i(T_{\text{HO}}) < 1$, $\Omega_i(T_j, R) = \Omega_1$ and $\Psi_i(T_k) = 1$.

Please note that (a) is out of the proposition conditions. Instead, (16) can be rewritten imposing (b), leading to:

$$\begin{aligned}
\Psi_i(T_j) &\geq \frac{\Omega_1}{\Omega_1} \cdot (\Psi_i(T_{\text{HO}}) - 1) + 1) \\
\Psi_i(T_j) &\geq \Psi_i(T_{\text{HO}})
\end{aligned}$$

which is true if, and only if, $\Delta_i(T_j) \leq$ *HO-Threshold*, which is indeed what the human operator would like to get from his criterion application to tasks within the *HO-Threshold*. Hence, following (15) under the [F/Co,Cl/NC] conditions it is always possible to compute an α_{HO}-value that makes the robots behave following the human-operator criterion. □

Likewise, it is important to highlight that the α_{HO}-value needs to be calculated every time a robot is ready to make a decision. This need for adaptation arises from $\Psi_i(T_{\text{HO}})$, which is not constant. Its value depends on the relation between the *HO-Threshold* and the relative distance to the current furthest task. That way, the robots can autonomously adapt the weights of the *task* utility function according to the changing conditions of the environment in order to be always consistent with the human-operator criterion.

4.3.2. [Cl/Co,F/Co] Case

This analysis is devoted to checking the applicability of the α_{HO} when the conditions to achieve a good trade-off between path cost and connectivity level are less demanding than in the [F/Co,Cl/NC] case. In the [Cl/Co,F/Co] case, although one task is closer than the other, the differences in the positive connectivity level offered by them could make the furthest task more attractive than the closest. From that, considering the connectivity level offered by the closest, two cases may be identified: (i) When T_j offers a higher level of connectivity than T_k. In such a case, there is no doubt that independently of the α_{HO} value, the selection would always favour the task T_j because it is the closest as well; (ii) On the contrary, when T_k offers a higher level of connectivity than T_j, the selection of T_k will depend on both how distant from robot it is and how much more connected would be the robot on T_k respect to T_j.

Finally, to show that the α_{HO} value does not introduce any unwanted side effect on the task selection process when tasks belong to the [Cl/Co,F/Co] case, it is needed to prove that it neither contradicts the first case nor restricts the occurrence of the second case.

Proposition 4. *In the presence of two tasks subject to the [Cl/Co,F/Co] case conditions, if T_j is the closest and simultaneously the one which provides the highest level of connectivity, then the application of the α_{HO} value will never result in the selection of T_k.*

Proof. By contradiction, it is assumed that under these conditions the selection could be in favour of T_k, implying that the following inequality holds:

$$\begin{aligned} &\Phi_i(\alpha, T_j, R) = \alpha \cdot \Psi_i(T_j) + \beta \cdot \Omega_i(T_j, R) \leq \alpha \cdot \Psi_i(T_k) + \beta \cdot \Omega_i(T_k, R) = \Phi_i(\alpha, T_k, R) \\ &\alpha \cdot (\Psi_i(T_j) - \Psi_i(T_k)) + \beta \cdot (\Omega_i(T_j, R) - \Omega_i(T_k, R)) \leq 0 \end{aligned} \tag{17}$$

Which implies that, independently of the α_{HO} value, the terms $(\Psi_i(T_j) - \Psi_i(T_k))$ and $(\Omega_i(T_j, R) - \Omega_i(T_k, R))$ should not be positive simultaneously. Thus, either $(\Psi_i(T_j) \leq \Psi_i(T_k))$ or $(\Omega_i(T_j, R) \leq \Omega_i(T_k, R))$. However, this contradicts the hypothesis where T_j is stated as the closest and the one which simultaneously provides the highest level of connectivity, and accordingly the proposition has been demonstrated. □

Proposition 5. *In the presence of two tasks subject to the [Cl/Co,F/Co] case conditions, if T_j is the closest and T_k the one which provides the highest level of connectivity, then the application of the α_{HO} value will never be conclusive concerning the task selection.*

Proof. The relation between the utility of tasks is written as follows in (18):

$$\begin{aligned} &\Phi_i(\alpha, T_j, R) = \alpha \cdot \Psi_i(T_j) + \beta \cdot \Omega_i(T_j, R) \lesseqgtr \alpha \cdot \Psi_i(T_k) + \beta \cdot \Omega_i(T_k, R) = \Phi_i(\alpha, T_k, R) \\ &\alpha \cdot (\Psi_i(T_j) - \Psi_i(T_k)) \lesseqgtr \beta \cdot (\Omega_i(T_k, R) - \Omega_i(T_j, R)) \\ &\alpha \cdot (\Psi_i(T_j) - \Psi_i(T_k)) \lesseqgtr (1 - \alpha) \cdot (\Omega_i(T_k, R) - \Omega_i(T_j, R)) \\ &\alpha \lesseqgtr \frac{\Omega_i(T_k, R) - \Omega_i(T_j, R)}{(\Omega_i(T_k, R) - \Omega_i(T_j, R)) + (\Psi_i(T_j) - \Psi_i(T_k))} \end{aligned} \tag{18}$$

Substituting $(\Omega_i(T_k, R) - \Omega_i(T_j, R)) = x$ and $(\Psi_i(T_j) - \Psi_i(T_k)) = u$, it is possible to state that in order to favour the selection of T_j the inequality (19) must be held, otherwise the (20):

$$\alpha \geq \frac{x}{u+x} \implies \alpha = sup\left(\frac{x}{u+x}\right) \tag{19}$$

$$\alpha \leq \frac{x}{u+x} \implies \alpha = inf\left(\frac{x}{u+x}\right) \tag{20}$$

$$\text{s.t. } 0 \leq x \leq 1$$
$$0 \leq u \leq 1$$
$$\text{given that: } \Omega_i(T_k, R) \geq \Omega_i(T_j, R)$$
$$\Psi_i(T_j) \geq \Psi_i(T_k)$$

On this domain, the function $\frac{x}{u+x}$ presents an absolute maximum equal to 1 in the point $(x,u) = (1,0)$, and absolute minima equal to 0 along the line segment defined by $(x,u) = (0,u)$. Assessing the α_{HO} expression derived in (15) with $(0,u)$ leads to (21) and (22), respectively:

$$0 = \frac{\Omega_1}{1 + \Omega_1 - \Psi_i(T_{\text{HO}})} \quad \therefore \quad 0 = \Omega_1 \tag{21}$$

$$\begin{aligned} 1 &= \frac{\Omega_1}{1 + \Omega_1 - \Psi_i(T_{\text{HO}})} \\ 1 - \Psi_i(T_{\text{HO}}) &= 0 \quad \therefore \quad \Psi_i(T_{\text{HO}}) = 1 \end{aligned} \tag{22}$$

From which, while the condition expressed in (21) is reached when $|R| \to \infty$, the one expressed in (22) is reached when *HO-Threshold* tends to 0. The condition (21) is unreachable in practice implying that no α_{HO} can make the task T_k always preferred over T_j. Conversely, the condition (22) is reachable if, and only if, the human operator deliberately does not want to care about connectivity. Otherwise, there is no positive α_{HO}-value that can make the task T_j always preferred over T_k.

Consequently, when $\alpha_{\text{HO}} \in (0..1]$ under the [Cl/Co,F/Co] conditions, it is not possible to hold a single preference over time. □

4.4. Considerations and Usefulness

In order to establish the task selection criterion, the human operator only needs to choose the extra distance *HO-Threshold*—according to his expertise and knowledge—he is willing to ask the robots to travel in order to keep or enlarge the connectivity level of the fleet. Once the *HO-Threshold* is set, robots are capable of selecting tasks consistently with the *HO* criterion following the Equation (15). Furthermore, it is important to note that the *HO-Threshold* value does not change along the exploration but, as was pointed out, the α_{HO} does, due to the dependency on the Ψ function. This explains the need for auto-adaptive capabilities concerning the multi-objective Φ function.

Additionally, it also worth noticing that setting *HO-Threshold*$= \infty$ it is a practical way to implement an event-based connectivity approach where the tasks that provide connectivity will always be preferred over the rest, no matter how close they are.

5. Task Allocation Scheme

The allocation scheme is founded on two pillars: the coordination method and the task selection algorithm.

5.1. Coordination Method

In order to take advantage of the individual computing power of the robots, to avoid the single point of failure, and to deal better with the presence of real communication constraints during the exploration, a decentralised approach is followed. Typically, estimation of travelling costs and target benefits, as well as mapping and localisation, are the tasks chosen to be made locally by the robots.

However, to achieve a cooperative behaviour, both the local map and localisation information must be shared among team members.

Additionally, the relation between $|T|$ and $|R|$ can result in two somewhat different behaviours: (i) If $|T| < |R|$, not all robots would be needed to reach all targets. Some robots may choose to keep quiet; (ii) When $|T| \geq |R|$ all robots would be needed in order to reach the maximum amount of targets at a time. When robots decide to explore, the task selection is made coordinately. Robots coordinate their actions implicitly, sharing specific information (such as locations, eventually already-done-selections, and local maps) and running the same selection algorithm. Thus, it is possible for the multi-robot system to compute a coordinated-tasks-to-robots distribution in a decentralised way [15,17,44].

To do so properly, the exchanging information time is carefully set up. The system is fully asynchronous, meaning that: (i) Robots do not wait for others; (ii) After selecting a task, the robots do exchange their selection in order to prevent future overlappings; (iii) Local maps and—by means of this —the sets of new available tasks are periodically exchanged, each time two conditions are met: (1) A waypoint of the planned path is reached; (2) New information has been gathered; (iv) Localisation data is exchanged at a higher rate than maps because its influence on the task selection algorithm is higher too.

While localisation data is exchanged periodically, the rest of data exchanging is triggered by events instead. These policies make the system more efficient and flexible because: (i) No data is transmitted when there is no new information to exchange; (ii) There is no need to set up any rate parameter when exploring different environments. The robot life-cycle algorithm is sketched in Algorithm 1.

Algorithm 1 Robotic Agent Life-cycle algorithm.

```
function EXPLORE(i,R,HO-Threshold)
  ▷ i stands for the robot position in vector R.
  ▷ R stands for the robots location vector.

    atT ← true                                        ▷ atTarget flag.
    pose ← R[i]
    gMap ← getMap(pose)                               ▷ Occupancy Grid map.
    while true do
        R* ← ∅                                        ▷ Vector of connected robot locations.
        for j ∈ R ∧ j ≠ i do
            if Γ_i(j, gMap) > 0 then                  ▷ Connected robot.
                R*[j] = rcvPose(j)                    ▷ Asking for localisation data.
                sndPose(pose, j)                      ▷ Sending own localisation data.
                gMap = mapMerge(gMap, rcvMap(j))      ▷ Asking for local maps.
            end if
        end for
        if atT then
            T = getFrontierTasks(gMap)                ▷ Tasks location vector.
            task ← getAssignment(i, R*, T, |R|, HO-Threshold)
            goto(task)
        end if
        pose = getPose()                              ▷ Global localisation.
        atT ← pose = task
        [gMap, ni] = mapMerge(gMap, getMap(pose))     ▷ Mapping.
        if atT ∨ ni then                              ▷ R_i arrives at task or new information was gathered.
            for j ∈ R* ∧ j ≠ i do
                sndMap(gMap, j)                       ▷ Sending local map.
            end for
        end if
    end while
end function
```

5.2. Task Selection Algorithm

The task selection process employs the multi-objective utility function Φ defined in (7) with α_{HO} values dynamically adapted by (15) to solve the *MRTA* problem stated in Secion 2.5. The corresponding algorithm is sketched in Algorithm 2.

Algorithm 2 Task selection algorithm.

1: **function** GETASSIGNMENT($i, R^*, T, M,$ HO-Threshold)
 ▷ i stands for the robot position in vector R^*.
 ▷ R^* stands for the robots location vector.
 ▷ T stands for the current tasks location vector.
 ▷ M stands for the fleet size.

2: $[T^u, T^a] \leftarrow T$
3: $R^u \leftarrow R^*$
4: $T^{\mathrm{HO}} \leftarrow \{T_j \mid T_j \in T^u, \Delta_k(T_j) \leq \text{HO-Threshold}, \forall R_k \in R^u\}$ ▷ Relative distance $\Delta_k(T_j)$ is defined in Section 3.2.
5: **for each** k in R^u **do**
6: **for each** j in T^{HO} **do**
7: $PU[k,j] = \Psi_k(T_j)$ ▷ Path utility matrix.
8: **end for**
9: **end for**
10: $\alpha_{\mathrm{HO}} \leftarrow \dfrac{\Omega_1}{1 - \Psi_i(\text{HO-Threshold}) + \Omega_1}$ ▷ (15)
11: $\beta = 1 - \alpha_{\mathrm{HO}}$
12: $N^{\mathrm{HO}} = |T^{\mathrm{HO}}|$
13: $M^u = |R^u|$
14: $T2RDist \leftarrow Ar^{N^{\mathrm{HO}}}_{M^u}$ ▷ Tasks-to-robots distributions $T2RDist \in \mathbb{N}^{|Ar^{N^{\mathrm{HO}}}_{M^u}| \times M^u}$
15: **for each** row r in $T2RDist$ **do**
16: $\Phi[r] = \sum_{k=1}^{M^u} \alpha_{\mathrm{HO}} \cdot PU[k,j] + \beta \cdot \Omega_k\left(T_j^{\mathrm{HO}}, [R^u, T^a]\right), j = T2RDist(r,k)$
17: **end for**
18: $T^* \leftarrow T2RDist[r,i] \mid \arg\max_r \Phi[r]$
19: **return** T^*
20: **end function**

Firstly, the input parameter R^* specifically corresponds to the locations of the teammates currently connected with the robot R_i. Next, in lines 2 and 3, both the *task* and *robot* location sets are split up into two subsets each one (assigned and unassigned items, respectively). Line 4 is in charge of taking only the unassigned tasks that are within the *HO-Threshold* scope from every robot. Afterwards, from lines 5 until 9, the path utility matrix is computed regarding all possible task-robot combinations. Next, lines 10 and 11 aim to compute the α_{HO} and β values according to (15). The set of tasks-to-robots distributions is calculated from line 12 to 14. Finally, from line 15 to 17 all possible assignments are evaluated using the Φ function while the task corresponding to robot i of the best assignment is selected in line 18.

Some considerations on Algorithm 2 are hereafter discussed. Concerning the computation of the set of tasks-to-robots distributions (lines 12 to 14), it provides a way to potentially avoid falling in local minima or even taking wrong decisions. Note that the *connectivity utility* function is subject to locality conditions and thus, it is not possible to compute optimal distributions from the application of iterative polynomial-time assignation algorithms such as the *Hungarian method* [14].

On the contrary, Algorithm 2 can choose the optimum tasks-to-robots distribution by evaluating all possible T^{HO}-to-R^u distributions. Nevertheless, this process may be potentially very hard since $|Ar^{N^{\mathrm{HO}}}_{M^u}| = \frac{N^{\mathrm{HO}}!}{(N^{\mathrm{HO}} - M^u)!} = \Pi_{m=1}^{n}(N^{\mathrm{HO}} - m + 1) = \Pi_{m=0}^{n-1}(N^{\mathrm{HO}} - m) \rightarrow O(N^{*M^u})$. Therefore, the smaller $|T^{\mathrm{HO}}|$ and $|R^u|$ the faster the algorithm will run. In the first case $|T^{\mathrm{HO}}|$ is bounded by pruning $|T^u|$ with the help of *HO-Threshold*.

On the contrary, even being naturally bounded ($|R| \geq |R^*| \geq |R^u|$), the set R could imply a large R^u. Besides, all efforts are to keep the fleet connected as much as possible, leading to $|R^*| \rightarrow |R|$. Fortunately, in a fully asynchronous multi-robot system the probability of two or more robots being simultaneously making a decision is negligible.

Finally, note that Algorithm 2 assumes $|T^{\text{HO}}| > |R^u|$; otherwise the tasks-to-robots distribution cannot be computed. In such a case, the input parameters are managed in order to conduct a robots-to-tasks distribution instead. In turn, $|Ar^{M^u}_{N^{\text{HO}}}|$ does not represent a significant effort since $M^u \geq N^{\text{HO}}$ holds for small values.

6. Baseline Statement and AAMO Approach Results

The aims of this section are: (i) To establish a baseline on the main figure of merits that will be defined to asses the benefits of different approaches; (ii) To assess and analyse the performance of different instances of the *Auto-Adaptive Multi-Objective (AAMO)* approach (different instances—from now on—refer to different *HO-Threshold* setup values) under non-ideal communication conditions; (iii) To compare *AAMO* instances against other approaches under non-ideal communication conditions.

Regarding the first purpose, the baseline is established regarding two state-of-art approaches so that the simulation runnings concern the comparison between a *Yamauchi*-based algorithm [5] and the *minPos* algorithm [17] under ideal communication conditions. These algorithms were chosen since they are decentralised, as are the author's proposal; while *Yamauchi* is a reference on exploration and typically serves itself as a comparison baseline, the *minPos* proposal has demonstrated very good performance, outperforming other important reference algorithms.

On the contrary, regarding the *AAMO* assessment and the comparison with other approaches, the simulation runnings concern exploration missions subject to non-ideal communication conditions. In this case, the primary purpose is to understand how compromised could be the exploration time performance when the connectivity level is prioritised and to reveal possible improvements concerning previous techniques. In consequence, there are experiments which compare only the performance achieved by different instances of *AAMO*, while in other experiments, where relevant, comparison with state-of-art performance is taken into account too.

6.1. Simulation Setup

All simulations were conducted over *MORSE* physics simulator (www.openrobots.org/morse/doc/stable/morse.html) using *ATRV*-like robots equipped with laser range sensors. The more relevant simulation parameters are shown in Table 3.

Furthermore, it is important to precise that except for *Communication range* that depends on the device, the rest of communication factors were taken from [38] regarding their strong dependency on the materials present in the environment. The values of *HO-Threshold* correspond to 66%, 50%, and 33% of the communication range c_i, respectively. In all simulations localisation and low level motion control are taken for granted.

Table 3. Simulation setup.

Robot Features & Capabilities	
Model	*ATRV*
Maximum speed (s_i)	1 (m/s)
Laser scan window	360 (°)
Laser range	6 (m)
Laser resolution	3 (°)
Communication parameters.	
Range (c_i)	30 (m)
Wall attenuation factor (W_{af})	3.1 (dBm)
Distance attenuation factor (D_{af})	1.523
C factor	4 (walls)
Fleet features.	
Heterogeneity	Homogeneous
Initial positions	Left Bottom corner
Environment features.	
Terrain	80×80 (m^2)
Wall height	2 (m)
Wall thickness	0.2 (m)
Corridor width	8 (m)
Grid Map features & parameters.	
Mesh	Cartesian grid
Cell side	$2r_i$
AAMO parameters.	
γ	3
ρ	$2 \cdot (\|R\| - 1)$
HO-Threshold	20, 15, 10 (m).

6.1.1. Scenarios

Simulations are conducted over synthetic scenarios (See Figure 10) where long distances and obstacle presence may offer similar challenging conditions that would be expected in the real world. The *Loop* and *Cross* scenarios (see Figure 10a,b) were mainly used to confirm the correctness of the implemented solutions and to show the advantages of using a multi-robot approach over a single one. Unfortunately, and caused by the shape and size of the free zones, on those scenarios there are nearly no possibilities to demonstrate any advantage of the proposed approaches over the others. Finally, the *Maze* scenario (Figure 10c), that represents the most challenging environment, was used to establish comparative results among the approaches. These results are further analysed below in Sections 6.3 and 6.4, respectively. Due to the big amount of collected data, only the values related to *Maze* runnings are summarised and discussed here. Even so, all charts and screen-shots generated from data concerning all of the three environments are available online: www.fing.edu.uy/~fbenavid/projects/MuRE/mure.html.

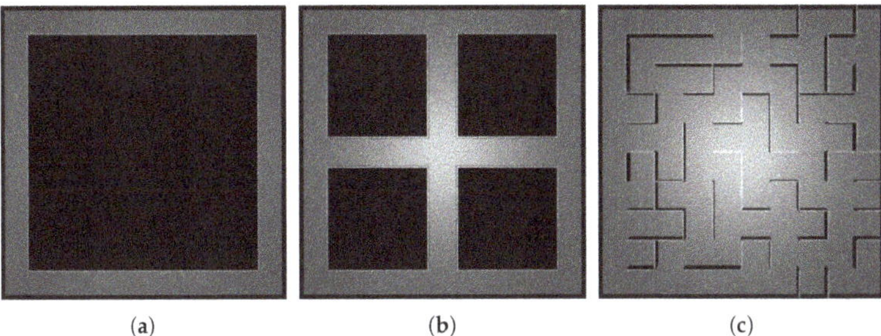

Figure 10. Benchmark scenarios. All terrains cover an 80 × 80 m^2 flat surface with static obstacles (walls including the outer perimeter). Proposed as benchmarks in [45]. (**a**) Loop-like scenario; (**b**) Cross-like scenario; (**c**) Maze-like scenario.

6.1.2. Robotic Agent Architecture

From a software architecture point of view, each robot is organised in three layers. In Figure 11 the main components are roughly depicted. Each layer is responsible for different aspects grouped by abstraction levels so that the higher layer, the more abstract are the issues which the software components are devoted to.

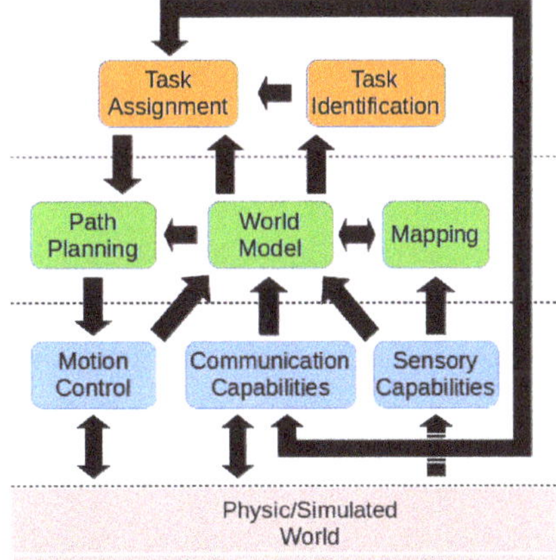

Figure 11. Robotic Agent Architecture. The first layer includes the software components that represent systems or devices through which the agent can interact with the environment. The second layer includes models and algorithms to keep the models up to date. The third layer includes the task identification and selection algorithms. Components on the shadowed zone were developed during this work.

Going bottom-up in the layer stack, in the first layer the components are in charge of the interaction between the robotic agent and the environment. The *Motion Control* component is taken from the *MORSE* (*MORSE* physics simulator www.openrobots.org/morse/doc/stable/morse.html) repository and is responsible for controlling the motors. Besides, in this work, the component follows a way-point-based motion strategy. In the *Sensory Capabilities* component all sensory systems in charge of gathering environmental information are grouped. The most relevant information comes from the *Pose* and the *Laser scanner* sensors, also taken from the *MORSE* repository. From the *Pose* sensor it is possible to know the robot configuration $X_i(t) = \{x_i(t), y_i(t), \theta_i(t)\}$ at any time—implementing the localisation capability—while the laser gives an array of distance measurements $z(t)$ from which is possible to build the map of the close surroundings. Finally, the *Communication Capabilities* component is asked to manage every aspect related to communications receiving/sending information from/to (see incoming/outgoing arrows) other team members. In this work, and since only the distance and wall attenuation effects (discarding other sources of perturbation) are considered, the communication is simulated in a very simple manner directly applying the communication model introduced in Section 2.3.

The second layer represents the core of the system where the models and algorithms that support the highest level functionalities—namely related to the exploration purpose of the system—are allocated. On the one side, the *World Model* component is in charge of modelling all physic interaction between the robotic agents and its surroundings. By keeping several structures up-to-date (e.g., occupancy grid map, the position of the fleet members, assignment of the fleet members), it is also able to support foretelling services that would be required for the highest level algorithms. On the other side, *Mapping* and *Path Planning* components are also supported by the *World Model* component since it gives an access point to the mapping structures and the kinematic models as well. The *Mapping* component implements a standard occupancy grid approach [46] where the posterior of the map is calculated from a collection of separate problems of estimating $p(m_k|z(t), X_i(t))$ for all grid cell m_i and where each m_i has attached to it one of the occupancy values $S = \{f, o, u\}$ (previously defined in Section 2.1). The *Path Planning* component implements the wave-front propagation approach introduced in [43].

Finally, high level decisions as coordination are taken in the third layer when the task allocation scheme is executed by the *Task Assignment* component. In particular, the arrow between *Task Assignment* and *Communication Capabilities* components represents the exchange of current positions and task assignments from the agent to the fleet and vice-versa.

6.2. Figure of Merits

The performance of approaches is assessed regarding the following figures of merit. The first three are the most popular and represent the strongest quality indicators [20]. The fourth has been taken from [47] and sometimes can be useful to explain the results concerning the first two. The fifth was inspired by [48] in order to measure the connectivity quality. Besides, a sixth indicator is proposed here in order to have a better qualitative analysis of the connectivity aspects. Moreover, the connected components of the topology along the exploration are also plotted. The indicators are defined as follows:

- *Total exploration Time (TT)*: time elapsed from the beginning until the end of exploration measured in seconds.
- *Path Length (PL)*: sum of the distance travelled by each robot measured in meters.
- *Coverage **R**atio (CR)*: percentage of the accessible terrain covered by the team. Calculated as: $\frac{\text{explored cells} \cdot 100}{\text{accessible cells}}$.
- *Over-Sensing cell **R**atio (OSR)*: percentage of cells sensed as new by more than one robot. Calculated as: $\frac{\text{over-sensed cells} \cdot 100}{\text{explored cells}}$.
- ***D**isconnection **L**ast **R**atio (DLR)*: percentage of TT where at least one robot is totally unconnected. Calculated from the Fiedler number corresponding to the network connectivity graph (see Section 2.6.3).
- ***M**aximum **D**isconnection **L**ast **R**atio (MDLR)*: calculated as: $\frac{\text{longest disconnection period} \cdot 100}{TT}$.

6.3. Baseline Statement

In this section, a baseline of performance on the main indicators is established from runnings of both *Yamauchi* and *minPos* approaches under ideal communication conditions. Since the exploration problem is expected to be more difficult under non-ideal communication conditions than otherwise [20], the obtained results may be considered as a baseline of the first four indicators—defined before in

Section 6.2—with respect to the corresponding performance achieved in runnings conducted under non-ideal communication conditions.

6.3.1. Collected Data

In order to conduct the assessment and comparison stated above, at least ten realistic software-in-the-loop simulations were executed on the *Maze* scenario presented in Figure 10. All collected data is presented in Table 4 and are organised obeying the following scheme. The columns refer to (from left to right): figure of merits (FM); approaches, where *Y* and *MP* stand for *Yamauchi* and *MinPos*, respectively; and the fleet size $|R|$. In each fleet size, the average *AVE* and standard deviation *StD* values are registered.

Table 4. Yamauchi and MinPos results under ideal communication conditions on Maze environment.

FM		$\|R\|$													
		1		2		3		4		5		8		10	
		AVE	**StD**	**AVE**	**StD**	**AVE**	**StD**	**AVE**	**StD**	**AVE**	**StD**	**AVE**	**StD**	**AVE**	**StD**
TT	Y	1958	121.8	1288	255.9	902	128.6	791	125.6	647	78.5	516	41.3	**459.5**	41.5
	MP	**1898**	148.0	**1044**	110.1	**779**	72.8	**615**	52.0	**505**	48.8	**496**	19.7	482	37.4
PL	Y	1308	94.0	1581	157.4	1665	158.9	1831	225.7	1896	186.2	2093	95.9	**2294**	173.9
	MP	**1268**	59.0	**1413**	124.1	**1420**	85.4	**1467**	94.8	**1592**	107.8	**2053**	50.3	2438	140.7
CR	Y	99.1	0.09	99.0	0.06	99.0	0.04	99.1	0.05	99.0	0.04	99.0	0.05	99.1	0.04
	MP	99.0	0.04	99.1	0.06	99.1	0.06	99.1	0.05	99.1	0.05	99.1	0.07	99.1	0.07
OSR	Y	0.0	0.00	1.23	0.57	2.22	0.48	3.68	0.70	5.25	0.81	5.45	0.21	7.00	0.12
	MP	0.0	0.00	**0.93**	0.19	**2.31**	0.38	**2.94**	0.25	**4.37**	0.50	5.45	0.11	7.00	0.08

6.3.2. Baseline Assessment

We start the analysis highlighting that both approaches can adequately explore all the environments presented above in Section 6.1.1. Coherently, both approaches achieve high levels of *CR*. This can be seen clearer in Figure 12.

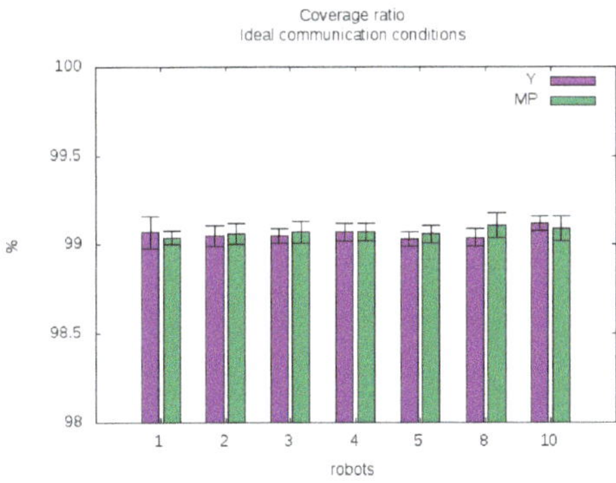

Figure 12. Coverage ratio (CR) under ideal communication conditions. Both approaches achieve a coverage bigger than 99% of the terrain regardless of the fleet size.

Furthermore, the *minPos* approach outperforms *Yamauchi* concerning TT as was expected. However, the most notorious differences of performance are observed on fleets which size is less than or equal to five robots, as can be seen in Figure 13.

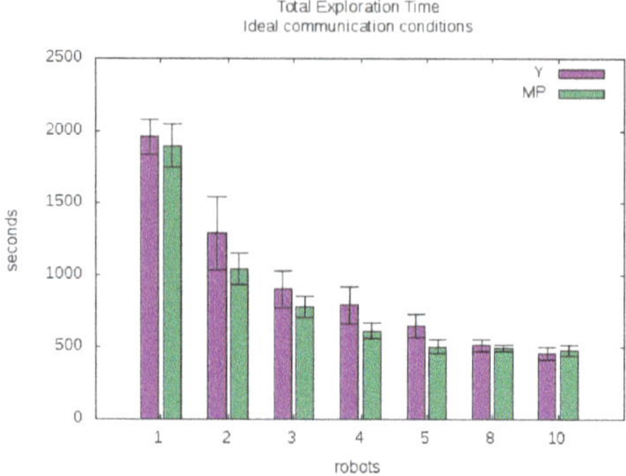

Figure 13. Total exploration time (TT) under ideal communication conditions. Both approaches show a decreasing trend of TT as the fleet size increase. Nevertheless, the fact that the performance improvements are decreasing suppose the existence of a limit on the benefit from robots adding.

In crowded environments, going from one location to another is often more difficult than in the presence of fewer robots. Therefore, due to collision avoidance manoeuvres, both approaches show an increasing *PL* when the fleet size increases. This behaviour may be observed in the corresponding chart in Figure 14. On the one hand, *Yamauchi* presents a trend with an almost invariant slope along the different fleet size values. On the other hand, under *MinPos*, the trend of *PL* presents a positive but minor slope from one to five-robot-sized fleet after what it becomes very steep.

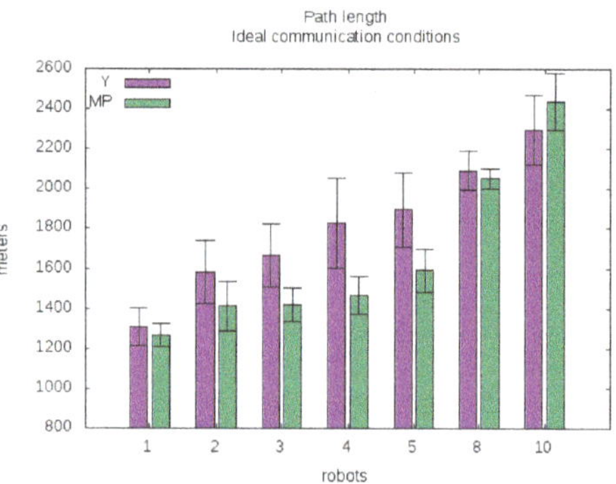

Figure 14. Path length (PL) under ideal communication conditions. The trend of PL is upward in both cases.

Hence, the analysis is divided into two cases. Firstly, when fleet size is less than or equal to five robots, *MinPos* is more efficient than *Yamauchi* since both approaches achieve very similar coverage ratios (see Figure 12) despite in the latter robots need to traverse longer distances than in the former, on average. That is expected since the *Yamauchi* approach does not take care about the dispersion of the fleet as the *MinPos* does and consequently, in the former robots are forced to deal with crowding more frequently than in the latter. This is a remarkable difference given that the energy needed to support an exploration mission will be closely related to the distance traversed by robots.

Contrarily, as the fleet size increase beyond five robots, the shape of the scenario and the peculiar wall distribution all together seem to make the crowding unavoidable for the *MinPos* approach, causing a severe worsening on its *PL* performance.

Finally, it is interesting to observe the over-sensing-cell phenomenon, because, by observing the amount of rework done by the fleet during exploration tasks, it also gives a good measure of the system efficiency.

In this case, we start the analysis pointing that in an ideal world—with perfect communications, perfect sensing and instantaneous actions—there would be no place for over-sensing. Nevertheless, in the real world, communications and sensing systems are not perfect and, more important, all actions take time. Even the ones which do not involve motion such as sensing, computing and communicating actions need some window time to be executed. Therefore, many things can happen simultaneously, e.g., sensing actions conducted on the same objects. In such a case, two or more robots might report the discovery of the same cells.

In conclusion, even under ideal communication conditions, it is possible to register some level of over-sensing, and this level is unavoidable because of the parallel nature of the system. However, it is equally interesting to analyse the over-sensing results: (i) When the fleets are obeying different policies; (ii) To have a baseline against which the results obtained under non-ideal communication conditions may be compared.

Backing to the experiments, during the simulation runnings we verify that the most significant over-sensing record is mainly generated at starting steps when all robots are very close to each other (recall that all robots start from the same corner of the scenario, see Table 3) and, in consequence, its sensing scopes overlap each other, significantly. In Figure 15 the robot placement setup at the starting time is shown.

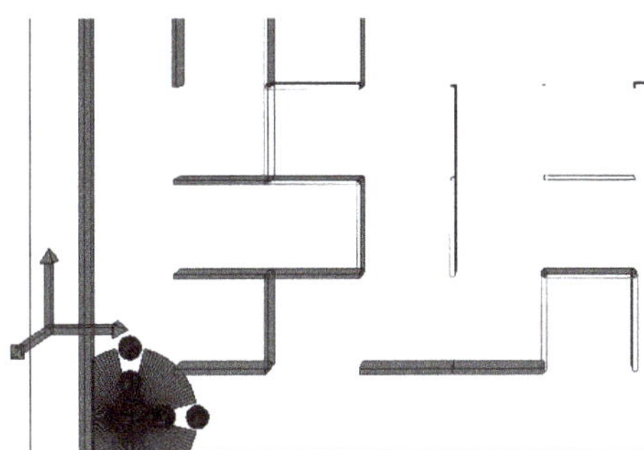

Figure 15. Robot placement setup at starting time. Robots are represented by black dots. The sensing scope of the robot placed right in the corner is represented by a grey area where it is possible to see the laser aces and the obstruction caused by some teammates. Robots are placed from the corner along the x and y axes. As the fleet size increase, new robots are placed next following the row of robots on each axis, alternately.

Conversely, after this initial period, the robots overlap each other less frequently, and hence the *OSR* remains almost unchangeable over time, in both approaches. Despite this, minor differences may be highlighted. Due to a better fleet distribution on the terrain—which decreases the probability of simultaneous sensing events—the fleet makes slightly less rework under *MinPos* approach than under *Yamauchi* approach (see Figure 16).

6.3.3. Conclusions

Concerning the maze scenario, the conclusions of the section are: (i) Regarding fleets integrated with at most five robots, the *MinPos* approach is clearly advantageous (outperforming the *Yamauchi* approach in all assessed figures of merit); (ii) The benefits of employing the *MinPos* approach are severely affected when fleet increase beyond five robots, decreasing quickly or even disappearing when it is about eight robots.

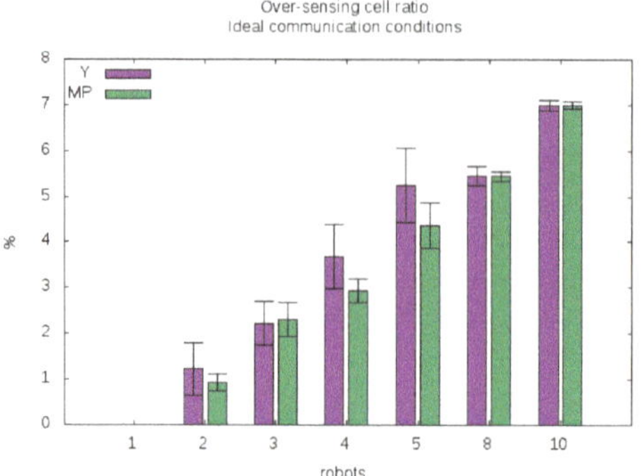

Figure 16. Over-sensing ratio (OSR) under ideal communication conditions. This shows how as fleet size increases the trend of OSR is upward as well. This is expected since the more robots sensing the environment the higher the probability of simultaneously sensing the same cells.

6.4. AAMO Assessment

This section aims to study the impact of using different *HO-Threshold* values on the performance of the proposed *AAMO* approach when the fleet is asked to explore an environment under non-ideal communication conditions. Moreover, these results are compared with the one achieved by other approaches like *Yamauchi* and *MinPos*—when they are subject to non-ideal communication conditions too—and also with an *event-based-connectivity* strategy that does make all efforts in favour of connectivity (regardless the total exploration time).

This last comparison is namely important because the performance of this kind of strategy may serve as an upper bound on the connectivity level over time and the total exploration time as well. To do so, typically two strategies (based on different connection requirements, see Section 1.1.1) can be considered: the ones which force the robots to be connected only on task-arrival time (kind of event-based connectivity) or the ones which force the robots to keep always connected—even during the path traversal periods (continuous connectivity). In the former, the robots are forced to select only between tasks which location would not cause isolation on arrival—regarding the current task assignment of the fleet. Nevertheless, it does not take into account the connectivity level along the path between the current robot location and the location of the task under consideration. Conversely, the latter imposes stronger restrictions on the fleet mobility in order to guarantee connectivity at all times. Consequently, depending on the application field the latter strategy would be recommended but is more complex to implement than the former. On the contrary, the former allows a simpler implementation but could lead to a lower level of connectivity along the exploration. Concerning this document, a connectivity-at-task-arrival-time based strategy is used for comparison purposes.

Besides, it is also important to highlight that, despite *Yamauchi* and *MinPos* assume ideal communication conditions, neither approach needs to be modified or adapted in order to properly run under non-ideal communication condition. Nevertheless, in the *MinPos* case, some severe degradation is expected because of the following working hypothesis are not guaranteed anymore: All robots share the same map and know the position of the other fleet members, at all times. This could lead to incoordinations that, in turn, would harm the dispersion strategy on which the approach is strongly based. Conversely, in the *Yamauchi* case, the level of expected degradation is fewer due to the coordination level between robots is fewer as well. Robots only try to avoid going to the same task simultaneously.

6.4.1. Collected Data

In order to conduct the assessment and comparison stated above, at least ten realistic software-in-the-loop simulations were executed on the *Maze* scenario presented in Figure 10. All collected data is presented in Table 5 and are organised obeying the following scheme. The columns refer to (from left to right): **F**igure of **M**erits (FM), *Approach*, where *Y*, *MP*, *EbC*, and *AAMO:HO-Th* stand for ***Y**amauchi*, ***M**in**P**os*, ***E**vent-**b**ased **C**onnectivity* (implemented by an *AAMO:∞* instance, as was mentioned above in Section 4.4) and ***A**uto-**A**daptive **M**ulti-**O**bjective:**H**uman**O**perator-**Th**reshold*, respectively; and fleet size $|R|$. The *HO-Threshold* values are 20 m, 15 m, and 10 m (equivalent to 66%, 50%, and 33% of the communication range c_i, respectively). Besides, since the communication conditions are non-ideal all runnings only concern fleets integrated with multiple robots (explicitly avoiding the single robot case because the communication conditions do not make any difference on it).

Table 5. *AAMO* Results obtained under non-ideal communication conditions on Maze environment.

FM	Approach	$\|R\|$									
		2		3		4		5		8	
		AVE	StD	AVE	StD	AVE	StD	AVE	StD	AVE	StD
TT	Y	1216	131.7	**904**	73.6	**759**	78.2	653	53.3	496	86.5
	MP	1201	118.2	945	95.4	801	90.2	683	57.9	**491**	29.2
	EbC	1751	131.0	1600	190.7	1339	291.7	1028	154.6	750	89.0
	AAMO:20	1292	87.8	1100	88.1	913	94.9	767	104.0	661	108.2
	AAMO:15	1222	73.6	1100	130.1	823	79.0	723	65.7	606	85.8
	AAMO:10	**1137**	85.3	960	123.1	774	94.3	**620**	76.9	514	23.1
PL	Y	1592	144.2	**1707**	173.8	**1842**	190.0	**1846**	139.0	**2216**	290.7
	MP	1583	134.9	1744	177.7	1911	200.2	1960	173.8	2375	159.9
	EbC	2215	154.0	2929	323.5	3416	618.6	3181	415.1	3412	296.3
	AAMO:20	1726	114.4	2086	222.5	2243	236.1	2394	252.8	2782	376.1
	AAMO:15	1669	86.9	2056	207.8	2106	204.8	2263	219.2	2536	222.6
	AAMO:10	**1542**	119.9	1859	225.6	1982	215.5	2007	221.7	2279	248.5
CR	Y	99.2	0.33	99.4	0.37	99.3	0.33	99.3	0.24	99.2	0.20
	MP	99.3	0.35	99.4	0.29	99.4	0.33	99.6	0.34	99.2	0.20
	EbC	99.1	0.04	99.0	0.05	99.1	0.09	99.1	0.11	99.2	0.30
	AAMO:20	99.1	0.26	99.2	0.28	99.2	0.28	99.3	0.35	99.4	0.37
	AAMO:15	99.2	0.35	99.3	0.32	99.3	0.31	99.2	0.12	99.4	0.39
	AAMO:10	99.3	0.32	99.3	0.40	99.3	0.29	99.3	0.29	99.1	0.05
OSR	Y	13.39	6.72	20.21	17.71	28.95	19.00	23.28	13.94	6.44	1.28
	MP	20.49	13.22	28.72	18.85	30.46	19.77	27.98	18.51	6.10	0.87
	EbC	2.47	2.98	3.99	3.01	3.94	1.96	5.04	2.34	5.44	0.01
	AAMO:20	**1.85**	1.57	2.68	1.12	3.53	2.29	**4.56**	1.46	5.92	0.83
	AAMO:15	1.94	1.50	**2.47**	1.07	**2.40**	0.10	4.81	2.04	**5.43**	0.02
	AAMO:10	2.47	2.98	3.99	3.01	3.94	1.96	5.04	2.34	5.44	0.01
DLR	Y	77.0	9.10	85.9	6.35	79.7	7.47	77.1	5.94	60.7	9.38
	MP	81.7	7.69	87.2	7.26	80.2	8.28	77.0	8.52	59.6	10.74
	EbC	**14.5**	2.60	**30.9**	3.83	**29.0**	9.02	41.8	11.16	**39.4**	8.64
	AAMO:20	40.6	9.32	54.0	9.43	44.1	13.29	51.8	10.11	46.5	13.92
	AAMO:15	45.6	12.41	53.9	15.36	55.4	14.85	**39.9**	11.22	47.8	12.30
	AAMO:10	61.3	9.49	68.6	13.24	54.9	9.53	62.8	11.32	46.6	14.28
MDLR	Y	34.8	14.30	58.0	26.30	41.8	18.06	49.3	15.41	25.0	6.66
	MP	33.8	10.88	55.5	16.44	40.6	17.52	44.6	18.38	27.2	8.40
	EbC	**3.3**	0.39	**6.6**	1.70	**6.7**	2.99	**12.7**	6.90	**11.0**	3.84
	AAMO:20	17.8	10.47	21.7	8.96	15.9	7.61	24.2	12.99	18.0	8.88
	AAMO:15	19.2	6.49	22.2	9.65	24.4	13.17	15.2	6.62	17.4	4.36
	AAMO:10	27.2	10.11	33.6	10.95	20.0	5.72	24.9	9.15	22.9	8.61

6.4.2. Effectiveness Assessment

We start the analysis highlighting that all implemented approaches—and particularly all *AAMO* instances—can adequately explore all the environments presented above in Section 6.1.1. Coherently, all instances achieve a high level of *CR* when exploring the *Maze* environment, as can be appreciated in Figure 17.

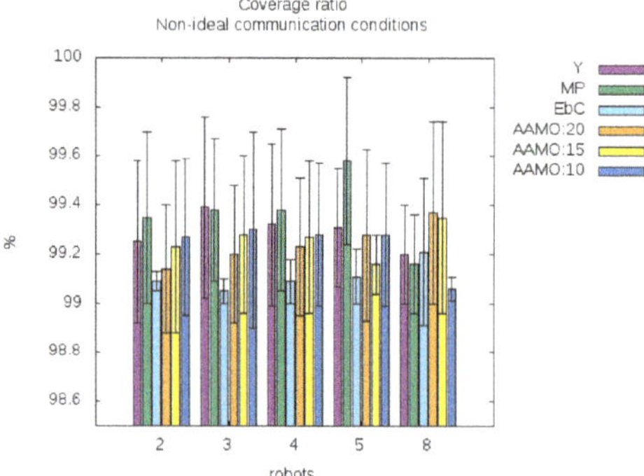

Figure 17. *AAMO* Coverage ratio. Regardless of how different the *HO-Threshold* values are, in all cases, the *AAMO* approach can cover more than 99% of the terrain.

6.4.3. *AAMO vs.* Baseline Comparison

Concerning *TT*, as was expected in multi-robot systems, all *AAMO* instances benefit from adding robots to the fleet. This result can be seen in Figure 18a. Nevertheless, compared to the baseline results all *AAMO* instances show performance degradation (see Figure 18b).

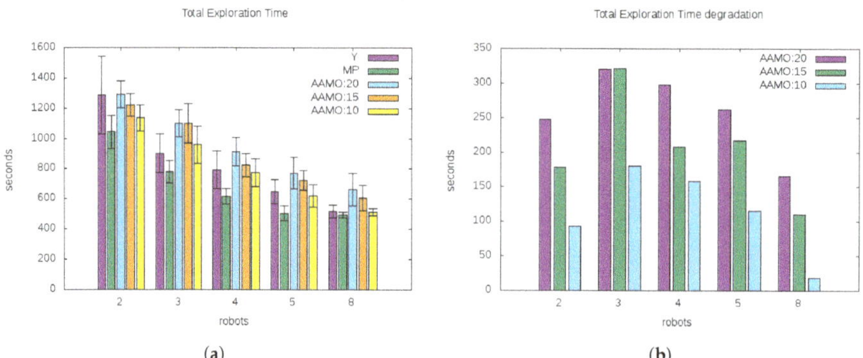

Figure 18. *AAMO* Total Exploration Time (TT) under non-ideal communication conditions and Degradation with respect to baseline results. (**a**) All *AAMO* instances show a decreasing trend of TT as the fleet size increase. The *Yamauchi* and *MinPos* approach results (coloured in purple and green, respectively) obtained under ideal communication conditions are placed together to make the comparison easier; (**b**) The degradation is expressed in terms of the difference between the TT achieved by each of the *AAMO* instances and the one achieved by the *MinPos* approach, for each fleet size.

The evidence indicates that the more efforts made in favour of connectivity (bigger HO-Threshold) the worst *TT*. In other words, not all *HO-Threshold* setup values produce the same level of performance degradation. Since the degradation of TT performance could be very problematic in many application fields, this subject is carefully analysed.

At first, the PL indicator can help to initially explain why the fleet spends more time under *AAMO* approach than under the *MinPos* approach, to explore the same environment. In Figure 19a it is possible to observe the same behaviour as in the baseline (see Section 6.3): larger fleets imply bigger PL; while Figure 19b shows the difference between the corresponding total length of the paths traversed by fleets.

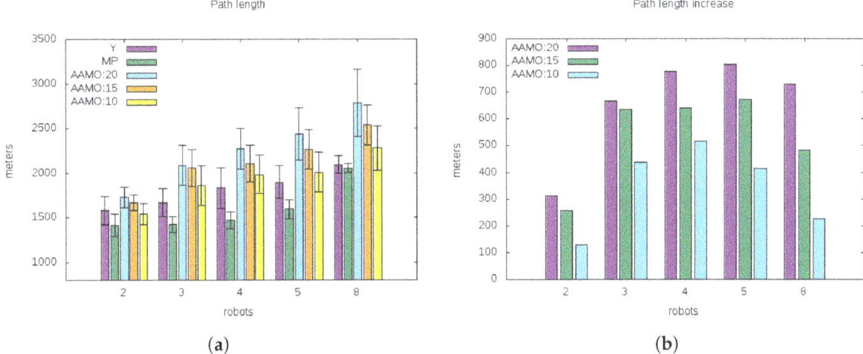

Figure 19. *AAMO* Path length (PL) under non-ideal communication conditions and Degradation with respect to baseline results. (**a**) An increasing trend of PL is shown by all *AAMO* instances as the fleet size increase. The *Yamauchi* and *MinPos* approach results (coloured in purple and green, respectively) obtained under ideal communication conditions are placed together to make the comparison easier; (**b**) The degradation is expressed in terms of the difference between the PL achieved by each of the *AAMO* instances and the one achieved by the *MinPos* approach, for each fleet size.

The similarity between Figures 18b and 19b is remarkable and could explain, to a large extent, the origin of TT degradation. Simply, under the *AAMO* approach, the robots are asked to invest some effort (translated as a distance using the HO-Threshold) in order to keep the fleet connected and hence it is logic to get a bigger PL as a result. Moreover, the tradeoff between path and connectivity utility discussed in Section 3.1 shows up through these results, reflecting that the price of connectivity maintenance is the inability to apply an optimal policy concerning path costs.

Nevertheless, there exists a small portion of the TT degradation that cannot be explained by the PL increasing. Therefore, the hypothesis assumed in the tractability analysis made at the end of Section 5.2 are compared here with the simulation results in order to add a complementary explanation on the TT degradation. Furthermore, this TT degradation shows a parabolic trend as the fleet size increase, reaching a maximum about three-sized fleets, independently of the *HO-Threshold* values. Thus, the analysis will be conducted observing what happens when the fleet size does change but the *HO-Threshold* does not (in order to explain the shape of the curve or the relative values), and the opposite conditions are imposed in order to explain the absolute values.

In any case, it is worth knowing that the *Task selection* algorithm is the most demanding software component in the software architecture of the robots. Hence, the overall performance of the multi-robot system is highly determined by the performance of this component. In turn—as was pointed out in Section 5.2—its performance is strongly influenced by the number of unassigned thresholded tasks $n = |T^{\mathrm{HO}}|$ and the number of unassigned robots in a connected component $m = |R^u|$ that are making a decision at the same time, in the following way: $|Ar^n_m| = \frac{n!}{(n-m)!} = \Pi_{m=0}^{n-1}(n-m) \rightarrow O(n^m)$. Therefore, the smaller $|T^{HO}|$ and $|R^u|$ the faster the algorithm will run. Please recall that $|T^{HO}|$ is upper bounded by the amount of unassigned tasks $|T^u|$.

Firstly, from Figures 20 and 21, it is possible to examine how $|R^u|$ and $|T^u|$ change along explorations depending on the fleet size. In all cases, both values show well defined patterns that are easily identifiable. Concerning $|R^u|$ (see Figure 20) it is possible to state that in all *AAMO* instances—working in a fully asynchronous modality—the probability of two or more robots simultaneously running a decision making process is negligible. Thus the majority of time either none robot is making a decision or at most one robot is evaluating the available tasks.

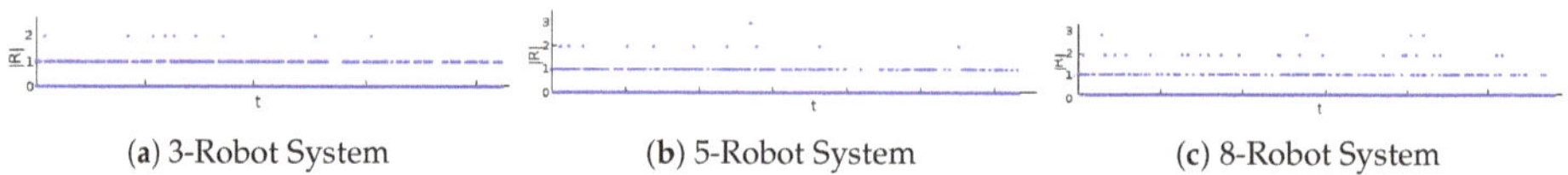

(**a**) 3-Robot System (**b**) 5-Robot System (**c**) 8-Robot System

Figure 20. The maximum amount of unassigned robots $|R^u|$ in any connected component over time under different sized multi-robot systems. All images concern instances of *AAMO* set with HO-Threshold = 15. Blue dots represent the $|R^u|$ (on average) that are simultaneously deciding along the exploration.

Results obtained during simulations are summarised in Table 6 and show a behaviour that is consistent with this last statement independently of the fleet size. The low ratio of robot coincidences is remarkable (e.g., for 3-sized fleets, about 96% of the decision making moments have only one robot participating on them).

Table 6. *AAMO* Robot Coincidence on Decision Making moments.

HO-Threshold	$\|R\|$	**$\|R^u\|$ that are simultaneously making a decision**							
		1		**2**		**3**		**4**	
		AVE	**StD**	**AVE**	**StD**	**AVE**	**StD**	**AVE**	**StD**
AAMO:10	3	0.959	0.01	0.041	0.01	$\simeq 0$	$\simeq 0$	n/a	n/a
	5	0.895	0.02	0.097	0.02	0.011	0.01	$\simeq 0$	$\simeq 0$
	8	0.807	0.02	0.153	0.04	0.037	0.02	$\simeq 0$	$\simeq 0$
AAMO:15	3	0.969	0.02	0.031	0.02	$\simeq 0$	$\simeq 0$	n/a	n/a
	5	0.929	0.02	0.068	0.03	0.003	0.01	$\simeq 0$	$\simeq 0$
	8	0.823	0.03	0.146	0.03	0.024	0.01	$\simeq 0$	$\simeq 0$
AAMO:20	3	0.968	0.02	0.032	0.02	$\simeq 0$	$\simeq 0$	n/a	n/a
	5	0.917	0.02	0.080	0.02	0.003	0.01	$\simeq 0$	$\simeq 0$
	8	0.838	0.02	0.148	0.03	0.018	0.01	$\simeq 0$	$\simeq 0$

In conclusion, in practice, the worsening of the TT performance is apparently only related to the incidence of the *HO-Threshold* on the $|T^{HO}|$ value. Next, this relation is carefully studied, and some answers are essayed.

The parabola described by the TT degradation values in Figure 18 suggests the presence of two factors impacting on this behaviour. One presses the trend upwards and the other in a counter sense. In the following, two particular factors are analysed: the fleet size and the bounded condition of the environment. (i) As the fleet size increase robots make progress faster, causing $|T^{HO}|$ to increase more quickly as well. When $|T^{HO}|$ rises, the task selection algorithm becomes slower, and thus the increase in the fleet size could explain the first increasing section of the trend; (ii) In bounded environments, the multi-robot exploration systems typically show two mobility patterns that characterise, in turn, two different exploration stages: (1) One is characterised by the dispersion of the fleet on the terrain. In such a stage, the new available tasks appear closer to each other, and its total amount $|T^u|$ is upward; (2) On the contrary, the second exploration stage is characterised by the convergence of the fleet to the remaining unexplored zones starting when it is no longer possible to disperse the fleet until the end of the exploration. In such a stage, the new available tasks generally appear further to each other and its total amount $|T^u|$ is decreasing. Therefore, since the tasks T^{HO} are the ones which are closer than a relative distance *HO-Threshold*, under the *AAMO* approach, it is statistically less demanding for the robots to select a task during the last exploration stage than in the initial one.

Additionally, either when the fleet size increase or the *HO-Threshold* decrease, the transition from the first to the second exploration stage is achieved faster. This fact can be corroborated in both Figures 21 and 22. For instance, concerning Figure 21, the 3-Robot system spends about 410 s to reach the end of the dispersion stage whereas the 5-Robot system and 8-Robot system spend about 320 s and 260 s, respectively. Likewise, from Figure 22, the AAMO:20 instance spends about 310 s to reach the end of the dispersion stage whereas the AAMO:15 and AAMO:10 spend about 260 s and 150 s, respectively.

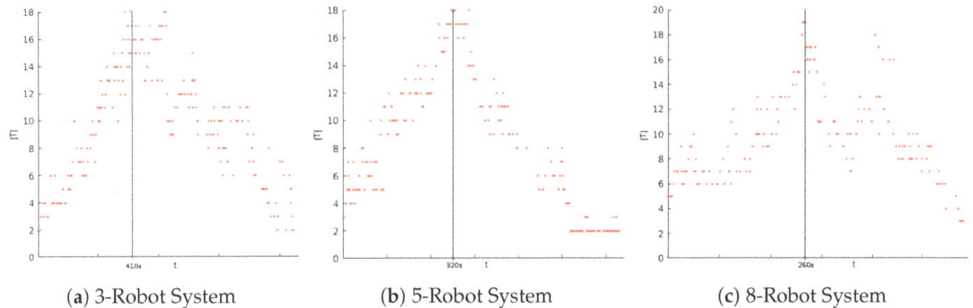

(**a**) 3-Robot System (**b**) 5-Robot System (**c**) 8-Robot System

Figure 21. Amount of unassigned tasks $|T^u|$ over time for different sized multi-robot systems. All images concern instances of *AAMO* set with HO-Threshold = 15. The maximum $|T^u|$ and the end of the dispersion stage are reached at the same time. Red dots represent the $|T^u|$ considered by robots (on average) along the exploration.

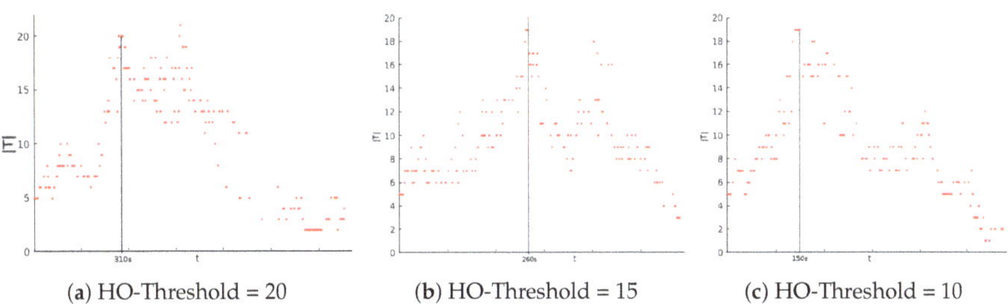

(**a**) HO-Threshold = 20 (**b**) HO-Threshold = 15 (**c**) HO-Threshold = 10

Figure 22. Number of unassigned tasks $|T^u|$ over time for different instances of the *AAMO* approach on 8-Robot systems. The maximum $|T^u|$ and the end of the dispersion stage are reached at the same time. Red dots represent the $|T^u|$ considered by robots (on average) along the exploration.

Hence, despite the fact the impact of the fleet size on the exploration stage transition appears to be higher than the one caused by the *HO-Threshold* value, both aspects contribute to reducing the task selection effort enabling robots to save time in the task allocation procedure anticipatedly.

In conclusion, when the *AAMO* is executed in bounded environments, the addition of robots and the decreasing of *HO-Threshold* can almost entirely mitigate the worsening in the total exploration time performance. Please note that the performance degradation of AAMO:10 instances is almost null for eight-sized fleets.

From these promising results, in the following, all *AAMO* instances are compared with the other approaches concerning non-ideal communication conditions.

6.4.4. *AAMO* Efficiency Assessment

In this section, several statistical analyses were performed on different indicators to demonstrate the efficiency of the proposed *AAMO* approach. Wilcoxon signed-rank tests were performed (a non-parametric test was chosen since data in each condition do not follow a normal distribution) to compare samples from two populations. More precisely, it tests the indicator differences between approaches for a given fleet size.

Firstly, in relation to TT (see Figure 23), the evidence confirms two expected results: (i) All approaches benefit from adding robots to the fleet. A Wilcoxon difference test was performed regarding TT and the fleet size for each approach. All comparisons present a significant decrease in TT when fleet size increases (p-value < 0.05); (ii) Since it only takes care of connectivity, the EbC approach shows the worst performance regardless the fleet size. Wilcoxon tests showed a significant result (p-value < 0.001) for all comparisons between approaches given a fleet size.

Additionally, all *AAMO* instances show competitive TT results even slightly outperforming other approaches in the case of *AAMO:10*. In particular, a Wilcoxon difference test showed that *AAMO:10* has a smaller TT than *MinPos* for 2 and 5 robots (resp. $W = 169$, p-value < 0.01, and $W = 159$, p-value < 0.05).

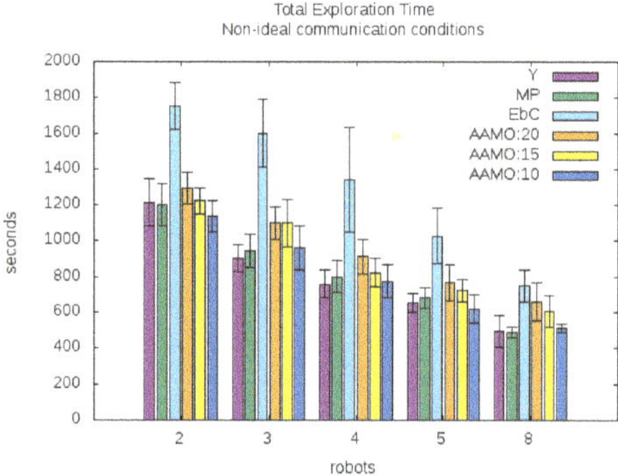

Figure 23. Total Exploration Time (TT) under non-ideal communication conditions.

Secondly, concerning the PL indicator (see Figure 24), the EbC approach present the worst performance, coherently. Again, the Wilcoxon test showed significant results (p-value < 0.001). Likewise, all *AAMO* instances show competitive results too.

Besides, and as was pointed above, the TT and PL results show that the lack of ideal communication conditions negatively affects the *MinPos* approach more than the *Yamauchi* approach. Wilcoxon tests showed a trend for 4 and 5 robots (p-value < 0.1) concerning TT, and a sigfinicant difference in PL for 4 robots (p-value < 0.05).

Up to this point, the *AAMO* approach has shown results as good as the *MinPos* approach. Next, the indicators related to connectivity are analysed in order to properly assess the potential advantages of the *AAMO* approach in the presence of more realistic communication conditions.

The DLR indicator trend is shown in Figure 25. As can be seen, while the performance of the *MinPos* and *Yamauchi* approaches are the worst, the EbC performance is remarkably the best. These visual results were confirmed by Wilcoxon tests between approaches for each fleet size. DLR indicator is significantly bigger (p-value < 0.001) for MinPos and Yamauchi than AAMO and EbC approaches, except for 8-sized fleets where these results are significant only when compared to EbC. Moreover EbC has a significant smaller DLR indicator (p-value < 0.05) than all the others approaches except for the 5 and 8 robots cases, in which any statistical difference can be found between AAMO approaches and EbC.

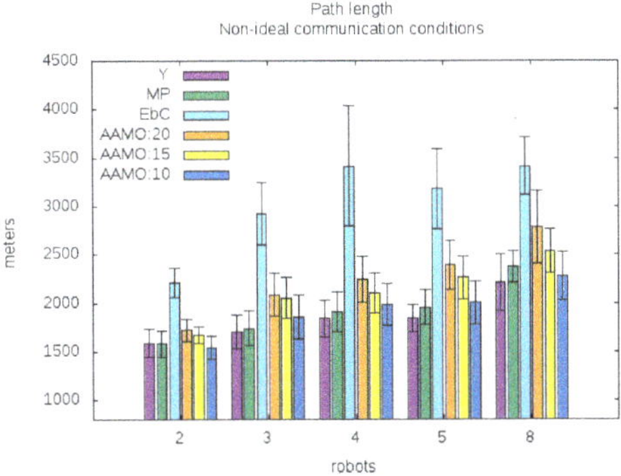

Figure 24. Path length (PL) under non-ideal communication conditions.

Similarly, the *AAMO* approach results represent a very good improvement with respect to both *MinPos* and *Yamauchi* approaches. The chart in Figure 25 reveals that our approach outperforms both Yamauchi and minPos approaches independently of the fleet size on average. Nevertheless, the smaller fleet, the greater outperforming. The explanation can arise correctly from intuition: when the environment is bounded, the probability of being disconnected tends to decrease as the fleet size increase. Therefore, the benefits of our approach tend to be smaller when the fleet size increases. Either way, it is always meaningful. Please note that even in the largest fleet size case, the DLR of *AAMO* represents an improvement of 20% on average compared to the corresponding *Yamauchi* or *MinPos*.

Furthermore, the relation between TT, DLR and *HO-Threshold* is noticeable. The more effort demanded by the human operator (higher threshold), the slower but higher connected the *AAMO* performs. This claim is confirmed by Wilcoxon tests that showed a significantly bigger (p-value < 0.05) TT indicator for AAMO:20 than for AAMO:10, and also show that the DRL indicator is significantly smaller (p-value < 0.05) for AAMO:20 than for AAMO:10, regardless the fleet size.

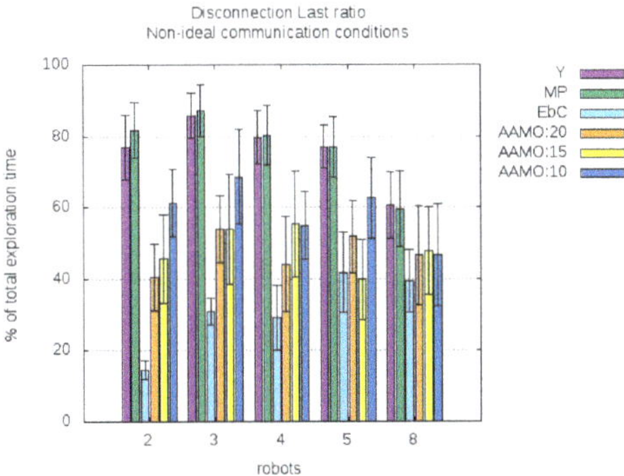

Figure 25. Disconnection Last Ratio (DLR) under non-ideal communication conditions. The bigger the *HO-Threshold*, the smaller DLR. This fact holds showing an oscillatory behaviour as the fleet size increase.

Regarding the oscillation registered, it could suggest the existence of the following rational pattern. When fleet size is even, the easier way to avoid isolation situations is keeping in pairs (connected with at least another teammate). Contrarily, when the fleet size is odd, not all robots can keep in pairs. In case the fleet has divided, at least one sub group must be composed of three robots. Therefore, this oscillatory behaviour could hint at the fact that odd-sized fleets need to make little more effort to avoid robot isolation situations and are consequently subject to bigger DLR results as well.

Likewise, it is interesting to analyse the DLR indicator and network topology together. This way it is possible to get a closer notion about the interaction between robots along the exploration. Figure 26 is devoted to showing the number of connected components present in the network, averaged over time.

Please note that for the AAMO:20 instance—run on 2-Robot fleet—the DLR is about 40% (see Figure 25), coinciding with the percentage achieved by the 2CC of the same fleet size in Figure 26. In other words, the fleet holds a network composed of one single connected component during 60% (100%–40%) of total exploration time. Consistently, this is equivalent to say that during this portion of the time none robot has been disconnected.

Additionally, and as a matter of fact, the chart shows that as the fleet size increase it is more challenging to keep the whole fleet connected: 1CC stack is decreasing in size as the fleet size increase. Nevertheless, it also shows that simultaneously with the adding of new robots, the fleet is more and more cohesive (in relative terms). This fact may be corroborated looking at the upper part of the chart where the stacks corresponding to the greatest number of connected components are plotted. The following pattern can be observed: the number of connected components (given by nCC) increase slower than the fleet size n. Again, the fact that the *Maze* scenario is bounded may explain this phenomenon to a large extent.

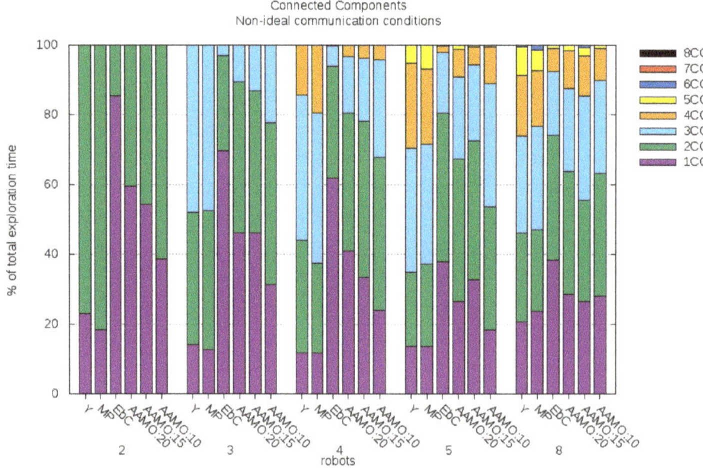

Figure 26. Network topology composition under non-ideal communication conditions averaged over time. Depending on the number of connected components and the fleet size, it is possible to study the existence of sufficient conditions to fall into isolation situations. For instance, for a 3-Robot fleet, the 2CC or 3CC topologies imply having at least one robot isolated while for a 5-Robot fleet this implication is related to 3CC, 4CC, or 5CC topologies, and so on.

Although all this information gives an approximated notion about how disconnected is the fleet (group perspective) along explorations, it is not enough to hint what is happening at the individual level. Thus, it is also interesting to study the worst case of the individual disconnections last. This way it is easier to evaluate both coordination capabilities (how long a robot is unable to coordinate its actions with any other teammates) and risky situations (how long the fleet present single points of failure). Recall that the key motivations in considering communication constraints are strongly related with the rework avoidance: (i) When robots are unconnected they have fewer possibilities to coordinate their actions hence they could visit the same regions unnecessarily. Hence, keep them connected is a way to favour the efficiency; (ii) In the presence of damages or inner failures the exploration strategy should take those events into account preventing the need of re-exploration.

In Figure 27 the trend followed by Maximum Disconnection Last Ratio MDLR indicator is depicted showing that the bigger *HO-Threshold*, the shorter disconnection periods (Wilcoxon tests showed that the MDLR indicator is significant smaller (p-value < 0.05) for AAMO:20 than for AAMO:10, for 2 and 3 robots, and tends to be smaller (p-value $= 0.09$) for 4 robots) and that the last of isolation situations is at most equivalent to half of the DLR values for every fleet size and *HO-Threshold* value as well. In other words, the isolation situations regard more than one single robot and this in turn, reveals that under the *AAMO* approach the robots often intent to rejoin each other.

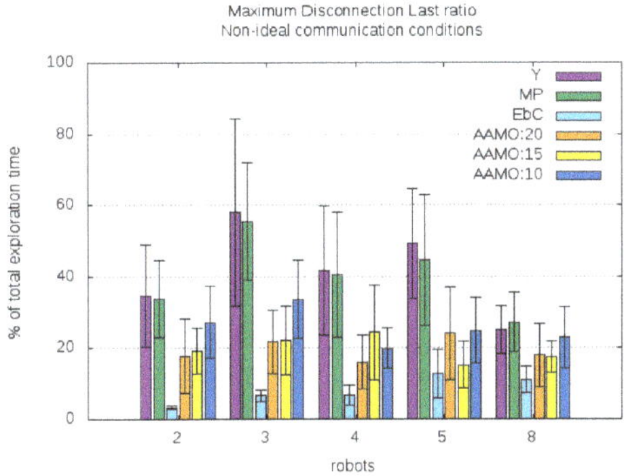

Figure 27. Maximum Disconnection Last Ratio (MDLR) under non-ideal communication conditions. MDLR shows the longest individual isolation period registered by some fleet member along the exploration. The trend is oscillatory following the same pattern as the DLR indicator.

At last but not least, it is worth to discuss the trend of OSR as the fleet size increase. The results obtained by the different *AAMO* instances are depicted in Figure 28. In Section 6.3 the OSR levels were achieved mostly thanks to simultaneous sensing actions, conversely, in these simulation runnings, the OSRs achieve higher levels due to non-ideal communication conditions. As was expected, the more the mapping information of the robots is out-of-date with respect to each other, the higher the OSR. However, in any communication conditions, the same upper bound is achieved. This suggests that the size and bounded condition of the *Maze* environment could be limiting the over-sensing phenomenon when fleet size increase beyond five robots.

To sum up and concerning the *Maze* scenario and the baseline stated in Section 6.3, the conclusions of this section are: (i) The *AAMO* approach can be employed as a strategy to coordinate multi-robot systems that are dedicated to exploration tasks; (ii) As was expected, the *HO-Threshold* value directly impacts on the connectivity level that the fleet is able to hold during the mission; (iii) Likewise, the relation between *HO-Threshold* values and the TT and DLR/MDLR indicators is the expected: the bigger the *HO-Threshold* value, the worse TT performance, but the better DLR/MDLR ratios; (iv) Although all instances of the *AAMO* approach present TT degradation with respect to the baseline, in any case it is not significantly due to the computation of the proposed task-to-robots distribution; (v) All *AAMO* instances outperform the baseline concerning the DLR and MDLR indicators; (vi) With the exception of DLR/MDLR, all instances of the *AAMO* approach outperform the *EbC* approach; (vii) The topology of the fleet networks shown during exploration is consistent with the *HO-Threshold* values, for all *AAMO* instances.

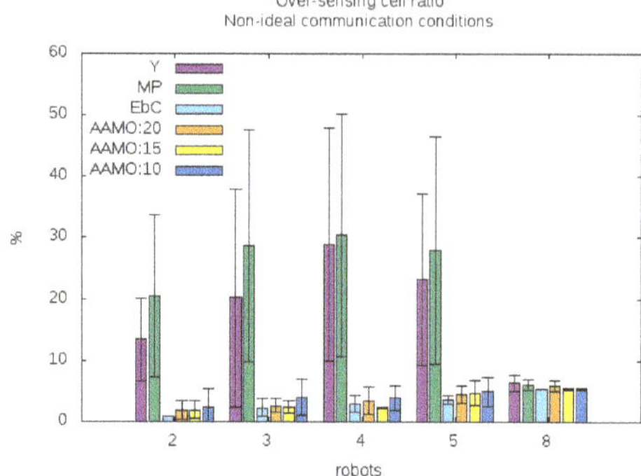

Figure 28. Over-sensing ratio (OSR) under non-ideal communication conditions. The *Yamauchi* and *MinPos* approach results (coloured in purple and green, respectively) obtained under ideal communication conditions are placed together to make the comparison easier.

The *AAMO* approach shows effectiveness and flexibility (through the *HO-Threshold* setup) to tackle the multi-robot exploration problem. Particularly concerning the efficiency related to both completion time and connectivity level maintenance, the approach appears as an intermediate solution that presents much better TT performance than the most restrictive approach *EbC* and better connectivity level along exploration than the approaches that do not take care about communication issues.

7. Conclusions

7.1. Task Assessment Problem

The proposed *Auto-Adaptive Multi-Objective (AAMO)* approach follows a multi-objective assessment strategy where the tasks under consideration are assessed regarding two objectives: the cost associated with the corresponding shortest path and the connectivity level each task location can offer to robots at arrival time. The multi-objective strategy is implemented employing a weighted sum that trades travelling cost off for connectivity levels. Up until now, all these concepts are quite standard being present in several state-of-art approaches.

Nevertheless, in this work: (i) Connectivity awareness ability is given to the robots by modelling attenuation effects that commonly affect the communication signal strength; (ii) the weights of these potentially conflicting objectives are derived from formal analysis instead of a training stage, making the system more adaptable to different environments; (iii) The human operator is asked to use his application-field expertise to play a part in the task assessment process by setting a distance threshold until which the tasks that preserve or enlarge connectivity are preferred over the rest. All this leads to a more flexible system where the robots can deal with communication constraints adjusting the weights of each objective independently of any scenario in a more intuitive manner, saving a lot of training time.

All existence and correctness proofs conducted on the task selection procedure support the fact that the robots are always capable of auto-adapting the objectives weights in order to select the tasks accordingly with the human-operator criterion. In conclusion, this task assessment approach may be very advantageous considering its ease of deployment.

7.2. Task Allocation Problem

Concerning the tasks-to-robots distribution algorithm, all previous proposals explicitly avoid the combinatorial blow-up of allocation complexity using different heuristics. Nevertheless, heuristic-based approaches make assumptions that cannot be verified at all times. In consequence, when the heuristic fails the robots choose suboptimal distributions.

Taking into account this limitation and since the number of possible distributions depends on both the number of available tasks and the number of robots making a decision, the proposal presented here computes optimal distributions based on more general assumptions such as: (i) Robots can implicitly coordinate their actions; (ii) Asynchronism may keep the number of simultaneous decision making at small values; (iii) Pruning the furthest tasks (out of the scope defined by the human operator –*HO-Threshold*) does not prevent the computation of optimal tasks-to-robots distribution.

7.3. Connectivity Maintenance Problem

While all event-based connectivity approaches consist in the execution of regaining-connectivity actions in the presence of specific events (e.g., typically disconnections, whenever it happens or periodically after a certain amount of time), the *AAMO* approach integrates a less restrictive connectivity strategy where the robots are motivated but not compelled to regain connectivity. When selecting their targets, the robots are always considering the opportunity cost of keeping connected or regaining connectivity, implicitly. Furthermore, in reconnection cases, the task location becomes the meeting point itself eliminating the need for rendezvous policy implementation and, maybe more important, avoiding deviations from natural paths. This way, the policy is utterly transparent to the eyes of the external observer: every time it is possible to explore and keep or enlarge connectivity level the robots will choose this option. On the contrary, when it is not, they merely behave guided by a pure path-cost exploration.

Particularly concerning the efficiency related to both completion time and connectivity level maintenance, the approach is capable of decreasing the last of disconnection periods without a noticeable degradation of the completion exploration time, appearing as an intermediate solution that presents much better completion time performance than the most restrictive event-based connectivity approaches and better connectivity level along exploration than the approaches that do not take care about communication issues.

7.4. Future Research Directions

New research questions have arisen along this work leaving, as a result, several opportunities to improve the developed system. Although the environments employed in simulations are proposed as benchmarks, it would be beneficial to check the validity and performance of the proposed approach on a broader variety of scenarios. Large office-like environments would be exciting to put the system into more realistic situations like mapping buildings where larger fleets could be employed, too. Since the robot model defined can support robots with different characteristics, exploiting heterogeneity could be a promising research direction. Integrating a fleet with heterogeneous robots (e.g., different in size, sensory, and motion capabilities) could enhance the skills of the fleet. For instance, given their greater mobility, UAVs could help the fleet to keep connected by playing a relay role, while small terrestrial robots could be the key to get into access-restricted spaces. At last but not least, executions on real systems are also planned. Despite the goodness of any simulator, many important details escape their scopes. The proposed approach is designed to serve as a solution for real-world applications so that it is imperative to verify its feasibility in real scenarios. In such a case, localisation and mapping errors cannot be ignored. Both simultaneous Localisation and Mapping (SLAM) algorithms and the sensory and motor devices should be carefully studied to limit the influence of this kind of errors on the high-level decision components. Regarding the equipment availability of the involved laboratories, the candidate platforms would be either IRobot or KheperaIII units.

Author Contributions: Conceptualisation, Methodology, Software, Validation, Analysis and Writing-original draft preparation, F.B.; Writing reviews and Supervision, E.G., C.C. and P.M.

Funding: This research was partially funded by the Comisión Sectorial de Investigación (CSIC-UdelaR) through its mobility program MIA.

Conflicts of Interest: The authors declare no conflict of interest.

Abbreviations

The following acronyms are used in this manuscript:

AAMO	Auto-Adaptive Multi-Objective approach
CC	Connected Component
CR	Coverage Ratio
EbC	Event-based Connectivity approach
DLR	Disconnection Last Ratio
DR	Dual Role approach
LoS	Line-of-sight
MANET	Mobile Ad-hoc NETwork
MDLR	Maximum Disconnection Last Ratio
MDP	Markov Decision Process
MORSE	Modular Open Robots Simulation Engine
MRS	Multi-Robot System
MP	MinPos approach
OSR	Over-Sensing Ratio
PL	Path length
POMDP	Partially Observable Markov Decision Process
SLAM	simultaneous Localisation and Mapping
TT	Total exploration Time
UAV	Unmanned Aerial Vehicle
Y	Yamauchi approach

References

1. Burgard, W.; Moors, M.; Stachniss, C.; Schneider, F. Coordinated multi-robot exploration. *IEEE Trans. Robot.* **2005**, *21*, 376–386. [CrossRef]
2. Wurm, K.; Stachniss, C.; Burgard, W. Coordinated multi-robot exploration using a segmentation of the environment. In Proceedings of the 2008 IEEE/RSJ International Conference on Intelligent Robots and Systems, Nice, France, 22–26 September 2008; pp. 1160–1165. [CrossRef]
3. Simmons, R.; Apfelbaum, D.; Burgard, W.; Fox, D.; Moors, M.; Thrun, S.; Younes, H. *Coordination for Multi-Robot Exploration and Mapping;* Aaai: Menlo Park, CA, USA; AAAI Press: Cambridge, MA, USA; MIT Press: London, UK, 1999/2000; pp. 852–858.
4. Elfes, A. Using Occupancy Grids for Mobile Robot Perception and Navigation. *Computer* **1989**, *22*, 46–57. [CrossRef]
5. Yamauchi, B. Frontier-based exploration using multiple robots. In Proceedings of the Second International Conference on Autonomous Agents AGENTS 98, Minneapolis, MN, USA, 10–13 May 1998; pp. 47–53. [CrossRef]
6. Keidar, M.; Kaminka, G.A. Efficient frontier detection for robot exploration. *Int. J. Robot. Res.* **2014**, *33*, 215–236. [CrossRef]
7. Keidar, M.; Kaminka, G.A. Robot Exploration with Fast Frontier Detection: Theory and Experiments. In Proceedings of the International Conference on Autonomous Agents and Multiagent Systems, Valencia, Spain, 4–8 June 2012; pp. 113–120.
8. Yuan, J.; Huang, Y.; Tao, T.; Sun, F. A cooperative approach for multi-robot area exploration. In Proceedings of the IEEE/RSJ 2010 International Conference on Intelligent Robots and Systems, IROS 2010—Conference Proceedings, Taipei, Taiwan, 18–22 October 2010; pp. 1390–1395. [CrossRef]
9. Korsah, G.A.; Stentz, A.; Dias, M.B. A comprehensive taxonomy for multi-robot task allocation. *Int. J. Robot. Res.* **2013**, *32*, 1495–1512. [CrossRef]
10. Gerkey, B.P. A Formal Analysis and Taxonomy of Task Allocation in Multi-Robot Systems. *Int. J. Robot. Res.* **2004**, *23*, 939–954. [CrossRef]
11. Sheng, W.; Yang, Q.; Tan, J.; Xi, N. Distributed multi-robot coordination in area exploration. *Robot. Auton. Syst.* **2006**, *54*, 945–955. [CrossRef]
12. Zlot, R.; Stentz, A.; Dias, M.; Thayer, S. Multi-robot exploration controlled by a market economy. In Proceedings of the 2002 IEEE International Conference on Robotics and Automation, Washington, DC, USA, 11–15 May 2002; Volume 3, pp. 3016–3023. [CrossRef]

13. Cavalcante, R.C.; Noronha, T.F.; Chaimowicz, L. Improving combinatorial auctions for multi-robot exploration. In Proceedings of the 2013 16th International Conference on Advanced Robotics (ICAR), Montevideo, Uruguay, 25–29 November 2013; pp. 1–6. [CrossRef]
14. Kuhn, H.W. The Hungarian method for the assignment problem. *Naval Res. Logist.* **1955**, *2*, 83–97. [CrossRef]
15. Hollinger, G.A.; Singh, S. Multirobot coordination with periodic connectivity: Theory and experiments. *IEEE Trans. Robot.* **2012**, *28*, 967–973. [CrossRef]
16. Pham, V.C.; Juang, J.C. A multi-robot, cooperative, and active slam algorithm for exploration. *Int. J. Innov. Comput. Inf. Control* **2013**, *9*, 2567–2583.
17. Bautin, A.; Simonin, O. MinPos: A Novel Frontier Allocation Algorithm for Multi-robot Exploration. *ICIRA* **2012**, *7507*, 496–508.
18. Valentin, L.; Murrieta-Cid, R.; Muñoz-Gómez, L.; López-Padilla, R.; Alencastre-Miranda, M. Motion strategies for exploration and map building under uncertainty with multiple heterogeneous robots. *Adv. Robot.* **2014**, *28*, 1133–1149. [CrossRef]
19. Banfi, J.; Li, A.Q.; Basilico, N.; Rekleitis, I.; Amigoni, F. Asynchronous multirobot exploration under recurrent connectivity constraints. In Proceedings of the IEEE International Conference on Robotics and Automation, Stockholm, Sweden, 16–21 May 2016; pp. 5491–5498. [CrossRef]
20. Amigoni, F.; Banfi, J.; Basilico, N. Multirobot Exploration of Communication-Restricted Environments: A Survey. *IEEE Intell. Syst.* **2017**, *32*, 48–57. [CrossRef]
21. Vazquez, J.; Malcolm, C. Distributed Multirobot Exploration Maintaining a Mobile Network. In Proceedings of the IEEE International Conference on Intelligent Systems, Varna, Bulgaria, 22–24 June 2004.
22. Rooker, M.N.; Birk, A. Multi-robot exploration under the constraints of wireless networking. *Control Eng. Pract.* **2007**, *15*, 435–445. [CrossRef]
23. Mosteo, A.R.; Montano, L.; Lagoudakis, M.G. Multi-robot routing under limited communication range. In Proceedings of the 2008 IEEE International Conference on Robotics and Automation, Pasadena, CA, USA, 19–23 May 2008; pp. 1531–1536. [CrossRef]
24. Le, V.T.; Bouraqadi, N.; Stinckwich, S.; Moraru, V.; Doniec, A. Making networked robots connectivity-aware. In Proceedings of the IEEE International Conference on Robotics and Automation, Kobe, Japan, 12–17 May 2009; pp. 3502–3507. [CrossRef]
25. Michael, N.; Zavlanos, M.M.; Kumar, V.; Pappas, G.J. Maintaining Connectivity in Mobile Robot Networks. *Springer Tracts Adv. Robot. (STAR)* **2009**, *54*, 117–126. [CrossRef]
26. Derbakova, A.; Correll, N.; Rus, D. Decentralized self-repair to maintain connectivity and coverage in networked multi-robot systems. In Proceedings of the IEEE International Conference on Robotics and Automation, Shanghai, China, 9–13 May 2011; pp. 3863–3868. [CrossRef]
27. Pei, Y.; Mutka, M.W. Steiner traveler: Relay deployment for remote sensing in heterogeneous multi-robot exploration. In Proceedings of the IEEE International Conference on Robotics and Automation, Saint Paul, MN, USA, 14–18 May 2012; pp. 1551–1556. [CrossRef]
28. Laëtitia Matignon, L.J.; Mouaddib, A.I. Coordinated Multi-Robot Exploration Under Communication Constraints Using Decentralized Markov Decision Processes. In Proceedings of the Twenty-Sixth AAAI Conference on Artificial Intelligence, Toronto, ON, Canada, 22–26 July 2012.
29. Couceiro, M.S.; Figueiredo, C.M.; Rocha, R.P.; Ferreira, N.M. Darwinian swarm exploration under communication constraints: Initial deployment and fault-tolerance assessment. *Robot. Autonomous Syst.* **2014**, *62*, 528–544. [CrossRef]
30. Pralet, C.; Lesire, C. Deployment of Mobile Wireless Sensor Networks for Crisis Management: A Constraint-Based Local Search Approach. In Proceedings of the International Conference on Principles and Practice of Constraint Programming, Toulouse, France, 5–9 September 2014; pp. 870–885.
31. Jensen, E.A.; Nunes, E.; Gini, M. *Communication-Restricted Exploration for Robot Teams*; Workshop on Multiagent Interaction without Prior Coordination; Number Association for the Advancement of Artificial Intelligence (AAAI): Palo Alto, CA, USA, 2014; pp. 15–21.
32. Cesare, K.; Skeele, R.; Yoo, S.H.; Zhang, Y.; Hollinger, G. Multi-UAV exploration with limited communication and battery. In Proceedings of the IEEE International Conference on Robotics and Automation, Seattle, WA, USA, 26–30 May 2015; Volume 2015, pp. 2230–2235. [CrossRef]
33. Magán-Carrión, R.; Camacho, J.; García-Teodoro, P.; Flushing, E.F.; Di Caro, G.A. A Dynamical Relay node placement solution for MANETs. *Comput. Commun.* **2017**, *114*, 36–50. [CrossRef]

34. Magán-Carrión, R.; Rodríguez-Gómez, R.A.; Camacho, J.; García-Teodoro, P. Optimal relay placement in multi-hop wireless networks. *Ad Hoc Netw.* **2016**, *46*, 23–36. [CrossRef]
35. Rahman, M.M.; Bobadilla, L.; Abodo, F.; Rapp, B. Relay vehicle formations for optimizing communication quality in robot networks. In Proceedings of the IEEE International Conference on Intelligent Robots and Systems, Vancouver, BC, Canada, 24–28 September 2017; pp. 6633–6639. [CrossRef]
36. Nestmeyer, T.; Robuffo Giordano, P.; Bülthoff, H.H.; Franchi, A. Decentralized simultaneous multi-target exploration using a connected network of multiple robots. *Autonomous Robots* **2017**, *41*, 989–1011, [CrossRef]
37. Banfi, J.; Quattrini Li, A.; Rekleitis, I.; Amigoni, F.; Basilico, N. Strategies for coordinated multirobot exploration with recurrent connectivity constraints. *Autonomous Robots* **2018**, *42*, 875–894. [CrossRef]
38. Bahl, P.; Padmanabhan, V.N. RADAR: An In-Building RF-based User Location and Tracking System. In Proceedings of the IEEE INFOCOM 2000, Conference on Computer Communications, Nineteenth Annual Joint Conference of the IEEE Computer and Communications Societies, Tel Aviv, Israel, 26–30 March 2000; pp. 775–784, [CrossRef]
39. Caccamo, S.; Parasuraman, R.; Freda, L.; Gianni, M.; Ogren, P. RCAMP: A resilient communication-aware motion planner for mobile robots with autonomous repair of wireless connectivity. In Proceedings of the IEEE International Conference on Intelligent Robots and System, Vancouver, BC, Canada, 24–28 September 2017; pp. 2010–2017, [CrossRef]
40. Fink, J.; Ribeiro, A.; Kumar, V. Robust Control of Mobility and Communications in Autonomous Robot Teams. *IEEE Access* **2013**, *1*, 290–309. [CrossRef]
41. MacQueen, J. Some Methods for classification and Analysis of Multivariate Observations. In Proceedings of the 5th Berkeley Symposium on Mathematical Statistics and Probability, Berkeley, CA, USA, 1967.
42. Frey, B.J.; Dueck, D. Clustering by passing messages between data points. *Science* **2007**, *315*, 972–976. [CrossRef] [PubMed]
43. Bautin, A.; Simonin, O.; Charpillet, F. SyWaP: Synchronized wavefront propagation for multi-robot assignment of spatially-situated tasks. In Proceedings of the 2013 16th International Conference on Advanced Robotics, ICAR, Montevideo, Uruguay, 25–29 November 2013. [CrossRef]
44. Bautin, A.; Simonin, O.; Charpillet, F. Towards a communication free coordination for multi-robot exploration. In Proceedings of the 6th National Conference on Control Architectures of Robots, CAR2011, Grenoble, France, 24–25 May 2011.
45. Yan, Z.; Fabresse, L.; Laval, J.; Bouraqadi, N. Metrics for performance benchmarking of multi-robot exploration. In Proceedings of the IEEE International Conference on Intelligent Robots and Systems, Hamburg, Germany, 28 September–2 October 2015; pp. 3407–3414. [CrossRef]
46. Thrun, S.; Burgard, W.; Fox, D. *Probabilistic Robotics (Intelligent Robotics and Autonomous Agents)*; The MIT Press: Cambridge, MA, USA, 2005.
47. Pal, A.; Tiwari, R.; Shukla, A. Communication constraints multi-agent territory exploration task. *Appl. Intell.* **2013**, *38*, 357–383. [CrossRef]
48. Satici, A.C.; Poonawala, H.; Eckert, H.; Spong, M.W. Connectivity preserving formation control with collision avoidance for nonholonomic wheeled mobile robots. In Proceedings of the IEEE/RSJ International Conference on Intelligent Robots and Systems, Chicago, IL, USA, 14–18 September 2013; pp. 5080–5086. [CrossRef]

Article

Robust Visual-Aided Autonomous Takeoff, Tracking, and Landing of a Small UAV on a Moving Landing Platform for Life-Long Operation

Pablo R. Palafox [1,*], Mario Garzón [2], João Valente [3] and Juan Jesús Roldán [4] and Antonio Barrientos [4]

1 Computer Vision & Artificial Intelligence Group, Technical University Munich, Boltzmannstrasse 3, 85748 Garching, Germany
2 Univ. Grenoble Alpes, INRIA, Grenoble INP, 38000 Grenoble, France
3 Information Technology Group, Wageningen University & Research, 6708 PB Wageningen, The Netherlands
4 Centro de Automática y Robótica (UPM-CSIC), Universidad Politécnica de Madrid, José Gutiérrez Abascal, 2, 28006 Madrid, Spain
* Correspondence: pablo.rodriguez-palafox@tum.de

Received: 31 May 2019; Accepted: 25 June 2019; Published: 29 June 2019

Featured Application: Autonomous takeoff, tracking, and landing maneuvers on a moving target with application to a fleet of robots with aerial and ground vehicles that need to operate for extended periods of time, as in Search and Rescue tasks.

Abstract: Robot cooperation is key in Search and Rescue (SaR) tasks. Frequently, these tasks take place in complex scenarios affected by different types of disasters, so an aerial viewpoint is useful for autonomous navigation or human tele-operation. In such cases, an Unmanned Aerial Vehicle (UAV) in cooperation with an Unmanned Ground Vehicle (UGV) can provide valuable insight into the area. To carry out its work successfully, such as multi-robot system requires the autonomous takeoff, tracking, and landing of the UAV on the moving UGV. Furthermore, it needs to be robust and capable of life-long operation. In this paper, we present an autonomous system that enables a UAV to take off autonomously from a moving landing platform, locate it using visual cues, follow it, and robustly land on it. The system relies on a finite state machine, which together with a novel re-localization module allows the system to operate robustly for extended periods of time and to recover from potential failed landing maneuvers. Two approaches for tracking and landing are developed, implemented, and tested. The first variant is based on a novel height-adaptive PID controller that uses the current position of the landing platform as the target. The second one combines this height-adaptive PID controller with a Kalman filter in order to predict the future positions of the platform and provide them as input to the PID controller. This facilitates tracking and, mainly, landing. Both the system as a whole and the re-localization module in particular have been tested extensively in a simulated environment (Gazebo). We also present a qualitative evaluation of the system on the real robotic platforms, demonstrating that our system can also be deployed on real robotic platforms. For the benefit of the community, we make our software open source.

Keywords: robust autonomous landing; unmanned aerial vehicle; unmanned ground vehicle; multi-robot systems; Kalman filter; PID controller; re-localization module

1. Introduction

Robotics is increasingly taking on greater importance in our lives. One of the main areas where this can be perceived is Search and Rescue (SaR) tasks [1]. Robots designed for this kind of task, known as SaR robots, must operate on many occasions in unknown environments, move over unstable surfaces, and face multiple difficulties in order to carry out their mission, e.g., obtaining a map of the environment to facilitate the subsequent intervention of the rescue brigades [2]. Using a single robot under such conditions poses big difficulties: whether it moves on the surface or flies nearby areas, there are intrinsic difficulties for each type of robot. Thus, by building heterogeneous teams of robotic platforms that can jointly operate in such scenarios, it is possible to bring about great benefits, since the shortcomings of each robot can be compensated with the strengths of the other [3,4].

Indeed, while aerial robots have the unique ability to obtain top views from the terrain and move without being hampered by the elements that may be found on the ground after a collapse, their reduced flight autonomy limits their operating time to a few tens of minutes. Moreover, their load capacity is generally less than 1 kg, which limits the type of sensors or equipment that can be deployed.

On the other hand, terrestrial robots are able to overcome, in general, the requirements of energy autonomy and payload. In addition, they can act as relays for communication systems, as well as provide high computing capabilities and data storage to the system. They have, however, limited mobility, especially in cluttered environments, such as narrow bridges or inclined planes. Additionally, their ability to obtain information about their environment may also be limited by their low height above the ground level and by the very elements of the scenario.

The literature contains multiple examples of successful collaboration of ground and aerial robots to carry out different missions: exploration in wide areas with obstacles [3]; precision farming for ground moisture sampling [5]; surveillance in complex environments using route optimization strategies [6]; and supporting aerial surveys in maritime environments [7], where the maritime robot acts as a mobile landing platform of the Unmanned Aerial Vehicle (UAV) when it has to perform an emergency landing, charge its batteries, or be picked up by an operator. All these examples prove the efficiency and benefit of building mixed robotic systems comprised of a terrestrial and an aerial robot for many different and complex tasks.

This work proposes a step towards obtaining such a joint team by developing a system that enables a UAV to: (1) take off autonomously from a landing platform attached to a Unmanned Ground Vehicle (UGV); (2) detect, localize, and follow the ground robot while in the air; and (3) land autonomously on the moving platform when required.

The proposed system differs from previous works by presenting a novel height-adaptive controller for tracking and landing. In essence, the behavior of a Proportional-Integral-Derivative (PID) controller is modified according to the UAV's distance to the landing platform along the vertical axis. By doing so, the performance and robustness of the system as a whole are largely increased.

Two different approaches to track and land robustly the UAV on the moving landing platform are presented: in the first variant, the aforementioned height-adaptive controller uses the current position of the landing platform as the target; the second approach extends the height-adaptive controller with a prediction algorithm (based on a Kalman filter) that predicts the future position of the platform and feeds it to the controller. This facilitates tracking and, more importantly, the autonomous landing of the aerial robot.

Another key novelty introduced in this work is the addition of a recovery and re-localization module for both tracking and landing. This further helps to increase the robustness of the system, because the UAV can re-detect the landing platform autonomously in case the latter disappears from the field of view of the UAV's camera or if the relative error between the landing platform and the UAV in the immediate moments before landing is greater than a threshold.

Furthermore, a novel finite state machine is presented in this work, which together with our re-localization module allows for life-long operation, as we will demonstrate in Section 4.

The proposed system in its two versions has been extensively tested on a realistic three-Dimensional (3D) simulated environment (Gazebo [8]) and deployed for qualitative evaluation on real robotic platforms. Figure 1 shows the proposed aerial-ground robot fleet. We employ Robotnik's Summit XL as the UGV and the Parrot's AR Drone 2.0 as the UAV. In the simulated environment, we use the corresponding Robot Operating System (ROS) [9] packages, namely the *summit_xl_sim* and *tum_simulator* packages.

Figure 1. Cooperation between Unmanned Aerial Vehicles (UAVs) and Unmanned Ground Vehicles (UGVs) can greatly benefit Search and Rescue (SaR) tasks where both long-term operation and a wide aerial view are required. The UAV can travel on top of the UGV (possibly recharging its battery) and, when needed, take off, inspect the area, and land again autonomously.

This paper is organized as follows. Section 2 analyzes previous work. In Section 3, an overview of our system is presented, and the robotic frameworks and platforms used are described. Section 4 presents an extensive evaluation of the system in the simulated environment and qualitative tests in the real robotic platforms. Section 5 concludes this work.

2. Related Work

The development of drone-related applications has exploded in recent years. In particular, there is enormous interest in using these robots for detecting and monitoring terrestrial mobile objects using their on-board cameras. However, as previously mentioned, their low autonomy renders them unable to perform tasks of long duration. That is why much of the research so far has focused on landing UAVs on mobile platforms, giving them greater versatility, since in many scenarios, it is impossible to ensure a stationary landing area. A good overview of the research done in the development of vision-based autonomous landing systems, as well as the challenges in this field, can be found in [10].

Ling et al. [11] tried to solve the problem that arises when taking pictures of icebergs using drones launched from a ship. Traditionally, the aerial vehicle had to be rescued semi-manually between two operators: one would pilot the drone until it was close enough to the boat for a second operator to manually recover it, with the danger that this action entailed. Ling proposed a precision landing algorithm to eliminate completely the human participation in this type of situation, which uses a downward-facing camera to track a target on the landing platform and generates high quality relative pose estimates.

Lee et al. [12] focused on the use of vertical cameras and Image-Based Visual Servoing (IBVS) algorithms to track a platform in a two-dimensional space and perform a Vertical Take-Off and Landing (VTOL). They obtained the speed at which the platform moved, and then they used this information as a reference to perform an adaptive control of sliding movement. Compared to other vision-based control algorithms that reconstruct a complete 3D representation of the objective (which requires accurate depth estimates), the IBVS algorithms are computationally less expensive.

Prior to these two works, Saripalli and Sukhatme [13] worked with vision algorithms for the autonomous landing of a helicopter on a mobile platform. They used Hu's moments of inertia [14] for an accurate detection of the objective and a Kalman filter for tracking. Based on the output of the tracking algorithm, it was possible for them to implement a trajectory controller that ensured the landing on the mobile target.

The literature also contains some proposals with Model Predictive Control (MPC). Maces et al. [15] considered a mission with three phases (target detection, target tracking, and autonomous landing) that were modeled in a state machine. During the last two phases, an MPC is used for position control, whereas a PID controller is employed for altitude control. The system we present extends the state machine proposed by Maces with a key additional phase, namely a *recovery mode*. This new state increases the system's robustness by allowing the UAV to re-locate the landing platform autonomously in case the latter accidentally leaves the field of view of the drone's camera. Feng et al. [16] combined a vision-based target position measurement, a Kalman filter for target localization, an MPC for the guidance of the UAV, and an integral control for robustness. They tested their algorithms on a DJI M100 quadcopter and reached a maximum error of 37 cm with a platform moving at up to $12\,\mathrm{m\,s^{-1}}$.

However, there are works that considered other techniques. Almeshal et al. [17] proposed a neural network to estimate the target position, as well as a PID controller to track it and perform landing, and validated it with a Parrot AR.Drone quadcopter. Finally, Yang et al. [18] developed a complete UAV autonomous landing system using a hybrid camera array (fish-eye and stereo cameras) and a state estimation algorithm based on motion compensation and tested them with multiple platforms (Parrot Bebop and DJI M100).

A common assumption in many of these systems is that the speed of the mobile target is low enough for the UAV to be able to land on it without compromising the integrity of both robotic platforms. However, experiments carried out by the German Aerospace Center (DLR) have demonstrated that it is possible to land a fixed wing drone on a net attached to the top of a car moving at $70\,\mathrm{km\,h^{-1}}$ [19]. Note, however, that in this experiment the ground vehicle followed a linear trajectory, which is not always possible in SaR missions, where the debris forces the UGV to make turns almost continuously.

Indeed, there are substantial differences when trying to land a UAV on a terrestrial moving platform describing either a linear or a circular trajectory. Most of the research so far focused solely on the former, without thoroughly considering that the movement of the target can also be circular or even a mixture of both, thus producing random trajectories. This is, therefore, an interesting line of work, since in SaR tasks we want to provide the terrestrial robot with complete freedom of movement. In such a scenario, the UAV has to adapt to the trajectory described by the ground robot for a successful landing. The work presented in this paper takes a step forward in this direction by demonstrating a system capable of autonomously landing a UAV for both a linear and a circular trajectory of the moving landing platform.

Finally, all of the works above described strategies towards precise landing in moving platforms, but none presented a full system capable of operating for extended periods of time. In our work, a robust state machine together with a recovery and re-localization module allows for life-long operation, as we show in Section 4.

3. Proposed Approach

This section describes the proposed approach: (1) a state machine to execute robustly the complete autonomous takeoff, tracking, and landing of a UAV on a moving landing platform (Section 3.1); (2) detection and localization of the mobile target using a downward-looking camera (Section 3.2); and (3) vision-based tracking of the mobile platform while in flight (Section 3.3).

3.1. State Machine

The autonomous takeoff-tracking-landing system proposed in this paper builds upon a finite state machine (Figure 2) with five states: *landed, taking off, tracking, landing,* and *re-localizing*.

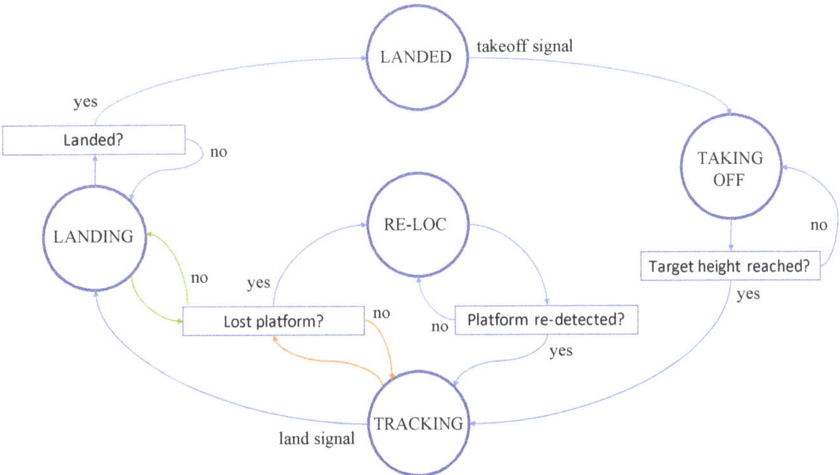

Figure 2. Finite state diagram that determines the behavior of the UAV.

Landed is the default state when launching the system and corresponds to the UAV resting on top of the landing platform, awaiting for a takeoff signal. As soon as this takeoff order is received, the drone's state changes to *taking off*, which represents the period during which the UAV is gaining altitude at a constant speed of $1\,\mathrm{m\,s^{-1}}$ along its z axis (please refer to Figure 1 for a view of the UAV's axis). The detection-localization algorithm (Section 3.2) is also launched at this point, as well as the tracking module (Section 3.3), so that the drone can start following the landing platform while ascending.

Once the nominal height has been reached (set to 4 m in our experiments), the state automatically changes to *tracking*, and the drone stops ascending. It will now follow the landing platform, keeping a constant altitude. To do so, a PID controller computes the necessary speed signals (both in the x and y axes) required to reduce the drone's distance to the landing platform's centroid in the xy-plane, which is parallel to the ground.

A land command (user-induced or automated) will trigger the start of the landing maneuver and shift the state to *landing*. The aerial robot will start its descent towards the moving landing platform at a constant downward speed of $-0.3\,\mathrm{m\,s^{-1}}$ along its z axis, while the height-adaptive PID controller provides the necessary speed commands along the UAV's x and y axis. Note that the PID gains are constantly being updated depending on the altitude (Section 3.3.1), thus height-adaptive.

Recovery module. The system can enter into recovery mode for either of the following two reasons: (1) the tracking algorithm registers when the landing platform was detected for the last time, and if more than 0.5 s pass without getting a new position, the state changes to *re-localizing*; (2) if the relative error between the landing platform's centroid and the UAV's body frame is bigger than a threshold (0.25 m in our experiments) at the final landing stages—when the sonar indicates values smaller than 0.7 m—the system will also enter recovery mode. In both cases, the drone will start gaining altitude at a speed of $1\,\mathrm{m\,s^{-1}}$, and the height-adaptive PID controller will be turned off so that no speed commands are sent along the x or y directions. Note that the second condition is designed to work as a more strict and early detection of potential landing failures.

The intuition behind ascending vertically is that the viewed area by the UAV's downward-looking camera is gradually increased. When the landing platform is viewed again, the state is changed to tracking and maintained so until the nominal tracking altitude of 4 m is reached again. At that point, new incoming landing signals may be processed again. As we will show in a set of extensive experiments (Section 4.2.3), this re-localization strategy will prove to be key when the landing platform moves faster than the nominal velocity, keeping the system alive and preventing failed landings.

3.2. Detection and Localization of the Mobile Platform

In this section, we describe the proposed method to *detect* and *localize* the landing platform relative to the UAV's coordinate frame. To this end, several standard computer vision techniques were used. It should be noted that this task is not the central topic of this work, but rather a required task to accomplish the rest of the steps that make up the autonomous landing of a UAV on a moving target. Thus, it was assumed that the landing platform had an easy to detect pattern (color-wise) or a marker on top of it. In either case, OpenCV [20] functions are readily available to accomplish shape- and color-based or more accurate marker-based detection.

The detection algorithm is summarized in Algorithm 1. First, the input frame is converted to the Hue, Saturation, Value (HSV) color model; second, a color mask is applied, where we keep all pixels within a certain HSV range, in particular that which corresponds to the red color; third, a Gaussian filter is applied to blur the image; finally, the Canny edge detection and Hough line transform algorithms are employed, followed by polygon fitting and computation of the centroid's coordinates in the image plane. An example result of this approach using the downward-looking camera of a real aerial vehicle (AR Drone 2.0) is shown in Figure 3.

Algorithm 1 Detection algorithm.

Input: *video_feed*

Output: *centroid*

```
while true do
    image ← getImage(video_frame)
    image_hsv ← rgb2hsv(image)
    masked_image ← hsvMask(image_hsv)
    blured_image ← gaussianBlur(masked_image)
    edges ← cannyEdgeDetector(blured_image)
    lines ← houghLineTransform(edges)
    polygon ← polygonFitting(lines)
    centroid ← computateFirstOrderMomentOfArea(polygon)
return centroid
```

(**a**) HSV img (**b**) masked img (**c**) blurred img (**d**) centroid

Figure 3. Detection of the landing platform and extraction of its centroid using the imagery from the downward-looking camera of an AR Drone 2.0.

To convert the previously calculated coordinates of the centroid in the image plane Ω to a three-space vector in the camera frame the inverse of the well-known pinhole camera model can be used. This process is depicted in Figure 4.

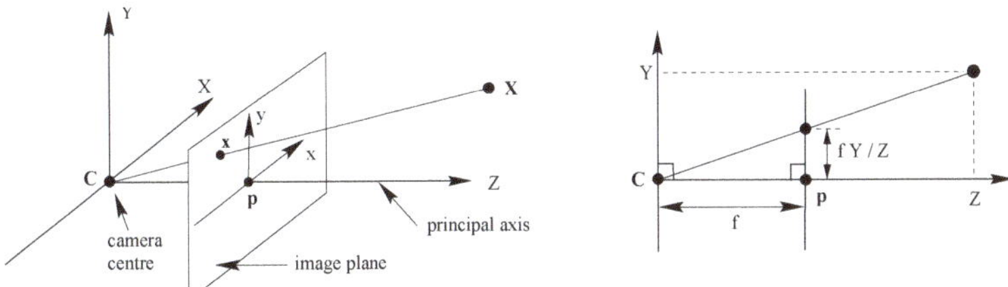

Figure 4. Pinhole camera geometry. C is the camera center and p the principal point [21].

Assuming radial distortion has been removed in a pre-processing step, Equations (1) and (2), i.e., the pinhole camera model equations, can be used to define the projection of a three-space vector in the camera frame into the image plane:

$$\lambda \begin{pmatrix} u' \\ v' \\ 1 \end{pmatrix} = \begin{pmatrix} f & 0 & c_x \\ 0 & f & c_y \\ 0 & 0 & 1 \end{pmatrix} \begin{pmatrix} X \\ Y \\ Z \end{pmatrix}, \tag{1}$$

$$\begin{aligned} u &= u'/\lambda \\ v &= v'/\lambda \end{aligned} \tag{2}$$

where $(u, v)^T$ is the projection of the point on the image plane expressed in pixels; f is the focal length in pixels; $(c_x, c_y)^T$ are the coordinates of the principle point of the camera; and (X, Y, Z) the three-space coordinates of the landing platform's centroid in the UAV's camera frame. Note that in the above equation we have assumed square and non-skewed pixels.

To obtain our desired 3D coordinates in the camera space, we need to invert the pinhole model. Additionally, if we know Z beforehand (in our case, Z corresponds to the vertical distance from the center of the UAV's camera frame to the center of the landing platform), we can easily compute X and Y:

$$\begin{pmatrix} X \\ Y \end{pmatrix} = Z \begin{pmatrix} \frac{u-c_x}{f_x} \\ \frac{v-c_y}{f_y} \end{pmatrix} \tag{3}$$

and thus obtain the 3D position of the landing platform's centroid with respect to the UAV's camera frame.

We discard measurements taken when the Inertial Measurement Unit (IMU) indicates an inclination bigger than a threshold. Therefore, in theory, relative positions of the landing platform with respect to the UAV computed by the detection-localization algorithm are always obtained without a major tilt, thus producing reliable estimates to a certain degree. Nonetheless, a minor level of noise is always present in the output of this algorithm. Leveraging a Kalman filter, as we will explain in Section 3.3.2, will prove to be an effective way to deal with such measurement noise.

3.3. Tracking the Mobile Platform

Two variants for the tracking algorithm were explored. The first one uses the currently estimated 3D position of the landing platform's centroid relative to the UAV's body frame as the input cue for a height-adaptive PID controller. The required 3D position is computed by transforming the output of the detection-localization algorithm presented in Section 3.2 to the UAV's body frame, as detailed in the following subsection. Note that by height-adaptive we refer to the fact that the PID gains are modified continuously depending on the UAV's flight altitude at every instant.

The second variant extends this height-adaptive PID with a Kalman filter to predict the future position of the landing platform. This prediction is then used as the target position for the same height-adaptive PID controller. The implementation of the Kalman filter is based on a previous work by the authors [22] where the prediction was used for tracking pedestrians.

Additionally, embedded in the tracking module, a *height control system* ensures that a proper descent speed is set in every instant depending on the current UAV's flight altitude. Furthermore, as mentioned above, this height-adaptive control system updates the PID gains depending on the current altitude: essentially, the goal is to have a faster response the closer the UAV is to the landing platform along the vertical axis (perpendicular to the ground).

After having presented the general scheme of both algorithms, i.e., with (w/) and without (w/o) the prediction of the future position of the moving platform, each variant is explained with more detail in Sections 3.3.1 and 3.3.2, respectively.

3.3.1. Height-Adaptive, Non-Predictive PID Controller

A PID controller is a control loop feedback mechanism to compute the necessary control variable $u(t)$ that makes the error $e(t)$ between the desired process value or setpoint $r(t)$ and the measured process value $y(t)$ converge to zero as fast as possible. To do so, it uses three different actions, namely a proportional, an integral, and a derivative action, each contributing differently to the control action. The mathematical expression of a PID controller is given by:

$$u(t) = K_p\, e(t) + K_i \int_0^t e(t')\, dt' + K_d\, \frac{\mathrm{d}e(t)}{\mathrm{d}t}, \tag{4}$$

where $e(t) = r(t) - y(t)$. In our system, the desired process value, $r(t)$, is the 2D zero vector, i.e., the origin of the UAV's coordinate frame relative to the coordinate frame itself. The measured process value, $y(t)$, is the two-vector containing the x and y coordinates from the 3D position of the landing platform's centroid relative to the UAV's body frame. Thus, $e(t)$ in our system is the distance in the xy-plane (parallel to the ground) between the UAV's frame and the centroid of the landing platform at time t.

Recall now that in the localization stage we compute the three-space coordinates of the landing platform in the drone's camera frame, which we denote by P_{cam}. If we are to use this position as a target value for the PID controller, the first step is to transform its coordinates to the UAV's body frame, P_{body}, since what we want is for the center of the UAV—and not for the center of its downward-looking camera—to get closer to the centroid of the landing platform. The transformation T_{body_cam} that maps a point from the camera frame to the body frame of the UAV is known from design and gives us:

$$P_{body} = T_{body\_cam}\, P_{cam}. \tag{5}$$

We can then compute the position error $e(t)$ using the expression:

$$e(t) = \sqrt{(P_{body_x})^2 + (P_{body_y})^2} \tag{6}$$

and finally, calculate the control variable $u(t)$.

Tuning. Finding the optimal PID gains (proportional gain K_p, integral gain K_i, and derivative gain K_d) was carried out through heuristic rules, i.e., looking first for the K_p that provided the desired response while keeping K_i and K_d at zero; then gradually increasing K_i to cancel the position error (and slightly decreasing the K_p so that the system did not become unstable); and, finally, increasing K_d so that the response of the system was faster while fixing the values of K_p and K_i obtained in the former steps. In practice, we found a PI controller, i.e., without the derivative action, to work better than a complete PID.

Height-adaptive PI controller. We also found a height-adaptive PI controller to perform better than a fixed-gain PI, since different responses are needed when the drone is hovering at 4 m above the landing platform than moments before landing. In particular, the UAV needs to react faster the closer it is to the landing platform along the vertical axis.

An initial tuning of the PI gains for an altitude of 4 and 2 m in the simulation is shown in Table 1. Note that no distinction has been made between the x and y axes of the UAV, since we can disregard the different dynamics of the UAV with respect to each of these axes.

Table 1. Height-dependent PI gains.

PID Gains	Altitude Range	
	4 m	2 m
K_p	0.694	0.697
K_i	0.198	0.199

We noticed that an exponential function of the form:

$$K_x = Q\, e^{-T\, p_z} \tag{7}$$

could nicely fit the values we had obtained manually, where Q and T are the parameters of the function, p_z the drone's altitude, and K_x the height-dependent PID gain we want to model. After further fine-tuning, the final expressions for both the proportional and integral gains are the following:

$$K_p = 0.7\, e^{-0.002\, p_z},$$
$$K_i = 0.2\, e^{-0.002\, p_z}.$$

Landing. The descent speed during landing remains at a constant value of 0.3 m s^{-1} when flying 0.7 m above the landing platform, i.e., when the sonar indicates more than 0.7 m. Below 0.7 m, the UAV increases its downward speed notably to 2.0 m s^{-1}. The intuition behind this design choice is that, when too close to the landing platform, the latter is not viewed completely by the UAV's downward-looking camera, and in turn the computed centroid might not represent the real center position of the platform. It is therefore a better option to rely on the correct measurements taken at altitudes higher than 0.7 m and then perform the final approach stages faster.

To determine whether the drone has successfully landed on the landing platform, we use the sonar measurements and the linear acceleration provided by the IMU. If there exists some linear acceleration, either in the x or y direction, that means the drone is moving. Moreover, if the sonar indicates a value smaller than a threshold for a certain period of time, the drone is assumed to have landed; this is not, however, a sufficient condition, since it might as well have landed on the ground. Therefore, to be certain that the UAV has actually landed on the moving platform and not on the ground, both conditions must be met, namely (1) the sonar measurement must be smaller than a threshold persistently and (2) the IMU must indicate a non-zero linear acceleration. Note that these assumptions are valid because the landing platform is assumed to be moving constantly.

In practice, the UAV mostly lands smoothly without major rebounds after the first contact with the landing platform, as we could verify in a set of preliminary experiments. After all, given that the height-adaptive PID controller ensures that the landing platform's centroid and the UAV's body frame are aligned along the vertical axis at all times, landing mostly occurs very close to the center of the platform. Therefore, in all experiments (Section 4) we can safely use the initial touch point between the landing platform and the UAV as stop signal for data recording.

3.3.2. Height-Adaptive, Predictive PID Controller

In this section, we describe how we integrated the pedestrian trajectory prediction algorithm developed in a previous work by the authors [22] into the height-adaptive PID controller described in the previous section. The resulting pipeline looks as follows:

1. The detection-localization algorithm (Section 3.2) outputs the centroid of the landing platform in the UAV's camera frame. The prediction algorithm requires as input a 3D point relative to an *inertial* reference system. Therefore, we must transform the centroid of the landing platform from the drone's camera frame into the the world's frame:

$$P_{world} = T_{world\_cam} \, P_{cam}. \tag{8}$$

2. The new position is then sent to the prediction algorithm (Kalman filter), which returns a vector of future positions of the centroid of the landing platform relative to the world frame $\hat{P}_{world}$. The first element in this vector (with index zero) corresponds to the current position of the landing platform. The next element (index one) corresponds to the next predicted position after a user-defined time step. Correspondingly, the element with index two corresponds to a prediction carried out with twice the defined time step. In general, the number of steps in this path of predicted positions is computed as the ratio between a user-provided *path time* and the *time step*.
3. Subsequently, the predicted future position of the landing platform is transformed from the world's frame into the UAV's body frame:

$$\hat{P}_{body} = T_{body\_world} \, \hat{P}_{world}. \tag{9}$$

4. Finally, the x and y coordinates of $\hat{P}_{body}$ are used to calculate the controller's error, i.e., the distance in the xy-plane between the UAV and the predicted position of the landing platform. Using this error we can now calculate the speed commands in x and y, i.e., $u(t)$ in (4), that make this error converge to zero.

Configuration of the Kalman filter. We use a value of 0.1 s as the time step for our predictions and a path time of 0.1 s, thus obtaining a vector of future predictions of size two, where the element with index one corresponds to the predicted position of the landing platform. (Recall that the element with index zero corresponds to the *current* position of the landing platform's centroid.) We studied the effect of using different path times and then accessing different future positions within this vector of predictions depending on the altitude, but in a set of preliminary experiments, we found that the performance improvement of this design choice was minor or even negative.

By using the time step and path time defined above, if the landing platform were to move, for instance, linearly at a speed of $0.5\,\mathrm{m\,s^{-1}}$, the prediction algorithm would estimate an increment of 0.05 m along the current trajectory of the moving target. We found that such a minor prediction (in this example, the predicted position differs only in 5 cm from the current position) benefits the landing accuracy notably, as presented in Section 4, when compared to the non-predictive system.

For completeness, we detail the co-variance matrices of the process noise, Q, and the observation noise, R, both diagonal matrices of the form:

$$Q = \begin{pmatrix} 2 & 0 & 0 & 0 \\ 0 & 2 & 0 & 0 \\ 0 & 0 & 2 & 0 \\ 0 & 0 & 0 & 2 \end{pmatrix}, \qquad R = \begin{pmatrix} 2 & 0 \\ 0 & 2 \end{pmatrix}, \tag{10}$$

which were obtained empirically.

As will become apparent in Section 4, employing a Kalman filter to predict the future positions of the landing platform not only provided the system with knowledge about the trajectory described by the UGV, but also helped further stabilize the measurements from the detection-localization algorithm (Section 3.2).

In this section, we described a system for the autonomous takeoff, tracking, and landing of a small UAV on a moving landing platform. We presented two variants for the tracking module: a non-predictive height-adaptive PID controller and its predictive counterpart, which leverages a Kalman filter to predict the future position of the landing platform. By doing so, not only do we filter out noise in the measured relative 3D position of the landing platform with respect to the UAV, but also allow the UAV itself to stay slightly ahead of the UGV by directly feeding this virtual future position to the height-adaptive PID. This approach will prove to be crucial to accomplish successful landings, especially when the landing platform describes non-linear trajectories.

4. Results

In this section, we present the obtained results both in the simulated and real environments. In Section 4.1, some general design considerations are given. Section 4.2 details the results obtained for an extensive set of experiments in the simulated environment, which serve as a validation of the system performance both in nominal and more demanding conditions. Finally, in Section 4.3, we present some qualitative real-world experiments to demonstrate that our system can be deployed in real robotic platforms.

4.1. Design of the Testing Environments

As previously mentioned, the scope of this work is focused on the development of a robust control algorithm for the autonomous takeoff, tracking, and landing of a UAV on a moving landing platform for life-long operation. Therefore, the experiments were designed so as to facilitate the detection task. Nevertheless, it should be noted that the control algorithm can operate together with any kind of detection method.

The initial approach was based on placing a visual marker on top of the landing platform, as shown in Figure 1. However, this approach was discarded after testing that from a height of 4 m the downward-looking camera of the real AR Drone 2.0 was unable to detect the marker robustly. Therefore, the visual marker was replaced by a red-colored, square-like landing platform like that shown in Figures 3 and 5 for the real and simulated environments, respectively. Such a landing platform can be easily detected by using standard computer vision techniques based on color and shape detection, as already described in Section 3.2. To further simplify this task, both simulated and real tests were performed on flat and feature-less terrains.

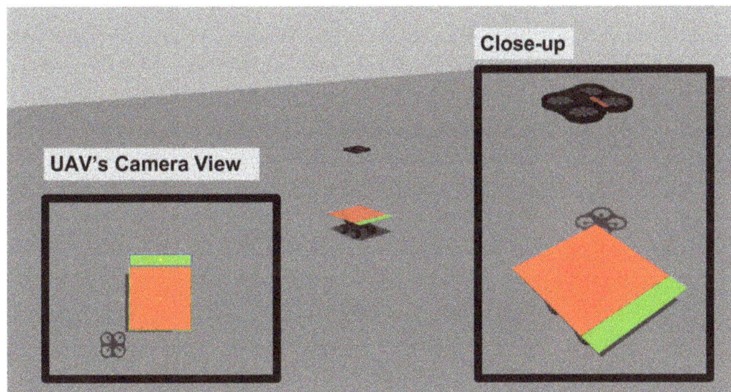

Figure 5. Simulated environment and robotic platforms.

4.2. Experiments in the Simulated Environment

The simulation experiments were conducted in the Gazebo simulator [8]. We designed two sets of experiments: one evaluated the system under the same initial conditions; a second set of tests aimed at evaluating the robustness of the system as a whole and the utility of our novel re-localization module in particular.

4.2.1. Experiments under the Same Initial Conditions

In this set of experiments, the system was re-launched every time so as to have the same initial conditions for every try. Moreover, the landing platform followed two different trajectories, namely linear and circular. A total of 20 takeoff-tracking-landing maneuvers were executed, 10 for each trajectory type, always re-initiating the whole system for every new test. Within a trajectory type, we tested the two variants of our tracking algorithm, i.e., w/ and w/o prediction. In all experiments, we computed the Euclidean distance between the landing platform's centroid and the UAV's body frame with respect to the latter. In the following, we will denote this as the *error*. Note that this error only coincides with the error fed to our height-adaptive PID controller in the mode w/o prediction. The details of the two types of trajectories together with the results obtained in each case are described next.

Linear trajectory. The UGV (and thus the landing platform) describes a linear trajectory at a speed of $0.5\,\mathrm{m\,s^{-1}}$ along its x axis. Initially, both the UGV and the UAV are at rest, with the latter lying on the landing platform. After a takeoff signal, the aerial vehicle begins its ascent phase to a pre-defined height of 4 m, and the UGV starts moving along a linear trajectory. Upon detection of the landing platform, the UAV automatically begins to follow the UGV by reducing its distance to the latter in the xy-plane, as explained in previous sections. After 30 s in tracking mode, a landing signal is sent automatically to the drone, which begins its descent towards the UGV.

Figure 6 shows the trajectory described by the UAV and the UGV for a single experiment of the linear trajectory. Results for both the non-predictive and predictive modes of the controller are presented. At first glance, no notable differences can be appreciated. The system w/ prediction is, however, more stable than its non-predictive counterpart as can be noted by visualizing the slightly more smooth curves in Figure 6b compared to those in Figure 6a. Importantly, note the scale of the y axis in Figure 6a,b and how the trajectory of the UAV is bounded within roughly 3 cm after the first 5 m in both cases (w/ and w/o prediction), following almost perfectly the linear trajectory described by the landing platform.

Figure 7 plots the error along both the x and y axis for a single experiment of the linear trajectory. As with the previous figures, results for both for the non-predictive and predictive variants are shown. Note that the error along the UAV's y axis is close to zero for both modes (w/ and w/o prediction), as was expected in the case of a linear trajectory. As for the error along the x axis, the general trend is similar in both modes, though more bounded for the algorithm w/ prediction. The big initial peak in Figure 7a is due to the fact that the UGV starts moving at the same time that the UAV takes off, thus generating a relatively large initial error along the x axis that is corrected after the first 10 s of simulation.

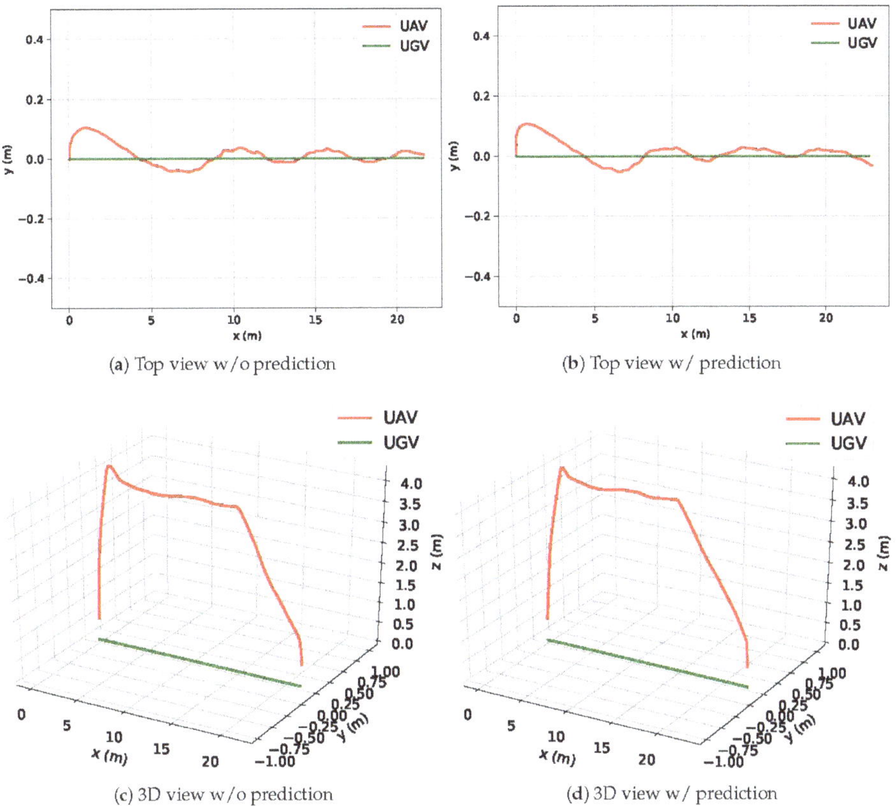

(a) Top view w/o prediction (b) Top view w/ prediction

(c) 3D view w/o prediction (d) 3D view w/ prediction

Figure 6. Views of the movements of the UGV and the UAV during a linear trajectory experiment.

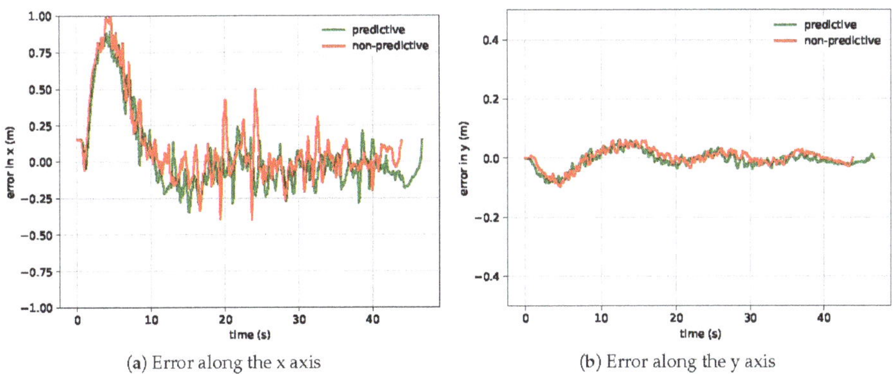

(a) Error along the x axis (b) Error along the y axis

Figure 7. Errors for a linear trajectory sequence.

Circular trajectory. In this experiment, the UGV describes circles with a forward speed along its x axis of $0.5\,\mathrm{m\,s^{-1}}$ and an angular velocity along its z axis of $0.05\,\mathrm{rad\,s^{-1}}$. In theory, this would result in a circular trajectory of radius 10 m. In practice, due to the friction coefficient of the UGV's wheels, it results in a circle of radius 12 m in this particular case. Note that the takeoff-tracking-landing procedure described for the linear trajectory is also followed here.

Figure 8 shows the trajectory described by the UAV and the UGV for a single experiment of the circular trajectory. As with previous figures, results for both the non-predictive and predictive variants are presented for comparison. Note how the UAV's trajectory resulting from the system w/ prediction matches that of the landing platform slightly better than in non-predictive mode, especially in the region between 6 and 8 m along the x axis.

Similarly, Figure 9 plots the error along both the x and y axis for a single experiment of the circular trajectory. Once again, non-predictive and predictive approaches are compared. The reader will again note that the system leveraging a predictive action outperforms its non-predictive counterpart in terms of a smaller error overall, both along the x and the y axis.

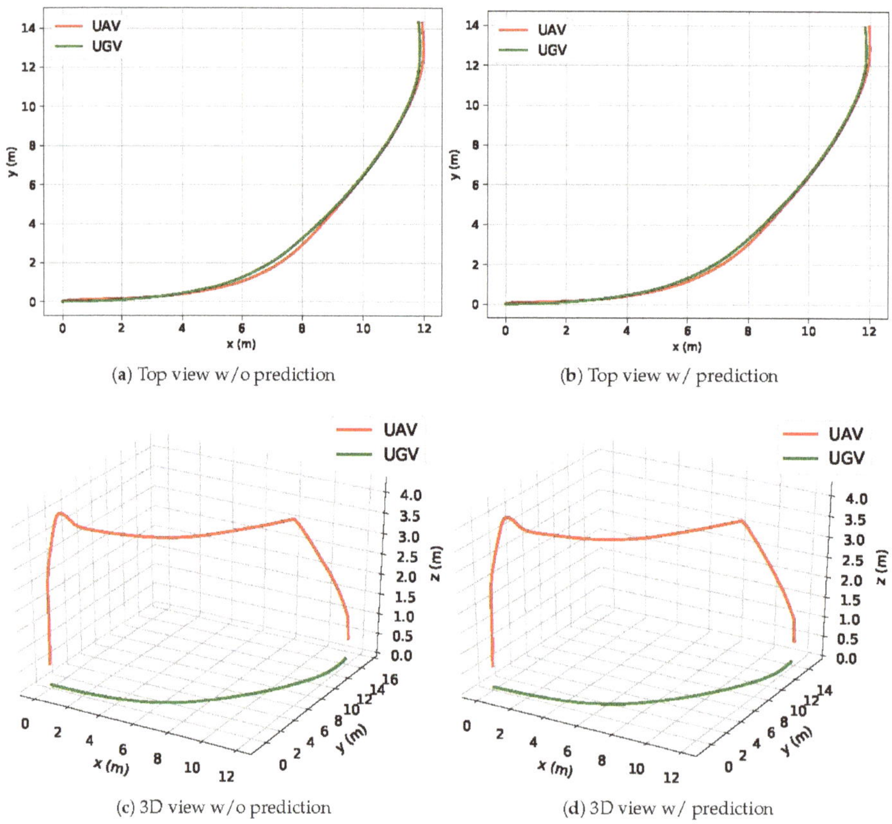

Figure 8. Views of the movements of the UGV and the UAV during a circular trajectory experiment.

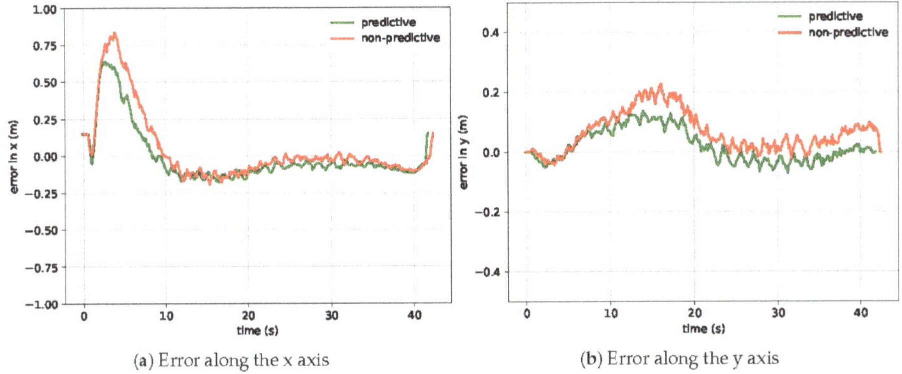

(a) Error along the x axis (b) Error along the y axis

Figure 9. Errors for a circular trajectory sequence.

4.2.2. Overall Results in the Simulated Environment for a Linear and Circular Trajectory

Figure 10 reports the errors in the x and y axis, respectively, for all 20 experiments. Each box within a boxplot represents five experiments, where for each experiment we computed the mean error value of the whole trajectory (from takeoff until landing).

Note that the mean error in all cases was always lower than 0.12 m. Note as well how for both trajectory types (linear and circular) and for both the error in the x and y axes, the variant w/ prediction achieved lower error than its non-predictive counterpart.

In particular, the error along the x axis (Figure 10a) for a linear trajectory was around 0.10 cm on average for the system w/o prediction and around 0.04 cm for the system w/ predictive action; for a circular trajectory, this error was smaller in general terms: around 0.04 cm for the system w/o prediction and close to zero when employing the predictive height-adaptive PI controller.

With regards to the error along the y axis (Figure 10b), for a linear trajectory the mean error for all ten experiments (five for each mode, i.e., w/o and w/ prediction) was very small with almost no dispersion of the data at all. For the circular trajectory, the error produced by the non-predictive tracking algorithm was in the range of 0.06 cm, while the system leveraging the Kalman filter achieved lower errors in the range of 0.03 cm.

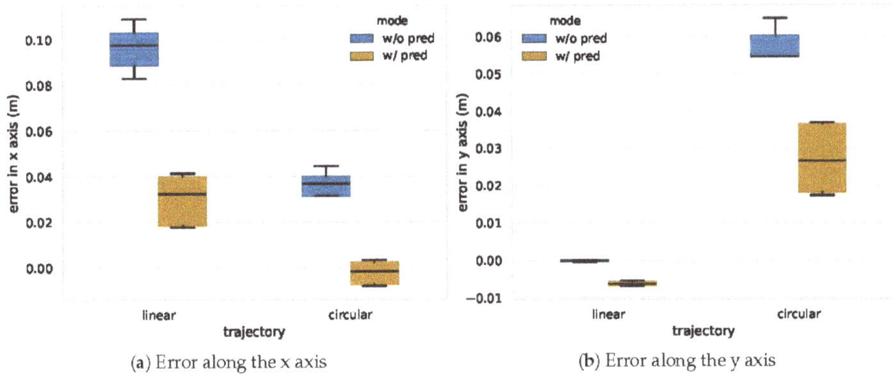

(a) Error along the x axis (b) Error along the y axis

Figure 10. Comparison of the errors between linear and circular trajectories, both for the non-predictive (w/o pred) and predictive (w/ pred) variants.

Overall, the system performed robustly in all tests, managing to land flawlessly in all cases. We have therefore demonstrated how an approach based on a predictive height-adaptive PI controller outperforms its non-predictive counterpart, allowing the UAV to follow and land on the moving landing platform consistently, performing a more stable and accurate flight.

4.2.3. Experiments to Test the Life-Long Operation Capabilities of the System

To further demonstrate the robustness of our algorithm and its life-long operation capability, we carried out the following experiment: we launched the system once and let the UAV perform up to a maximum of 50 takeoff-tracking-landing maneuvers continuously. For every iteration, 10 s pass from the moment of takeoff before an automated landing signal is sent. Once the UAV lands, it rests for 1 s on top of the landing platform before taking off again to complete a new takeoff-tracking-landing maneuver.

This test was carried out for a circular trajectory and for two different velocity conditions of the landing platform, resulting in a total of 100 attempted maneuvers: on the one hand, nominal conditions ($v_x = 0.5\,\mathrm{m\,s^{-1}}, w_z = 0.05\,\mathrm{rad\,s^{-1}}$); and, on the other hand, more demanding conditions ($v_x = 0.7\,\mathrm{m\,s^{-1}}, w_z = 0.07\,\mathrm{rad\,s^{-1}}$). Note that, in practice, the UGV model, i.e., the Summit XL, rarely reaches linear speeds higher than $v_x = 0.7\,\mathrm{m\,s^{-1}}$, commonly operating at a nominal speed of $0.5\,\mathrm{m\,s^{-1}}$ along its x axis. However, we wanted to evaluate our re-localization module thoroughly under more challenging conditions. We used our best-performing system, namely our system based on a predictive, height-adaptive PID controller. Figure 11 visualizes the trajectories described by both the UAV and the UGV for the two speed conditions mentioned above.

Note that, in practice, the trajectory followed by the UGV is never exactly a circle due to the wheels' friction coefficients. Moreover, the UGV also drifted in time, thus resulting in various circles centered in different locations, as can be seen in Figure 11.

As gathered in Table 2, for a velocity of ($v_x = 0.5\,\mathrm{m\,s^{-1}}$, $w_z = 0.05\,\mathrm{rad\,s^{-1}}$) the system managed to land the aerial vehicle successfully in all 50 consecutive maneuvers with a Mean Absolute Error (MAE) along the UAV's x axis of 0.127 m and 0.245 m along its y axis. Moreover, the maximum absolute error along the x and y axes was 0.734 m and 0.653 m, respectively. When operating at a higher velocity ($v_x = 0.7\,\mathrm{m\,s^{-1}}$, $w_z = 0.07\,\mathrm{rad\,s^{-1}}$), the re-localization module had to be launched 16 times, leaving a total of 34 successful landings. In this case, the MAE along the UAV's x axis was 0.245 m and 0.232 m along the y axis, and the maximum absolute errors were 1.526 m and 1.441 m for the x and y axes, respectively.

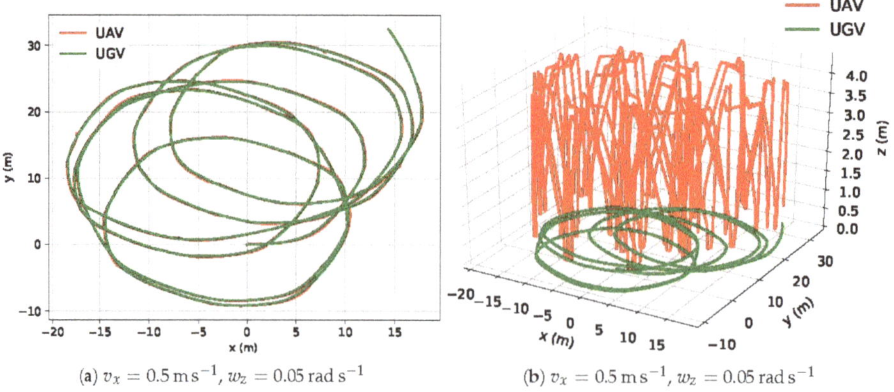

(a) $v_x = 0.5\,\mathrm{m\,s^{-1}}$, $w_z = 0.05\,\mathrm{rad\,s^{-1}}$ (b) $v_x = 0.5\,\mathrm{m\,s^{-1}}$, $w_z = 0.05\,\mathrm{rad\,s^{-1}}$

Figure 11. *Cont.*

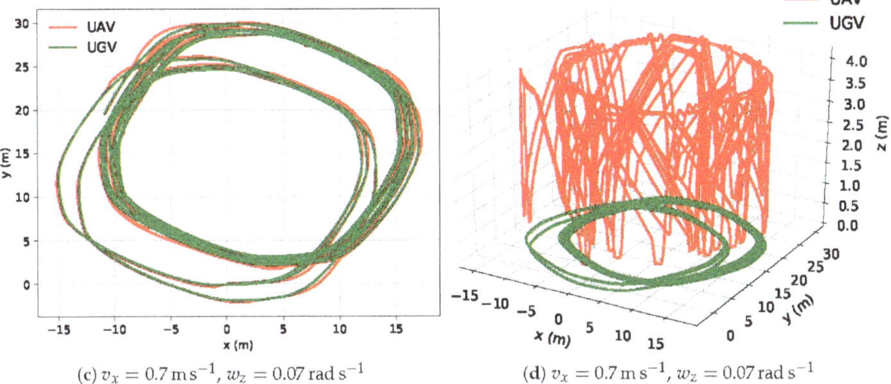

(c) $v_x = 0.7\,\mathrm{m\,s^{-1}}$, $w_z = 0.07\,\mathrm{rad\,s^{-1}}$ (d) $v_x = 0.7\,\mathrm{m\,s^{-1}}$, $w_z = 0.07\,\mathrm{rad\,s^{-1}}$

Figure 11. Trajectories described by the UAV (red) and the UGV (green) when performing a total of 50 consecutive takeoff-tracking-landing maneuvers continuously for two different linear and angular speeds of the UGV.

Table 2. Statistics for the life-long capability experiments. Linear velocities (v_x) are given in $\mathrm{m\,s^{-1}}$ and angular velocities (w_z) in $\mathrm{rad\,s^{-1}}$.

Study Variables	Landing Platform Velocities (v_x, w_z)	
	$(0.5, 0.05)$	$(0.7, 0.07)$
Total test time	1072.41 s	1249.09 s
Successful landings (num)	50/50	34/50
Successful landings (%)	100%	68%
Re-localization maneuvers	0	16
$\max(\lvert error\_x \rvert)$	0.734 m	1.526 m
MAE_x	0.127 m	0.245 m
$\max(\lvert error\_y \rvert)$	0.653 m	1.441 m
MAE_y	0.103 m	0.232 m

Note that all 16 recovery maneuvers under the more challenging conditions were triggered not because the UAV lost sight of the landing platform, but due to a too large error—greater than 0.25 m in our experiments—in the moments before landing. Such an early detection of potential failed landings allows the system to avoid accidents and flawlessly continue functioning even under challenging velocity conditions of the landing platform. An example of a relocalization maneuver can be viewed in Figure 12. Moreover, Figure 13 shows the error along the UAV's z axis during the first 200 s of the experiment for both velocity conditions studied. Overall, the rate of successful landings was 100% for the nominal velocity of the UGV and 68% in the case of more challenging (and rather unusual) conditions.

In this subsection, we demonstrated how our system performs flawlessly at the nominal speed of the UGV. More importantly, we have shown that by leveraging our novel recovery module, potential failed landing maneuvers can be detected and avoided, thus demonstrating the great benefits that such a system brings about, both in terms of robustness and reliability.

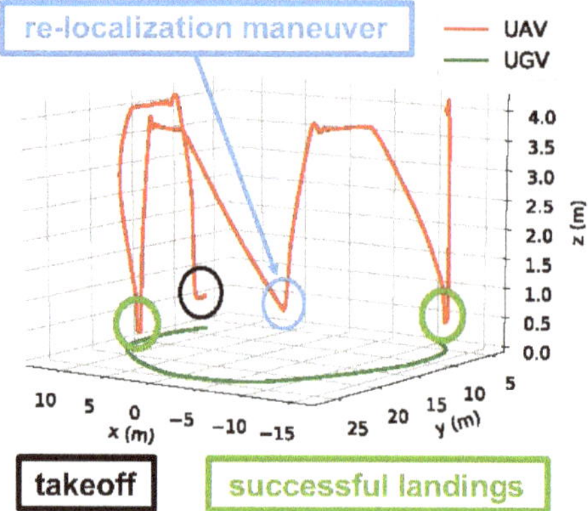

Figure 12. Detail of the 3D trajectory for a velocity of ($v_x = 0.7\,\mathrm{m\,s^{-1}}$, $w_z = 0.07\,\mathrm{rad\,s^{-1}}$) where a re-localization maneuver takes place. The run-time interval represented is [47.79 s, 99.84 s].

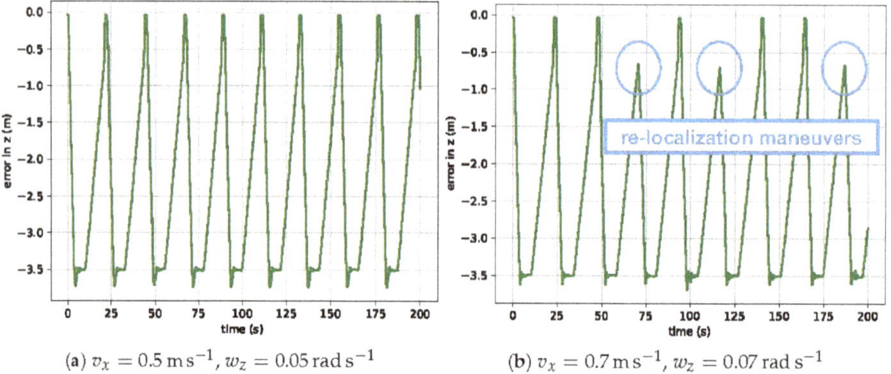

(**a**) $v_x = 0.5\,\mathrm{m\,s^{-1}}, w_z = 0.05\,\mathrm{rad\,s^{-1}}$ (**b**) $v_x = 0.7\,\mathrm{m\,s^{-1}}, w_z = 0.07\,\mathrm{rad\,s^{-1}}$

Figure 13. Error along the z axis (relative UAV-UGV distance along the z axis) during the first 200 s for both velocity conditions studied. (**b**) shows three re-localization maneuvers.

4.3. Experiments in the Real Environment

On the real robotic platforms we only tested the height-adaptive PID w/o predictive action, since for the predictive system to have worked we would have needed an additional means to localize the landing platform's position in global coordinates, as described in Section 3.3.2. In the simulated environment, transforming positions to a fixed global frame was straightforward. In the real world, however, this is more complex; implementing a Visual Inertial Odometry (VIO) or even a full visual Simultaneous Localization and Mapping (vSLAM) system would have been required in order to localize the drone in the scene with respect to a fixed frame.

What we did, however, was localize the landing platform relative to the UAV's coordinate frame, which is the only input required by the non-predictive approach. Therefore, on the real robotic platforms we qualitatively tested the system that employs our novel height-adaptive PID w/o prediction.

In particular, we performed five takeoff-tracking-landing sequences both for the linear and circular trajectory, following the same strategy as that described in Section 4.2.1. The UAV landed successfully in all five experiments for the linear trajectory and only failed once for the circular trajectory. The landing quality is visualized in Figure 14. We therefore demonstrate that the system presented in this work can be deployed on real robotic platforms. Figure 15 visualizes one of the linear trajectory experiments, and Figure 16 shows a sequence of a recovery maneuver. The complete sequences can be found in the video provided as Supplementary Material, or at https://youtu.be/CCrPBw_we2E.

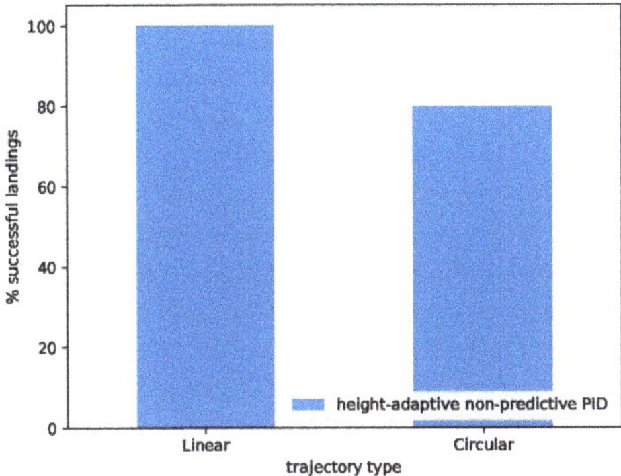

Figure 14. Percentage of successful landings in the real environment when using the height-adaptive, non-predictive PID controller for a linear and circular trajectory of the UGV. Note that these experiments were obtained by re-launching the system from scratch for every new test, as depicted in Section 4.2.1.

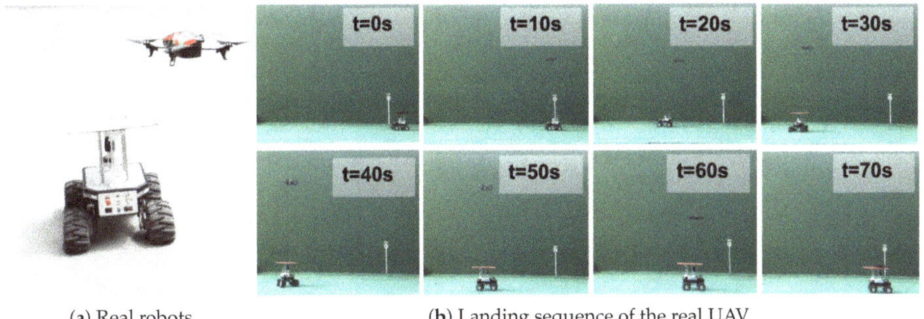

(**a**) Real robots (**b**) Landing sequence of the real UAV

Figure 15. Real robotic platforms (**a**) and landing sequence (**b**).

Figure 16. Re-localization maneuver in the real environment.

The reader must note that the real experiments were targeted as a qualitative demonstration of how our system can be integrated into real robotic platforms. We believe that the numerous quantitative experiments presented for the simulated environment (where we have used the same UAV model as in the real tests, as well as the same UGV) can serve to demonstrate the robustness and accuracy of the system, while the qualitative tests performed on the real robots can demonstrate that our system can be deployed on the real world.

5. Conclusions

In this work, we proposed a ROS-based system that enables a UAV to take off, track, and land autonomously on a moving landing platform. A novel height-adaptive PID controller suffices to operate the UAV satisfactorily when the landing platform describes either a linear or a circular trajectory at a speed of $0.5\,\mathrm{m\,s^{-1}}$ along its x axis. Introducing a Kalman filter to predict the future position of the landing platform further improves the overall performance of the system, reducing the position error in comparison to the non-predictive approach.

Furthermore, we proposed a finite state machine architecture to keep track of different stages robustly. Together with a novel recovery module, they enable our system to operate in a continuous manner, providing it with life-long operation capability.

We extensively tested the system in the simulated environment (Gazebo), executing a total of 120 takeoff-tracking-landing sequences and reporting detailed results that validate the system's performance. We also implemented our algorithms on real robotic platforms and carried out qualitative evaluations, thus demonstrating that our system can be deployed in the real world.

Regarding future work, using a UAV with a better downward-looking camera would allow leveraging a marker detection system instead of the current color- and shape-based detection algorithm. By doing so, the whole system could be deployed in any kind of environment, regardless of the terrain's texture. Furthermore, a module could be added to localize the UAV in global coordinates, e.g., VIO or visual SLAM. This would allow implementing the predictive variant of our system in real platforms, which has demonstrated to outperform its non-predictive counterpart in the simulated environment.

Supplementary Materials: The software presented in this work is publicly available at https://github.com/pablorpalafox/uav-autonomous-landing. A video demonstrating the system can be viewed at https://youtu.be/CCrPBw_we2E. Furthermore, we also provide as Supplementary Material all our log files as raw CSV files (plus several Python scripts) so that the results presented in this work can be reproduced.

Author Contributions: Conceptualization, P.R.P. and M.G.; methodology, P.R.P., M.G., and J.V.; software, P.R.P. and M.G.; validation, P.R.P. and M.G.; formal analysis, P.R.P., J.V., and J.J.R.; investigation, P.R.P.; resources, P.R.P. and A.B.; data curation, P.R.P.; writing, original draft preparation, P.R.P.; writing, review and editing, M.G., J.V., and J.J.R.; visualization, P.R.P. and J.V., ; supervision, M.G.; project administration, A.B.; funding acquisition, J.J.R. and A.B.

Funding: The research leading to these results received funding from RoboCity2030-DIH-CM, Madrid Robotics Digital Innovation Hub, S2018/NMT-4331, funded by "Programas de Actividades I+Den la Comunidad de Madrid" and cofunded by Structural Funds of the EU, and from the project DPI2014-56985-R (Robotic protection of critical infrastructures), financed by the Ministry of Economy and Competitiveness of the Government of Spain.

Conflicts of Interest: The authors declare no conflict of interest.

Abbreviations

The following abbreviations are used in this manuscript:

UAV	Unmanned Aerial Vehicle
UGV	Unmanned Ground Vehicle
PID	Proportional-Integral-Derivative
3D	Three-Dimensional
IBVS	Image-Based Visual Servoing
ROS	Robot Operating System
MPC	Model Predictive Control
VTOL	Vertical Take-Off and Landing
HSV	Hue, Saturation, Value
IMU	Inertial Measurement Unit
MAE	Mean Absolute Error
CSV	Comma-Separated Values

References

1. Liu, Y.; Nejat, G. Robotic Urban Search and Rescue: A Survey from the Control Perspective. *J. Intell. Robot. Syst.* **2013**, *72*, 147–165. [CrossRef]
2. Cardona, G.A.; Calderon, J.M. Robot Swarm Navigation and Victim Detection Using Rendezvous Consensus in Search and Rescue Operations. *Appl. Sci.* **2019**, *9*, 1702. [CrossRef]
3. Garzón, M.; Valente, J.; Zapata, D.; Barrientos, A. An aerial-ground robotic system for navigation and obstacle mapping in large outdoor areas. *Sensors* **2013**, *13*, 1247–1267. [CrossRef] [PubMed]
4. Polvara, R.; Sharma, S.; Wan, J.; Manning, A.; Sutton, R. Vision-Based Autonomous Landing of a Quadrotor on the Perturbed Deck of an Unmanned Surface Vehicle. *Drones* **2018**, *2*, 15. [CrossRef]
5. Roldán, J.; Garcia-Aunon, P.; Garzón, M.; de León, J.; del Cerro, J.; Barrientos, A. Heterogeneous multi-robot system for mapping environmental variables of greenhouses. *Sensors* **2016**, *16*, 1018. [CrossRef] [PubMed]
6. Reardon, C.; Fink, J. Air-ground robot team surveillance of complex 3D environments. In Proceedings of the 2016 IEEE International Symposium on Safety, Security, and Rescue Robotics (SSRR), Lausanne, Switzerland, 23–27 October 2016; pp. 320–327.
7. Borreguero, D.; Velasco, O.; Valente, J. Experimental Design of a Mobile Landing Platform to Assist Aerial Surveys in Fluvial Environments. *Appl. Sci.* **2018**, *9*, 38. [CrossRef]
8. Koenig, N.; Howard, A. Design and use paradigms for gazebo, an open-source multi-robot simulator. In Proceedings of the 2004 IEEE/RSJ International Conference on Intelligent Robots and Systems (IROS) (IEEE Cat. No. 04CH37566), Sendai, Japan, 28 September–2 October 2004; Volume 3, pp. 2149–2154.
9. Quigley, M.; Conley, K.; Gerkey, B.; Faust, J.; Foote, T.; Leibs, J.; Wheeler, R.; Ng, A.Y. ROS: An open-source Robot Operating System. In Proceedings of the ICRA Workshop on Open Source Software, Kobe, Japan, 17 May 2009; Volume 3, p. 5.
10. Kong, W.; Zhou, D.; Zhang, D.; Zhang, J. Vision-based autonomous landing system for unmanned aerial vehicle: A survey. In Proceedings of the 2014 International Conference on Multisensor Fusion and Information Integration For Intelligent Systems (MFI), Beijing, China, 28–29 September 2014; pp. 1–8.
11. Ling, K. Precision Landing of a Quadrotor UAV on a Moving Target Using Low-Cost Sensors. Master's Thesis, University of Waterloo, Waterloo, ON, Canada, 2014.
12. Lee, D.; Ryan, T.; Kim, H.J. Autonomous landing of a VTOL UAV on a moving platform using image-based visual servoing. In Proceedings of the 2012 IEEE International Conference on Robotics and Automation, Saint Paul, MN, USA, 14–18 May 2012; pp. 971–976.
13. Saripalli, S.; Sukhatme, G.S. Landing on a moving target using an autonomous helicopter. In *Field and Service Robotics*; Springer: Mt Fuji, Japan, 2006; pp. 277–286.
14. Hu, M.K. Visual pattern recognition by moment invariants. *IRE Trans. Inf. Theory* **1962**, *8*, 179–187.
15. Macés-Hernández, J.A.; Defaÿ, F.; Chauffaut, C. Autonomous landing of an UAV on a moving platform using Model Predictive Control. In Proceedings of the 2017 11th Asian Control Conference (ASCC), Gold Coast, QLD, Australia, 17–20 December 2017; pp. 2298–2303.
16. Feng, Y.; Zhang, C.; Baek, S.; Rawashdeh, S.; Mohammadi, A. Autonomous Landing of a UAV on a Moving Platform Using Model Predictive Control. *Drones* **2018**, *2*, 34. [CrossRef]
17. Almeshal, A.; Alenezi, M. A Vision-Based Neural Network Controller for the Autonomous Landing of a Quadrotor on Moving Targets. *Robotics* **2018**, *7*, 71. [CrossRef]
18. Yang, T.; Ren, Q.; Zhang, F.; Xie, B.; Ren, H.; Li, J.; Zhang, Y. Hybrid Camera Array-Based UAV Auto-Landing on Moving UGV in GPS-Denied Environment. *Remote Sens.* **2018**, *10*, 1829. [CrossRef]
19. German Aerospace Center. Autonomous Landing at Full Speed. Available online: https://www.dlr.de/dlr/en/desktopdefault.aspx/tabid-10080/150_read-16413/#/gallery/21679 (accessed on 29 May 2019).
20. Bradski, G. The OpenCV Library. 2000. Available online: http://www.drdobbs.com/open-source/the-opencv-library/184404319 (accessed on 29 May 2019).
21. Hartley, R.; Zisserman, A. *Multiple View Geometry in Computer Vision*; Cambridge University Press: Cambridge, UK, 2003.
22. Garzón, M.; Garzón-Ramos, D.; Barrientos, A.; Cerro, J.D. Pedestrian Trajectory Prediction in Large Infrastructures. In Proceedings of the 13th International Conference on Informatics in Control, Automation and Robotics, Lisbon, Portugal, 29–31 July 2016; pp. 381–389.

Article

On Sharing Spatial Data with Uncertainty Integration Amongst Multiple Robots Having Different Maps

Abhijeet Ravankar [1,*,†], Ankit A. Ravankar [2,†], Yohei Hoshino [1] and Yukinori Kobayashi [2]

[1] School of Regional Innovation and Social Design Engineering, Faculty of Engineering, Kitami Institute of Technology, Kitami, Hokkaido 090-8507, Japan

[2] Division of Human Mechanical Systems and Design, Faculty of Engineering, Hokkaido University, Sapporo, Hokkaido 060-8628, Japan

* Correspondence: aravankar@mail.kitami-it.ac.jp

† These authors contributed equally to this work.

Received: 30 May 2019; Accepted: 4 July 2019; Published: 8 July 2019

Abstract: Information sharing is a powerful feature of multi-robot systems. Sharing information precisely and accurately is important and has many benefits. Particularly, smart information sharing can improve robot path planning. If a robot finds a new obstacle or blocked path, it can share this information with other remote robots allowing them to plan better paths. However, there are two problems with such information sharing. First, the maps of the robots may be different in nature (e.g., 2D grid-map, 3D semantic map, feature map etc.) as the sensors used by the robots for mapping and localization may be different. Even the maps generated using the same sensor (e.g., Lidar) can vary in scale or rotation and the sensors used might have different specifications like resolution or range. In such scenarios, the 'correspondence problem' in different maps is a critical bottleneck in information sharing. Second, the transience of the obstacles has to be considered while also considering the positional uncertainty of the new obstacles while sharing information. In our previous work, we proposed a 'node-map' with a confidence decay mechanism to solve this problem. However, the previous work had many limitations due to the decoupling of new obstacle's positional uncertainty and confidence decay. Moreover, the previous work applied only to homogeneous maps. In addition, the previous model worked only with static obstacles in the environment. The current work extends our previous work in three main ways: (1) we extend the previous work by integrating positional uncertainty in the confidence decay mechanism and mathematically model the transience of newly added or removed obstacles and discuss its merits; (2) we extend the previous work by considering information sharing in heterogeneous maps build using different sensors; and (3) we consider dynamic obstacles like moving people in the environment and test the proposed method in complex scenarios. All the experiments are performed in real environments and with actual robots and results are discussed.

Keywords: information sharing; multi-robot systems; positional uncertainty; path planning; mapping

---

1. Introduction

Mobile robots are increasingly being used to automate many tasks; tasks which are mostly dull, dangerous, or demanding are a good fit for autonomous robots. The industrial sector has already benefited a lot from 'factory robots'. Recently, a new class of robots called 'service robots' have been increasing. These robots are used to provide several common services like cleaning and delivering, dispatching and moving items. These service robots are also used for specific tasks like patrolling and escorting people. Generally, multiple robots are used for such tasks in large service areas as there are several advantages. One of the major advantages of using multiple robots is wide area coverage. Multiple robots can cover a large area and perform several tasks simultaneously. Task parallelism is possible as different robots can perform different tasks at the same time. Some robots may be cleaning, some patrolling, while others may be delivering items to specific locations. Fault tolerance is another advantage of multi-robot systems. Even if one of the robots goes out of service the entire service does not stop as other robots can finish the task. Moreover, with task coordination, multiple robots can perform the task efficiently and quickly.

However, with the introduction of multiple robots in a system, there are several challenges which need to be addressed. Among these problems, effective communication between the multiple robots is a major challenge. Communication forms the basis of other major modules like task coordination, task distribution and collective execution. Accurate and content rich information is important for the successful execution of many tasks.

Although there are many benefits of sharing spatial information in a multi-robot system, in this paper, we consider the case of sharing obstacle information in a multi-robot system. The environments at many service places, like hospitals and warehouses, are very dynamic with moving entities and new obstacles. To navigate autonomously in such environments, robots need a map of the environment and need to localize themselves within it. This is generally achieved through a SLAM (Simultaneous Localization and Mapping) [1] module. Generally, if one robot finds a new obstacle in the environment, it only updates its own map. The other robots do not benefit from this knowledge. However, if the robot shares this knowledge with other robots along with updating its map, other robots can update their maps and plan better paths with the real-time information. This is shown in Figure 1, in which, Robot R1 finds a new obstacle and blocked path at the center passage of the service area and shares the spatial coordinates of the obstacle with other robots R2, $\cdots$, R5 which can use this information in generating optimal trajectories. The extension of use-case scenarios other than obstacle information sharing is straightforward.

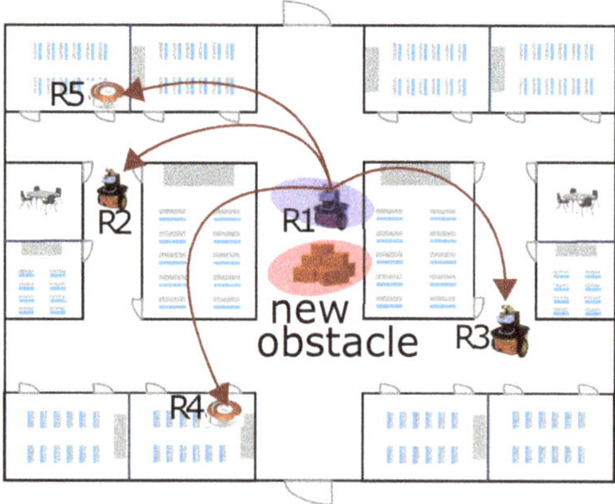

Figure 1. Robot R_1 finds a new obstacle blocking the path and shares this information with other robots ($R_1, \cdots, R_5$). The blue and red ellipses represents the robot's and obstacle's positional uncertainty, respectively.

Related Works

There is a plethora of previous works related to sharing information in multi-robot systems. Sharing corresponding matches of an object by two robots to calculate an accurate relative localization over time is proposed in Reference [2]. Work in Reference [3] proposes sharing visual information. In Reference [4], task negotiation between multiple robots by sharing information is proposed to decide the sequence in which the tasks should be performed by different robots. Work in Reference [5,6] proposes a protocol to share the region of interest between robots for efficient task cooperation. In-fact, multi-robot sport activities like Robo-soccer [7,8] heavily relies on meaningful information sharing between robots to achieve a common goal. Virtual pheromones have been proposed to be used for coordinating master-slave robots in References [9,10]. Path planning of multiple robots using information from external security cameras is proposed in Reference [11]. In addition, a direct obstacle coordinate information sharing was proposed in our previous work [12] without considering the uncertainty. However, this is a limitation as in practical systems there is always some uncertainty associated with robot's localized information and mapped obstacle's position due to sensor errors. RoboEarth [13–15] is another platform which heavily uses information exchange through cloud.

Such information sharing has huge merits in robot path planning. Path planning is an active area of research and in the context of multi-robot systems path-planning has shown promising advantages through information sharing between robots. Multi-robot collision avoidance has been discussed in Reference [16]. Work in Reference [17] presents a mechanism in which robots share information about their remaining battery power and accordingly avoid collision by giving priority to a robot with less battery power over the shortest path. An interesting approach of collaborative navigation through visual-servoing is presented in Reference [18,19] which heavily relies on reliable and efficient inter-robot communication to share information. The proposed work focuses on multiple robots sharing information about the dynamic changes in the remote area of the environment. This enables the robots to use updated and timely information to efficiently plan their paths. Information sharing among multiple robots for efficient path planning usually involves a decentralized approach [20] in which each robot calculates its path individually and decisions to change paths or avoid obstacles is done later based on the received messages from other robots. This is unlike centralized path planners [21] in which all the paths of all the robots are calculated simultaneously. In Reference [22], a motion planner is proposed for multiple robots with limited ranges of sensing and communication to reach the goal in dynamic environments. In Reference [23], a navigational technique for multiple service robots in a robotic wireless network (RWN) is presented in which robots download map information from map servers for safe navigation. Semantic information is used among multiple robots for efficient task coorindation in Reference [24].

In Reference [25], a practical case of multi-robot navigation in warehouse has been discussed. The proposed work also deals with the positional uncertainty of robots and obstacles. In this context, a decentralized approach for collaboration between multiple robots in presence of uncertainty are considered for robot action in Reference [26]. A review of multi-robot navigation strategies can be found in References [27–29].

The proposed work is an extension of our previous work [12]. Our previous work proposed the idea of a 'Node-Map' and obstacle's confidence decay mechanism. However, there were many limitations which are addressed in this extended work. The new major contributions are:

1. **Uncertainty Integration in the Improved Confidence Decay Mechanism:** The previous work [12] did not consider the amount of estimated positional uncertainty of obstacles in the confidence decay. Both were decoupled entities. However, this was a serious drawback in the previous work because irrespective of the amount of positional uncertainty, confidence of all the obstacles decayed at the same rate. This caused several false map updates corresponding to dynamic obstacles which generally have large uncertainty associated. In the extended work, we have mathematically modeled the integration of positional uncertainty in the confidence decay mechanism. This is discussed in 'Section 4.1 Integrating Uncertainty in Confidence Decay Mechanism'.
2. **New Experiments with Heterogeneous Maps with Different Sensors:** Another shortcoming of the previous work was that it only worked with the same type of 2D grid-maps made with the same type of sensors. However, in the extended work, we include new experiments with heterogeneous maps (3 dimensional RGBD map and 2D grid-map) made from different sensors. In this regard, the merits of using the 'node-map' as a means of smoothly sharing information

coherently between heterogeneous maps are also discussed. This is discussed in 'Section 6.1 Experiments with Heterogeneous Maps'.

3. **New Experiments in Dynamic Environment with Moving People and Testing Under Pressure:** The previous work only worked with static obstacles. In the extended work, new experiments have been performed to test the method when people are randomly moving in the vicinity of the robot and obstructing its navigation. In this regard, the tight coupling of new obstacle's uncertainty in the confidence decay mechanism plays a vital role to avoid false map-updates corresponding to the dynamic obstacles. This is discussed in 'Section 6.2. Results with Dynamic Entities (Moving Obstacle)'.

The comparison of the previous work with the extended work is summarized in Table 1. In addition, the proposed work discusses the algorithm to generate the T-node map.

Table 1. Comparison of this extended work with the previous work [12].

Feature	Previous Work [12]	Extended Work
Sharing New Obstacle's Position Information	Yes	Yes
Consideration of Positional Uncertainty of Obstacles	No	Yes
Confidence Decay Mechanism	Yes	Yes
Uncertainty Influence Over Confidence Decay	No	Yes
Experiments in Very Dynamic Environment (e.g., Moving People)	No	Yes
Robots have Different Types of Sensors	No	Yes
Tests with Heterogeneous Maps	No	Yes

The paper starts by first explaining the correspondence problem in different maps in Section 2. The node-map representation is explained in Section 3. Section 4 briefly explains obstacle removal and update in the nodemap and Section 4.1 explains the integration of positional uncertainty in the confidence decay mechanism. Further, using this coupling with Extended Kalman Filter is explained in Section 5 with detailed algorithm. The experimental results are discussed in Section 6. Section 6.1 explains about the experiments with heterogeneous maps and Section 6.2 discusses the results with dynamic entities (moving people). Finally, Section 7 concludes the paper.

2. Correspondence Problem in Different Maps

In dynamic environments, the new objects in the environment could be the temporary or new permanent obstacles. Both needs to be estimated in the map for correct path planning. A robot estimates the absolute position (x_{obs}, y_{obs}) of an obstacle in its map through its SLAM module. This estimation also has an uncertainty (Σ_{obs}) associated with it which arises mainly from sensor errors. This information about the new obstacle $(x_{obs}, y_{obs}, \Sigma_{obs})$ is difficult to be directly shared with other robots.

A common problem occurring in multi-robot systems is information sharing in different types of maps (e.g., 2D grid-map, 3D semantic map, feature map etc.) made from different sensors used by the robots for mapping and localization. Even the maps generated using the same sensor (e.g., Lidar) can vary in scale or rotation and the sensors used might have different specifications like resolution or range. In such scenarios, the 'correspondence problem' in different maps is a critical bottleneck in information sharing. Moreover, the uncertainty of localization also adversely affects the information sharing. In other words, it is important to consider how to easily correspond local spatial information in one map to spatial information in a separate map of different type or scale while considering the uncertainty.

This is graphically explained in Figure 2. There is a scale difference between Map1 and Map2. Whereas, Map2 and Map3 differ by a rotation factor. A spatial obstacle information, for example, position (x^1, y^1) will correspond to different spatial coordinates in Map2 and Map3. In most real world scenarios, these scale and rotation differences are generally not known. Some previously proposed techniques [30] to find the necessary translation and rotation can be applied to transform the spatial information in one map to another. However, the computation costs are expensive and could introduce undesired delays.

Although the example in Figure 2 is simplified for illustration, in actual scenarios different maps may have different levels of noise and even feature dimensions. Moreover, some robots may only have a partial map information. Similarly, Figure 3 discusses the problem of robots having different types of maps [31]. Figure 3a is a map in the form of a graph, Figure 3b is a dense 3D map, while Figure 3c is the gridmap of the same environment. It is difficult to for the robots to correlate spatial information in such different types of maps.

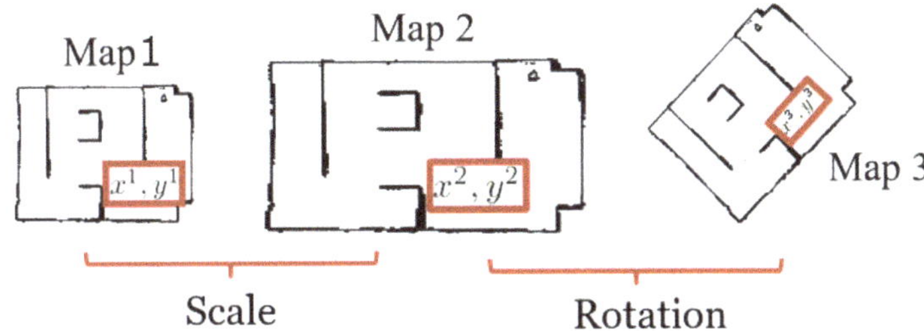

Figure 2. Correspondence problem due to the scale and rotational differences between maps.

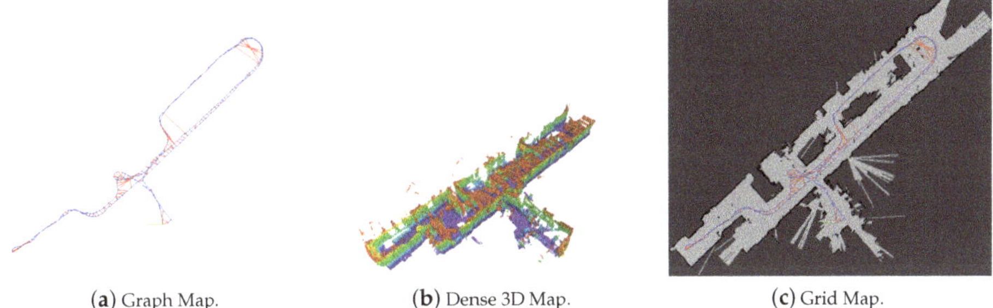

(**a**) Graph Map. (**b**) Dense 3D Map. (**c**) Grid Map.

Figure 3. Correspondence problem due to the different types of maps [31] of the same environment. (**a**) Graph map. (**b**) Dense 3D map. (**c**) Grid map.

In the proposed work, it is assumed that the robots work in the common service area whose map is available to the robots. This map itself could be heterogenous, for example, grid-map, RGBD map and so forth, which is built using different sensors mounted on different robots. Moreover, the maps could be build from different anchor points. Thus, different robots could have heterogeneous maps.

3. 'T-Node' Map Representation

A T-node representation of the map has been proposed in our previous work [12]. We briefly explain the T-node map and how obstacles are represented in it. Later, we describe how path planning is done on the node map. It is assumed that each robot is also assigned a unique robot-id (R_{id}) and the robots are on the same network to exchange messages with each other.

A node is defined as a point of turn in a path of the map. The paths are represented as a network of these nodes in the map. Figure 4a shows the node representation of the environment shown in Figure 1. Notice that, the nodes $n_1, n_2, \cdots, n_{12}$ are the points of turns in the map. The terminal nodes are shown in red color in Figure 4a. Nodes are connected to each other through edges. Figure 4b shows the T-node map with an obstacle placed between the nodes n_9 and n_3. The distance between the nodes n_9n_3 is L and the distance of the obstacle from node n_3 is x, which can easily be estimated using an on-board distance sensor.

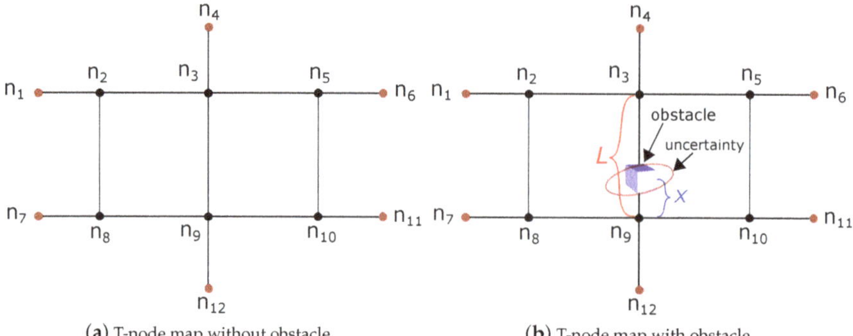

(a) T-node map without obstacle.

(b) T-node map with obstacle.

Figure 4. T-node representation of the environment shown in Figure 1. (a) T-node map without obstacle. (b) T-node map with obstacle between nodes n_9n_3.

A table stores T-node map's information viz. traversable/blocked edges (paths) and any changes at the edges. All the robots have access to this table. In the context of Figure 4b, the corresponding information is shown in Table 2. The table contains a set of four information about each path: (1) binary information of whether a new obstacle is found on an edge, (2) a binary information if the path is blocked and cannot be traversed, (3) details of the obstacle if the path is changed and (4) the timestamp (T_s) when the information was updated. The details of the information will vary according to the type of the sensor used. For example, in case of Lidar, the obstacle information will contain: the obstacle coordinates from the node (d_x, d_y), dimensions of the obstacle like width (w_{obs}) and height (h_{obs}) and the positional uncertainty associated in estimating the obstacle (σ_x, σ_y). The uncertainty information comes from the SLAM module used in the robot. As shown in T-node map of Figure 4b, only one of the edges n_3n_9 is obstructed. This information is reflected in Table 2. It is possible that a new obstacle is found on a path, however the path could be still be traversed. A blocked path cannot be traversed by the robot.

Table 2. T-node map information corresponding to Figure 4b.

Node Path	New Obstacle	Path Blocked	Meta-Data
n_1n_2	0	0	-
...	...	...	...
n_3n_9	1	1	{ d:(d_x, d_y),w:w_{obs},h:h_{obs},Σ:(σ_x, σ_y), T_s }
n_9n_{12}	0	0	-

Each robot has a copy of this table which has small memory requirement as the meta-data for only the changed paths are required. Moreover, information is communicated to other robots only when some path information is changed. A T-node representation makes it easier for a robot to share information with other robots. The local maps maintained by the two robots might differ by some rotation, translation or scale. As an example, Figure 5a shows the section of the map of Figure 4b with obstacles. Figure 5b shows a scaled version, Figure 5c a rotated version and Figure 5d a scaled and rotated version of Figure 5a. However, the nodes on the paths remains the same and information that there is an obstacle on one of the edges is still conveyed clearly from Table 2 which maintains the details of the obstacles. In addition, with a T-node representation a global map is not required.

Even for a large number of nodes, only those edges which are changed is communicated to the robots. The small data size ensures fast and reliable communication with small communication bandwidth.

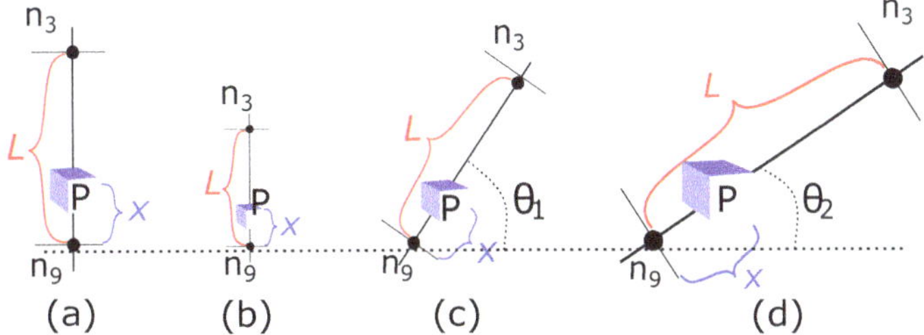

Figure 5. Scale and rotation effects on the T-node map. In all cases, meaningful information can still be shared between robots. (**a**) Original section n_9n_3 of Figure 4b. (**b**) Figure 5a scaled down. (**c**) Figure 5a rotated by angle θ_1. (**d**) Figure 5a scaled up and rotated by angle θ_2.

The T-node map can be generated by maneuvering the robot in the environment and setting points of turns (where the robot turns by around 90 degrees) as nodes. Automatic generation of T-node map is also possible if a map is available. For example, if there is a grid-map with obstacles (black), open (white) and unknown (grey) areas, the first step is to generate a binary image of the grid-map which is done by turning all unknown cells to blocked (black) value. This is shown in Figure 6a. Noise is removed by successively applying morphological erode and dilate operations [32,33]. The next step is to apply skeletonization algorithm [34,35]. Many skeletonization and thinning algorithms generate unnecessary tentacles which needs to be removed using pruning algorithm [36]. Result of showing skeletonization on binary map of Figure 6a is shown in Figure 6b. Line segments are then detected using techniques like SVD and Hough Transform [1]. The end-points of segments which are within a small threshold distance (δ) can be clustered [37] using k-means [38,39], fuzzy c-means [40] or density based clustering methods [41] into a single node as shown in Figure 6c,d. A graph 'N' of these nodes $\{n_1, n_2, \cdots, n_m\}$ form the T-node map of the environment. The pseudo-code is given in Algorithm 1.

Algorithm 1: T-node-map Generation

Data: m : Gridmap, m_height : map height, m_width : map width
Function node_mapping(m)
    for *each* row in m_height **do**
        for *each* col in m_width **do**
            if cell m[row][col] is unknown **then**
                $m[row][col] \leftarrow l_{occupied}$
Successively erode and dilate binary image multiple times [32,33]
Apply skeletonization algorithm [34,35]
Apply prunning algorithm [36]
Detect lines segments and their endpoints using algorithm [1]
Cluster nearby endpoints in range δ with k means algorithm. [37]
Mark clustered points as nodes N $\leftarrow$ $\{n_1, n_2, \cdots, n_m\}$
return(N)

It should be noted that a 'node' is merely a point of turn in the navigational graph. It does not include any feature information (e.g., corners, line, color, etc.) of the map. Hence, map-merger on T-Node map is not possible. However, traditional methods [42,43] can be used to first merge feature-rich maps and thereby T-node map. Moreover, since the characteristics of the navigational paths are different for unmanned ground vehicles (UGVs) and unmanned aerial vehicles (UAVs), this work does not consider the case of heterogeneous robots. Only ground robots are considered and the proposed method will work well for both differential drive robots and skid-steer drive robots.

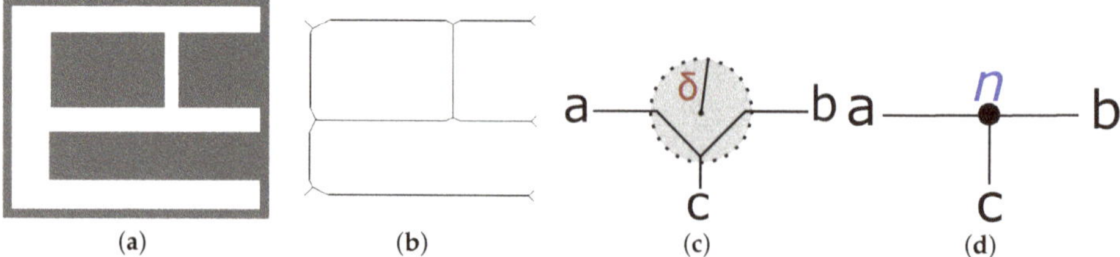

(a) (b) (c) (d)

Figure 6. T-node map generation. (**a**) Binary grid-map. (**b**) Skeleton map. (**c**) Clustering within δ distance. (**d**) Clustered node n.

4. Obstacle Removal and Update in T-Node Map

This section briefly discusses about modeling the transience of obstacles first proposed in our previous work [12]. The new contribution of this extended work lies in integrating the positional uncertainty in the decay mechanism which is discussed in the next Section 4.1. The obstacles in the passages may be permanent or temporary and removed after some time. Regardless of the transience of the obstacles, all the robots update their respective T-node map. Newly added obstacles can only be re-confirmed if a robot actually visits the area near the obstacle. Hence, once an obstacle information on a particular node has been updated and communicated to other robots, robots update their map and if that obstacle is found again, the timestamp is updated. However, there is no upper time bound of when a robot would actually visit the particular location and update its map. The obstacle might already have been removed by that time. This problem needs to be modeled mathematically.

A timestamp (T_s) is maintained for each obstacle representing the time at which the obstacle was last seen. If an obstacle has recently been added to the map and a short time has elapsed since its addition, then the probability that it has not been removed is high. On the contrary, if a lot of time has elapsed since the addition of the obstacle, the probability that it still exists in the map is less. We model a confidence (c) measure which represents this probability $0 \leq c_{th} \leq 1$. The maximum value of confidence is 1 and its value decreases with time. The robots assumes that the obstacle still exists in the map until the corresponding confidence has not dropped to below a threshold confidence (C_{th}). Depending on the nature of the environment, a threshold time (t_{th}) is chosen in which the confidence decays to c_{th} value and the time in which the confidence decays to zero is (t_z). To model the confidence decay, the following family of curves are chosen.

$$c = 1 - \left(\frac{t_{th}}{t_z}\right)^n \tag{1}$$

The curves given by Equation (1) have the desired characteristic that for higher values of n, the curve flattens out more and delays confidence decay until the threshold time (t_{th}) and after that it decays quickly to zero in t_z time. For a given c_{th}, t_{th} and t_z, the value of the degree of the curve (n) can be found by solving Equation (1) as,

$$\begin{aligned} \left(\frac{t_{th}}{t_z}\right)^n &= 1 - c_{th}, \quad (0 \leq c_{th} \leq 1), \\ n \log\left(\frac{t_{th}}{t_z}\right) &= \log(1 - c_{th}), \\ \implies n &= \frac{\log(1 - c_{th})}{\log\left(\frac{t_{th}}{t_z}\right)}. \end{aligned} \tag{2}$$

Figure 7 shows the curves for the decay function given by Equation (1) for various values of n. The Ufactor in Figure 7 shows the uncertainty factor which is discussed in Section 4.1. The various curves have been generated for $c_{th} = 0.55$ and $t_z = 600$ s, for varying values of t_{th} between 300 s to

600 s. It can be seen that the corresponding values of n can be found for different t_{th} according to Equation (2). Moreover, as the value of n increases, the decay curves flatten out more taking more time to reach the threshold time and then quickly decrease to zero.

For a given instantaneous value of confidence c, the elapsed time t is calculated from Equation (1) as,

$$t = e^{\frac{1}{n}\log(1-c)+\log(t_z)}. \tag{3}$$

The time remaining (t_{rem}) to reach the threshold time (t_{th}) is,

$$t_{rem} = t_{th} - e^{\frac{1}{n}\log(1-c)+\log(t_z)}. \tag{4}$$

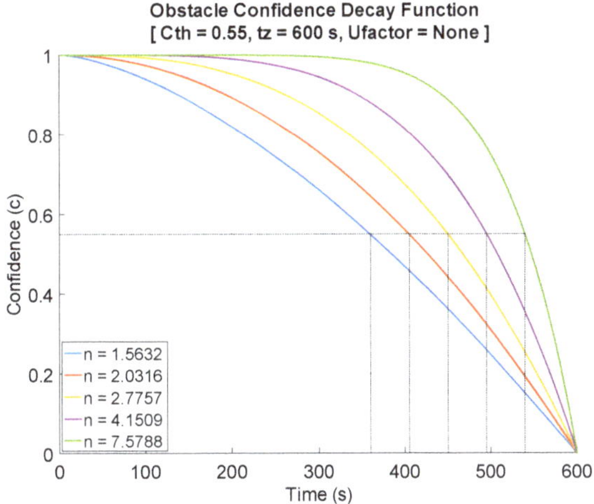

Figure 7. Obstacle confidence decay function. $c_{th} = 0.55$, $t_z = 600$ s. Effects of uncertainty are not considered and Ufactor = 0.

4.1. Integrating Uncertainty in Confidence Decay Mechanism

Uncertainty in obstacle's position affects the rate of confidence decay. If there is a large uncertainty in obstacle's spatial position, the threshold time t_{th} is reduced by an uncertainty factor. In case of no uncertainty, Equations (1)–(4) are used. In SLAM, the obstacle's uncertainty in state is generally represented by the covariance matrix (Σ_t),

$$\Sigma_t = \begin{bmatrix} \sigma_x^2 & \sigma_{xy} & \sigma_{x\theta} \\ \sigma_{xy} & \sigma_y^2 & \sigma_{y\theta} \\ \sigma_{x\theta} & \sigma_{y\theta} & \sigma_\theta^2 \end{bmatrix} \tag{5}$$

If the uncertainty given by Σ_t is large, the confidence falls down faster and vice-versa. Hence, the confidence decay is modeled as,

$$\text{Confidence decay} \propto \frac{1}{\text{Uncertainty given by } \Sigma_t}. \tag{6}$$

The eigenvalues ($\lambda_1, \cdots \lambda_n$) and eigenvectors ($\vec{v_1}, \cdots \vec{v_n}$) of the matrix Σ_t denotes the magnitude of the variance. Two largest eigenvalues λ_1 and λ_2 control the decay of confidence. The threshold time after uncertainty integration $t_{th}^{'}$ is given as,

$$t_{th}^{'} = t_{th} - \frac{\Psi}{\sqrt{\lambda_1^2 + \lambda_2^2}}, \tag{7}$$

where, Ψ is a controlling factor. The new confidence $c^{'}$ is given as,

$$\begin{aligned} c^{'} &= 1 - \left(\frac{t^{'}_{th}}{t_z}\right)^n \\ &= 1 - \left(\frac{t_{th} - \frac{\Psi}{\sqrt{\lambda_1^2 + \lambda_2^2}}}{t_z}\right)^n \\ &= t_z^{-n}\left(\lambda_1^2 + \lambda_2^2\right)^{\frac{-n}{2}} \left[t_z^n \left(\lambda_1^2 + \lambda_2^2\right)^{\frac{n}{2}} - \left\{t_{th}\left(\lambda_1^2 + \lambda_2^2\right)^{\frac{1}{2}} - \Psi\right\}^n\right]. \end{aligned} \tag{8}$$

The degree of the curve is given as,

$$\begin{aligned} n^{'} &= \frac{\log(1 - c^{'}_{th})}{\log\left(\frac{t^{'}_{th}}{t_z}\right)}, \\ \implies n^{'} &= \frac{\log\left(1 - t_z^{-n}\left(\lambda_1^2 + \lambda_2^2\right)^{-\frac{n}{2}} \left[t_z^n \left(\lambda_1^2 + \lambda_2^2\right)^{\frac{n}{2}} - \left\{t_{th}\left(\lambda_1^2 + \lambda_2^2\right)^{\frac{1}{2}} - \Psi\right\}^n\right]\right)}{\log\left(t_{th}\left(\lambda_1^2 + \lambda_2^2\right)^{\frac{1}{2}} - \Psi\right) - \log\left(t_z\left(\lambda_1^2 + \lambda_2^2\right)^{\frac{1}{2}}\right)}. \end{aligned} \tag{9}$$

Figure 8 shows the results of integrating the spatial uncertainty in the obstacle confidence decay time for various values $c_{th} = 0.55$, $t_z = 600$ s and different values of the uncertainty factory (Ufactor). In Figure 8, Ufactor represents,

$$\text{Ufactor} = \frac{\Psi}{\sqrt{\lambda_1^2 + \lambda_2^2}}, \tag{10}$$

where, values λ_1 and λ_2 capture the amount of estimated uncertainty. In Figure 8, Ufactor is given as a factor of threshold time. It can be seen that for more uncertainty, the curve starts to fall faster to the threshold time. Appropriate values of Ufactor can be chosen depending on different scenarios. Moreover, this value can also be changed dynamically.

The obstacle confidence decay mechanism ensures a smooth robot operation in multi-robot system where multiple robots frequently inform each other about the new obstacle information. If a robot receives an obstacle information update from another robot while it is navigating towards its goal location, then it would have to stop and update its map information which consumes time and computation. To avoid this, a check is performed to see if the information received affects the current navigation towards the goal. This is easily achieved by checking the blocked flag of the corresponding edge. If the blocked flag is set to 1 and the current navigational path is affected, the timestamp and other meta-data for the blocked edge are checked. Based on the obstacle's confidence value, path re-planning or continuation on the same path can be decided according to the priority of the task at hand.

A major benefit of tightly coupling the obstacle's uncertainty with confidence decay mechanism is minimizing the false map updates corresponding to the dynamic obstacles in vicinity. Generally, the uncertainty of dynamic obstacles is larger than that of static obstacles estimated by the underlying SLAM module. In the absence of uncertainty integration, confidence all the obstacles irrespective of their positional uncertainty decays at the same rate. Therefore, if there is a false map update corresponding to a dynamic obstacle (like moving people), it decays at the same rate like other fixed obstacles. This increases the chances of false map updates due to dynamic obstacles. However, with the uncertainty integration, the confidence of obstacles with larger positional uncertainty decay faster than those with less uncertainty. In effect, this allows minimizing false map updates, as they decay out quickly. This also prevents false notifications to other robots.

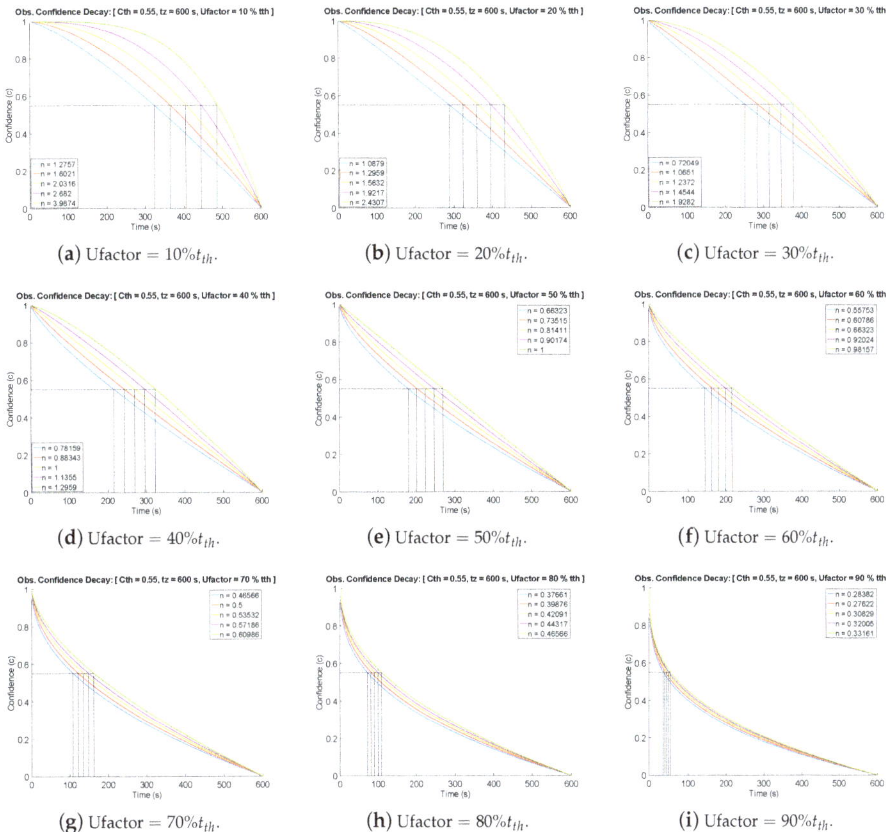

(a) Ufactor $= 10\%t_{th}$. (b) Ufactor $= 20\%t_{th}$. (c) Ufactor $= 30\%t_{th}$.

(d) Ufactor $= 40\%t_{th}$. (e) Ufactor $= 50\%t_{th}$. (f) Ufactor $= 60\%t_{th}$.

(g) Ufactor $= 70\%t_{th}$. (h) Ufactor $= 80\%t_{th}$. (i) Ufactor $= 90\%t_{th}$.

Figure 8. Obstacle confidence decay function with uncertainty integration. With more positional uncertainty, the confidence falls below the threshold confidence fast. In all the cases, $c_{th} = 0.55$ and $t_z = 600$ s.

5. Uncertainty Integrated Confidence Decay Mechanism with Extended Kalman Filter

The integration of confidence decay meachanism in EKF is given in Algorithm 2. The algorithm is straightforward and estimates the Kalman gain (K_t), robot's pose (μ_t) and the covariance (Σ_t) at time t until step 12. Later, Eigen values ($\lambda_1 \cdots \lambda_n$) are extracted from the covariance matrix (Σ_t) by applying Singular Value Decomposition. The degree of the confidence curve is then determined using the amount of uncertainty represented by the Eigen values in steps 14 and 15 of Algorithm 2 as explained in the previous section. Essentially, the degree of the curve is chosen to fasten the confidence decay inversely proportional to the positional uncertainty.

6. Experimental Results

This section presents the results of the experiments. The extended work discuss information sharing in heterogeneous maps made with different sensors and tests the proposed method under pressure with dynamic obstacles in the vicinity of robots.

We used Pioneer-P3DX [44] and Kobuki Turtlebot [45] robot shown in Figure 9a. Both the robots are wheeled differential drive robots and the motion model is explained in our previous work [12]. Both the robots used ROS [46] on Ubuntu computer and were on the same network to communicate with each other.

Algorithm 2: Uncertainty Integrated Confidence Decay with Extended Kalman Filter

1 # x_t: robot state, v_t, ω_t: translation and rotational velocity.

$$x_t = [x \;\; y \;\; \theta]^T$$

2 # EKF uses Jacobian to handle non-linearity. G_t: Jacobian of motion function w.r.t state

$$G_t \leftarrow \begin{bmatrix} 1 & 0 & -\frac{v_t}{\omega_t}cos\theta + \frac{v_t}{\omega_t}cos(\theta + \omega_t \Delta t) \\ 0 & 1 & -\frac{v_t}{\omega_t}sin\theta + \frac{v_t}{\omega_t}sin(\theta + \omega_t \Delta t) \\ 0 & 0 & 1 \end{bmatrix}$$

3 # V_t: Jacobian of motion w.r.t control

$$V_t = \begin{bmatrix} \frac{-sin\theta + sin(\theta + \omega_t \Delta t)}{\omega_t} & \frac{v_t(sin\theta - sin(\theta+\omega_t\Delta t))}{\omega_t^2} + \frac{v_t(cos(\theta+\omega_t\Delta t)\Delta t)}{\omega_t} \\ \frac{cos\theta - cos(\theta+\omega_t\Delta t)}{\omega_t} & -\frac{v_t(cos\theta - cos(\theta+\omega_t\Delta t))}{\omega_t^2} + \frac{v_t(sin(\theta+\omega_t\Delta t)\Delta t)}{\omega_t} \\ 0 & \Delta t \end{bmatrix}$$

4 # M_t: Covariance of noise in control space. $\alpha_1, \cdots, \alpha_4$: Error-specific parameters.

$$M_t = \begin{bmatrix} \alpha_1 v_t^2 + \alpha_2 \omega_t^2 & 0 \\ 0 & \alpha_3 v_t^2 + \alpha_4 \omega_t^2 \end{bmatrix}$$

5 # $\bar{\mu}_t$: Prediction updates in state.

$$\bar{\mu}_t = \mu_{t-1} + \begin{bmatrix} \frac{-v_t}{\omega_t}sin\theta + \frac{v_t}{\omega_t}sin(\theta + \omega_t\Delta t) \\ \frac{v_t}{\omega_t}cos\theta - \frac{v_t}{\omega_t}cos(\theta + \omega_t \Delta t) \\ \omega_t \Delta t \end{bmatrix}$$

6 # $\bar{\Sigma}_t$: Prediction updates in covariance.

$$\bar{\Sigma}_t = G_t \Sigma_{t-1} G_t + V_t M_t V_t^T$$

7 # $\bar{Q}_t$: Covariance of the sensor noise.

$$Q_t = \begin{bmatrix} \sigma_r^2 & 0 & 0 \\ 0 & \sigma_\phi^2 & 0 \\ 0 & 0 & \sigma_s^2 \end{bmatrix}$$

8 # $[m_{ix} \;\; m_{iy}]^T$: coordinates of the ith landmark. z_t^i: measurement. q: squared distance.

$$q = (m_{k,x} - \bar{\mu}_{t,x})^2 + (m_{k,y} - \bar{\mu}_{t,y})^2$$

$$\hat{z_t}^k = \begin{bmatrix} \sqrt{q} \\ atan2(m_{k,y} - \bar{\mu}_{t,y}, m_{k,x} - \bar{\mu}_{t,x}) - \bar{\mu}_{t,\theta} \\ m_{k,s} \end{bmatrix}$$

9 # H_t: Jacobian of measurement with respect to state.

$$H_t^k = \begin{bmatrix} -\frac{m_{k,x} - \bar{\mu}_{t,x}}{\sqrt{q}} & -\frac{m_{k,y} - \bar{\mu}_{t,y}}{\sqrt{q}} & 0 \\ \frac{m_{k,y} - \bar{\mu}_{t,y}}{q} & -\frac{m_{k,x} - \bar{\mu}_{t,x}}{q} & -1 \\ 0 & 0 & 0 \end{bmatrix}$$

10 # S_t: Measurement covariance matrix.

$$S_t^k = H_t^k \bar{\Sigma}_t [H_t^k]^T + Q_t.$$

11 # $j(i)$: likely correspondence after applying maximum likelihood estimate.

$$j(i) = \text{argmax} \; \frac{1}{\sqrt{\det(2\pi S_t^k)}} e^{-\frac{1}{2}(z_t^i - \hat{z}_t^k)^T [S_t^k]^{-1} (z_t^i - \hat{z}_t^k)}$$

12 # K_t: Kalman gain, μ_t: state, Σ_t: covariance.

$$K_t^i = \bar{\Sigma}_t [H_t^{j(i)}]^T [S_t^{j(i)}]^{-1}$$

$$\mu_t = \bar{\mu}_t + K_t^i (z_t^i - \hat{z}_t^{j(i)})$$

$$\Sigma_t = (I - K_t^i H_t^{j(i)}) \bar{\Sigma}_t$$

13 # Apply Singular Value Decomposition and get Eigen-values λ_i:

$$\lambda_1, \cdots \lambda_n = \text{svd}(\Sigma_t) = \text{svd}\left(\begin{bmatrix} \sigma_x^2 & \sigma_{xy} & \sigma_{x\theta} \\ \sigma_{xy} & \sigma_y^2 & \sigma_{y\theta} \\ \sigma_{x\theta} & \sigma_{y\theta} & \sigma_\theta^2 \end{bmatrix} \right)$$

14 # n: degree of decay curve, t_{th}: threshold time, c_{th}: threshold confidence, t_z: time to decay to zero.

$$n = \frac{\log(1 - c_{th})}{\log\left(\frac{t_{th}}{t_z}\right)}.$$

15 # n': degree of decay curve with uncertainty integrated, Ψ: decay control factor.

$$n' = \frac{\log\left(1 - t_z^{-n} \left(\lambda_1^2 + \lambda_2^2\right)^{-\frac{n}{2}} \left[t_z^n \left(\lambda_1^2 + \lambda_2^2\right)^{\frac{n}{2}} - \left\{ t_{th} \left(\lambda_1^2 + \lambda_2^2\right)^{\frac{1}{2}} - \Psi \right\}^n \right] \right)}{\log\left(t_{th} \left(\lambda_1^2 + \lambda_2^2\right)^{\frac{1}{2}} - \Psi\right) - \log\left(t_z \left(\lambda_1^2 + \lambda_2^2\right)^{\frac{1}{2}}\right)}$$

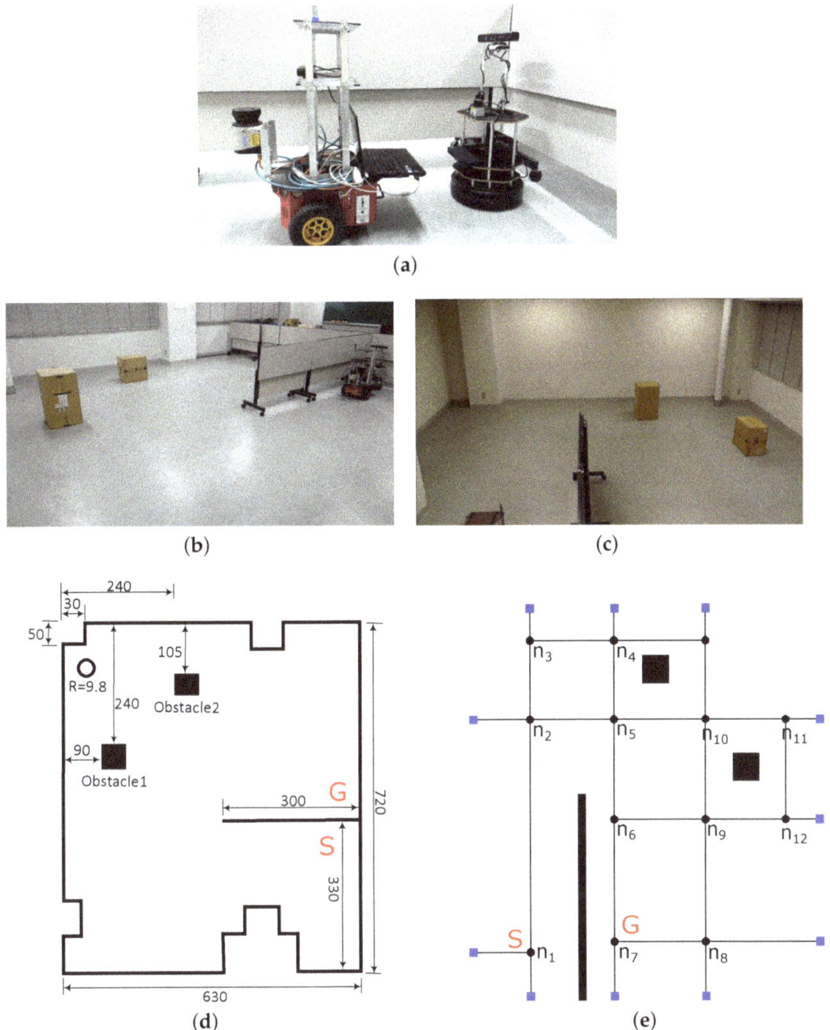

(a)

(b) (c)

(d) (e)

Figure 9. Experiment setup. (**a**) Differential drive robots Kobuki-Turtlebot2 and Pioneer-P3Dx. (**b**) Environment with initial position of robots. (**c**) Another view of the environment. (**d**) Environment dimensions. (**e**) Node-map of the environment where S and G are the start and goal points.

6.1. Experiments with Heterogeneous Maps

In this section, we describe the results of the proposed method with heterogeneous maps. The environment for experiments is shown in Figure 9b,c. The dimensions of the environment are shown in Figure 9d. The environment had two static obstacles 'Obstacle1' and 'Obstacle2' marked in Figure 9d. The start and goal positions are marked as 'S' and 'G', respectively, in Figure 9d. The T-node map is shown in Figure 9e.

The two robots used in the experiment were both equipped with 2D Lidar and RGBD sensors. As shown in Figure 9a, the Pioneer P3DX robot was equipped with a Sick-Lidar of 10 m range and ASUS Xtion-Pro RGBD camera. Turtlebot was equipped with a Hokuyo Lidar of 20 m range and a Kinect RGBD camera.

To test the proposed method with heterogeneous maps, Pioneer P3DX robot was programmed to use only the RGBD sensor to build a 3D map, and navigate in the environment. Pioneer P3DX first started navigation from location 'S' to the goal location 'G' as shown in Figure 10a. A* algorithm [47] and SHP algorithm [48,49] were used for path planning and path smoothing, respectively. As soon as the P3DX robot started moving, a long new obstacle was placed in the environment as shown in Figure 10b, well outside the range of the sensor. As shown in Figure 10c, the person moves in front of the robot and blocks its way purposefully. The details of dynamic obstacle are discussed in the next Section 6.2. P3DX perceives the new obstacle and alters its path towards the goal as shown in Figure 10d–f, while also updating the map with the newly added obstacle. P3DX was programmed to come back to its initial position 'S' and the navigation in shown in Figure 10g,h.

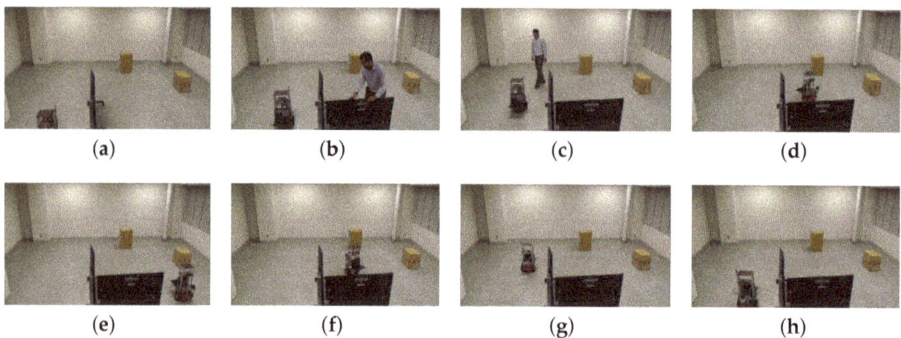

Figure 10. Timely snapshots of the experiment. (**a**) P3DX starts moving with old map. (**b**) Person adds a new obstacle. (**c**) Person moves in front of P3DX. (**d**,**e**) P3DX observes the new obstacle, changes trajectory and updates map. (**f**–**h**) P3DX return to the starting position. ***(Supplementary Materials)***

The updated 3D map build by P3DX robot is shown in Figure 11a. In this experiment, Turtlebot used only the 2D Lidar sensor with 2D gridmap. Hence, P3DX could not directly share the 3D point-cloud information due to the heterogeneous maps used by the two robots. By using the T-node map, P3DX blocked the path between nodes n_6 and n_7 and shared this information with the Turtlebot to plan appropriate path. More information regarding the dimensions of the new obstacle could also be shared for better path planning. Hence, the 3D information was converted to a 2D information to be shared with Turtlebot. Grid maps are the most commonly used 2D maps in which each grid-value represents whether the grid is occupied, free or unknown. The 3D point-cloud were projected to the ground which was detected using a RANSAC based plane detection [50]. This 2D information was shared by P3DX robot with Turtlebot and the updated T-node map is shown in Figure 11b. In the updated T-node map of Figure 11b, the obstacle is placed between the nodes n_6 and n_7 blocking it.

Turtlebot was programmed to navigate from the same start location 'S' to the goal location 'G'. In the absence of the proposed information sharing mechanism, the path planned by Turtlebot would be (Figure 11b),

$$S \to n_2 \to n_5 \to n_6 \to n_7$$

The Turtlebot would encounter a new obstacle between the nodes n_6 and n_7 and would have to re-plan a new path towards the goal. However, with the proposed information sharing mechanism, Turtlebot could directly plan a path considering the newly added obstacle and the planned path was (Figure 11b),

$$S \to n_2 \to n_5 \to n_6 \to n_9 \to n_8 \to n_7.$$

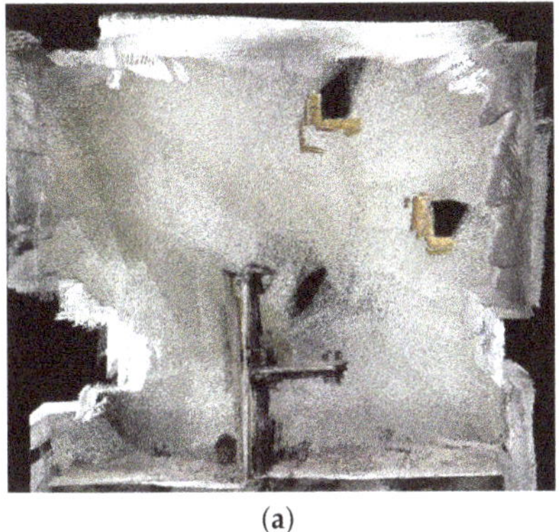

(a)

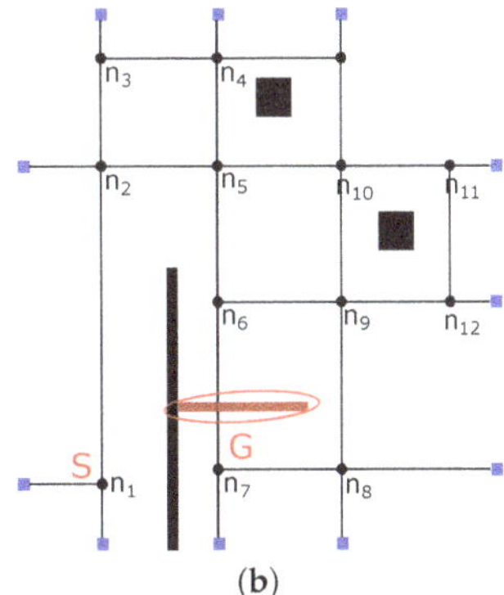

(b)

Figure 11. (**a**) RGBD map updated by P3DX. (**b**) Node-map. S and G are the start and goal points. The new obstacle is shown on nodes n_6n_7. Ellipse represents positional uncertainty.

Notice that, this appropriate path was generated by Turtlebot 'remotely' before actually encountering the new obstacle. Figure 12 shows the navigation of Turtlebot after considering the new obstacle. The entire navigation is illustrated between Figure 12a–h. In particular, it can be seen from Figure 12e–g, that Turtlebot maintains a safe threshold from the start itself. Turtlebot itself updated its map using the attached Lidar and the updated grid-map is shown in Figure 13.

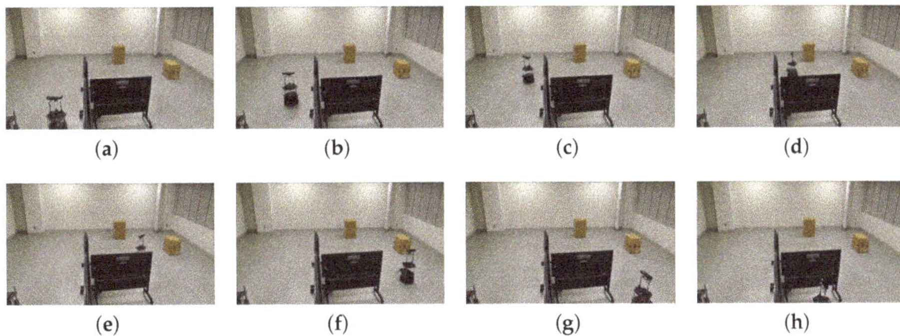

(a) (b) (c) (d)

(e) (f) (g) (h)

Figure 12. Timely snapshots of the experiment. (**a**–**h**) Turtlebot starts navigation with the updated information and plans a trajectory considering the new obstacle. (***Supplementary Materials***)

The dimensions of the obstacles in the experiment are given in Table 3.

The decay curve is shown in Figure 14. In the experiment, C_{th} was set to 0.45 and t_z to 20 min. Based on the uncertainty of the obstacle, the Ufactor was calculated as approximately 15% of t_z. Figure 14 shows the decay of confidence considering the uncertainty of the obstacles.

Figure 15 shows different decay curves for different amounts of estimated positional uncertainties of the new obstacle. Although Figure 14 shows the actual decay curve of the experiment, Figure 15 shows theoretical values for different values of uncertainty. Figure 15a–d shows the confidence decay with increasing uncertainty of 35%, 45%, 55% and 65%, respectively. It can be seen that, for increasing uncertainty, the curve decays much faster, as desired.

Thus, the T-node enables robots to share information across heterogeneous maps. Indeed, there is a need to transform the newly added obstacle's information to spatial coordinates but it can easily be achieved in real-time. Moreover, to avoid the overheads of such computation for time-critical applications, only the blocked/un-blocked information could also be shared. Using the same approach, information among other type of maps could be shared effectively.

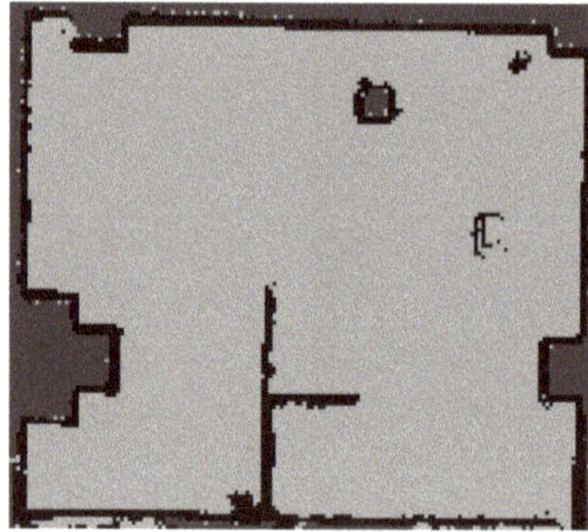

Figure 13. 2D gridmap updated by Turtlebot during navigation.

Table 3. Obstacle Dimensions in the Experiment.

Obstacle	Length × Width × Height
Obstacle1	40 cm × 40 cm × 68 cm
Obstacle2	50 cm × 35 cm × 50 cm
Newly Added Obstacle	300 cm × 5 cm × 100 cm

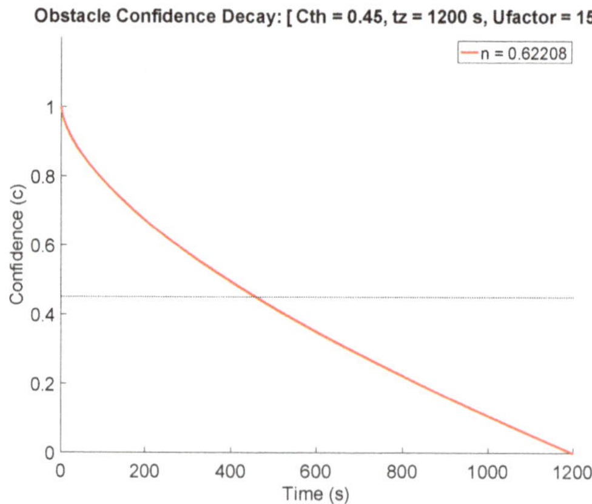

Figure 14. Confidence decay curve in the new experiment. C_{th} = 0.45, t_z = 1200 s, Ufactor = 15.

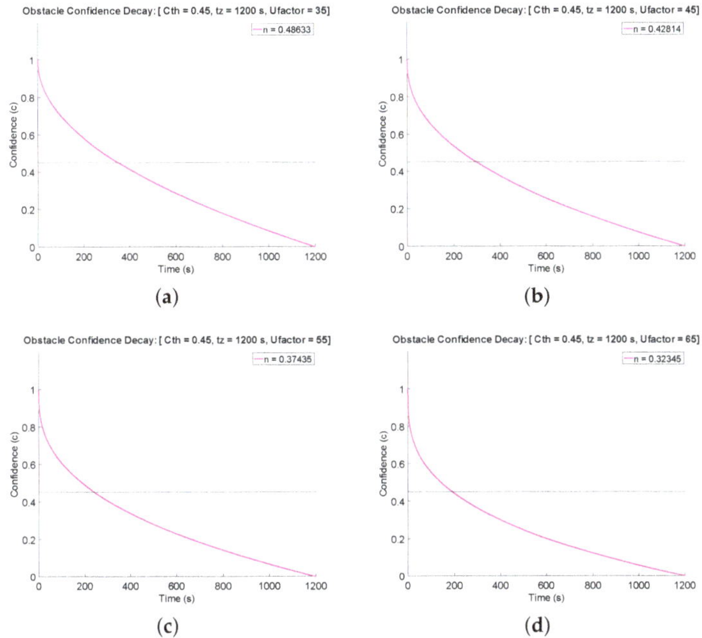

Figure 15. Estimation of confidence decay curve for different values of positional uncertainty with C_{th} = 0.45 and t_z = 1200 s. (**a**) Ufactor = 35%. (**b**) Ufactor = 45%. (**c**) Ufactor = 55%. (**d**) Ufactor = 65%. It can be seen that for higher uncertainties, the curve decays faster as desired.

6.2. Results with Dynamic Entities (Moving Obstacle)

The proposed method was tested under complex scenarios by purposefully moving a person in front of the robot and blocking its way. This is shown in Figure 16. As P3DX robot started navigation from start location 'S' to goal location 'G', a person blocked its way by randomly moving in front of the robot. This is shown in Figure 16a–j.

Similarly, the path of P3DX robot was blocked again while it was navigating back from the goal location 'G' to its start location 'S'. This is shown in Figure 17. The person randomly moved in front of the robot blocking its path as shown in Figure 17a–j.

In both the cases of Figures 16 and 17, the robot attempted to avoid collision and planned alternate trajectories or stopped if the person stands dangerously close to the robot. Moreover, in both the cases, the robot did not update the map corresponding to the person as a new obstacle in the map. This is because the positional uncertainty corresponding to the moving obstacle was large as calculated by Algorithm 2. Even if the person is falsely identified as a new obstacle and the map is updated, it has no adverse effects in the proposed method, as uncertainty is integrated in the confidence decay mechanism. Any wrong map update corresponding to dynamic obstacles has high probability of larger positional uncertainty corresponding to the dynamic obstacle and therefore a quicker decay given by Equation (6), (7) and (9). On the other hand, for static new obstacles in the map, the underlying SLAM (Algorithm 2) algorithm estimates smaller positional uncertainty and therefore a larger decay time, ensuring its permanence in the map.

Thus, uncertainty integration has two merits in the information sharing scheme. First, it acts a filter for wrong map updates corresponding to the dynamic obstacles in the environment through a quick confidence decay. Second, it ensures that only the correct information is shared with other robots corresponding to the new static obstacles. It should be noted that the dynamic detection of moving people can be done using image processing for camera-based sensors [51], RGBD sensors [52] or leg detector for Lidar-based sensors [53] and integration of such approaches [54] will increase the robustness of the system.

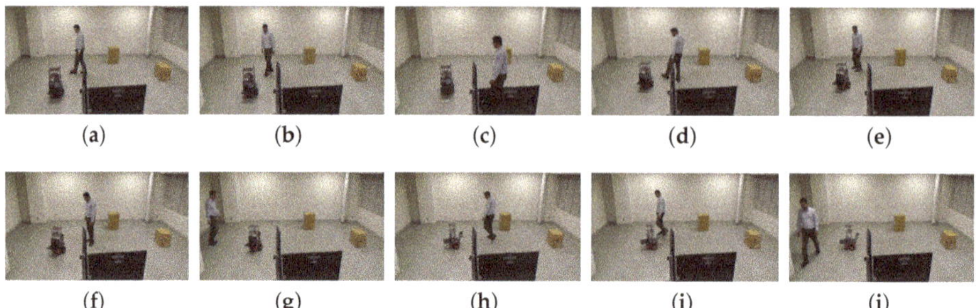

Figure 16. Dynamic obstacle experiment with P3DX navigation from position S to G in Figure 9. (**a–j**) Person moved randomly in front of P3DX for a long time moving in and out of the range of sensors. P3DX changed trajectories or stopped if the obstacle was dangerously close. (***Supplementary Materials***)

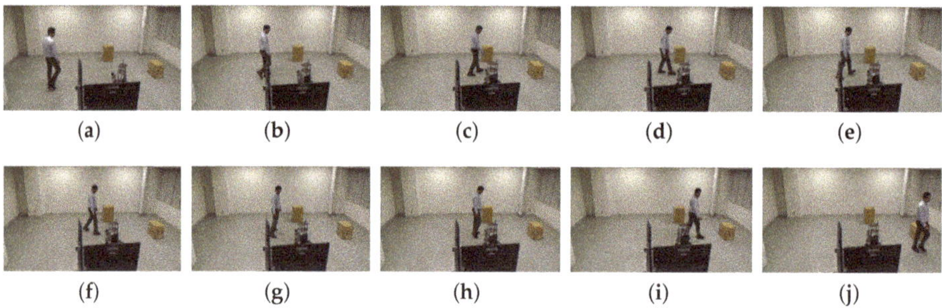

Figure 17. Dynamic obstacle experiment with P3DX navigation from position G to S in Figure 9. (**a–j**) To further test the method under pressure, a person moved randomly near P3DX. (***Supplementary Materials***)

7. Conclusions

Information sharing is a powerful technique which has many potential benefits in path planning of multi-robot systems. A node-map was proposed in our previous work to solve the problems of information sharing in different robots. We extended our previous work by integrating the positional uncertainty of the new obstacles in the confidence decay mechanism which models the transience of the obstacles. This minimizes false map updates and notifications in the system. New experiments were performed to share information about new obstacles in heterogeneous maps. The results shown that using the nodemap allows the robots to smoothly share the information. Moreover, since path planning is also done using the nodemap, efficient trajectories considering the position of new obstacles can be done in real-time. The information sharing mechanism allows the robots to obtain timely information about remote obstacles in the map without having to explicitly visit those areas. In addition, new experiments were performed to test the proposed mechanism in complex environments with moving people in the vicinity of the robots. Due to the tight coupling of uncertainty and decay mechanism, the dynamic obstacles could be filtered and avoided false update of the map. Even if there is some false update, the confidence corresponding to them decays fast due to larger uncertainty. Experiment results confirm that, in the long run in large environments employing multiple robots, the proposed method can improve the efficiency of the system in terms of shorter distance traveled by the robots and shorter planning time by eliminating path re-planning. In future, we will continue to test the robustness of the proposed method in more complex and realistic environments such as cafeterias and offices.

Supplementary Materials: The supplementary materials are available online at http://www.mdpi.com/2076-3417/9/13/2753/s1.

Author Contributions: A.R. and A.A.R. conceived the idea, designed, performed experiments, and summarized the research; Y.K. made valuable suggestions to analyze the data and improve the manuscript. Y.H. provided important feedback to improve the manuscript. The manuscript was written by A.R.

Funding: This research received no external funding.

Conflicts of Interest: The authors declare no conflict of interest.

References

1. Ravankar, A.; Ravankar, A.A.; Hoshino, Y.; Emaru, T.; Kobayashi, Y. On a Hopping-points SVD and Hough Transform Based Line Detection Algorithm for Robot Localization and Mapping. *Int. J. Adv. Robot. Syst.* **2016**, *13*, 98. [CrossRef]
2. Wang, R.; Veloso, M.; Seshan, S. Multi-robot information sharing for complementing limited perception: A case study of moving ball interception. In Proceedings of the 2013 IEEE International Conference on Robotics and Automation, Karlsruhe, Germany, 6–10 May 2013; pp. 1884–1889. [CrossRef]
3. Riddle, D.R.; Murphy, R.R.; Burke, J.L. Robot-assisted medical reachback: using shared visual information. In Proceedings of the ROMAN 2005, IEEE International Workshop on Robot and Human Interactive Communication, Nashville, TN, USA, 13–15 Augest 2005; pp. 635–642. [CrossRef]
4. Cai, A.; Fukuda, T.; Arai, F. Cooperation of multiple robots in cellular robotic system based on information sharing. In Proceedings of the IEEE/ASME International Conference on Advanced Intelligent Mechatronics, Tokyo, Japan, 20–20 June 1997; p. 20. [CrossRef]
5. Rokunuzzaman, M.; Umeda, T.; Sekiyama, K.; Fukuda, T. A Region of Interest (ROI) Sharing Protocol for Multirobot Cooperation With Distributed Sensing Based on Semantic Stability. *IEEE Trans. Syst. Man Cybern. Syst.* **2014**, *44*, 457–467. [CrossRef]
6. Samejima, S.; Sekiyama, K. Multi-robot visual support system by adaptive ROI selection based on gestalt perception. In Proceedings of the 2016 IEEE International Conference on Robotics and Automation (ICRA), Stockholm, Sweden, 16–21 May 2016; pp. 3471–3476. [CrossRef]
7. Özkucur, N.E.; Kurt, B.; Akın, H.L. A Collaborative Multi-robot Localization Method without Robot Identification. In *RoboCup 2008: Robot Soccer World Cup XII*; Springer: Berlin/Heidelberg, Germany, 2009; pp. 189–199. [CrossRef]
8. Sukop, M.; Hajduk, M.; Jánoš, R. Strategic behavior of the group of mobile robots for robosoccer (category Mirosot). In Proceedings of the 2014 23rd International Conference on Robotics in Alpe-Adria-Danube Region (RAAD), Smolenice, Slovakia, 3–5 September 2014; pp. 1–5. [CrossRef]
9. Ravankar, A.; Ravankar, A.A.; Kobayashi, Y.; Emaru, T. Avoiding blind leading the blind. *Int. J. Adv. Robot. Syst.* **2016**, *13*. [CrossRef]
10. Ravankar, A.; Ravankar, A.A.; Kobayashi, Y.; Emaru, T. On a bio-inspired hybrid pheromone signalling for efficient map exploration of multiple mobile service robots. *Artif. Life Robot.* **2016**, 221–231. [CrossRef]
11. Ravankar, A.; Ravankar, A.A.; Kobayashi, Y.; Emaru, T. Intelligent Robot Guidance in Fixed External Camera Network for Navigation in Crowded and Narrow Passages. *Proceedings* **2017**, *1*, 37. [CrossRef]
12. Ravankar, A.; Ravankar, A.; Kobayashi, Y.; Emaru, T. Symbiotic Navigation in Multi-Robot Systems with Remote Obstacle Knowledge Sharing. *Sensors* **2017**, *17*, 1581. [CrossRef] [PubMed]
13. Hunziker, D.; Gajamohan, M.; Waibel, M.; D'Andrea, R. Rapyuta: The RoboEarth Cloud Engine. In Proceedings of the 2013 IEEE International Conference on Robotics and Automation, Karlsruhe, Germany, 6–10 May 2013; pp. 438–444. [CrossRef]
14. Waibel, M.; Beetz, M.; Civera, J.; D'Andrea, R.; Elfring, J.; Gálvez-López, D.; Häussermann, K.; Janssen, R.; Montiel, J.M.M.; Perzylo, A.; et al. RoboEarth. *IEEE Robot. Autom. Mag.* **2011**, *18*, 69–82. [CrossRef]
15. Tenorth, M.; Perzylo, A.C.; Lafrenz, R.; Beetz, M. Representation and Exchange of Knowledge About Actions, Objects, and Environments in the RoboEarth Framework. *IEEE Trans. Autom. Sci. Eng.* **2013**, *10*, 643–651. [CrossRef]

16. Stenzel, J.; Luensch, D. Concept of decentralized cooperative path conflict resolution for heterogeneous mobile robots. In Proceedings of the 2016 IEEE International Conference on Automation Science and Engineering (CASE), Fort Worth, TX, USA, 21–25 Augest 2016; pp. 715–720. [CrossRef]
17. Ravankar, A.; Ravankar, A.A.; Kobayashi, Y.; Jixin, L.; Emaru, T.; Hoshino, Y. An intelligent docking station manager for multiple mobile service robots. In Proceedings of the Control, Automation and Systems (ICCAS), 2015 15th International Conference on, Busan, Korea, 13–16 October 2015; pp. 72–78. [CrossRef]
18. Ravankar, A.; Ravankar, A.; Kobayashi, Y.; Emaru, T. Hitchhiking Robots: A Collaborative Approach for Efficient Multi-Robot Navigation in Indoor Environments. *Sensors* **2017**, *17*, 1878. [CrossRef] [PubMed]
19. Ravankar, A.; Ravankar, A.; Kobayashi, Y.; Hoshino, Y.; Peng, C.C.; Watanabe, M. Hitchhiking Based Symbiotic Multi-Robot Navigation in Sensor Networks. *Robotics* **2018**, *7*, 37. [CrossRef]
20. Guo, Y.; Parker, L. A distributed and optimal motion planning approach for multiple mobile robots. In Proceedings of the 2002 IEEE International Conference on Robotics and Automation (Cat. No.02CH37292), Washington, DC, USA, 11–15 May 2002; Volume 3, pp. 2612–2619. [CrossRef]
21. Svestka, P.; Overmars, M.H. *Coordinated Path Planning for Multiple Robots*; Technical Report UU-CS-1996-43; Department of Information and Computing Sciences, Utrecht University: Utrecht, The Netherlands, 1996.
22. Clark, C.M.; Rock, S.M.; Latombe, J. Motion planning for multiple mobile robots using dynamic networks. In Proceedings of the 2003 IEEE International Conference on Robotics and Automation (Cat. No.03CH37422), Taipei, Taiwan, 14–19 September 2003; Volume 3, pp. 4222–4227. [CrossRef]
23. Teng, R.; Yano, K.; Kumagai, T. Efficient Acquisition of Map Information Using Local Data Sharing over Hierarchical Wireless Network for Service Robots. In Proceedings of the 2018 Asia-Pacific Microwave Conference (APMC), Kyoto, Japan, 6–9 November 2018; pp. 896–898. [CrossRef]
24. Ravankar, A.A.; Ravankar, A.; Peng, C.; Kobayashi, Y.; Emaru, T. Task coordination for multiple mobile robots considering semantic and topological information. In Proceedings of the 2018 IEEE International Conference on Applied System Invention (ICASI), Chiba, Japan, 13–17 April 2018; pp. 1088–1091. [CrossRef]
25. Pinkam, N.; Bonnet, F.; Chong, N.Y. Robot collaboration in warehouse. In Proceedings of the 2016 16th International Conference on Control, Automation and Systems (ICCAS), Gyeongju, Korea, 16–19 October 2016; pp. 269–272. [CrossRef]
26. Regev, T.; Indelman, V. Multi-robot decentralized belief space planning in unknown environments via efficient re-evaluation of impacted paths. In Proceedings of the 2016 IEEE/RSJ International Conference on Intelligent Robots and Systems (IROS), Daejeon, Korea, 9–14 October 2016; pp. 5591–5598. [CrossRef]
27. Gasparetto, A.; Boscariol, P.; Lanzutti, A.; Vidoni, R. Path Planning and Trajectory Planning Algorithms: A General Overview. In *Motion and Operation Planning of Robotic Systems: Background and Practical Approaches*; Springer International Publishing: Cham, Switzeland, 2015; pp. 3–27. [CrossRef]
28. Tang, S.H.; Kamil, F.; Khaksar, W.; Zulkifli, N.; Ahmad, S.A. Robotic motion planning in unknown dynamic environments: Existing approaches and challenges. In Proceedings of the 2015 IEEE International Symposium on Robotics and Intelligent Sensors (IRIS), Langkawi, Malaysia, 18–20 October 2015; pp. 288–294. [CrossRef]
29. Ravankar, A.; Ravankar, A.; Kobayashi, Y.; Hoshino, Y.; Peng, C.C. Path Smoothing Techniques in Robot Navigation: State-of-the-Art, Current and Future Challenges. *Sensors* **2018**, *18*, 3170. [CrossRef] [PubMed]
30. Montijano, E.; Aragues, R.; Sagüés, C. Distributed Data Association in Robotic Networks With Cameras and Limited Communications. *IEEE Trans. Robot.* **2013**, *29*, 1408–1423. [CrossRef]
31. Ravankar, A. Probabilistic Approaches and Algorithms for Indoor Robot Mapping in Structured Environments. Ph.D. Thesis, Hokkaido University, Sapporo, Japan, 2015.
32. Ravankar, A.; Kobayashi, Y.; Ravankar, A.; Emaru, T. A connected component labeling algorithm for sparse Lidar data segmentation. In Proceedings of the 2015 6th International Conference on Automation, Robotics and Applications (ICARA), Queenstown, New Zealand, 17–19 Feburary 2015; pp. 437–442. [CrossRef]
33. Ravankar, A.; Ravankar, A.A.; Kobayashi, Y.; Jixin, L.; Emaru, T.; Hoshino, Y. A novel vision based adaptive transmission power control algorithm for energy efficiency in wireless sensor networks employing mobile robots. In Proceedings of the 2015 Seventh International Conference on Ubiquitous and Future Networks, Sapporo, Japan, 7–10 July 2015; pp. 300–305. [CrossRef]
34. Zhang, T.Y.; Suen, C.Y. A Fast Parallel Algorithm for Thinning Digital Patterns. *Commun. ACM* **1984**, *27*, 236–239. [CrossRef]

35. Yang, D.H.; Hong, S.K. A roadmap construction algorithm for mobile robot path planning using skeleton maps. *Adv. Robot.* **2007**, *21*, 51–63. [CrossRef]
36. Bai, X.; Latecki, L.; Yu Liu, W. Skeleton Pruning by Contour Partitioning with Discrete Curve Evolution. *Pattern Anal. Mach. Intell. IEEE Trans.* **2007**, *29*, 449–462. [CrossRef] [PubMed]
37. Ravankar, A.A.; Hoshino, Y.; Ravankar, A.; Jixin, L.; Emaru, T.; Kobayashi, Y. Algorithms and a framework for indoor robot mapping in a noisy environment using clustering in spatial and Hough domains. *Int. J. Adv. Robot. Syst.* **2015**, *12*. [CrossRef]
38. MacQueen, J.B. Some Methods for Classification and Analysis of MultiVariate Observations. In *Proceedings of the Fifth Berkeley Symposium on Mathematical Statistics and Probability*; Cam, L.M.L., Neyman, J., Eds.; University of California Press: Berkeley, CA, USA, 1967; Volume 1, pp. 281–297.
39. Lloyd, S.P. Least squares quantization in pcm. *IEEE Trans. Inf. Theory* **1982**, *28*, 129–137. [CrossRef]
40. Jain, A.K.; Dubes, R.C. *Algorithms for Clustering Data*; Prentice-Hall, Inc.: Upper Saddle River, NJ, USA, 1988.
41. Ester, M.; Kriegel, H.P.; Sander, J.; Xu, X. *A Density-Based Algorithm for Discovering Clusters in Large Spatial Databases with Noise*; AAAI Press: Menlo Park, CA, USA, 1996; pp. 226–231.
42. Fox, D.; Ko, J.; Konolige, K.; Limketkai, B.; Schulz, D.; Stewart, B. Distributed Multirobot Exploration and Mapping. *Proc. IEEE* **2006**, *94*, 1325–1339. [CrossRef]
43. Ravankar, A.A.; Ravankar, A.; Emaru, T.; Kobayashi, Y. A hybrid topological mapping and navigation method for large area robot mapping. In Proceedings of the 2017 56th Annual Conference of the Society of Instrument and Control Engineers of Japan (SICE), Kanazawa, Japan, 19–22 September 2017; pp. 1104–1107. [CrossRef]
44. Pioneer P3-DX. Pioneer P3-DX Robot. 2018. Available online: https://www.robotshop.com/community/robots/show/pioneer-d3-px (accessed on 11 Janurary 2019).
45. TurtleBot 2. TurtleBot 2 Robot. 2018. Available online: http://turtlebot.com/ (accessed on 11 Janurary 2019).
46. Quigley, M.; Conley, K.; Gerkey, B.P.; Faust, J.; Foote, T.; Leibs, J.; Wheeler, R.; Ng, A.Y. ROS: An Open-Source Robot Operating System. In Proceedings of the ICRA Workshop on Open Source Software, Kobe, Japan, 12–15 May 2009.
47. Hart, P.; Nilsson, N.; Raphael, B. A Formal Basis for the Heuristic Determination of Minimum Cost Paths. *Syst. Sci. Cybern. IEEE Trans.* **1968**, *4*, 100–107. [CrossRef]
48. Ravankar, A.; Ravankar, A.A.; Kobayashi, Y.; Emaru, T. Path smoothing extension for various robot path planners. In Proceedings of the 2016 16th International Conference on Control, Automation and Systems (ICCAS), Gyeongju, Korea, 16–19 October 2016; pp. 263–268. [CrossRef]
49. Ravankar, A.; Ravankar, A.A.; Kobayashi, Y.; Emaru, T. SHP: Smooth Hypocycloidal Paths with Collision-Free and Decoupled Multi-Robot Path Planning. *Int. J. Adv. Robot. Syst.* **2016**, *13*, 133. [CrossRef]
50. Xu, B.; Jiang, W.; Shan, J.; Zhang, J.; Li, L. Investigation on the Weighted RANSAC Approaches for Building Roof Plane Segmentation from LiDAR Point Clouds. *Remote Sens.* **2015**, *8*, 5. [CrossRef]
51. Enzweiler, M.; Gavrila, D.M. Monocular Pedestrian Detection: Survey and Experiments. *IEEE Trans. Pattern Anal. Mach. Intell.* **2009**, *31*, 2179–2195. [CrossRef] [PubMed]
52. Spinello, L.; Arras, K.O. People detection in RGB-D data. In Proceedings of the 2011 IEEE/RSJ International Conference on Intelligent Robots and Systems, San Francisco, CA, USA, 25–30 September 2011; pp. 3838–3843. [CrossRef]
53. Pantofaru C. Leg Detector. 2019. Available online: http://wiki.ros.org/leg_detector (accessed on 3 May 2019).
54. Moeslund, T.B.; Granum, E. A Survey of Computer Vision-Based Human Motion Capture. *Comput. Vis. Image Understand.* **2001**, *81*, 231–268. [CrossRef]

Article

Contingent Task and Motion Planning under Uncertainty for Human–Robot Interactions

Aliakbar Akbari *, Mohammed Diab and Jan Rosell

Institute of Industrial and Control Engineering, Universitat Politècnica de Catalunya, 08034 Barcelona, Spain; mohammed.diab@upc.edu (M.D.); jan.rosell@upc.edu (J.R.)

* Correspondence: ali.akbari101@gmail.com

Received: 8 February 2020; Accepted: 24 February 2020; Published: 1 March 2020

Abstract: Manipulation planning under incomplete information is a highly challenging task for mobile manipulators. Uncertainty can be resolved by robot perception modules or using human knowledge in the execution process. Human operators can also collaborate with robots for the execution of some difficult actions or as helpers in sharing the task knowledge. In this scope, a contingent-based task and motion planning is proposed taking into account robot uncertainty and human–robot interactions, resulting a tree-shaped set of geometrically feasible plans. Different sorts of geometric reasoning processes are embedded inside the planner to cope with task constraints like detecting occluding objects when a robot needs to grasp an object. The proposal has been evaluated with different challenging scenarios in simulation and a real environment.

Keywords: task and motion planning; manipulation planning; robot-human interactions; perception

---

1. Introduction

Robotic manipulation tasks become highly challenging when a mobile manipulator is required to obtain a feasible plan to solve a given problem under potential uncertainties. Uncertainty shall be viewed in the initial state of the robot environment, e.g., objects may rest in different positions or some object features (like color) could be initially unknown for a robot. Uncertainty, moreover, must be considered in the result of manipulation actions (as nondeterministic effects) since there could be different action outcomes. To deal with such uncertainties, robots generally look for a sequence of actions to satisfy the goal of a task and perform replanning in the case of action execution failure or uncertain situations. This process may be costly while a robot requires repetition of expensive replanning.

To tackle those challenging issues, these problems can rely on contingent task planning which plans in belief space and can generate conditional plans under uncertainty in terms of initial state and action effects. Contingent-based task planners can provide a tree of plans rather than a single sequence of executive actions. Therefore, uncertainty is observed during the plan execution, and the tree of plans is followed according to the binary observation values.

Other challenges are related to some demanding or difficult tasks which are either not performable easily by robots or are out of their reach, but that can be done in collaboration with a human operator. In these cases, the robot can ask the human operator to do some particular difficult actions, to transfer some objects located in the human workspace or to share knowledge that is initially incomplete to the robot. Moreover, there could be some geometric constraints imposed in the environment, e.g., lack of space for placing objects, occlusions, kinematic issues, etc., and the finding of the geometric values for each manipulation action becomes substantial in order to make a manipulation plan feasible. Therefore, the way of combining task and motion planning plays a significant role when the manipulation task is highly constrained in terms of geometric information and there is amount of uncertainty.

In this paper, we are going to deal with manipulation tasks carried out by a mobile manipulator assisted by a human operator. The mobile manipulator will be responsible to execute the main task, while the human operator will be responsible for some difficult actions (like to open some box-like containers which cannot be opened by the robot), to share knowledge with the robot, and to transfer objects to the robot when they are not reachable. Uncertainty in the initial state and in some action effects are considered. Some manipulation and sensing actions are considered in the current proposal, which allow illustration of the approach, and that can be extended to handle a broader set of manipulation tasks. No geometric uncertainty is considered, e.g., in the robot motion or the object poses.

Contributions: To deal with the aforementioned challenges, we propose a contingent-based task and motion planner based on *Contingent-FF* [1] that works under uncertainty and considers human–robot collaboration. The *Contingent-FF* includes two main components, *heuristic evaluation* and *search space*, and results in a tree-shaped set of plans involving sensing actions. Three main contributions extend the basic *Contingent-FF* planner:

- *Robot action reasoning.* Two types of geometric reasoning are proposed and integrated with the basic planner: *relaxed geometric reasoning* and *lazy motion evaluation*. The former refers to *Reachability*, *Spatial*, and *Manipulation* reasoning. This reasoning process is embedded within the heuristic computation of the planner. Motion paths are lazily evaluated when actions are selected by the state space search. If the reasoning processes fail, geometric constraints are fed back to the planner. This part of the computation is done offline and aims to prune infeasible actions due to geometric constraints and to obtain a feasible set of actions in the tree of plans. As the basic contingent planner considers only symbolic reasoning, this module enables it to incorporate geometric reasoning to deal with practical applications. The reasoning process provides feasible initial and goal configurations for motion planning queries, improving its success rate and thus the overall performance of the planner in the generation of a feasible manipulation plan.
- *Human–robot collaboration.* There are some actions which can be executed by the robot and others that require the collaboration of a human operator. The proposed relaxed geometric reasoning is extended to inform the planner about which actions cannot be executed by the robot, and hand over them to the human operator, allowing the planner to handle those cases where the selection of actions to be performed by a human operator is required. In these cases, the geometric world resulting by these actions is simulated and used in the planning system for further geometric reasoning evaluation. This step makes the basic planner flexible to consider the result of human actions, extending its performance to situations it is not able to be handled autonomously.
- *State observation.* To observe the binary outcomes of actions, two modules are proposed: *perception* and *human knowledge*. Perception is used to detect, e.g., the actual locations of the objects or some objects feature like color. The knowledge provided by the operator is required for more difficult observations like determining if a can is filled or empty, or if a glass contains a given drink. Action observation takes place at execution time. The combining of both modules widens the capacity of the planner to identify the current situation of the robot's world and decide the best course of actions in execution, thus improving the planner performance in finding feasible solutions.

One of the main advantages of the proposed framework is that the offline computation is valid and works despite the actual values of the uncertainty variables or the actual outcomes of the executable actions.

The rest of the paper is structured as follows. First, Section 2 summarizes some related work and Section 3 explains a proposal for contingent task and motion planning. Afterwards, Section 4 presents and illustrates the proposed relaxed geometric reasoning for mobile manipulators, Section 5 demonstrates contingent heuristic computation using relaxed information, Section 6 details tree-based planning using search space, and Section 7 presents manipulation plan execution using sensing and human interaction. Finally, Section 8 shows some implementation issues as well as empirical results, and Section 9 sketches the conclusions and future works.

2. Related Work

Manipulation problems of different nature have been tackled in the literature with different strategies, e.g., the manipulation problem of Navigation Among Movable Obstacles (NAMO) has been addressed in [2,3] using a backward search algorithm, and dual-arm table-top manipulation problems by combining motion planning and task assignment [4]. These robotic applications, like many others, must deal with different sources of uncertainty and the use of sensors and perception strategies may be required, e.g., the studies in [5,6] have investigated the machine robotic cell scheduling problem for manufacturing systems with or without sensor inspection. The following sections classify more approaches in the field of task and motion planning with and without uncertainty.

2.1. Task and Motion Planning without Uncertainty

Recently, much study has been centered to solve robotics manipulation tasks by combining task and motion planning problems with no consideration on uncertainty. It is assumed that the initial state of the environment is perfectly known, and actions are deterministic, i.e., state of planning is only changed by the selected action. There is a huge number of task planners being able to solve manipulation problems under perfect information [7].

In principle, two methods of combining task and motion planning have been explored: interleaved or simultaneously. Several studies call first task planning, and then motion planning to determine whether a plan is feasible or not such as [8–12]. In the case of failure, geometric constraints are identified and reported to task planning and the procedure continues. This might be costly as a number of times the process could be repeated in order to find a geometrically feasible plan.

On the other hand, other approaches enable task planning to incorporate geometric reasoning within the task planning process [13–18]. Hence, in this case, task planning results in a feasible manipulation plan. In this line, we recently proposed a heuristic-based task and motion planner [19] to deal with constrained table-top problems for bi-manual robots by offering different type of geometric reasoners that can be used in heuristic computation or when an action is selected. Our previous approach does not consider any uncertainty, human actions, and reasoning about mobile manipulation problems which are the subjects of this paper.

The way of integrating task and motion planning information in the current proposal is based on the simultaneous approach in order to generate feasible plans, and is an extension of [19] that copes with mobile manipulators, uncertainty, and to consider collaborative tasks with human operators.

2.2. Task and Motion Planning under Uncertainty

There are some situations in which a robot has incomplete information about its manipulation environment; therefore, it needs to plan under uncertainty. Task planning under uncertainty is a well-established field in Artificial Intelligence. Conditional-based task planners can provide conditional plan to cope with uncertain information when either the initial state is not completely known, or the result of actions are nondeterministic. There are various classes of planning in this field like conformant, contingent, or probabilistic planning.

Conformant planning looks for plans under given uncertainty concerning the start state and the effects of symbolic actions, assuming no sensing capabilities during the execution of the plan. The plan should be successful regardless of which is the start state. Contingent planning also considers uncertainty regarding the start state and the effects of actions. However, it can provide some sort of observation over a conditional plan in execution. Probabilistic planning does planning under probabilistic uncertainty regarding the start state and the effects of actions.

More details on some approaches following conditional-based task planning are commented next as we are interested in this type of planner due to its feature of providing observation over a conditional plan. Some conditional task planners are *Contingent-FF* [1], *POND* [20], and *PKS* [21]. They plan in the belief space and compute conditional plans in the offline mode, which are guided by the result of sensing actions. On the other hand, there are some conditional task planners like *K-Planner* [22], *SDR* [23], and *HCP* [24] solving conditional plans online. Although these planners can prune some branches by considering online sensing actions, satisfying the goal of task may not be possible and the planners may face with dead-end even if there is a solution.

The concept of contingent-based task and motion planning has also emerged. For instance, the *Planning with Knowledge and Sensing (PKS)* planner considers incomplete information and performs contingent planning [25] in two main scenarios, using *force sensing* and *visual sensing*. In a similar direction, offline-based hybrid conditional task and motion planning has been proposed [26], i.e., task planning is foremost performed, and then geometric evaluation is considered by incorporating low-level feasibility checks inside conditional planning (assuming that actuation actions are deterministic). On the contrary, the approach proposed here interweaves simultaneously efficient geometric reasoning inside the task planning process to provide geometrically feasible plans. The approach also copes with collaboration between the mobile robot and a human operator to perform a manipulation task.

3. A Proposal for Contingent Task and Motion Planning

This section first presents a brief overview of the original *Contingent-FF* task planning, and the modifications introduced in the present proposal to compute geometrically feasible manipulation conditional plans.

3.1. Contingent-FF Overview

The *Contingent-FF* task planner [1] handles uncertainty in the initial state and in the result of actions. The task planner has two main components which are heuristic computation and search space. For the heuristic computation, the planner uses a modified version of the *Relaxed Planning Graph (RPG)* used in the *Fast-Forward (FF)* planner [27]. The relaxed plan including a number of relaxed actions is computed from the *RPG*, and the heuristic value is the length of this relaxed plan. Also, promising actions (called helpful actions in *FF*) are extracted from the relaxed plan as a pruning technique in the search space, as discussed in *FF*. The *Contingent-FF* planner extends the *RPG* process, called *CRPG*, by adding unknown facts in an additional layer in the heuristic phase. Known facts are basically those which do not have uncertainty and unknown facts are the ones which could be the result of nondeterministic actions or uncertain in the initial state. It introduces reasoning about unknown facts that allows such facts to become known in the *RPG* process. Once *CRPG* is successfully built, the relaxed plan is extracted.

In *Contingent-FF*, belief states including known and unknown facts are considered. The search space starts from the initial belief state and applies an *And-Or* search. The search space progress is guided by the heuristic value and helpful actions. The result of planning provides conditional plans that may involve a variety of sensing actions whose outcome causes different plan branches.

3.2. Planning Formulation

Our planning system domain $\mathcal{D}$ is a tuple $\langle \mathcal{A}, \Omega, \mathcal{F}, \mathcal{W}, S_g \rangle$ where $\mathcal{A}$ is the action space, Ω is the sensing action space, $\mathcal{F}$ is a set of literals, $\mathcal{W}$ is a workspace involving a mobile manipulator $\mathcal{R}$ (described by the pose of the base Pos_{rob} along the arm configuration Q_{rob}) and a number of objects $\mathcal{O}$, and, S_g is a set of grasping poses described for objects. Objects are denoted as: $\mathcal{O}$=$\{\mathcal{O}_1^m(pos,fe) \ldots \mathcal{O}_j^m(pos,fe), \mathcal{O}_1^f(pos,fe) \ldots \mathcal{O}_k^f(pos,fe)\}$, where j and k are the number of *Movable* and *Fixed* objects respectively, whose initial position and orientation are denoted by *pos*, and whose features are denoted by *fe*.

An action $a \in \mathcal{A}$ is a tuple $\langle$*name(a), pre(a), effect(a), coneffect(a), geom(a)*, $\mathcal{Q}(a)\rangle$, where *name(a)* is the action symbolic name, *pre(a)* is a propositional formula which must hold for the action to be applied, *geom(a)* is the numerical counterpart of an action containing geometric information, *effect(a)* represents the negative and positive effects of a on the state it is applied to, and $\mathcal{Q}(a)$ is a query function to the motion planner which computes a motion between two robot configurations and stores the solution if any. A relaxed action $a' \in \mathcal{A}'$ (where $\mathcal{A}'$ is the relaxed action space) is similar to the action despite it does not consider any negative effects. Actions refer to executable actions, i.e., requiring motion, and can be done by either the robot or a person. The following actions types are considered to deal with some examples of mobile robot manipulation:

- *Transit*: an action done by the robot to travel from one configuration to another one without an attached object.
- *Transfer*: an action done by the robot to move an attached object from one pose to another one.
- *Push*: an action done by the robot to push an object from one pose to another one.
- *Open*: an action done by the robot to open a box-like container (articulated cap with prismatic joint is assumed with two positions corresponding to fully closed and fully opened, the state being stored in the containers objects features).
- *HumanTransfer*: an action done by a person to transfer/push an object to the robot workspace.
- *HumanOpen*: an action done by a person to open a box-like container.

Each sensing action is a tuple $\langle pre(a), o(a) \rangle$, where *o(a)* is a literal with uncertainty. These are actions not involving motion, devoted to observing the value of *o(a)*. The observation is done in run-time. Some sample sensing actions are considered in the proposed planning system. They are the following:

- *SenseColor*: a sensing action is done by a perception module to determine the color of an object.
- *SensePose*: a sensing action is done by a perception module to determine the pose of an object.
- *CheckContainer*: a sensing action is done by a person to evaluate whether a container is open or not.
- *CheckCan*: a sensing action is done by a person to evaluate whether can-like objects are filled or not.

A belief state S is a tuple $S = \langle \mathcal{P}, \mathcal{V} \rangle$ where $\mathcal{P}$ includes a set of known literals which hold in that state and a set of uncertain literals which may hold or not in the state, and $\mathcal{V}$ represents a full geometric description of the scene, i.e., configurations of robots and poses of objects corresponding to certain and uncertain literals. An executable action from a state S_1 results in a new world state using the state transition functions $S_2.\mathcal{P} := S_1.\mathcal{P} - effect^-(a) + effect^+(a)$ and $S_2.\mathcal{V} := S_1.\mathcal{V} - geom^-(a) + geom^+(a)$. A sensing action splits a belief state and introduces two branches into the plan marked with *o(a)*, and $\sim$ *o(a)*.

The planning problem $\mathcal{T}$ is expressed by a tuple $\langle \mathcal{D}, \mathcal{S}_0, \mathcal{G} \rangle$ where $\mathcal{D}$ is a domain, $\mathcal{S}_0$ consists of a set of literals representing the initial symbolic state $\mathcal{I}$ such that $\mathcal{I} \subseteq \mathcal{F}$ along their geometric assignments regarding the initial state of the world $\mathcal{W}_0$, and $\mathcal{G} \subseteq \mathcal{F}$ is the set of symbolic goal conditions. The solution of a *Combined Task and Motion Planning (TAMP)* problem under uncertainty, which we denote by π, is a tree-shaped conditional plan, i.e., a sequence of symbolic actions achieving $\mathcal{G}$, along with a feasible motion for each action.

3.3. Geometric Constraint Predicates

Basically, three general predicates, evaluated by geometric reasoning, are allocated that set constraints to the task states: *isCrit*($\mathcal{O}_j^m$, $\mathcal{O}'$, *Pos*), *infeasByRob*($\mathcal{R}$, $\mathcal{O}'$, *Pos*), and *assist*(*Human*, $\mathcal{O}'$, *Pos*). The first predicate indicates that there is a blocking object $\mathcal{O}_j^m$ which is located towards the target object $\mathcal{O}'$ placed in the pose *Pos*. The second one shows that the target object cannot be manipulated by the robot $\mathcal{R}$ in the corresponding pose *Pos*. The last predicate shows the manipulation action with the target object $\mathcal{O}'$ and the corresponding *Pos* must be done by a human operator *Human*.

The proposed predicates are interleaved inside the pre- and post-conditions of the actions. Concerning the actions done by the robot, the predicates ∼*isCrit* and ∼*infeasByRob* are inserted within the preconditions of the actions *Transit*, *Open*, *Transfer*, and *Push* in order to avoid moving the robot to any unreachable or infeasible configuration. Referring to the post-conditions, the last two actions may include the negation of the predicate *isCrit* if they are moving a blocking object to a placement where the obstruction does no longer hold. With respect to the actions performed by a human, the preconditions of the actions *HumanTransfer* and *HumanOpen* may include the predicate *assist* in order to indicate the requirement of a human operator.

To illustrate the use of these predicates, the actions *Transit* and *HumanOpen* are described next. The action template *Transit*($\mathcal{R}$, $\mathcal{O}_i^m$, *Surface*, *Pos*) is designed to move the robot $\mathcal{R}$ (arm and base), without holding any object, towards a grasp configuration of a manipulatable object $\mathcal{O}_i^m$ located on a surface *Surface* at pose *Pos*. In the final configuration, the robot holds the target object. The action is applicable if the following preconditions hold: the object is located on a surface, top of the object is clear, the robot arm is empty, it can reach the grasp configuration if there is no movable objects blocking its way to $\mathcal{O}_i^m$. The last precondition is represented by fact *isCrit*($\mathcal{O}_j^m$, $\mathcal{O}_i^m$, *Pos*); objects that make this fact to hold are called *Critical Objects* which are the objects blocking the way of reaching the object. As a result of the action, the robot holds an object.

Transit($\mathcal{R}$, $\mathcal{O}_i^m$, *Surface*, *Pos*):

Pre: *onSurface*($\mathcal{O}_i^m$, *Surface*, *Pos*), *armEmpty*($\mathcal{R}$), *clear*($\mathcal{O}_i^m$), ∼*infeasByRob*($\mathcal{R}$, $\mathcal{O}_i^m$, *Pos*),
$\forall\mathcal{O}_j^m$ ∼*isCrit*($\mathcal{O}_j^m$, $\mathcal{O}_i^m$, *Pos*)

Effect: *holding*($\mathcal{R}$, $\mathcal{O}_i^m$, *Pos*), ∼*clear*($\mathcal{O}_i^m$), ∼*armEmpty*($\mathcal{R}$), ∼*onSurface*($\mathcal{O}_i^m$, *Surface*, *Pos*)

The action template *HumanOpen*(*Human*, $\mathcal{O}_i^m$, *Pos*, *Closed*, *Open*) is used to open a box-like container $\mathcal{O}_i^m$ when it is closed. The action is applicable if the robot needs the assistance from a human operator, represented by the *assist* predicate, and its status is closed, shown by the predicate *status*. These conditions are introduced in the action preconditions. As a result of the action, the corresponding container will be open.

HumanOpen(*Human*, $\mathcal{O}_i^m$, *Pos*, *Closed*, *Open*):

Pre: *assist*(*Human*, $\mathcal{O}_i^m$, *Pos*), *status*($\mathcal{O}_i^m$, *Closed*)

Effect: *status*($\mathcal{O}_i^m$, *Open*), ∼*status*($\mathcal{O}_i^m$, *Closed*)

3.4. The Proposed Framework

The proposed framework for task and motion planning under uncertainty extends the basic *Contingent-FF* planner, aiming to incorporate different geometric reasoning procedures, observations on sensing actions, as well as human–robot collaboration within planning. The overview of the system is sketched in Figure 1. It involves three main parts: *Heuristic Computation*, *Space Search*, and *Conditional Plans Evaluation*.

Heuristic Computation basically provides a value which is distance to goal and promising actions for each belief state. The basic *CRPG* is initially computed and the associated relaxed plan is obtained. This plan is forwarded to the relaxed geometric reasoner determining the feasibility of actions in terms of reachability, collisions, manipulation constraints, and graspability. The heuristic value is returned along with helpful actions if such constraints are met. If a constraint is violated, the associated belief state is updated with facts describing the cause of failure, and an alternative relaxed plan is looked for. Hence, the heuristic function is informative both in terms of symbolic and geometric constraints.

Space Search maintains the basic algorithm of the *Contingent-FF* planner that is based on the *And-Or* search strategy. From each belief state, the action resulting in the state with lowest heuristic value is selected and is a candidate to be added to the conditional plans. The only difference is that the heuristic value now accounts for geometric constraints.

Conditional Plans Evaluation tries to solve motion planning for an action if possible. It considers sensing procedure to evaluate sensing actions and may assign actions, which are infeasible for robots, to operators.

If motion planning fails, the current belief state is updated with the cause of failure and the search resumes. In general, the geometric failure could be due to collisionable objects, so the literal *isCrit* is added to the belief state. If the failure is because of fixed obstacles blocking the way of reaching an object, inverse kinematic problems, or motion planning time-out problem, the literal *infeasByRob* is added to the state. In such failure, if the type of the evaluated action is either transit or open, the literal *assist* is also inserted.

Otherwise, when motion planning succeeds, the action is added to the tree-shaped conditional plans at hand. After finding the complete conditional plans, feasible actions are executed by a robot or a human operator in the real world and sensing actions are observed either using a perception module or the information provided by the human operator in run-time. Therefore, the robot plan can determine the correct branch to follow up its plan.

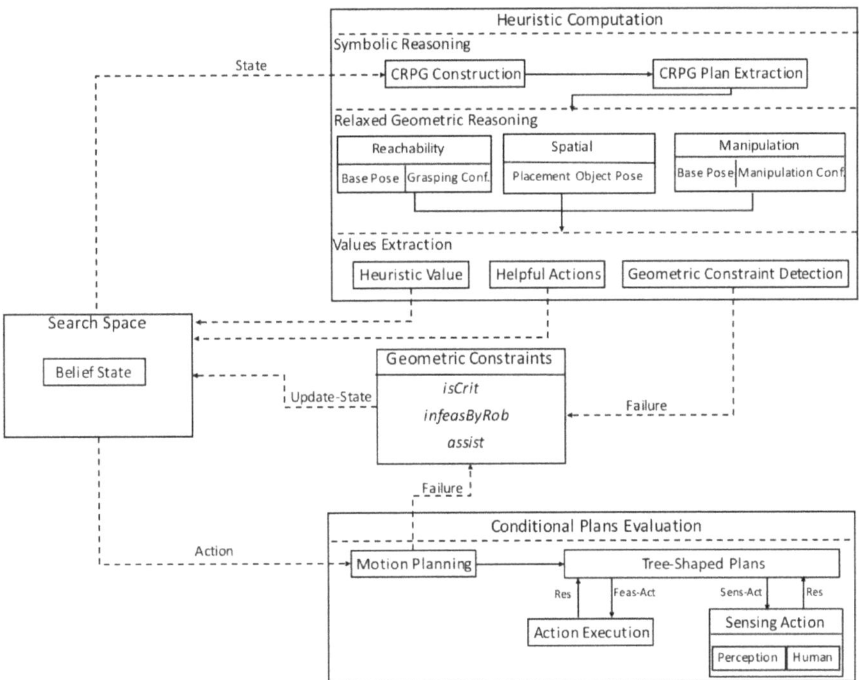

Figure 1. The proposed system overview of contingent task and motion planning using the extended version of *Contingent-FF*.

4. Relaxed Geometric Reasoning for Mobile Manipulators

Relaxed geometric reasoning is the evaluation of geometric conditions of actions with no call to motion planning. It indicates that a feasible motion is likely to be obtained for the selected actions if certain task constraints are satisfied. Therefore, the relaxed geometric reasoning process contains three modules: *reachability reasoning*, *spatial reasoning*, and *manipulation reasoning*. This set of reasoning extends our previous relaxed geometric reasoning process [19] to consider reasoning on mobile manipulation along human actions.

Reachability reasoning ($\mathcal{R}_{rch}$): This reasoning is applied for only *transit* action. To transit the robot to a target position, a feasible arm configuration and robot base pose must be first obtained. A set of robot poses is considered in the workspace of the robot. From each pose, an *Inverse Kinematic (IK)* solver is called for each candidate grasping pose, and moreover the result of *IK* is determined whether it is collision-free or not. The first collision-free *IK* solution along the corresponding grasping pose is reported if possible. Otherwise, failure is reported if there is neither *IK* solution nor collision-free configuration among a set of robot poses. Accordingly, the reasoner returns the collisionable objects.

Spatial reasoning ($\mathcal{R}_{sp}$): This reasoning is applied for the *transfer, push, open, humanOpen*, and *humanTransfer* actions. This is considered to find a valid placement for an object within a given region with no consideration of the robot. A pose is sampled, i.e., an object lies in the target region and the initial stable posture is maintained. The feasibility of the sampled pose is also determined through a collision-checking procedure to verify whether there are any collisions with other objects in the robot environment or not. If it is valid, the sampled pose is stored in the geometry details of the action which transfers the object. Otherwise, another sample will be attempted. In the case that all tried samples are not valid, failure occurs and the collisionable objects are reported. Moreover, some constraints are taken into consideration while the sample placement is accomplished. For example, in the case of the *push* action, the sample is considered in the direction in which the object is being pushed. In the case of *humanOpen* and *Open*, the valid object placement is extracted from the object feature (the box cap is assumed to have a single full-open position).

Manipulation reasoning ($\mathcal{R}_{mnp}$): This reasoning is considered to evaluate the compatibility of the grasp poses to move an object from the initial position to the final one (using $\mathcal{R}_{sp}$). The process applies $\mathcal{R}_{rch}$ reasoning to return on of the feasible ways to transfer an object from initial to final position in terms of collision-free *IK* solution. In the case that there is no possible solution meeting these conditions because of collisions, then the collisionable objects are returned. In this way, it can obtain the valid robot pose and grasping configuration when the robot manipulates an object.

Algorithm 1 describes the relaxed geometric function when applying the actions. Algorithm is detailed below:

- Reasoning about the robot actions [lines 5–15]: The *transit* action calls the reachability reasoning by the function $\mathcal{R}_{rch}$ [line 6]. The *transfer, push* and *open* actions call the spatial reasoning by the function $\mathcal{R}_{sp}$ [line 11], and then call the function $\mathcal{R}_{mnp}$ [line 13]. If the reasoning processes are successfully done, the corresponding response is set to feasible and geometric details are appended to the evaluated action [line 8] and [line 15]. On the contrary, if the failure is due to manipulatable objects, the response is set to *infeasible-criticalObjects* and the collisionable objects are stored in $\mathcal{CO}$. In other cases of failure, the response is set to *infeasible-infeasByRob* that could be because of collisions with fixed obstacles or because the *IK* module is not able to find a configuration.
- Reasoning and finding the geometric values of the human actions [lines 16–23]: The *humanTransfer* action calls the spatial reasoning by the function $\mathcal{R}_{sp}$ [line 17]. This function is responsible to find the pose of the object placement for the human action and inserts it to the action [line 19]. For the *humanOpen* action, the pose of the container object being opened is extracted from the object feature by the spatial reasoner function $\mathcal{R}_{sp}$ [line 21] and is stored into the action details [line 23].

Algorithm 1: *RelaxGeomReas(a)*

```
CO ← ∅
a.geom⁺ ← ∅
i ← 0
Res = False
if a.name = Transit then
    {Res, Q_rob, Pos_rob, CO, g} ← R_rch(a)
    if Res = feasible then
        a.geom⁺.add(Q_rob, Pos_rob, g)
else if a.name = Transfer or Push or Open then
    while i < Max do
        {Res_sp, O_j^m(pos_goal), CO} ← R_sp(a)
        if Res_sp = feasible then
            {Res, Q_rob, Pos_rob, CO, g} ← R_mnp(a, O_i^m(pos_goal))
            if Res = feasible then
                a.geom⁺.add(Q_rob, Pos_rob, O_j^m(pos_goal), g)
else if a.name = HumanTransfer then
    {Res, O_j^m(pos_goal)} ← R_sp(a)
    if Res = feasible then
        a.geom⁺.add(O_j^m(pos_goal))
else if a.name = HumanOpen then
    {Res, O_j^m(pos_open)} ← R_sp(a)
    if Res = feasible then
        a.geom⁺.add(O_j^m(pos_open))
else
    //a is not required to be checked;
    return Null
return {Res, CO}
```

5. Contingent Heuristic Computation using Relaxed Information

Heuristic computation returns the heuristic value as well as helpful actions using relaxed symbolic along geometric reasoning from each state. Algorithm 2 explains the modified version of *Contingent-FF* heuristic computation for a given belief state S and goal $\mathcal{G}$ by taking into account geometric information. This involves three steps: computing the *CRPG* and the relaxed plan π', determining π', and computing the heuristic value and the helpful actions, as follows.

Computing the CRPG and π' [lines 1–2]: The *CRPG* graph $CRPG_{gr}$ involving state layers and action layers is built by the function CRPGConst [line 1]. The function CRPGPlan extracts π' from that graph [line 2]. The process is performed in a similar way to the standard *Contingent-FF*.

Evaluating π' [lines 3–13]: Actions in π' are sent to the relaxed geometric reasoning for the feasibility evaluation [line 5]. Basically, this process tries to figure out whether there is any feasible world to meet the action conditions or not as we proposed in [19]. Upon failure, the function MaxUp [line 9] determines whether a predefined maximum number of trials is reached or not to update the belief state and find another relaxed plan. If updating the state is required, the feedback of the geometric reasoner is evaluated. In the case of failure because of *infeasible-criticalObjects*, the literal *isCrit(*$\mathcal{CO}$*,* $\mathcal{O}'$*, Pos)* with critical objects is added to the current belief state. Otherwise, the failure is because of *infeasible-infeasByRob* and the literal *infeasByRob(*$\mathcal{R}$*,* $\mathcal{O}'$*, Pos)* is added to the state. In this case, if the type of action is either transit or open, the literal *assist(Human, $\mathcal{O}'$, Pos)* is also added to the state.

Computing the heuristic value and with helpful actions [lines 14–15]: In the case that the relaxed plan is geometrically feasible with respect to the geometric reasoning evaluation, the heuristic value along the helpful actions are achieved. The function HValue extracts the heuristic value $h(S)$ [line 14] and the function HelpAct reports helpful actions $H(S)$ [line 15] as the generic *Contingent-FF*.

Algorithm 2: $CRPG(S, \mathcal{G})$

1 $CRPG_{gr} \leftarrow \text{CRPGConst}(\mathcal{G})$
2 $\pi' \leftarrow \text{CRPGPlan}(CRPG_{gr})$
3 **foreach** $\{a' \in \pi'\}$ **do**
4   **while** *True* **do**
5         $\{Res, \mathcal{CO}\} \leftarrow \text{RelaxGeomReas}(a')$
6   **if** $Res = True$ **then**
7   break
8   **else**
9   **if** $\text{MaxUp}(S) < Max$ **then**
10                 $S \leftarrow \text{UpState}(S, Res, \mathcal{CO})$
11   **return** $\text{CRPG}(S, \mathcal{G})$
12   **else**
13   **return** $\{\infty, \emptyset\}$
14 $h(S) \leftarrow \text{HValue}()$
15 $H(S) \leftarrow \text{HelpAct}()$
16 **return** $\{h, H(S)\}$

An example is considered to show how geometric constraints are captured and handled during the heuristic computation. The initial scene of the example is shown in Figure 2 where the robot is required to move *Can A* inside *Box*. To make the problem challenging, it is assumed that top grasps are not allowed and some side grasping poses are considered for each can. Several task constraints are imposed, e.g., there is no direct collision-free motion to reach *Can A*, and also the robot is not able to open *Box* and needs an operator assistant.

The computation of the heuristic process in terms of geometric feasibility is represented in Figure 3. The corresponding physical world for each relaxed plan has been shown also. Figure 3a shows the initial relaxed plan extracted. When the first action is forwarded for the relaxed geometric reasoning, the reachability reasoner fails. This is because when the inverse kinematic module checks side grasping poses considered for the box cap, all retrieved joint configurations have collisions with the box object. The geometric reasoning process is done by the proposed function RelaxGeomReas in Algorithm 2 that is added to the basic planner.

To handle this task constraint, the predicates *infeasByRob(tiago, box, posBox)* and *assist(person, box, posBox)* are asserted to the planning state by the associated reasoning process. The updating sate step is done by the proposed function UpState as it lets the planner know the detected constraints of the environment. Figure 3b shows the next heuristic computation taking into account the task constraint. In this case, the spatial reasoner module successfully finds the geometric state of *Box* after applying the action *humanOpen*. However, the reachability reasoner reports a failure for evaluating the *transit* action for the object *Can A* due to collision between the robot arm and *Can B*. This object is marked as a critical object, so the state is updated with the predicate *isCrit(can B, can A, pos A)*. The heuristic computation is again repeated, and finally the reasoning processes can correctly find feasible geometric details for the actions. This process results in geometrically feasible heuristic computation.

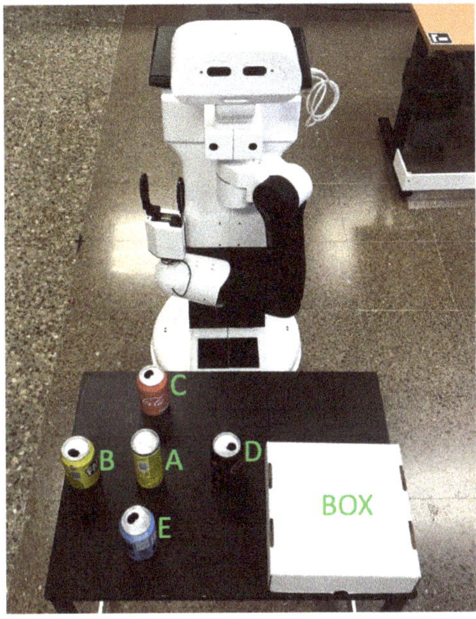

Figure 2. The initial scene where the robot requires placement of the object A within the box in the presence of geometric constrains.

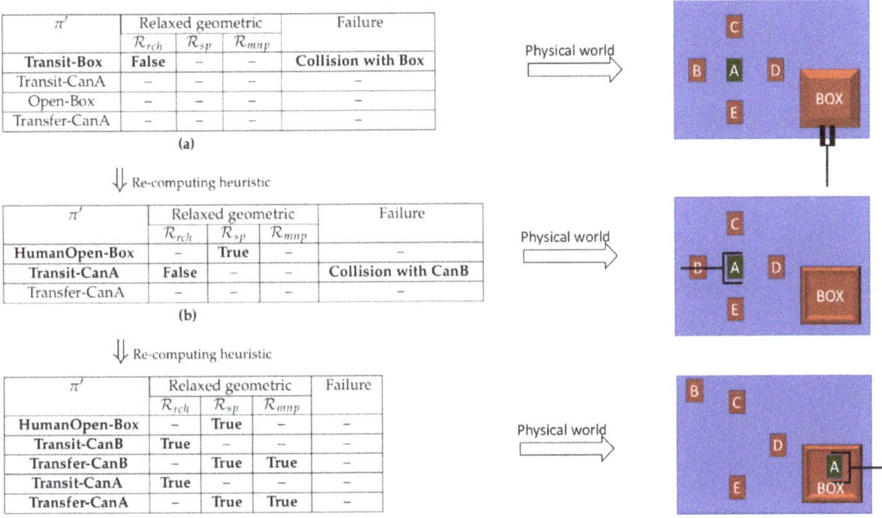

π'	Relaxed geometric			Failure
	$\mathcal{R}_{rch}$	$\mathcal{R}_{sp}$	$\mathcal{R}_{mnp}$	
Transit-Box	**False**	–	–	**Collision with Box**
Transit-CanA	–	–	–	–
Open-Box	–	–	–	–
Transfer-CanA	–	–	–	–

(a)

⇓ Re-computing heuristic

π'	Relaxed geometric			Failure
	$\mathcal{R}_{rch}$	$\mathcal{R}_{sp}$	$\mathcal{R}_{mnp}$	
HumanOpen-Box	–	**True**	–	–
Transit-CanA	**False**	–	–	**Collision with CanB**
Transfer-CanA	–	–	–	–

(b)

⇓ Re-computing heuristic

π'	Relaxed geometric			Failure
	$\mathcal{R}_{rch}$	$\mathcal{R}_{sp}$	$\mathcal{R}_{mnp}$	
HumanOpen-Box	–	**True**	–	–
Transit-CanB	**True**	–	–	–
Transfer-CanB	–	**True**	**True**	–
Transit-CanA	**True**	–	–	–
Transfer-CanA	–	**True**	**True**	–

(c)

Figure 3. The steps of the computation of heuristic using the relaxed geometric reasoning and the corresponding physical world. The information highlighted in bold shows the relaxed planning actions which have been currently tested by the proposed relaxed geometric reasoning. Others are those which have not been tested yet. True and false values show whether the reasoner is successful or failed. (**a**) The *transit* action to reach the Box fails. (**b**) The *transit* action for the Can A fails due to collision with other objects. (**c**) The final geometrically feasible heuristic computation.

6. Tree-Based Planning using Search Space

The *And-Or* search procedure as considered in the *Contingent-FF* planner is used to result in a feasible manipulation plan. The heuristic computation has been modified to incorporate geometric check and, moreover, selected actions must be evaluated using motion planning. The process is represented in Algorithm 3.

The algorithm gets $\mathcal{T}$ as input and outputs π if possible. First, the trials counter *trial* is set [line 1] and the state S_i is the initial belief state [line 4]. The function Search performs the standard search mechanism as *Contingent-FF* does [line 6]: it provides the next state using the transit function to visit S_{i+1} along the promising applicable action(s) with H_{S_i}. This step is done with the modified CRPG function (see Algorithm 2), by taking geometric constraints into account.

In the case that H_{S_i} does not exist [line 7], the algorithm performs another search from the beginning. Until the maximum number of iterations is not reached [line 9], the process is repeated with the initial state updated by the function UpdateInitState [line 11]. If the maximum number of trials is reached, the process returns failure [line 14].

Algorithm 3: The Proposed Planning Algorithm

inputs : $\mathcal{T}$=$\langle\mathcal{D},\mathcal{S}_0,\mathcal{G}\rangle$, $\mathcal{D}$=$\langle\mathcal{A},\Omega,\mathcal{F},\mathcal{W},S_g\rangle$
output: π
1 $trial \leftarrow 0$
2 $i \leftarrow 0$
3 $\pi \leftarrow \emptyset$
4 $\mathrm{S_i} \leftarrow S_{\mathrm{init}}$
5 **while** $\mathcal{G} \not\subseteq S_i$ **do**
6 $\quad \{H_{S_i}, S_{i+1}\} \leftarrow$ Search$(S_i, \mathcal{G}, \mathcal{A}, \Omega)$
7 $\quad$ **if** $H_{S_i} = \emptyset$ **then**
8 $\quad\quad trial \leftarrow trial + 1$
9 $\quad\quad$ **if** $trial < Max$ **then**
10 $\quad\quad\quad i \leftarrow 0$
11 $\quad\quad\quad S_i \leftarrow$ UpdateInitState()
12 $\quad\quad\quad$ Continue
13 $\quad\quad$ **else**
14 $\quad\quad\quad$ **return** fail
15 $\quad$ **else**
16 $\quad\quad$ **if** $H_{S_i} \notin \Omega$ ***And*** $H_{S_i}.name \neq$ HumanTransfer ***And*** $H_S.name \neq$ HumanOpen **then**
17 $\quad\quad\quad \{\mathcal{Q}, Res, \mathcal{CO}\} \leftarrow$ MotionPlanner(H_{S_i})
18 $\quad\quad$ **if** $H_{S_i} \in \Omega$ ***Or*** $H_{S_i}.name =$ HumanTransfer ***Or*** $H_S.name =$ HumanOpen ***Or*** $Res = feasible$ **then**
19 $\quad\quad\quad \pi$.append(H_{S_i})
20 $\quad\quad$ **else**
21 $\quad\quad\quad S_i \leftarrow$ UpdateState$(Res, \mathcal{CO})$
22 $\quad\quad\quad$ Continue
23 $\quad i \leftarrow i + 1$
24 **return** π

For those actions that either do not belong to the set of sensing actions and are not assigned to human, the MotionPlanner function is used to compute a collision-free path for the currently selected action(s) [line 17]. If a path is found, *Res* is set to *feasible* and the path $\mathcal{Q}$ is returned. Afterwards, π is appended with the sensing, human, or normal action(s) [line 19]. In the case of failure due to *infeasible-criticalObjects*, the literal *isCrit*($\mathcal{CO}$, $\mathcal{O}'$, *Pos*) with critical objects is added to the current belief state. Otherwise, the failure is because of motion planning time-out problem or collisionable fixed obstacles, the type of failure is *infeasible-infeasByRob* and the literal *infeasByRob*($\mathcal{R}$, $\mathcal{O}'$, *Pos*) is added to the state. In this case, if the type of action is either transit or open, the literal *assist(Human, $\mathcal{O}'$, Pos)* is also added to the state.

An example is considered to illustrate how the geometrically feasible conditional plan is obtained offline under belief information of the initial state. The scene depicted in Figure 4 shows the initial belief state of the mobile manipulation problem, where the color of the gray cylinder is uncertain (it could be red or green), it is not known whether the can is filled or not, nor if the containers are open or closed. The goal is to transfer the cylinder A to either the red or the green tray. Some particular placements regions allocated for the manipulatable objects if required:

- The green cylinders must be placed on the green tray.
- The red cylinders must be placed over the red tray.
- The blue cylinders must be placed within the containers.
- The can objects may be optionally placed anywhere over the table.

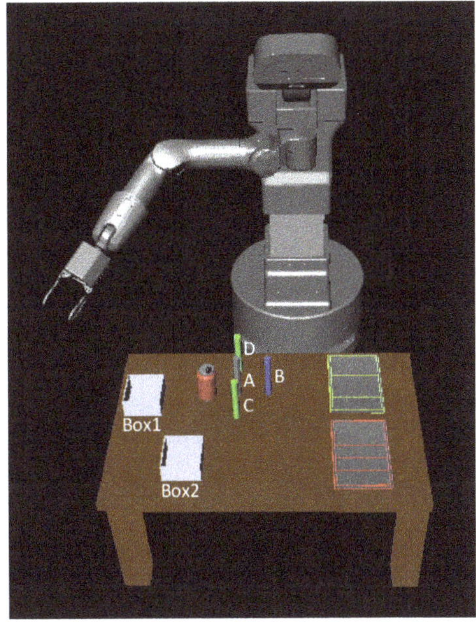

Figure 4. The manipulation example where the goal is to transfer cylinder A to one of the trays with respect to its color.

The complete conditional plan is represented in Figure 5 that is obtained by Algorithm 3. It is briefly discussed how this geometrically feasible plan is obtained. While the planning process is taking place, there are several challenges in terms of geometric constraints which are captured and handled by the proposed geometry reasoner. These steps are mainly done using the Search function which internally calls Algorithm 2. To reach the target object, the reachability reasoning process, place in the function RelaxGeomReas, detects cylinder B and reports that the object is blocking the way of reaching object A in the heuristic computation. Therefore, the predicate *isCrit(cylinderB, cylinderA, posA)* is inserted to the initial belief state of the planner using the function updating the belief state. This predicate says that cylinder B blocks the way of reaching cylinder A.

Furthermore, when the robot attempts to find a feasible configuration for opening box 1 in the case that the box is closed, the reachability reasoner fails due to colliding with the box. Here, it is the case that robot needs to ask a human operator for collaboration. Accordingly, the reasoner appends the predicates *infeasByRob(tiago, box1, posBox1)* and *assist(person, box1, posBox1)* to the corresponding state. The *humanOpen* action then appears.

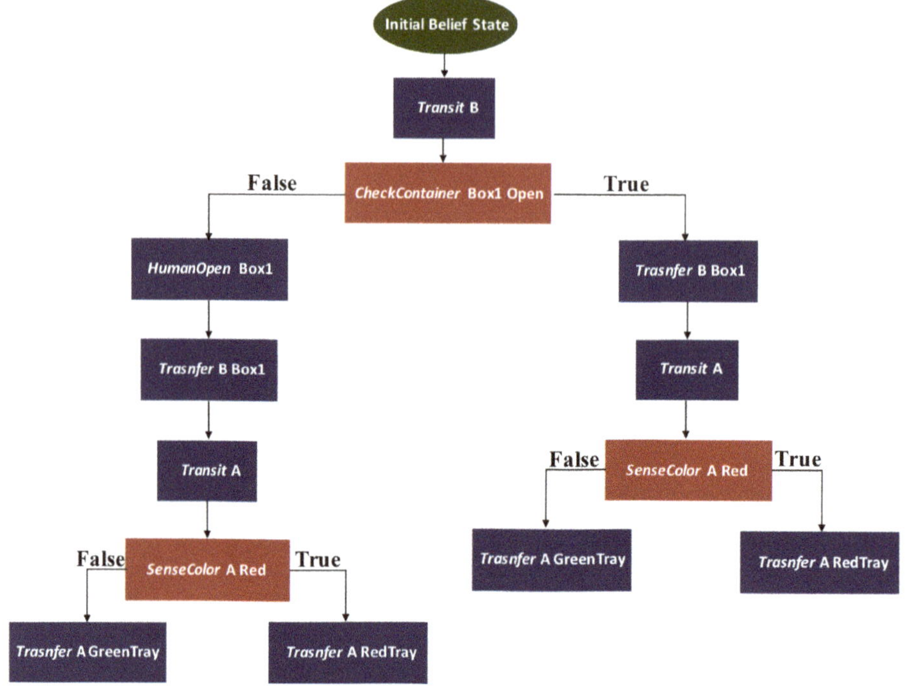

Figure 5. The conditional plan results from the proposed planning process. The actions highlighted with the blue color are the ones assigned to the robot or a human operator. Some important actions parameters are represented. The actions specified by the red color are sensing actions.

7. Manipulation Plan Execution using Sensing and Human Interaction

When the manipulation conditional plan is achieved, it will be forwarded to for the execution module. Algorithm 4 outlines the process of actions execution performed by the robot or human, and calls to the sensing actions. The conditional plan is initialized from its root [line 1]. For each action of the plan, its type first identified whether it is execution or sensing one. In the case of execution action, if it has to be executed by human, the function executeByHuman asks a person to do the corresponding action [line 5] and an operator then sends a command to the robot that the action has been done successfully. Otherwise, the action is executed by the robot [line 7].

On the other hand, if the type of action becomes sensing, the function senseAct determines the binary value of the sensing action which is *True* or *False*. This is done using the perception module allocated for the robot. Depending on the type of uncertainty, the function may request to human or activate a sensing module to observe the action value. Regarding the *CheckCan* or *CheckOpen* sensing actions, human information is used, while a sensing module is used for the *SensePose* and *SenseColor* sensing actions.

Algorithm 4: Manipulation Plan Execution

```
inputs : π
initializePlan(π)
foreach {H_S ∈ π} do
    if H_S ∈ A then
        if H_S.name = HumanTransfer Or HumanOpen then
            executeByHuman(H_S)
        else
            executeByRob(H_S)
    else
        Res = senseAct(H_S)
        selectBranch(π, H_S, Res)
```

8. Empirical Results and Discussion

This section describes some manipulation problem solved using the proposed framework, and the implementation issues. The mobile robot considered is *TIAGo*. It has 7 degrees of freedom arm, equipped with a gripper, mounted on a mobile platform through a lift torso.

The executive simulated result of the manipulation problem represented in Figure 4, called *Problem-1*, is shown in Figure 6. For the domain of the problem, a number of actions is considered for the robot being *transit, transfer, open,* and *push* along with some actions for an operator that are *humanTransfer* and *humanOpen*. Actions are selected according to the planning mechanism in terms of symbolic and geometric reasoning. We assume that the values of the uncertainty information are provided in run-time in simulation. Therefore, the executive plan is provided below:

Executive Plan: { *Transit-B, CheckContainer-Box1-Open (False), HumanOpen-Box1, Transfer-B-Box1, Transit-A, SenseColor-A-Red (True), Transfer-A-RedTray* }

The states represented in the figure are classified as follows:

(a) is the initial belief state of the robot and environment.
(b) is the state where the robot applies *transit* action to reach cylinder B.
(c) is the state where the sensing action *CheckContainer-Box1-Open* showed that Box 1 is currently closed.
(d) is the state where the *HumanOpen* action is executed as the robot is not capable enough to open the box.
(e) is the state where the robot places cylinder B within box 1.
(f) is the state where the robot transits to cylinder A.
(g) is the state where the sensing action *SenseColor-A-Red* showed that the cylinder is actually red.
(h) is the state where the robot moves its base and the arm configuration to place cylinder A over the red tray.

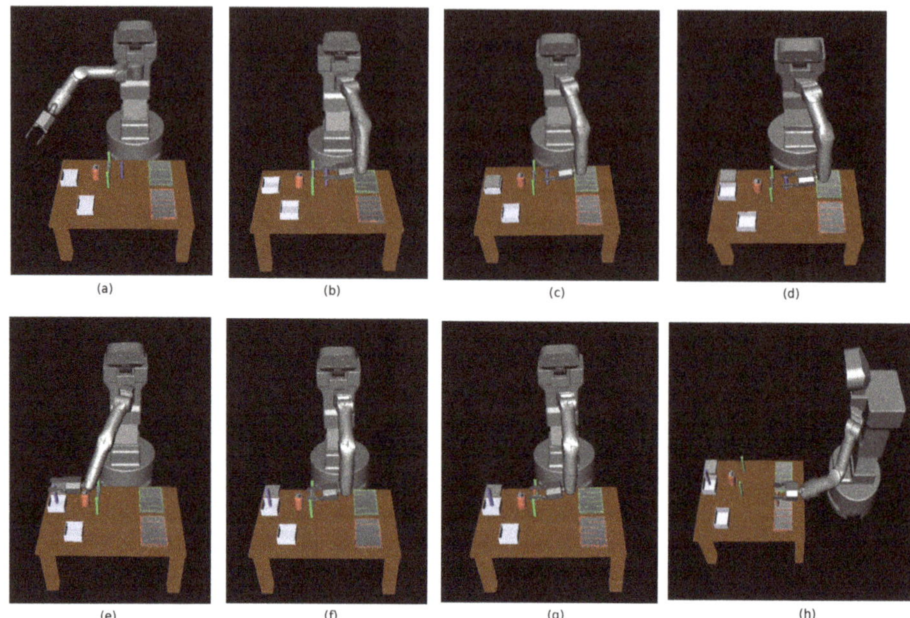

Figure 6. The simulation results of the executive plan performed by the *TIAGo* robot: (**a**) is the initial robot and environment state, (**b**) is the state after *Transit* action towards cylinder B, (**c**) is the result of the sensing action *CheckContainer-Box1-Open*, (**d**) is the state after applying *HumanOpen* action, (**e**) is the state after the robot executes the *Transfer* action for cylinder B, (**f**) is the state when the robot transits to cylinder A, (**g**) is the state resulting from the sensing action *SenseColor-A-Red*, and (**h**) is the state when the robot place cylinder A on the associated tray.

In addition, the proposal has been evaluated for other cluttered problems where the robot needs to sort objects according to the colors. Regarding the action domain, Robot actions are *transit* and *transfer*, and the action template *humanTransfer* is considered for an operator. The problem represented in Figure 7, called *Problem-2*, shows the initial and goal states of manipulation where the green and red objects must be located on the green and red regions respectively. The red object is not initially located on the table. The pink region is considered on the robot workspace where an operator can transfer objects. The planning uncertainties are the color of the green object which could be actually green or red and the location of the red object which could be on the robot table or in the human workspace. Therefore, the *humanTransfer* action is applied to transfer the object to the robot workspace as the robot is not allowed to move to the human workspace. The final executable plan would be to transfer the green object to the target placement region by the robot. It then looks for the red object and figures out the object is not located on the table and asks an operator to transfer the object. The *humanTransfer* action is selected in the conditional plan, so the requested object is transferred to the robot workspace. The operator updates the robot knowledge through the robot system terminal. The robot is aware that the human action has been successfully performed, and afterwards it travels to grasp the object. Eventually, the robot transfers the object to the target region.

The proposed approach has been tested for similar problems by increasing the number of objects and varying color and/or location uncertainties. The problems performance are represented in Table 1 in terms of conditional and executive plan length, and moreover planning time. *Problems-3* includes a cluttered problem where there are nine objects and three of them need to be sorted. Similar uncertainty of *Problem-2* is considered regarding the color and location of objects. *Problem-4* is the one where 12 objects exist and four of them must be sorted. In this case, the uncertainty information like the objects color and locations are considered for more objects.

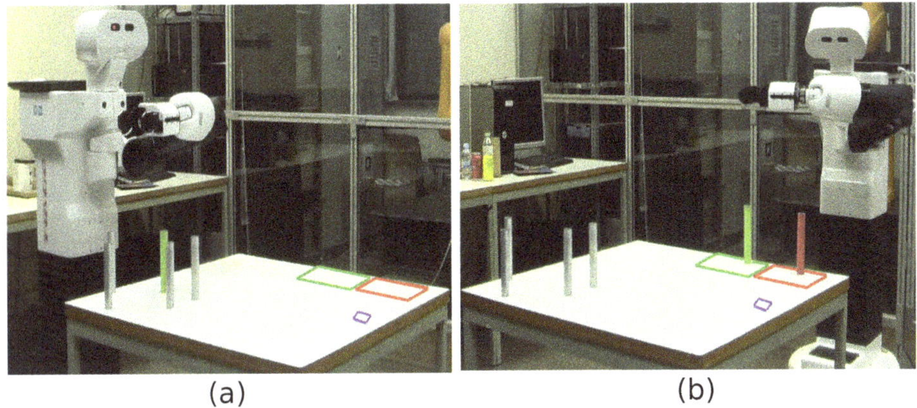

Figure 7. The manipulation example where green and red objects must be placed in the green and red regions. (**a**) shows the initial state of the problem. (**b**) shows the final state of the problem. The red object is not initially located in the robot workspace. The pink region is the place where human can transfer objects to the robot workspace. The solution can be visualized here: https://sir.upc.es/projects/ontologies/GreenRedHuman.mp4. The solution for the case that the red object is initially located on the table is visualized here: https://sir.upc.es/projects/ontologies/TiagoRedGreenRob.mp4.

Table 1. The conditional plan and executive plan length in terms of number of sensing and executive actions and planning time in seconds for the evaluated problems.

Problem	Conditional Plan		Executive Plan		Planning Time
	Sensing	Executive	Sensing	Executive	
Problem-1	3	10	2	5	35
Problem-2	3	13	2	5	59
Problem-3	3	19	2	8	163
Problem-4	7	41	3	12	449

Concerning the implementation framework, four components are considered: task planning, relaxed geometric reasoning, motion planning, and executive module. Task planning is developed using a modified version of the *Contingent-FF* planner coded in C++. All the action templates are described using *PDDL* by considering *ADL (Action Description Language*, ref. [28]) enabling us to define operators in a more compact way, using quantifiers and conditional effects. There is not any pre-processing step to compute geometric details of actions and they are computed and assigned during the manipulation planning process.

We use *The Kautham Project* [29], a C++-based open-source tool for motion planning that enables planning under geometric and kinodynamic constraints for relaxed geometric reasoning and motion planning. It uses the *Open Motion Planning Library (OMPL)* [30] as a core set of sampling-based planning algorithms. In this work, the *RRT-Connect* [31] motion planner is used for motion planning. This planner is one of the most efficient motion planners, but it does not guarantee optimal motions. *The Kautham Project* involves different collision checking modules to detect robot-object and object-object collisions, and features a placement sampling mechanism to find feasible object poses in the workspace. Relaxed geometric reasoning uses these modules to find feasible sample geometric instances for symbolic actions. The executive module uses a sensing module which uses the 3D camera mounted inside the *TIAGo* robot, and also some components provided by *PAL Robotics* to send a motion path to the robot. The communication between task, relaxed geometric reasoning, motion planning, and executive modules is done via *Robotic Operating System (ROS)* [32].

9. Conclusions

This paper has proposed a contingent-based task and motion planning approach able to cope with high-dimension mobile manipulation problems in the presence of high-level uncertainty and human interactions (referred to the sharing of knowledge and to collaborative actions which are out of the robot capabilities). For this purpose, the basic *Contingent-FF* planner has been modified to include robot action reasoning, human–robot collaboration, and state observation. A set of geometric reasoning processes has been offered to the planning process to capture the task constraints imposed in the robot environment and to update belief state while task planning is done. Moreover, some modules linked with the human knowledge along with the perception system, have been also designed to observe the binary outcomes of actions. It is worth noting that the proposed approach results in a tree-shaped conditional plan which is geometrically feasible regardless of the values of sensing actions.

To evaluate the proposed approach, several manipulation tasks have been executed in simulation and real environments to show the way of tackling human–robot interactions, and identifying and handling both geometric constraints and high-level uncertainty. Problems performance has been reported in terms of the length of the manipulation plan and planning time, considering an increasing number of objects. In all the cases, the robot in collaboration with the human operator has been able to solve the tasks despite the uncertainty and the constraints.

Future work will concentrate on manipulation tasks also subject to low-level geometric uncertainty, its effects in sensing and how it is transferred to task planning.

Author Contributions: A.A. and J.R. conceived the theoretical contributions, A.A. wrote the paper and implemented the whole framework, being assisted by M.D. for the perception part. All authors have read and agreed to the published version of the manuscript.

Funding: This work was partially supported by the Spanish Government through the project DPI2016-80077-R. Mohammed Diab is supported by the Spanish Government through the grant 2017.

Conflicts of Interest: The authors declare no conflict of interest.

References

1. Hoffmann, J.; Brafman, R. Contingent planning via heuristic forward search with implicit belief states. In Proceedings of the International Conference on Automated Planning and Scheduling, Monterey, CA, USA, 5–10 June 2005.
2. Stilman, M.; Schamburek, J.U.; Kuffner, J.; Asfour, T. Manipulation planning among movable obstacles. In Proceedings of the 2007 IEEE International Conference on Robotics and Automation, Roma, Italy, 10–14 April 2007; pp. 3327–3332.
3. Stilman, M.; Kuffner, J. Planning among movable obstacles with artificial constraints. *Int. J. Robot. Res.* **2008**, *27*, 1295–1307. [CrossRef]
4. Rodríguez, C.; Suárez, R. Combining motion planning and task assignment for a dual-arm system. In Proceedings of the 2016 IEEE/RSJ International Conference on Intelligent Robots and Systems (IROS), Daejeon, Korea, 9–14 October 2016; pp. 4238–4243.
5. Zarandi, M.F.; Mosadegh, H.; Fattahi, M. Two-machine robotic cell scheduling problem with sequence-dependent setup times. *Comput. Oper. Res.* **2013**, *40*, 1420–1434. [CrossRef]
6. Foumani, M.; Smith-Miles, K.; Gunawan, I.; Moeini, A. A framework for stochastic scheduling of two-machine robotic rework cells with in-process inspection system. *Comput. Ind. Eng.* **2017**, *112*, 492–502. [CrossRef]
7. Ghallab, M.; Nau, D.; Traverso, P. *Automated Planning: Theory & Practice*; Elsevier: San Francisco, CA, USA, 2004.
8. Lagriffoul, F.; Andres, B. Combining task and motion planning: A culprit detection problem. *Int. J. Robot. Res.* **2016**, *35*, 890–927. [CrossRef]

9. Dantam, N.; Kingston, Z.K.; Chaudhuri, S.; Kavraki, L.E. Incremental Task and Motion Planning: A Constraint-Based Approach. In Proceedings of the Robotics: Science and Systems, Cambridge, MA, USA, 18–22 June 2016.
10. He, K.; Lahijanian, M.; Kavraki, L.E.; Vardi, M.Y. Towards manipulation planning with temporal logic specifications. In Proceedings of the International Conference on Robotics and Automation, Seattle, WA, USA, 26–30 May 2015; pp. 346–352.
11. Akbari, A.; Muhayyudin; Rosell, J. Task and Motion Planning Using Physics-based Reasoning. In Proceedings of the IEEE International Conference on Emerging Technologies and Factory Automation, Luxembourg, 8–11 September 2015.
12. Srivastava, S.; Fang, E.; Riano, L.; Chitnis, R.; Russell, S.; Abbeel, P. Combined task and motion planning through an extensible planner-independent interface layer. In Proceedings of the IEEE International Conference on Robotics and Automation Robotics and Automation, Hong Kong, China, 31 May–5 June 2014; pp. 639–646.
13. Diab, M.; Akbari, A.; Muhayyuddin; Rosell, J. PMK—A Knowledge Processing Framework for Autonomous Robotics Perception and Manipulation. *Sensors* **2019**, *19*, 1166. [CrossRef] [PubMed]
14. Cambon, S.; Alami, R.; Gravot, F. A hybrid approach to intricate motion, manipulation and task planning. *Int. J. Robot. Res.* **2009**, *28*, 104–126. [CrossRef]
15. Garrett, C.R.; Lozano-Pérez, T.; Kaelbling, L.P. FFRob: An efficient heuristic for task and motion planning. In *Algorithmic Foundations of Robotics XI*; Springer: Cham, Switzerland, 2015; pp. 179–195.
16. Akbari, A.; Muhayyuddin; Rosell, J. Reasoning-based Evaluation of Manipulation Actions for Efficient Task Planning. In Proceedings of the ROBOT2015: Second Iberian Robotics Conference, Lisbon, Portugal, 19–21 November 2015.
17. Akbari, A.; Muhayyudin; Rosell, J. Task Planning Using Physics-based Heuristics on Manipulation Actions. In Proceedings of the IEEE International Conference on Emerging Technologies and Factory Automation, Berlin, Germany, 6–9 September 2016.
18. Akbari, A.; Muhayyuddin; Rosell, J. Knowledge-oriented task and motion planning for multiple mobile robots. *J. Exp. Theor. Artif. Intell.* **2019**, *31*, 137–162. [CrossRef]
19. Akbari, A.; Lagriffoul, F.; Rosell, J. Combined heuristic task and motion planning for bi-manual robots. *Auton. Robot.* **2018**, 1–16. [CrossRef]
20. Bryce, D.; Kambhampati, S.; Smith, D.E. Planning graph heuristics for belief space search. *J. Artif. Intell. Res.* **2006**, *26*, 35–99. [CrossRef]
21. Petrick, R.P.; Bacchus, F. Extending the Knowledge-Based Approach to Planning with Incomplete Information and Sensing. In Proceedings of the International Conference on Automated Planning and Scheduling, Whistler, BC, USA, 3–7 June 2004; pp. 2–11.
22. Bonet, B.; Geffner, H. Planning under partial observability by classical replanning: Theory and experiments. In Proceedings of the IJCAI Proceedings-International Joint Conference on Artificial Intelligence, Barcelona, Catalonia, Spain, 16–22 July 2011; Volume 22, p. 1936.
23. Shani, G.; Brafman, R.I. Replanning in domains with partial information and sensing actions. *IJCAI* **2011**, *2011*, 2021–2026.
24. Maliah, S.; Brafman, R.I.; Karpas, E.; Shani, G. Partially Observable Online Contingent Planning Using Landmark Heuristics. In Proceedings of the International Conference on Automated Planning and Scheduling, Portsmouth, NH, USA, 21–26 June 2014.
25. Gaschler, A.; Petrick, R.; Kröger, T.; Knoll, A.; Khatib, O. Robot task planning with contingencies for run-time sensing. In Proceedings of the International Conference on Robotics and Automation Workshop on Combining Task and Motion Planning, Karlsruhe, Germany, 6–10 May 2013.
26. Nouman, A.; Yalciner, I.F.; Erdem, E.; Patoglu, V. Experimental evaluation of hybrid conditional planning for service robotics. In Proceedings of the International Symposium on Experimental Robotics, Tokyo, Japan, 3–6 October 2016; pp. 692–702.
27. Hoffmann, J.; Nebel, B. The FF planning system: Fast plan generation through heuristic search. *J. Artif. Intell. Res.* **2001**, 253–302. [CrossRef]

28. Pednault, E.P.D. ADL: Exploring the Middle Ground Between STRIPS and the Situation Calculus. In Proceedings of the First International Conference on Principles of Knowledge Representation and Reasoning, Toronto, ON, Canada, 15–18 May 1989; Morgan Kaufmann Publishers Inc.: San Francisco, CA, USA, 1989; pp. 324–332.
29. Rosell, J.; Pérez, A.; Aliakbar, A.; Muhayyuddin; Palomo, L.; García, N. The Kautham Project: A teaching and research tool for robot motion planning. In Proceedings of the IEEE International Conference on Emerging Technologies and Factory Automation, Barcelona, Spain, 16–19 September 2014.
30. Sucan, I.; Moll, M.; Kavraki, L.E.; others. The open motion planning library. *Robot. Autom. Mag.* **2012**, *19*, 72–82. [CrossRef]
31. Kuffner, J.J.; LaValle, S.M. RRT-connect: An efficient approach to single-query path planning. In Proceedings of the International Conference on Robotics and Automation, San Francisco, CA, USA, 24–28 April 2000; Volume 2, pp. 995–1001.
32. Quigley, M.; Conley, K.; Gerkey, B.; Faust, J.; Foote, T.; Leibs, J.; Wheeler, R.; Ng, A.Y. ROS: An open-source Robot Operating System. In Proceedings of the International Conference on Robotics and Automation Workshop on Open Source Software, Kobe, Japan, 12–17 May 2009; Volume 3, p. 5.

Article

Automatic Design of Collective Behaviors for Robots that Can Display and Perceive Colors

David Garzón Ramos * and Mauro Birattari *

Institut de Recherches Interdisciplinaires et de Développements en Intelligence Artificielle (IRIDIA), Université Libre de Bruxelles, 1050 Brussels, Belgium

* Correspondence: dgarzonr@ulb.ac.be (D.G.R.); mbiro@ulb.ac.be (M.B.); Tel.: +32-02-6502729 (D.G.R.)

Received: 16 June 2020; Accepted: 1 July 2020; Published: 6 July 2020

Abstract: Research in swarm robotics has shown that automatic design is an effective approach to realize robot swarms. In automatic design methods, the collective behavior of a swarm is obtained by automatically configuring and fine-tuning the control software of individual robots. In this paper, we present `TuttiFrutti`: an automatic design method for robot swarms that belongs to AutoMoDe—a family of methods that produce control software by assembling preexisting software modules via optimization. The peculiarity of `TuttiFrutti` is that it designs control software for e-puck robots that can display and perceive colors using their RGB LEDs and omnidirectional camera. Studies with AutoMoDe have been so far restricted by the limited capabilities of the e-pucks. By enabling the use of colors, we significantly enlarge the variety of collective behaviors they can produce. We assess `TuttiFrutti` with swarms of e-pucks that perform missions in which they should react to colored light. Results show that `TuttiFrutti` designs collective behaviors in which the robots identify the colored light displayed in the environment and act accordingly. The control software designed by `TuttiFrutti` endowed the swarms of e-pucks with the ability to use color-based information for handling events, communicating, and navigating.

Keywords: swarm robotics; automatic design; AutoMoDe; evolutionary robotics

---

1. Introduction

A robot swarm [1,2] is a group of robots that operate autonomously without relying on a leader robot or on external infrastructures. By cooperating, the robots of a swarm can collectively accomplish missions that individual robots could not accomplish alone. The collective behavior of a robot swarm—and hence its ability to accomplish a particular mission—is the result of the interactions that the robots have with the environment and with their peers [3].

Unfortunately, conceiving and implementing a collective behavior for a robot swarm is particularly challenging. Indeed, to obtain a collective behavior, one must conceive and implement the control software of the individual robots. The problem is that no generally applicable method exists to tell what an individual robot should do so that the desired behavior is obtained [4]. Automatic design is a promising approach to address this problem. An automatic design method produces control software via an optimization algorithm that maximizes an appropriate mission-dependent objective function. For a recent literature review on the automatic design of robot swarms, see Francesca et al. [5].

Traditionally, research on the automatic design of robot swarms adopts the neuro-evolutionary approach [6,7]. Design methods based on neuro-evolution produce control software in the form of artificial neural networks. The architecture and parameters of the network are selected by an evolutionary algorithm. As an alternative to neuro-evolution, some modular methods have been proposed [8–14]. In the modular approach, preexisting software modules are combined and tuned by an optimization algorithm. Results show that modular methods are more suitable to produce communication-based behaviors [14] and are more robust to the so-called reality gap [8,15], that is, the possibly subtle but unavoidable differences between reality and the simulation models used in the design process.

In this paper, we present TuttiFrutti: a method for the automatic design of swarms of e-pucks (extended with an omnidirectional vision turret [16]) that can display and perceive colors. TuttiFrutti designs control software for the individual robots in the swarm by selecting, tuning, and assembling preexisting software modules into probabilistic finite state machines. TuttiFrutti is an instance of AutoMoDe [8]—a family of modular methods for the realization of robot swarms. TuttiFrutti differentiates from previous instances of AutoMoDe by enabling the production of control software that operates with information expressed in the form of colors. More precisely, TuttiFrutti is intended to solve classes of missions in which robots shall act according to colors displayed by objects in their environment and/or their peers. With TuttiFrutti, we significantly enlarge the variety of collective behaviors that can be obtained by AutoMoDe. The study we present in this paper is framed within the tenets of the automatic off-line design of robot swarms, as recently defined by Birattari et al. [17]: (i) TuttiFrutti is not intended to solve a specific design problem but rather a class thereof, without the need to undergo any problem-specific modification or adjustment; (ii) once a design problem is specified, human intervention is not provided for in any phase of the design process.

In our research, we address the following questions: Is TuttiFrutti capable of deciding whether a color displayed in the environment provides information useful to accomplish a mission? Can TuttiFrutti produce collective behaviors that exhibit color-based communication between robots? Do the extended capabilities of the e-puck increase the difficulty of automatically designing control software for the robot swarm? How could these new resources be used to create more complex missions?

We consider a model of the e-puck that can use its RGB LEDs for emitting color signals, and its omnidirectional vision turret [16] for detecting robots or other objects that display colors in the environment. We conduct our study to demonstrate that e-pucks that display and perceive colors enable the automatic design of collective behaviors with event-handling, communication and navigation properties. As a proof of concept, we assess TuttiFrutti in three missions in which colors displayed in the environment play a different role: STOP, AGGREGATION, and FORAGING. In STOP, the robots must stop moving as soon as a color signal appears in the environment. In AGGREGATION, the robots must aggregate in a region where a specific color is displayed. In FORAGING, the robots must forage in an environment that has two sources of items—the sources differ in the profit they provide and in the color displayed at their location. We report a statistical analysis of results obtained with realistic computer simulations and with a swarm of e-puck robots.

Alongside the results of TuttiFrutti, we report the results obtained by EvoColor—a design method based on neuro-evolution. EvoColor is a straightforward implementation of the neuro-evolutionary approach that, likewise TuttiFrutti, produces control software for swarms of e-pucks that can display and perceive colors. We report these results as a reference for appraising the complexity of the missions considered in our study. In the absence of a well established state-of-the-art off-line design method for the missions proposed here, we consider EvoColor as a reasonably appropriate yardstick against which we can assess the performance of TuttiFrutti. A thorough comparison of TuttiFrutti against any possible declination of the neuro-evolutionary approach is well beyond the scope of this paper.

The paper is structured as follows: Section 2 discusses previous related work; Section 3 introduces TuttiFrutti; Section 4 describes the experimental set-up; Section 5 presents the results; and Section 6 concludes the paper and highlights future work.

2. Related Work

In this section, we first introduce studies in which the robots of a swarm have the capabilities of displaying and perceiving colors. After, we revise related works on automatic design of robot swarms. Finally, we compare TuttiFrutti with the other existing instances of AutoMoDe.

Robots that can display or perceive colors have been largely used to demonstrate collective behaviors in swarm robotics. The literature on robot swarms that use visual information is extensive and it is not our intention to provide an exhaustive review. We exclude from our discussion any system in which robots only perceive visual information but do not display it—such as [18–22]. Instead, we focus on studies in which the robots both display and perceive colors to achieve collective behaviors [23–47].

Designers of robot swarms commonly use color lights to represent specific information that robots in the swarm must identify, process and/or transmit—the nature of the information varies from one study to another and is used ad hoc to obtain a particular behavior. For example, Nouyan et al. [35] designed a swarm that connects locations by establishing a chain of robots that act as waypoints for their peers. They conducted two experiments in which robots use colors differently: in the first experiment, robots repeat a pattern of 3 colors along the chain to indicate the sense in which the peers should move; in the second one, robots use colors to inform their peers about a location of interest. Mathews et al. [25] designed a swarm in which robots self-organize in mergeable structures. In their experiments, robots react to colored objects in their environment and display color signals that indicate their location. Garattoni and Birattari [31] designed a robot swarm that autonomously identifies and perform sequences of tasks. In their experiments, robots emit color signals to coordinate their collective action and associate each task in a sequence with objects that display a particular color. In a more general sense, one can find a similar approach in swarms that exhibit self-assembly and morphogenesis [23–26], collective fault detection [25,27], collective exploration [28–31], collective transport [28,30], coordinated motion [25,26,32], human-swarm interaction [33,34], chain formation [28,31,35], group size regulation [36], task allocation [31,37–40], object clustering [39,41], and foraging [42]—according to the taxonomy proposed by Brambilla et al. [4]. In these studies [23–42], designers manually established ad hoc relationships between the colors that a robot can perceive and the corresponding behavior that a robot must adopt when it perceives them. The research question we address in the present paper is whether automatic design methods [5,17] can establish similar relationships.

It is our contention that classes of missions that require the robots to communicate and react to color-based information are an appropriate benchmark to assess automatic design methods. First, the capability of displaying and perceiving colors is platform-independent and generalizes across different design methods—robot platforms used in swarm robotics often include LEDs and cameras [48]. Second, colors facilitate the conception and realization of complex missions—colored environments can be created in various manners [40,41,49–51]. Finally, colors simplify the visualization and monitoring of robot swarms [48]—a property relevant to the human understandability of collective behaviors, an open research area in swarm robotics [52]. Yet, no existing method for the automatic design of robot swarms targets classes of missions that require the robots to operate with color-based information. Specifically, we refer to methods that can be framed within the tenets of the automatic off-line design of robot swarms [17].

Few related studies have been conducted following the neuro-evolutionary approach [43–47]. Floreano et al. [43] evolved communication behaviors for a swarm performing a foraging mission. Ampatzis et al. [44] evolved self-assembly behaviors with a team of two robots. Sperati et al. [45,46] evolved behaviors for coordinated motion with a group of three robots, and later, evolved a dynamic chain of robots that perform a foraging-like mission. Trianni and López-Ibañez [47] used multi-objective optimization to evolve flocking and a two-robot collaborative behavior within a robot swarm. In the studies mentioned above, the methods under analysis have not been tested for their ability to generate control software autonomously. As a consequence, these studies belong in semi-automatic design rather than in automatic design [17]. Indeed, in these studies, researchers either focused a single mission [43,44,46] or modify the design methods and/or robot platform when they applied the methods under analysis to more than one [45,47]. In addition, we argue that mission-specific bias was manually introduced in the way the robots display and perceive colors: robots display colors that are manually defined at the beginning of the experiment [44]; robots display color-based information that is encoded by the researchers [46,47]; and the perception capabilities of the robots are adjusted to ease the emergence of a specific behavior [45,46]. We contend that these studies do not expose the full potential of using color-based information in the automatic design of collective behaviors. As a matter of fact, previous works were limited to produce control software for robots than can display and perceive a single [43–45] or at most two simultaneous colors [46,47].

Our research belongs in the family of modular design methods known as AutoMoDe [8]. In the rest of the section, we restrict our attention to this family. Francesca and Birattari [5] discussed how the capabilities of robot platforms limit the variety of collective behaviors that automatic design methods can produce. Methods conceived for e-puck robots with equal capabilities—such as `Vanilla` [8], `Chocolate` [9], `Maple` [10], `Waffle` [11], `Coconut` [12] and `IcePop` [13]—are restricted to address the same class of missions: robots in the swarm must position themselves regarding their peers [9,15] or few static environmental features [8–13,15,53], and they can only use a single global reference for navigation [8–13,15,53]. In contrast, Hasselmann et al. [14] obtained a larger variety of collective behaviors by considering an extended set of capabilities with respect to those considered in `Chocolate`. They introduced `Gianduja`—a method for the automatic design of swarms of e-pucks that can selectively broadcast binary messages. Hasselmann et al. showed that by broadcasting and reacting to messages, the robots can perform missions that `Chocolate` cannot address—for example, missions that require event-handling collective behaviors.

The approach we follow in our research is similar to the one of Hasselmann et al. [14]. We conceived `TuttiFrutti` as an instance of AutoMoDe that designs control software for e-pucks with the extended capability of communicating by displaying and perceiving colors. As we will see in Section 3, `TuttiFrutti` can address missions that require the robots to act according to color-based information—a class of missions that existing instances of AutoMoDe can not address.

3. AutoMoDe-`TuttiFrutti`

`TuttiFrutti` is a modular automatic off-line design method; it produces control software for swarms of e-pucks that can display and perceive colors. More precisely, `TuttiFrutti` is an instance of AutoMoDe specialized in the design of robot swarms that act according to color-based information. The variety of collective behaviors produced by previous instances of AutoMoDe have been so far restricted by the limited capabilities of the robots—see Section 2. We conceive `TuttiFrutti` to overcome this restriction from a twofold perspective: on the one hand, e-pucks that display and perceive colors could enable the design of robot swarms in which the individuals exhibit color-based communication; on the other hand, these swarms could perform missions in complex and time-varying environments. By introducing communication capabilities and missions with complex environments, we meant to enlarge significantly the variety of collective behaviors designed with AutoMoDe.

Three fundamental components characterize `TuttiFrutti`: the robot platform, the set of preexisting software modules, and the optimization process that produces the control software. In the following sub-sections we describe each of these components, their relationship, and how they differentiate from other instances of AutoMoDe.

3.1. Robot Platform

`TuttiFrutti` produces control software for an extended version of the e-puck [54,55]—see Figure 1. The e-puck is a two wheeled, small, educational robot often used in swarm robotics research [8–11,14,45,46]. We consider a model of the e-puck endowed with a set of sensors and actuators defined by the reference model RM 3—see Table 1. We adopt the concept of *reference model* [5] to formally characterize the platform for which `TuttiFrutti` can produce control software. RM 3 represents the capabilities of the robot both in real and simulated environments.

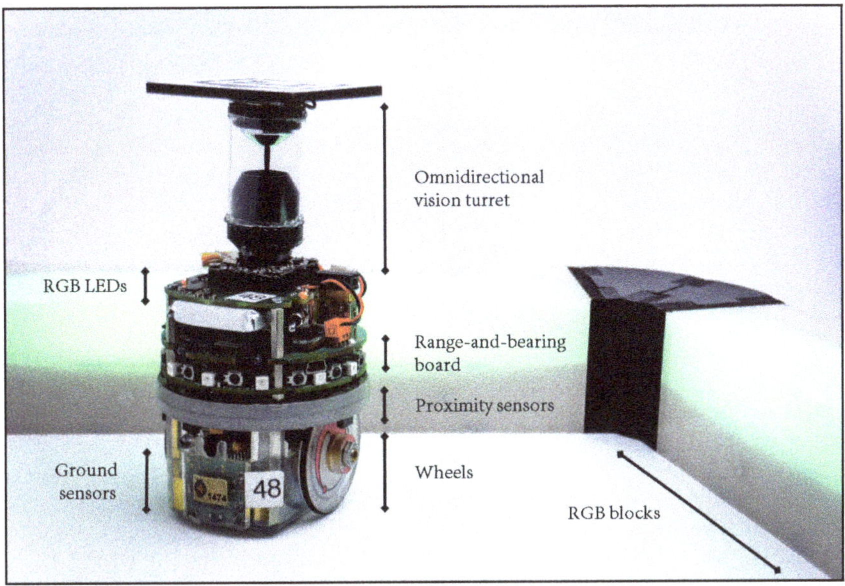

Figure 1. Extended version of the e-puck. The picture indicates the set of sensors and actuators defined by RM 3. Alongside, we show the RGB blocks that we use in our experiments with `TuttiFrutti`.

Table 1. Reference model RM 3. Novelties with respect to RM 1.1 are highlighted in gray. They concern the capabilities of displaying and perceiving colors. Robots can perceive: red (R); green (G); blue (B); cyan (C); magenta (M); and yellow (Y). Robots can display no color $(\varnothing)$; cyan (C); magenta (M); and yellow (Y). V_c is calculated likewise V_n—for each perceived color, the positions of color signals are aggregated into a unique attraction vector.

Input	Value	Description
$prox_{i\in\{1,\ldots,8\}}$	$[0,1]$	reading of proximity sensor i
$gnd_{j\in\{1,\ldots,3\}}$	$\{black, gray, white\}$	reading of ground sensor j
n	$\{0,\ldots,20\}$	number of neighboring robots detected
V_n	$([0.5,20];[0,2]\ \pi\,\text{rad})$	their relative aggregate position
$cam_{c\in\{R,G,B,C,M,Y\}}$	$\{yes, no\}$	colors perceived
$V_{c\in\{R,G,B,C,M,Y\}}$	$(1.0;[0,2]\ \pi\,\text{rad})$	their relative aggregate direction
Output	**Value**	**Description**
$v_{k\in\{l,r\}}$	$[-0.12, 0.12]$ m/s	target linear wheel velocity
LEDs	$\{\varnothing, C, M, Y\}$	color displayed by the LEDs

Period of the control cycle: 0.1 s.

The sensors and actuators available to the e-puck are proximity and ground sensors, a range-and-bearing board [56], an omnidirectional vision turret [16], right and left wheels, and RGB LEDs. The e-puck can detect nearby obstacles by its eight proximity sensors $(prox_i)$ distributed around its chassis. Three infrared ground sensors (gnd_j) allow the e-puck to differentiate between black, gray and white floor. By means of its range-and-bearing board, the e-puck knows the number of neighboring peers (n) in a range of 0.5 m. A vector (V_n) represents an attraction force to the neighboring e-pucks—to which the robot is subject—following the framework of virtual physics [57].

The omnidirectional vision turret allows the e-puck to perceive red, blue, green, cyan, magenta and yellow lights (cam_c) in a 360° field of view and within a range of about 0.5 m. For each color perceived, a unit vector (V_c) is associated, which represents a steady attraction to robots or objects that display the color. Finally, the control software of the robot can adjust independently the velocity of each wheel (v_k) between -0.12 and $0.12\,\text{m/s}$, and, using the three RGB LEDs placed on the top of the e-puck, can display cyan, magenta or yellow.

RM 3 is the first reference model adopted in the definition of a design method of the AutoMoDe family that includes the omnidirectional vision turret and the RGB LEDs of the e-puck platform. `Vanilla`, `Chocolate` and `Maple` are based on the simpler RM 1.1 [58]—the differences between RM 3 and RM 1.1 are highlighted in Table 1. `Gianduja` introduced the reference model RM 2 [58], associated to e-pucks that can exchange binary messages using their range-and-bearing board. An important difference between `TuttiFrutti` and other instances of AutoMoDe is that in RM 3 we removed the capability of the e-puck of measuring ambiance light. Although present in RM 1.1 and RM 2, this capability is incompatible with the RGB LEDs we added in RM 3.

3.2. Set of Preexisting Modules

The major characteristic of AutoMoDe is that it produces control software by assembling preexisting software modules. In `TuttiFrutti`, the modules are combined into probabilistic finite state machines—as in `Vanilla`, `Chocolate`, and `Gianduja` [8,9,14]. We conceived a set of modules that comprises six low-level behaviors—actions that a robot can take, and seven transition conditions—situations that trigger the change from one low-lever behavior to another. The set of modules of `TuttiFrutti` adapts and extends the modules originally conceived for `Vanilla`. We designed the modules to operate with RM 3 and they provide the e-puck different means to interact with robots and objects that display colors. Table 2 lists the low-level behaviors and transition conditions of `TuttiFrutti`. We further describe the modules in the following.

Table 2. Set of `TuttiFrutti`'s modules. Novelties with respect to `Vanilla` are highlighted in gray. They concern to the capability of acting upon perceived colors. The modules operate according to RM 3, see Table 1.

Low-Level Behavior *	Parameters	Description
EXPLORATION	$\{\tau, \gamma\}$	movement by random walk
STOP	$\{\gamma\}$	standstill state
ATTRACTION	$\{\alpha, \gamma\}$	physics-based attraction to neighboring robots
REPULSION	$\{\alpha, \gamma\}$	physics-based repulsion from neighboring robots
COLOR-FOLLOWING	$\{\delta, \gamma\}$	steady movement towards robots/objects of color δ
COLOR-ELUSION	$\{\delta, \gamma\}$	steady movement away from robots/objects of color δ
Transition Condition	**Parameters**	**Description**
BLACK-FLOOR	$\{\beta\}$	black floor beneath the robot
GRAY-FLOOR	$\{\beta\}$	gray floor beneath the robot
WHITE-FLOOR	$\{\beta\}$	white floor beneath the robot
NEIGHBOR-COUNT	$\{\xi, \eta\}$	number of neighboring robots greater than ξ
INVERTED-NEIGHBOR-COUNT	$\{\xi, \eta\}$	number of neighboring robots lower than ξ
FIXED-PROBABILITY	$\{\beta\}$	transition with a fixed probability
COLOR-DETECTION	$\{\delta, \beta\}$	robots/objects of color δ perceived

* All low-level behaviors display a color $\gamma \in \{\varnothing, C, M, Y\}$ alongside the action described.

3.2.1. Low-Level Behaviors

In EXPLORATION, the robot moves straight until it detects an obstacle in front ($prox_i$). Then, it rotates for a number of control cycles determined by the integer parameter τ, in a range of

$\tau \in \{0, \ldots, 100\}$. STOP maintains the robot standing still. In ATTRACTION and REPULSION, the robot moves closer (V_d) or farther from ($-V_d$) neighboring peers, respectively. In both cases, the velocity of the robot is a function of the number of robots detected (n) and the parameter $\alpha \in [0, 5]$. If the robot does not detect other robots, it moves straight. COLOR-FOLLOWING and COLOR-ELUSION move the robot with constant velocity towards (V_c) or away ($-V_c$) from robots or objects displaying specific colors (cam_c). The parameter $\delta \in \{R, G, B, C, M, Y\}$ determines the color to which the behaviors react. Robots can display the colors $\delta \in \{C, M, Y\}$, and other objects that might populate the environment can display the colors $\delta \in \{R, G, B\}$. If the robot does not perceive the color determined by δ, it moves straight. ATTRACTION, REPULSION, COLOR-DETECTION and COLOR-ELUSION incorporate a physics-based obstacle avoidance [59]. In all the low-level behaviors, the parameter $\gamma \in \{\varnothing, C, M, Y\}$ determines the color displayed by the RGB LEDs of the robot. The parameters τ, α, δ, and γ are tuned by the automatic design process.

3.2.2. Transition Conditions

BLACK-FLOOR, GRAY-FLOOR and WHITE-FLOOR trigger a transition when the robot steps on a portion of the floor (gnd_j) that is, respectively, black, gray or white. The parameter $\beta \in [0, 1]$ determines the probability of transitioning. NEIGHBOR-COUNT and INVERTED-NEIGHBOR-COUNT are transition conditions that consider the number of neighboring robots (n). NEIGHBOR-COUNT triggers a transition with a probability $z(n) \in [0, 1]$, with $z(n) = \frac{1}{1+e^{\eta(\xi - n)}}$. Conversely, INVERTED-NEIGHBOR-COUNT triggers a transition with a probability of $1 - z(n)$. The parameter $\xi \in \{0, \ldots, 10\}$ determines the inflection point of the probability function $z(n)$, the parameter $\eta \in [0, 20]$ determines its steepness. FIXED-PROBABILITY triggers a transition with a fixed probability determined by $\beta \in [0, 1]$. COLOR-DETECTION is based on the colors perceived by the robot (cam_c). The parameter $\delta \in \{R, G, B, C, M, Y\}$ defines the color that triggers a transition with a probability $\beta \in [0, 1]$. Robots in the swarm can display the colors $\delta \in \{C, M, Y\}$, and other objects that might populate the environment can display the colors $\delta \in \{R, G, B\}$. The parameters β, ξ, η, and δ are tuned by the automatic design process.

EXPLORATION, STOP, ATTRACTION and REPULSION are modified versions of the original low-level behaviors of `Vanilla`. In `TuttiFrutti`, we add the ability to control the color displayed by the LEDs. All the transition conditions, with exception of COLOR-DETECTION, are implementations of the original modules of `Vanilla`. COLOR-FOLLOWING, COLOR-ELUSION, and COLOR-DETECTION are modules we introduce here for the first time.

3.3. Design of Control Software

`TuttiFrutti` produces control software following the automatic design process proposed in `Chocolate` [9] and further studied in `Gianduja` [14]. The design of the control software is translated into an optimization problem—an optimization algorithm selects an appropriate combination of modules and parameters that, when uploaded to each robot, lead the swarm to exhibit a specific collective behavior. The collective behavior results then from an optimization process that maximizes the performance of the swarm, measured by an appropriate mission-specific performance measure. In `TuttiFrutti`, the architecture of the control software is a probabilistic finite state machine with a maximum of four states—each of which is a low-level behavior, and a maximum of four outgoing edges—to each of which a transition condition is associated. Edges always originate and end in different states—self-transitions are not allowed. The modules included in the finite state machine and the values of their parameters are selected off-line—that is, before the swarm is deployed on its target environment. To that purpose, `TuttiFrutti` uses Iterated F-race [60]—a multipurpose optimization method based on F-race [61]—to search the design space for effective control software configurations. The performance of the configurations is estimated through simulations performed in ARGoS3 [62], version beta 48, together with the argos3-epuck library [55]. The duration of the optimization process is determined by an *a priori* defined budget of simulations. Once the budget is exhausted, the design process terminates and Iterated F-race returns the best configuration found. This configuration is then uploaded to the robots and assessed in the target environment.

4. Experimental Set-Up

In this section, we describe our experiments in the design of collective behaviors for robots that can display and perceive colors. The study evaluates the capabilities of TuttiFrutti to address a class of missions in which colors displayed by objects in the environment provide relevant information to the robots. In the following, we first introduce the missions we consider in our study. Then, we present EvoColor—a baseline design method that serves as a reference to appraise the complexity of the missions. Finally, at the end of the section we describe the protocol we follow to assess TuttiFrutti.

4.1. Missions

We conceived three missions in which robot swarms operate in arenas surrounded by modular RGB blocks: STOP, AGGREGATION, and FORAGING. STOP and AGGREGATION are adaptations we make from the equivalent missions proposed by Hasselmann et al. to study Gianduja [14]. FORAGING is an abstraction of a foraging task, in a *best-of-n* fashion [63]—similar to the experiment of Valentini et al. [64]. The performance of the swarm is evaluated according to a mission-dependent objective function. We selected these missions because we conjecture that, to successfully perform them, the robots need to identify, process and/or transmit color-based information. In all missions, the time available to the robots is $T = 120$ s. The RGB blocks are arranged in walls that display colors on a per-mission basis—each RGB block is 0.25 m length and can display the colors red, green and blue $\{R, G, B\}$. In the context of these missions, when we reference to colored walls we imply that the RGB blocks arranged in the wall display the named color—for example, *"the green wall"* stands for a wall in which the RGB blocks composing it display the color green. In the following, we describe the scenario, the objective function, and the role of the colors for each mission. Figure 2 shows the arenas for the three missions.

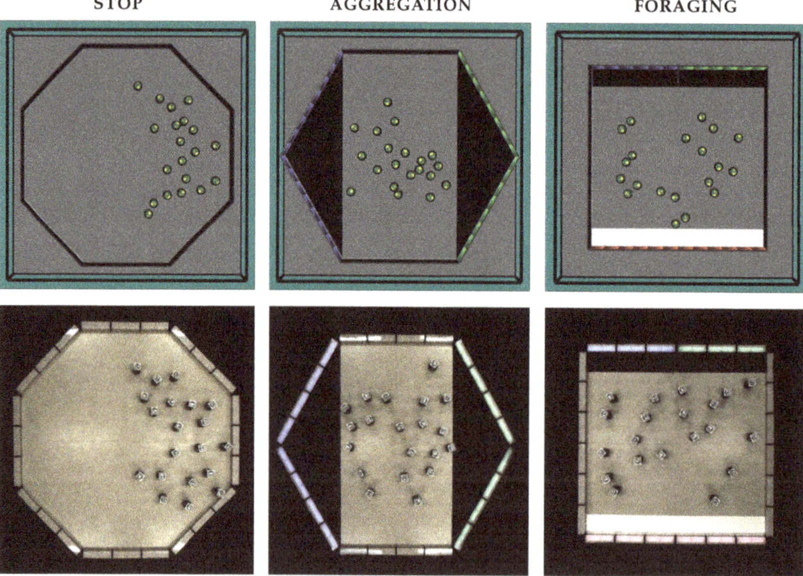

Figure 2. Set-up of the simulated (top) and real arena (bottom) for the missions STOP (left), AGGREGATION (center), and FORAGING (right). The images show an example of the initial position of the robots.

4.1.1. STOP

The robots must move until one of the walls that surrounds the arena emits a stop signal by turning green. Once the wall turns green, all the robots in the swarm must stop moving as soon as possible. The swarm operates in an octagonal arena of $2.75\,\mathrm{m}^2$ and gray floor. The wall that emits the stop signal is selected randomly. At the beginning of each run, the robots are positioned in the right side of the arena. Figure 2 (left) shows the arena for STOP.

The performance of the swarm (C_s) is measured by the objective function described by Equation (1); the lower the better.

$$C_s = \sum_{t=1}^{\bar{t}} \sum_{i=1}^{N} \bar{I}_i(t) + \sum_{t=\bar{t}+1}^{T} \sum_{i=1}^{N} I_i(t); \tag{1}$$

$$I_i(t) = \begin{cases} 1, \text{if robot } i \text{ is moving at time } t; \\ 0, \text{otherwise}; \end{cases} \qquad \bar{I}_i(t) = 1 - I_i(t).$$

C_s measures the amount of time during which the robots do not perform the intended behavior—before and after the stop signal. N and T represent respectively the number of robots and the duration of the mission. $\bar{t}$ indicates the time at which the stop signal is displayed. The time $\bar{t}$ is uniformly sampled between $[40, 60]$ s. We expect that TuttiFrutti produces collective behaviors with event-handling capabilities that allow the swarm to react when the stop signal appears.

4.1.2. AGGREGATION

The robots must aggregate in the left black region of the arena as soon as possible. The swarm operates in a hexagonal arena of about $2.60\,\mathrm{m}^2$ and gray floor. Triangular black regions of about $0.45\,\mathrm{m}^2$ are located at the left and right side of the arena. The walls lining the left black region are blue and those lining the right black region are green—the colors do not change during the mission. Each black region is characterized by the color of the walls that lines it. That is, the *blue zone* refers to the black region lined by blue walls and the *green zone* refers to the black region lined by green walls. At the beginning of each run, the robots are randomly positioned in the center of the arena—between the black regions. Figure 2 (center) shows the arena for AGGREGATION.

The performance of the swarm (C_a) is measured by the objective function described by Equation (2); the lower the better.

$$C_a = \sum_{t=1}^{T} \sum_{i=1}^{N} I_i(t) \tag{2}$$

$$I_i(t) = \begin{cases} 1, \text{if robot } i \text{ is not in the aggregation area at time } t; \\ 0, \text{otherwise.} \end{cases}$$

C_a indicates the time that the robots spend outside of the blue zone. N and T represent the number of robots and the duration of the mission, respectively. We expect that TuttiFrutti produces collective behaviors in which the swarm uses the blue walls as a reference to navigate and aggregate in the blue zone.

4.1.3. FORAGING

The robots must select and forage from the most profitable of two sources of items. The swarm operates in a squared arena of $2.25\,\mathrm{m}^2$ and gray floor. A rectangular white region of about $0.23\,\mathrm{m}^2$ is located at the bottom of the arena and represents the nest of the swarm. A rectangular black region of $0.23\,\mathrm{m}^2$ is located at the top of the arena and represents the two sources of items—the sources are separated by a short wall segment that does not display any color. This wall segment divides the black region in half. We account that an item is transported and successfully delivered when a robot travels from any of the sources to the nest. The walls lining the nest are red, the walls lining the left source are blue, and the walls lining the right source are green—the colors do not change during the mission. We consider then two types of sources of items: the *blue source*—the black region lined by blue walls; and the *green source*—the black region lined by green walls. At the beginning of each run, the robots are randomly positioned in the center of the arena—between the white and black areas. Figure 2 (right) shows the arena for FORAGING.

The performance of the swarm (C_f) is measured by the objective function described by Equation (3); the higher the better.

$$C_f = (\kappa)I_b + (-\kappa)I_g; \qquad (3)$$
$$\kappa = 1.$$

C_f indicates the aggregate profit of the total of items collected from the two sources. I_b corresponds to the number of items collected from the blue source, and I_g corresponds number of items collected from the green one. We added the factor κ to balance the profit of the items available in each source. In our study $\kappa = 1$. Items from the blue source account for a profit of $+1$ and items from the green source account for a penalization of -1. We expect that TuttiFrutti produces collective behaviors in which swarms use the blue walls as a reference to navigate towards the blue source, the green walls for avoiding the green source, and the red walls to navigate towards the nest.

4.2. Baseline Method: EvoColor

No standard design method exists to address the class of missions we consider in this study. Little related work exists—see Section 2—and refers only to mission-specific methods that follow the neuro-evolutionary approach. Indeed, as no extensive comparison has ever been performed between neuro-evolutionary methods across multiple missions, a state of the art in neuro-evolutionary robotics has not been identified, yet. Together with the results obtained with TuttiFrutti, in the following we will present also those obtained by EvoColor—a method based on neuro-evolution for the automatic design of swarms of e-pucks that can display and perceive colors. The results we will present should not therefore be considered as a comparison between TuttiFrutti and the state of the art in neuro-evolutionary robotics. In this context, our results should rather be considered as a comparison between TuttiFrutti and a reasonable instance of the neuro-evolutionary approach.

EvoColor is an adaptation of EvoStick [65]—a standard neuro-evolutionary method previously used as a yardstick in studies on the automatic design of robot swarms [8,10,15,53]. To the best of our knowledge, EvoStick is the only neuro-evolutionary method that has been tested via simulations and robot experiments on multiple missions without undergoing any per-mission modification. EvoColor produces control software for swarms of e-pucks that operate with RM 3—see Section 3.1. The control software has the form of a fully connected feed-forward artificial neural network with 41 input nodes (in), 8 output nodes (out) and no hidden layers. In this topology, the input nodes and output nodes are directly connected by synaptic connections (*conn*) with weights (ω) in a range of $[-5, 5]$. The activation of each output node is determined by the weighted sum of all inputs nodes filtered through a standard logistic function. EvoColor selects appropriate synaptic weights using an evolutionary process based on elitism and mutation. Just as in TuttiFrutti, the evolutionary process is conducted through simulations performed in ARGoS3, version beta 48, together with the argos3-epuck library. The evolution ends when an *a priori* defined budget of simulations is exhausted. Table 3 summarizes the topology of the neural network, the novelties with respect to EvoStick and the parameters used in the evolutionary process.

Table 3. Network topology and parameters of the evolutionary process in `EvoColor`. Novelties with respect to `EvoStick` are highlighted in gray. They concern the capability of displaying and perceiving colors. The neural network operates according to RM 3, see Table 1.

Input Node	Description
$in_{a\in\{1,\ldots,8\}}$	readings of proximity sensors $prox_{i\in\{1,\ldots,8\}}$
$in_{a\in\{9,\ldots,11\}}$	readings of ground sensors $gnd_{j\in\{1,\ldots,3\}}$
$in_{a\in\{12\}}$	value of the density function $z'(n)$
$in_{a\in\{13,\ldots,16\}}$	scalar projections of V_n
$in_{a\in\{17,\ldots,40\}}$	scalar projections of $V_{c\in\{R,G,B,C,M,Y\}}$
$in_{a\in\{41\}}$	bias input
Output Node	**Description**
$out_{b\in\{1,\ldots,4\}}$	tuples v' to map each velocity in the set $v_{k\in\{l,r\}}$
$out_{b\in\{5,\ldots,8\}}$	activation of each color in the set $\{\varnothing, C, M, Y\}$
Connection	**Description**
$conn_{s\in\{1,\ldots,328\}}$	synaptic connections with weights $\omega\in[-5,5]$
Number of generations *	—
Population size	100
Elite individuals	20
Mutated individuals	80
Evaluations per individual	10
Post-evaluation per individual **	100

* The number of generations is computed according to the budget of simulations. ** The population obtained in the last generation is post-evaluated to select the best individual.

The readings of the proximity $(prox)$ and ground (gnd) sensors are passed directly to the network. Information about the number of neighboring peers (n) is provided through the function $z'(n) \in [0,1]$, with $z'(n) = 1 - \frac{2}{1+e^{(n)}}$. The vector V_n and each vector in $V_{c\in\{R,G,B,C,M,Y\}}$ are translated into scalar projections onto four unit vectors that point at 45°, 135°, 225°, and 315° with respect to the front of the robot. Then, each projection is passed to the network through an independent input node. The last input of the network comes from a bias node. Four output nodes encode tuples (v') of negative and positive components of the velocity of the wheels. Each tuple is obtained from two independent output nodes and is defined as $v' = ([-12,0],[0,12])$. The velocity of a wheel (v) is computed as the sum of the two elements in a tuple (v'). Similarly, the color displayed by the RGB LEDs of the robot is selected by comparing the value of the output nodes that correspond to colors in the set $\{\varnothing, C, M, Y\}$. The color displayed corresponds to the maximum value found across the four colors.

`EvoColor` differs from `EvoStick` in two aspects: the reference model and how the output of the neural network is mapped to the velocity of the robots. First, `EvoColor` is based on RM 3 and `EvoStick` on RM 1.1. In accordance to RM 3, `EvoColor` does not integrate the capability of the e-pucks for sensing the intensity of ambiance light—originally integrated in `EvoStick`. The second difference between `EvoColor` and `EvoStick` is how the output of the neural network is mapped to the velocity of the e-pucks. In `EvoColor`, we introduce a velocity mapping based on tuples to facilitate the evolution of standstill behaviors—as we expect robots need them to perform STOP and AGGREGATION.

In `EvoStick`, the control software maps directly a single output node of the neural network into velocity commands $(v = [-12,12])$ for each wheel $(v_{k\in\{l,r\}})$—a robot can stand still only if the velocity of the two wheels is set exactly to 0. A standstill behavior is then difficult to achieve since only one pair of values in the output nodes maps exactly to $v_l = 0$ and $v_r = 0$; Moreover, the output nodes can not maintain a steady value because they are subject to the injection of sensory noise. In `EvoColor`, the control software maps the sum of elements of a tuple (v') into velocity commands for each wheel $v_{k\in\{l,r\}}$—each tuple is defined by two output nodes and provides a negative and a positive component to compute the velocity. We expect that this mapping facilitates the evolution of standstill behaviors: first, robots can stand still if the elements of each tuple (v') are any pair of values of equal magnitude—steady values are not required provided that the output nodes that encode the same tuple vary proportionally; second, the sum of the positive and negative components can cancel out the sensory noise injected in the output nodes that encode a tuple—given a proper tuning of the synaptic weights. If one compares EvoColor with EvoStick, the first has more freedom to tune neural networks that lead to standstill behaviors.

4.3. Protocol

We conduct experiments with twenty e-pucks on the missions described in Section 4.1. For each mission, we produce ten designs of control software with TuttiFrutti and ten with EvoColor. We assess the effectiveness of the methods by testing each design once in simulation and once with physical robots.

Statistics

We use box-plots to represent the performance of the control software we produce. For each method, we report the performance obtained in simulation (thin boxes) and with physical robots (thick ones). In all cases, we support comparative statements with an exact binomial test, at 95% confidence [66]: statements like "*A* performs *significantly* better/worse than *B*" imply that the comparison is supported by a an exact binomial test, at 95% confidence. In addition, we estimate the overall performance of TuttiFrutti with respect to EvoColor. To this purpose, we aggregate the results by comparing the performance of the two design methods across each mission. In the context of the overall performance of the design methods, any statement like "*A* performs *significantly* better/worse than *B*" also implies that the comparison is supported by an exact binomial test, at 95% confidence [66].

5. Results and Discussion

We present the qualitative and quantitative analysis of the results obtained with TuttiFrutti and EvoColor. We discuss first the behavior and performance of the swarms on a per-mission basis. Then, we elaborate on the aggregate performance across the three missions. In the end, we address the research questions presented in Section 1 and we discuss our findings. The code, control software and demonstrative videos are provided as Supplementary Material [67]. In the context of these results, references to colored robots imply that the robots display the named color—for example, "*cyan robots*" stands for robots that display the color cyan.

5.1. STOP

Figure 3 (left) shows the performance of TuttiFrutti and EvoColor in STOP. In this mission, TuttiFrutti performs significantly better than EvoColor.

From visual inspection, TuttiFrutti produced control software that effectively uses the capabilities of the robots for displaying and perceiving colors. The swarm first disperses and homogeneously covers the arena—aiming to rapidly detect the stop signal. If a robot detects the stop signal, it stands still and disseminates the information by emitting a signal of an arbitrary color. When other robots perceive the signal emitted by their peer, they also transition to a standstill behavior and relay the signal. The process continues until all robots in the swarm are standing still. We consider that this behavior shows the potential of TuttiFrutti for producing event-handling collective behaviors. The swarm collectively transitions from coverage to standstill when the stop signal appears. As we expected, TuttiFrutti produces control software that establishes communication protocols by correctly pairing the color of the signals that robots emit and the behavior other robots must adopt when they perceive them—similarly to the results obtained by Hasselmann et al. [14] with Gianduja.

EvoColor, unlike TuttiFrutti, designed collective behaviors that do not respond to the stop signal. The swarm adopts a rather simplistic behavior in which robots move until stopped by the walls. They remain then in a standstill behavior because they persistently push against the walls—no reaction can be appreciated in the swarms when the stop signal appears. This behavior was observed too in the experiments with physical robots, and in many cases, robots maintained standing-still behaviors by pushing against other robots too.

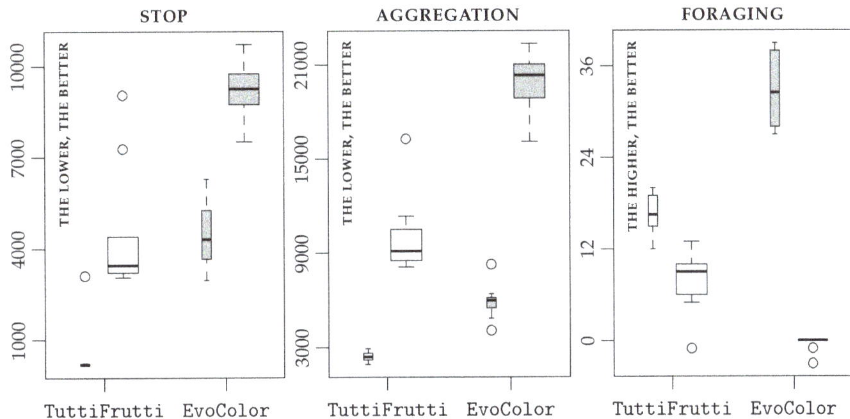

Figure 3. Performance obtained in the missions STOP (left), AGGREGATION (center), and FORAGING (right). The performance of TuttiFrutti is shown in white and the one of EvoColor in gray. Thin boxes represent results obtained in simulation and thick boxes the ones obtained with physical robots.

In the experiments with physical robots, both TuttiFrutti and EvoColor showed a significant drop in performance with respect to the simulations. However, the difference in mean performance between simulations and experiments with physical robots is larger for EvoColor than TuttiFrutti. Swarms deployed with the control software produced by TuttiFrutti showed the same collective behavior observed in simulation, although the rapidness in discovering the stop signal and disseminating the information decreased. In the case of EvoColor, robots often do not reach the walls and they push against each other to remain still in place.

Figure 4 shows an example of the control software produced by TuttiFrutti for STOP. Robots start in REPULSION with no color displayed ($\gamma = \varnothing$). They transition to STOP and turn yellow ($\gamma = Y$) when COLOR-DETECTION is triggered either by a green wall ($\delta = G$) or by yellow robots ($\delta = Y$). In this sense, robots change their behavior when they either perceive the stop signal or the yellow signals that other robots emit.

5.2. AGGREGATION

Figure 3 (center) shows the performance of TuttiFrutti and EvoColor in AGGREGATION. In this mission, TuttiFrutti performs significantly better than EvoColor.

Also in this case, from visual inspection, TuttiFrutti produced control software that effectively uses the capabilities that the robots have of displaying and perceiving colors. As we expected, TuttiFrutti designs collective behaviors in which robots reach and remain in the blue zone by moving towards blue walls. This behavior is often complemented with navigation or communication strategies that boost the efficiency of the swarm. For example, some instances of control software include a repulsion behavior that drives robots away from the green walls—robots reach the blue zone faster by avoiding unnecessary exploration in the green zone. In other instances, robots that step in the blue zone, or perceive the blue walls, emit a signal of an arbitrary color—other robots then follow this signal to reach the blue zone. In this sense, robots communicate and collectively navigate to aggregate faster. Finally, some instances combine the two strategies.

EvoColor designed collective behaviors in which robots use the colors displayed in the arena. Robots explore the arena until they step in one of the black regions—either at the blue or green zone. If robots step in the green zone, they move away from the green walls and reach the blue zone. If robots step in the blue zone, they attempt to stand still. In this sense, robots react and avoid the green walls as a strategy to aggregate in the blue zone.

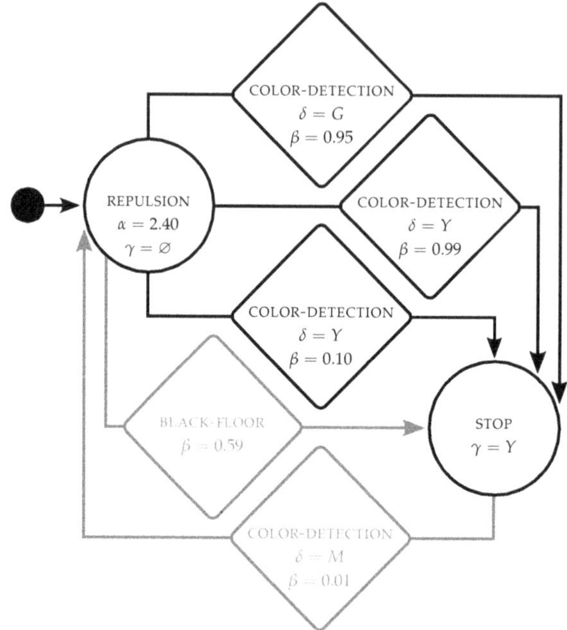

Figure 4. Instance of control software produced by `TuttiFrutti` for STOP. The probabilistic finite state machine shows the effective modules in black and non-reachable modules in light gray. Circular modules represent the low-level behaviors and rhomboid modules represent transition conditions.

The control software produced by `TuttiFrutti` and `EvoColor` showed a significant drop in performance when ported to the physical robots. As observed in STOP, the difference in mean performance between simulations and experiments with physical robots is larger for `EvoColor` than `TuttiFrutti`. Robot swarms that use the control software produced by `TuttiFrutti` display the same collective behaviors observed in simulation. The decrease in performance occurs because few robots that leave the blue zone do not return as fast as observed in the simulations. The control software produced by `EvoColor` does not port well to the physical robots—that is, robots appear to be unable to reproduce the behaviors observed in the simulation. Robots ramble in the arena and seem to react to the presence of their peers, however, no specific meaningful behavior could be identified by visual inspection.

Figure 5 shows an example of the control software produced by `TuttiFrutti` for AGGREGATION. Robots start in COLOR-FOLLOWING displaying yellow ($\delta = Y$) and move towards cyan robots ($\gamma = C$). When they perceive the blue walls ($\delta = B$), COLOR-DETECTION triggers and the robots transition to a second module COLOR-FOLLOWING in which they move towards the blue walls ($\delta = B$) while emitting a cyan signal ($\gamma = C$). By cycling in these behaviors, robots can navigate to the blue zone either by moving towards the blue walls or by following the cyan signals that other robots emit. The transition conditions FIXED-PROBABILITY, GRAY-FLOOR and NEIGHBOR-COUNT trigger the COLOR-FOLLOWING behavior that allows the robot to return to the aggregation area.

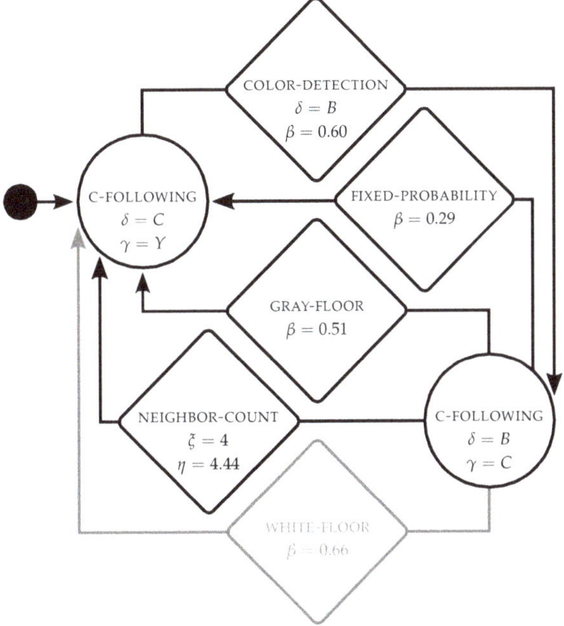

Figure 5. Instance of control software produced by `TuttiFrutti` for AGGREGATION. The probabilistic finite state machine shows the effective modules in black and non-reachable modules in light gray. Circular modules represent the low-level behaviors and rhomboid modules represent the transition conditions. Modules labeled as C-FOLLOWING stand for the low-level behavior COLOR-FOLLOWING.

5.3. FORAGING

Figure 3 (right) shows the performance of `TuttiFrutti` and `EvoColor` in FORAGING. In this mission, `EvoColor` performs significantly better than `TuttiFrutti` in simulation. However, `TuttiFrutti` performs significantly better than `EvoColor` in the experiments with physical robots.

As in the other missions, from visual inspection, `TuttiFrutti` produced control software that effectively uses the capabilities that the robots have of displaying and perceiving colors. Robots explore the arena and forage only from the profitable source. However, contrary to what we expected, `TuttiFrutti` designed collective behaviors that do not use all the colors displayed in the arena. In fact, robots mostly forage by randomly exploring the arena while moving away from the green wall—in other words, they only avoid to step in the green source. Although the swarm can perform the mission with this behavior, we expected that robots could navigate faster by moving towards the blue and red walls. Still, `TuttiFrutti` produced only few instances of control software in which robots react to more than one color—see Figure 6. We conjecture that `TuttiFrutti` exploits the convex shape of the arena to produce solutions that are effective at the minimal complexity—that is, the performance of a swarm in this mission might not improve even if robots react to all three colors.

`EvoColor` designed collective behaviors in which the swarm does not react to the colors displayed in the arena. Robots forage from the blue source by following the walls of the arena in a clockwise direction. This behavior efficiently drives the robots around the arena and across the blue source. When the robots reach the intersection that divides the blue and green source, they continue moving straight and effectively reach the nest. By cycling in this behavior, the swarm maintains an efficient stream of foraging robots.

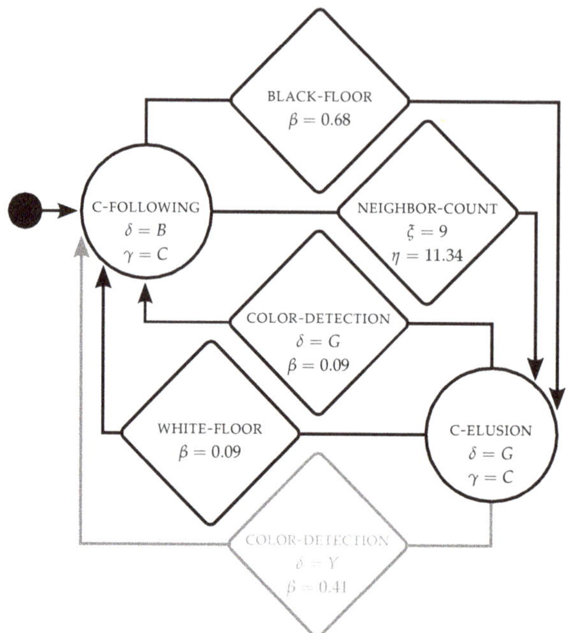

Figure 6. Instance of control software produced by `TuttiFrutti` for FORAGING. The probabilistic finite state machine shows the effective modules in black and non-reachable modules in light gray. Circular modules represent the low-level behaviors and rhomboid modules represent the transition conditions. Modules labeled as C-FOLLOWING and C-ELUSION stand for the low-level behaviors COLOR-FOLLOWING and COLOR-ELUSION, respectively.

`TuttiFrutti` and `EvoColor` showed a significant drop in performance in the experiments with physical robots, in comparison to the performance obtained in the simulations. Likewise the other two missions, the difference in mean performance between simulations and experiments with physical robots is larger for `EvoColor` than `TuttiFrutti`. In the case of `TuttiFrutti`, we did not observe any difference in the behavior of the swarms with respect to the simulations. Conversely, the collective behaviors designed by `EvoColor` are affected to the point that the swarm is unable to complete the mission. In the control software produced by `EvoColor`, the ability of the robots to follow the walls strongly depends on the fine-tuning of the synaptic weights in the neural network—more precisely, it requires a precise mapping between the proximity sensors and wheels of the robots. In the physical robots, the noise of the proximity sensors and wheels differs from the original design model, and a fine-tuned neural network is less effective. Indeed, the swarm is not any more able to maintain the stream of foraging robots, and on the contrary, robots stick to each other and to the walls.

We also observe a *rank inversion* of the performance of the two methods in this mission. As defined by Ligot and Birattari [53], a rank inversion is a phenomenon that manifests when an instance of control software outperforms another in simulation, but it is outperformed by the latter when it is evaluated on physical robots. In our experiments, `TuttiFrutti` is outperformed by `EvoColor` in simulation, but it outperforms `EvoColor` when it is ported to the physical robots. These results are consistent with the ones reported by Francesca et al. [8], and further discussed by Birattari et al. [68] and Ligot and Birattari [53], for comparisons between the modular and the neuro-evolutionary approach to the automatic design of robot swarms.

Figure 6 shows an example of the control software produced by `TuttiFrutti` for FORAGING. Robots start in COLOR-FOLLOWING displaying cyan ($\gamma = C$) and moving towards the blue wall ($\delta = B$). If a robot steps in one of the two sources, BLACK-FLOOR triggers and the robot transitions to COLOR-ELUSION—it then becomes cyan ($\gamma = C$) and moves away from the green wall ($\delta = G$). When the robot steps in the nest, WHITE-FLOOR triggers and the robot transitions back to COLOR-FOLLOWING. By cycling this behavior, robots move back and forth between the blue source and the nest. When robots are in COLOR-ELUSION, COLOR-DETECTION can trigger with a low probability ($\beta = 0.09$) if robots perceive the green wall ($\delta = G$). This transition mitigates the penalty caused by robots that step in the green source. If a robot steps in the green source, it transitions back to COLOR-FOLLOWING and moves towards the blue wall. Finally, the transition condition NEIGHBOR-COUNT can trigger when the robot perceives more than four neighboring robots. Yet, we do not find a clear effect of this transition in the overall behavior of the robots.

5.4. Aggregate Results

TuttiFrutti and EvoColor obtain similar results when the control software is evaluated with simulations. On the other hand, TuttiFrutti is significantly better than EvoColor when the control software is ported to the physical robots. It has already been pointed out that when control software developed in simulation is ported to a real-world platform, due to the reality gap one might observe both a drop in performance [53] and a substantial modification of the collective behavior [69]. The entity of these effects might depend on the design method, and some design methods might be more robust than others [53]. Our results indicate that EvoColor is more affected by the reality gap than TuttiFrutti across the three missions considered. This is apparent both in the entity of the performance drop we measured and in the fact that the collective behaviors of the control software generated by EvoColor are often dramatically differently in simulation and in the real world, while the ones of the control software generated by TuttiFrutti are essentially unchanged.

By introducing TuttiFrutti, we also investigated the impact of an extended design space in the optimization process of AutoMoDe. The size of the design space in Vanilla and Chocolate is $O(|B|^4\,|C|^{16})$, as estimated by Kuckling et al. [70]. B and C represent, respectively, the number of modules in low-level behaviors and transition conditions. Using the same method as Kuckling et al., we estimate the design space in TuttiFrutti to be $O(|4B|^4\,|C|^{16})$—that is, 256 times larger than the one searched by Chocolate. Notwithstanding the larger design space, we do not find evidence that TuttiFrutti is affected by the increased number of parameters to tune. Indeed, TuttiFrutti produced effective control software for all missions considered.

5.5. Discussion

In the following, we first address the research questions defined in Section 1 and then we discuss our findings.

TuttiFrutti selects, tunes and assembles control software that operates with information that is available in the environment in the form of colors. In the three missions, the robot swarm reacts to these colors and act according to the information they provide in each case—both for handling events and navigating. Additionally, we observed that TuttiFrutti can design collective behaviors that exhibit color-based communication between robots. For example, TuttiFrutti designed collective behaviors with color-based communication in STOP and AGGREGATION—missions in which communication can influence the performance of the robot swarm. These collective behaviors are feasible thanks to the extended capabilities of the e-puck, capabilities that translated into a larger space of possible control software than the one considered by Vanilla and Chocolate—early versions of AutoMoDe. As the design space of TuttiFrutti is larger than the one of Vanilla and Chocolate, one could have expected that the automatic design process would have difficulties in producing meaningful control software. Still, we did not find evidence that TuttiFrutti suffers from an increased difficulty to design collective behaviors for robot swarms. The reference model RM 3 and the set of modules introduced with TuttiFrutti allowed it to conceive STOP and AGGREGATION—variants of missions already studied with AutoMoDe, and FORAGING—a new mission framed within the *best-of-n* problem.

By introducing TuttiFrutti, we enlarged the variety of collective behaviors that can be produced with the AutoMoDe family.

We argue that the experiments we conducted with TuttiFrutti show evidence that automatic modular design methods can establish a mission-specific relationship between the colors that the robots perceive and the behavior that they must adopt. In Section 2, we described experiments in which this relationship enabled the design of complex collective behaviors [25,31,35]. We find that these collective behaviors have similarities with those designed by TuttiFrutti—for example, robots react to colored objects in the environment and use colors signals to communicate with their peers. We conjecture that TuttiFrutti, or design methods that might share its characteristics, can produce a wider range and more complex collective behaviors than those described in this paper. In this sense, we believe that research with robot swarms that can perceive and display colors has the potential to close the gap between the complexity of the missions performed with manual design methods, and those performed with automatic design.

6. Conclusions

In this paper, we introduced AutoMoDe-TuttiFrutti—an automatic method to design collective behaviors for robots that can perceive and communicate color-based information. We designed control software for swarms of e-pucks that comply with RM 3—e-pucks can use their LEDs to display colors and their omnidirectional vision turret to perceive them. The capability of the robots to act upon different colors translated into an increased variety of collective behaviors compared to previous instances of AutoMoDe. We assessed TuttiFrutti on a class of missions in which the performance of the swarm depends on its ability to use color-based information for handling events, communicating, and navigating.

We conducted experiments in simulation and with physical robot swarms performing three missions: STOP, AGGREGATION and FORAGING. In all cases, TuttiFrutti designed collective behaviors that effectively use color-based information. In STOP, the swarm collectively changes its behavior when a specific color signal appears. In STOP and AGGREGATION, the swarm exhibits communication behaviors in which robots pair the color signals they emit and the colors to which they react. In AGGREGATION and FORAGING, robots use the colors they perceive as a reference to navigate the environment. In FORAGING, swarms differentiate two sources of items and forage from the profitable one. Alongside the results obtained with TuttiFrutti, we assessed a method based on neuro-evolution: EvoColor. In STOP and FORAGING, EvoColor designed collective behaviors that do not use color-based information. In AGGREGATION, EvoColor designed collective behaviors in which robots use the colors they perceive to navigate the environment—likewise TuttiFrutti. The aggregated results showed that TuttiFrutti performs better than EvoColor in the class of missions we considered. Results with physical robots suggest that TuttiFrutti crosses the reality gap better than EvoColor—result partially sustained by the visual inspection of the behavior of the robots.

Automatic design methods can effectively produce control software for swarms of robots that can display and perceive colors. We demonstrated that TuttiFrutti establishes an appropriate relationship between the colors that the robots perceive and the behavior they must adopt. In our experiments, this relationship was established on a per-mission basis and responded to the specifications of each mission. Yet, the set of missions on which we assess TuttiFrutti is far from being exhaustive, and more research work is needed to define the limitations of the design method. Future work will be devoted to assess TuttiFrutti in a larger and more complex class of missions. It is our contention that TuttiFrutti can design collective behaviors to address missions that involve a larger number of features in the environment and time-varying conditions. As observed in STOP, robots can effectively transition between two collective behaviors. We foresee that this ability enables the design of swarms that can perform missions with two or more sequential tasks. To the best of our knowledge, the design of collective behaviors to address this class of missions has not been studied in the context of automatic off-line design of robot swarms.

Supplementary Materials: The code, control software and demonstrative videos are available online at http://iridia.ulb.ac.be/supp/IridiaSupp2019-008.

Author Contributions: Implementations and experiments were done by D.G.R. The paper was drafted by D.G.R. and refined by M.B. The research was directed by M.B. All authors have read and agreed to the published version of the manuscript.

Funding: The project has received funding from the European Research Council (ERC) under the European Union's Horizon 2020 research and innovation programme (grant agreement No 681872). M.B. acknowledges support from the Belgian *Fonds de la Recherche Scientifique*—FNRS. D.G.R. acknowledges support from the Colombian Administrative Department of Science, Technology and Innovation—COLCIENCIAS.

Acknowledgments: The authors thank Federico Pagnozzi and Jonas Kuckling for reading a preliminary version of this paper.

Conflicts of Interest: The authors declare no conflict of interest. The funders had no role in the design of the study; in the collection, analyses, or interpretation of data; in the writing of the manuscript, or in the decision to publish the results.

References

1. Beni, G. From swarm intelligence to swarm robotics. In *Swarm Robotics, Proceedings of the SAB 2004 International Workshop, Santa Monica, CA, USA, 17 July 2004*; Şahin, E., Spears, W.M., Eds.; LNCS; Springer: Berlin, Germany, 2004; Volume 3342, pp. 1–9. [CrossRef]
2. Şahin, E. Swarm robotics: From sources of inspiration to domains of application. In *Swarm Robotics, Proceedings of the SAB 2004 International Workshop, Santa Monica, CA, USA, 17 July 2004*; Şahin, E., Spears, W.M., Eds.; LNCS; Springer: Berlin, Germany, 2004; Volume 3342, pp. 10–20. [CrossRef]
3. Dorigo, M.; Birattari, M.; Brambilla, M. Swarm robotics. *Scholarpedia* **2014**, *9*, 1463. [CrossRef]
4. Brambilla, M.; Ferrante, E.; Birattari, M.; Dorigo, M. Swarm robotics: A review from the swarm engineering perspective. *Swarm Intell.* **2013**, *7*, 1–41. [CrossRef]
5. Francesca, G.; Birattari, M. Automatic design of robot swarms: Achievements and challenges. *Front. Robot. AI* **2016**, *3*, 1–9. [CrossRef]
6. Nolfi, S.; Floreano, D. *Evolutionary Robotics: The Biology, Intelligence, and Technology of Self-Organizing Machines*, 1st ed.; A Bradford Book; MIT Press: Cambridge, MA, USA, 2000.
7. Trianni, V. *Evolutionary Swarm Robotics*; Springer: Berlin, Germany, 2008. [CrossRef]
8. Francesca, G.; Brambilla, M.; Brutschy, A.; Trianni, V.; Birattari, M. AutoMoDe: A novel approach to the automatic design of control software for robot swarms. *Swarm Intell.* **2014**, *8*, 89–112. [CrossRef]
9. Francesca, G.; Brambilla, M.; Brutschy, A.; Garattoni, L.; Miletitch, R.; Podevijn, G.; Reina, A.; Soleymani, T.; Salvaro, M.; Pinciroli, C.; et al. AutoMoDe-Chocolate: Automatic design of control software for robot swarms. *Swarm Intell.* **2015**, *9*, 125–152. [CrossRef]
10. Kuckling, J.; Ligot, A.; Bozhinoski, D.; Birattari, M. Behavior trees as a control architecture in the automatic modular design of robot swarms. In *Swarm Intelligence, Proceedings of the 11th International Conference, ANTS 2018, Rome, Italy, 29–31 October 2018*; Dorigo, M., Birattari, M., Blum, C., Christensen, A.L., Reina, A., Trianni, V., Eds.; LNCS; Springer: Cham, Switzerland, 2018; Volume 11172, pp. 30–43. [CrossRef]
11. Salman, M.; Ligot, A.; Birattari, M. Concurrent design of control software and configuration of hardware for robot swarms under economic constraints. *PeerJ Comput. Sci.* **2019**, *5*, e221. [CrossRef]
12. Spaey, G.; Kegeleirs, M.; Garzón Ramos, D.; Birattari, M. Comparison of different exploration schemes in the automatic modular design of robot swarms. In Proceedings of the Reference AI & ML Conference for Belgium, Netherlands & Luxemburg, BNAIC/BENELEARN, Brussels, Belgium, 6–8 November 2019; CEUR Workshop Proceedings; Beuls, K., Bogaerts, B., Bontempi, G., Geurts, P., Harley, N., Lebichot, B., Lenaerts, T., Gilles, L., Van Eecke, P., Eds.; CEUR-WS.org: Aachen, Germany, 2019; Volume 2491.
13. Kuckling, J.; Ubeda Arriaza, K.; Birattari, M. Simulated annealing as an optimization algorithm in the automatic modular design of robot swarms. In Proceedings of the Reference AI & ML Conference for Belgium, Netherlands & Luxemburg, BNAIC/BENELEARN, Brussels, Belgium, 6–8 November 2019; CEUR Workshop Proceedings; Beuls, K., Bogaerts, B., Bontempi, G., Geurts, P., Harley, N., Lebichot, B., Lenaerts, T., Gilles, L., Van Eecke, P., Eds.; CEUR-WS.org: Aachen, Germany, 2019; Volume 2491.

14. Hasselmann, K.; Robert, F.; Birattari, M. Automatic design of communication-based behaviors for robot swarms. In *Swarm Intelligence, Proceedings of the 11th International Conference, ANTS 2018, Rome, Italy, 29–31 October 2018*; Dorigo, M., Birattari, M., Blum, C., Christensen, A.L., Reina, A., Trianni, V., Eds.; LNCS; Springer: Cham, Switzerland, 2018; Volume 11172, *LNCS*, pp. 16–29. [CrossRef]
15. Francesca, G.; Brambilla, M.; Brutschy, A.; Garattoni, L.; Miletitch, R.; Podevijn, G.; Reina, A.; Soleymani, T.; Salvaro, M.; Pinciroli, C.; et al. An experiment in automatic design of robot swarms: AutoMoDe-Vanilla, EvoStick, and human experts. In *Swarm Intelligence, Proceedings of the 9th International Conference, ANTS 2014, Brussels, Belgium, 10–12 September 2014*; Dorigo, M., Birattari, M., Garnier, S., Hamann, H., Montes de Oca, M., Solnon, C., Stützle, T., Eds.; LNCS; Springer International Publishing: Berlin, Germany, 2014; Volume 8667, pp. 25–37. [CrossRef]
16. École Polytechnique Fédérale de Lausanne. Omnidirectional Vision Turret for the e-Puck. 2010. Available online: http://www.e-puck.org/index.php?option=com_content&view=article&id=26&Itemid=21 (accessed on 3 July 2020).
17. Birattari, M.; Ligot, A.; Bozhinoski, D.; Brambilla, M.; Francesca, G.; Garattoni, L.; Garzón Ramos, D.; Hasselmann, K.; Kegeleirs, M.; Kuckling, J.; et al. Automatic off-line design of robot swarms: A manifesto. *Front. Robot. AI* **2019**, *6*, 59. [CrossRef]
18. Waibel, M.; Keller, L.; Floreano, D. Genetic team composition and level of selection in the evolution of multi-agent systems. *IEEE Trans. Evol. Comput.* **2009**, *13*, 648–660. [CrossRef]
19. Gauci, M.; Chen, J.; Li, W.; Dodd, T.J.; Groß, R. Self-organized aggregation without computation. *Int. J. Robot. Res.* **2014**, *33*, 1145–1161. [CrossRef]
20. Chen, J.; Gauci, M.; Li, W.; Kolling, A.; Groß, R. Occlusion-based cooperative transport with a swarm of miniature mobile robots. *IEEE Trans. Robot.* **2015**, *31*, 307–321. [CrossRef]
21. Lopes, Y.K.; Trenkwalder, S.M.; Leal, A.B.; Dodd, T.J.; Groß, R. Supervisory control theory applied to swarm robotics. *Swarm Intell.* **2016**, *10*, 65–97. [CrossRef]
22. Jones, S.; Winfield, A.; Hauert, S.; Studley, M. Onboard evolution of understandable swarm behaviors. *Adv. Intell. Syst.* **2019**, *1*, 1900031. [CrossRef]
23. O'Grady, R.; Christensen, A.L.; Dorigo, M. SWARMORPH: Multirobot morphogenesis using directional self-assembly. *IEEE Trans. Robot.* **2009**, *25*, 738–743. [CrossRef]
24. O'Grady, R.; Groß, R.; Christensen, A.L.; Dorigo, M. Self-assembly strategies in a group of autonomous mobile robots. *Auton. Robot.* **2010**, *28*, 439–455. [CrossRef]
25. Mathews, N.; Christensen, A.L.; O'Grady, R.; Mondada, F.; Dorigo, M. Mergeable nervous systems for robots. *Nat. Commun.* **2017**, *8*, 439. [CrossRef] [PubMed]
26. Mathews, N.; Christensen, A.L.; Stranieri, A.; Scheidler, A.; Dorigo, M. Supervised morphogenesis: Exploiting morphological flexibility of self-assembling multirobot systems through cooperation with aerial robots. *Robot. Auton. Syst.* **2019**, *112*, 154–167. [CrossRef]
27. Christensen, A.L.; O'Grady, R.; Dorigo, M. From fireflies to fault-tolerant swarms of robots. *IEEE Trans. Evol. Comput.* **2009**, *13*, 754–766. [CrossRef]
28. Nouyan, S.; Groß, R.; Bonani, M.; Mondada, F.; Dorigo, M. Teamwork in self-organized robot colonies. *IEEE Trans. Evol. Comput.* **2009**, *13*, 695–711. [CrossRef]
29. Ducatelle, F.; Di Caro, G.A.; Pinciroli, C.; Gambardella, L.M. Self-organized cooperation between robotic swarms. *Swarm Intell.* **2011**, *5*, 73–96. [CrossRef]
30. Dorigo, M.; Floreano, D.; Gambardella, L.M.; Mondada, F.; Nolfi, S.; Baaboura, T.; Birattari, M.; Bonani, M.; Brambilla, M.; Brutschy, A.; et al. Swarmanoid: A novel concept for the study of heterogeneous robotic swarms. *IEEE Robot. Autom. Mag.* **2013**, *20*, 60–71. [CrossRef]
31. Garattoni, L.; Birattari, M. Autonomous task sequencing in a robot swarm. *Sci. Robot.* **2018**, *3*, eaat0430. [CrossRef]
32. Ferrante, E.; Turgut, A.E.; Mathews, N.; Birattari, M.; Dorigo, M. Flocking in stationary and non-stationary environments: A novel communication strategy for heading alignment. In *Parallel Problem Solving from Nature, PPSN XI*; Schaefer, R., Cotta, C., Kołodziej, J., Rudolph, G., Eds.; Lecture Notes in Computer Science; Springer: Berlin/Heidelberg, Germany, 2010; Volume 6239, pp. 331–340. [CrossRef]

33. Giusti, A.; Nagi, J.; Gambardella, L.M.; Di Caro, G.A. Distributed consensus for interaction between humans and mobile robot swarms (demonstration). In *AAMAS '12: Proceedings of the 11th International Conference on Autonomous Agents and Multiagent Systems—Volume 3*; International Foundation for Autonomous Agents and Multiagent Systems: Richland, SC, USA, 2012; pp. 1503–1504.
34. Podevijn, G.; O'Grady, R.; Dorigo, M. Self-organised feedback in human swarm interaction. In Proceedings of the Workshop on Robot Feedback in Human-Robot Interaction: How to Make a Robot Readable for a Human Interaction Partner, Ro-Man 2012, Paris, France, 9 September 2012.
35. Nouyan, S.; Campo, A.; Dorigo, M. Path formation in a robot swarm: Self-organized strategies to find your way home. *Swarm Intell.* **2008**, *2*, 1–23. [CrossRef]
36. Pinciroli, C.; O'Grady, R.; Christensen, A.L.; Dorigo, M. Self-organised recruitment in a heteregeneous swarm. In Proceedings of the 2009 International Conference on Advanced Robotics, (ICAR), Munich, Germany, 22–26 June 2009; IEEE: Piscataway, NJ, USA, 2009; pp. 1–8.
37. Pini, G.; Brutschy, A.; Birattari, M.; Dorigo, M. Task partitioning in swarms of robots: Reducing performance losses due to interference at shared resources. In *Informatics in Control Automation and Robotics*; Andrade Cetto, J., Filipe, J., Ferrier, J.L., Eds.; Lecture Notes in Electrical Engineering; Springer: Berlin/Heidelberg, Germany, 2009; Volume 85, pp. 217–228. [CrossRef]
38. Pini, G.; Brutschy, A.; Frison, M.; Roli, A.; Dorigo, M.; Birattari, M. Task partitioning in swarms of robots: An adaptive method for strategy selection. *Swarm Intell.* **2011**, *5*, 283–304. [CrossRef]
39. Pini, G.; Brutschy, A.; Scheidler, A.; Dorigo, M.; Birattari, M. Task partitioning in a robot swarm: Object retrieval as a sequence of subtasks with direct object transfer. *Artif. Life* **2014**, *20*, 291–317. [CrossRef]
40. Brutschy, A.; Garattoni, L.; Brambilla, M.; Francesca, G.; Pini, G.; Dorigo, M.; Birattari, M. The TAM: Abstracting complex tasks in swarm robotics research. *Swarm Intell.* **2015**, *9*, 1–22. [CrossRef]
41. Allwright, M.; Bhalla, N.; El-faham, H.; Antoun, A.; Pinciroli, C.; Dorigo, M. SRoCS: Leveraging stigmergy on a multi-robot construction platform for unknown environments. In *Swarm Intelligence, Proceedings of the 9th International Conference, ANTS 2014, Brussels, Belgium, 10–12 September 2014*; Dorigo, M., Birattari, M., Garnier, S., Hamann, H., Montes de Oca, M., Solnon, C., Stützle, T., Eds.; Lecture Notes in Computer Science; Springer: Cham, Switzerland, 2014; Volume 8667, pp. 158–169. [CrossRef]
42. Brambilla, M.; Brutschy, A.; Dorigo, M.; Birattari, M. Property-driven design for swarm robotics: A design method based on prescriptive modeling and model checking. *ACM Trans. Auton. Adapt. Syst.* **2014**, *9*, 17:1–17:28. [CrossRef]
43. Floreano, D.; Mitri, S.; Magnenat, S.; Keller, L. Evolutionary conditions for the emergence of communication in robots. *Curr. Biol.* **2007**, *17*, 514–519. [CrossRef]
44. Ampatzis, C.; Tuci, E.; Trianni, V.; Christensen, A.L.; Dorigo, M. Evolving self-assembly in autonomous homogeneous robots: Experiments with two physical robots. *Artif. Life* **2009**, *15*, 465–484. [CrossRef]
45. Sperati, V.; Trianni, V.; Nolfi, S. Evolving coordinated group behaviours through maximisation of mean mutual information. *Swarm Intell.* **2008**, *2*, 73–95. [CrossRef]
46. Sperati, V.; Trianni, V.; Nolfi, S. Self-organised path formation in a swarm of robots. *Swarm Intell.* **2011**, *5*, 97–119. [CrossRef]
47. Trianni, V.; López-Ibáñez, M. Advantages of task-specific multi-objective optimisation in evolutionary robotics. *PLoS ONE* **2015**, *10*, e0136406. [CrossRef]
48. Nedjah, N.; Silva Junior, L. Review of methodologies and tasks in swarm robotics towards standardization. *Swarm Evol. Comput.* **2019**, *50*, 100565. [CrossRef]
49. Mayet, R.; Roberz, J.; Schmickl, T.; Crailsheim, K. Antbots: A feasible visual emulation of pheromone trails for swarm robots. In *Swarm Intelligence, Proceedings of the 7th International Conference, ANTS 2010, Brussels, Belgium, 8–10 September 2010*; Dorigo, M., Birattari, M., Di Caro, G.A., Doursat, R., Engelbrecht, A.P., Floreano, D., Gambardella, L.M., Groß, R., Şahin, E., Sayama, H., et al., Eds.; Lecture Notes in Computer Science; Springer: Berlin/Heidelberg, Germany, 2010; Volume 6234, pp. 89–94. [CrossRef]
50. Brutschy, A.; Pini, G.; Decugnière, A. *Grippable Objects for the Foot-Bot*; Technical Report TR/IRIDIA/2012-001; IRIDIA, Université Libre de Bruxelles: Brussels, Belgium, 2012.
51. Soleymani, T.; Trianni, V.; Bonani, M.; Mondada, F.; Dorigo, M. Bio-inspired construction with mobile robots and compliant pockets. *Robot. Auton. Syst.* **2015**, *74*, 340–350, doi:10.1016/j.robot.2015.07.018. [CrossRef]
52. Kolling, A.; Walker, P.; Chakraborty, N.; Sycara, K.; Lewis, M. Human interaction with robot swarms: A survey. *IEEE Trans. Hum.-Mach. Syst.* **2016**, *46*, 9–26. [CrossRef]

53. Ligot, A.; Birattari, M. Simulation-only experiments to mimic the effects of the reality gap in the automatic design of robot swarms. *Swarm Intell.* **2019**, 1–24. [CrossRef]
54. Mondada, F.; Bonani, M.; Raemy, X.; Pugh, J.; Cianci, C.; Klaptocz, A.; Magnenat, S.; Zufferey, J.C.; Floreano, D.; Martinoli, A. The e-puck, a robot designed for education in engineering. In Proceedings of the 9th Conference on Autonomous Robot Systems and Competitions, Castelo Branco, Portugal, 7 May 2009; Gonçalves, P., Torres, P., Alves, C., Eds.; Instituto Politécnico de Castelo Branco: Castelo Branco, Portugal, 2009; pp. 59–65.
55. Garattoni, L.; Francesca, G.; Brutschy, A.; Pinciroli, C.; Birattari, M. *Software Infrastructure for E-Puck (and TAM)*; Technical Report TR/IRIDIA/2015-004; IRIDIA, Université libre de Bruxelles: Brussels, Belgium, 2015.
56. Gutiérrez, Á.; Campo, A.; Dorigo, M.; Donate, J.; Monasterio-Huelin, F.; Magdalena, L. Open e-puck range & bearing miniaturized board for local communication in swarm robotics. In Proceedings of the IEEE International Conference on Robotics and Automation, ICRA, Kobe, Japan, 12–17 May 2009; Kosuge, K., Ed.; IEEE: Piscataway, NJ, USA, 2009; pp. 3111–3116. [CrossRef]
57. Spears, W.M.; Gordon, D.F. Using artificial physics to control agents. In Proceedings of the 1999 International Conference on Information Intelligence and Systems, Bethesda, MD, USA, 31 October–3 November 1999; IEEE Computer Society Press: Los Alamitos, CA, USA, 1999; pp. 281–288. [CrossRef]
58. Hasselmann, K.; Ligot, A.; Francesca, G.; Birattari, M. *Reference Models for AutoMoDe*; Technical Report TR/IRIDIA/2018-002; IRIDIA, Université libre de Bruxelles: Brussels, Belgium, 2018.
59. Borenstein, J.; Koren, Y. Real-time obstacle avoidance for fast mobile robots. *IEEE Trans. Syst. Man Cybern.* **1989**, *19*, 1179–1187. [CrossRef]
60. López-Ibáñez, M.; Dubois-Lacoste, J.; Pérez Cáceres, L.; Birattari, M.; Stützle, T. The irace package: Iterated racing for automatic algorithm configuration. *Oper. Res. Perspect.* **2016**, *3*, 43–58. [CrossRef]
61. Birattari, M.; Stützle, T.; Paquete, L.; Varrentrapp, K. A racing algorithm for configuring metaheuristics. In Proceedings of the GECCO 2002: Proceedings of the Genetic and Evolutionary Computation Conference, New York, NY, USA, 9–13 July 2002; Langdon, W.B., Cantú-Paz, E., Mathias, K., Roy, R., Davis, D., Poli, R., Balakrishnan, K., Honavar, V., Rudolph, G., Wegener, J., et al., Eds.; Morgan Kaufmann Publishers: San Francisco, CA, USA, 2002; pp. 11–18.
62. Pinciroli, C.; Trianni, V.; O'Grady, R.; Pini, G.; Brutschy, A.; Brambilla, M.; Mathews, N.; Ferrante, E.; Di Caro, G.A.; Ducatelle, F.; et al. ARGoS: A modular, parallel, multi-engine simulator for multi-robot systems. *Swarm Intell.* **2012**, *6*, 271–295. [CrossRef]
63. Valentini, G.; Ferrante, E.; Dorigo, M. The best-of-n problem in robot swarms: Formalization, state of the art, and novel perspectives. *Front. Robot. AI* **2017**, *4*, 9. [CrossRef]
64. Valentini, G.; Hamann, H.; Dorigo, M. Efficient decision-making in a self-organizing robot swarm: On the speed versus accuracy trade-off. In Proceedings of the AAMAS '15: Proceedings of the 2015 International Conference on Autonomous Agents and Multiagent Systems, Istanbul, Turkey, 4–8 May 2015; International Foundation for Autonomous Agents and Multiagent Systems: Richland, SC, USA, 2015; pp. 1305–1314.
65. Ligot, A.; Hasselmann, K.; Delhaisse, B.; Garattoni, L.; Francesca, G.; Birattari, M. *AutoMoDe, NEAT, and EvoStick: Implementations for the e-Puck Robot in ARGoS3*; Technical Report TR/IRIDIA/2017-002'; IRIDIA, Université libre de Bruxelles: Belgium, Brussel, 2017.
66. Conover, W.J. *Practical Nonparametric Statistics*, 3rd ed.; Wiley Series in Probability and Statistics, Applied Probability and Statistics Section; John Wiley & Sons: New York, NY, USA, 1999.
67. Garzón Ramos, D.; Birattari, M. Automatic Design of Collective Behaviors for Robots That Can Display and Perceive Colors: Supplementary Material. 2019. Available online: http://iridia.ulb.ac.be/supp/IridiaSupp2019-008 (accessed on 3 July 2020).
68. Birattari, M.; Delhaisse, B.; Francesca, G.; Kerdoncuff, Y. Observing the effects of overdesign in the automatic design of control software for robot swarms. In *Swarm Intelligence, Proceedings of the 10th International Conference, ANTS 2016, Brussels, Belgium, 7–9 September 2016*; Dorigo, M., Birattari, M., Li, X., López-Ibáñez, M., Ohkura, K., Pinciroli, C., Stützle, T., Eds.; Lecture Notes in Computer Science; Springer: Cham, Switzerland, 2016; Volume 9882, pp. 45–57. [CrossRef]

69. Floreano, D.; Husbands, P.; Nolfi, S. Evolutionary robotics. In *Springer Handbook of Robotics*, 1st ed.; Siciliano, B., Khatib, O., Eds.; Springer Handbooks; Springer: Berlin/Heidelberg, Germany, 2008; pp. 1423–1451. [CrossRef]
70. Kuckling, J.; Ligot, A.; Bozhinoski, D.; Birattari, M. *Search Space for AutoMoDe-Chocolate and AutoMoDe-Maple*; Technical Report TR/IRIDIA/2018-012; IRIDIA, Université Libre de Bruxelles: Brussels, Belgium, 2018.

Article

An Amalgamation of Hormone Inspired Arbitration Systems for Application in Robot Swarms

James Wilson [1,*], Jon Timmis [1,2] and Andy Tyrrell [1]

1 Department of Electronic Engineering, University of York, York YO10 5DD, UK
2 Faculty of Technology, University of Sunderland, Sunderland SR1 3SD, UK
* Correspondence: james.s.wilson@york.ac.uk

Received: 31 July 2019; Accepted: 22 August 2019; Published: 27 August 2019

Abstract: Previous work has shown that virtual hormone systems can be engineered to arbitrate swarms of robots between sets of behaviours. These virtual hormones act similarly to their natural counterparts, providing a method of online, reactive adaptation. It is yet to be shown how virtual hormone systems could be used when a robotic swarm has a large variety of task types to execute. This paper details work that demonstrates the viability of a collection of virtual hormones that can be used to regulate and adapt a swarm over time, in response to different environments and tasks. Specifically, the paper examines a new method of hormone speed control for energy efficiency and combines it with two existing systems controlling environmental preference as well as a selection of behaviours that produce an effective foraging swarm. Experiments confirm the effectiveness of the combined system, showing that a swarm of robots equipped with multiple virtual hormones can forage efficiently to a specified item demand within an allotted period of time.

Keywords: swarm; robotics; hormone; behaviour; arbitration; demand

---

1. Introduction

In nature, hormones provide an adaptation technique that cues behavioural change through chemical processing. As stimuli reach cells or organs hormone chemicals are produced and diffused throughout the body. The build up and gradual decay of these hormones as they are metabolised gives an organism contextual information based on how frequently stimuli are received. The balance and concentration of various hormones can then influence behaviour of the organism. These hormone induced changes to behaviour have been observed in a variety of natural examples [1–3].

In the context of robotics, previous work has shown that virtual hormones can be engineered to control, arbitrate and adapt swarms of robots amongst a small set of behaviours in a similar manner to the examples seen in nature [4–6]. However, it is yet to be shown how hormone systems could be used when a large array of behaviours and task types are available to a swarm. Evidence of virtual hormones being used to control such systems in simulation would provide evidence of their viability in non-abstracted tasks and support virtual hormone implementation in physical systems. This paper identifies for the first time, the viability of combining multiple hormone systems at once, each regulating a separate function or feature of the swarm. The primary goal of this amalgamation of hormone systems will be to ensure that the benefits of each system can provide improvements to the energy efficiency of a foraging swarm when combined, without disrupting the performance of other systems.

Having already explored several applications for hormone inspired systems in previous work [5,7] in which virtual hormone systems have effectively regulated behaviours and preference, respectively selecting appropriate states in a dynamic environment and allocating robots to environments based on their performance across different terrains. The work in this paper combines these applications to create an energy efficient foraging swarm regulated by numerous, simultaneously functioning hormones.

The hormones comprising the amalgamation operate at different levels of a behavioural hierarchy (illustrated in Figure 1), controlling preference, behavioural control and actuator control. Combining systems acting at these different levels of behaviour allows for the swarm to be controlled by hormones at every stage of operation, truly testing the combined systems capabilities and compatibility. This, alongside the fact that more than three times the number of individual hormone types previously studied have been used in these experiments means that the number of hormones used in this amalgamation can be considered numerous.

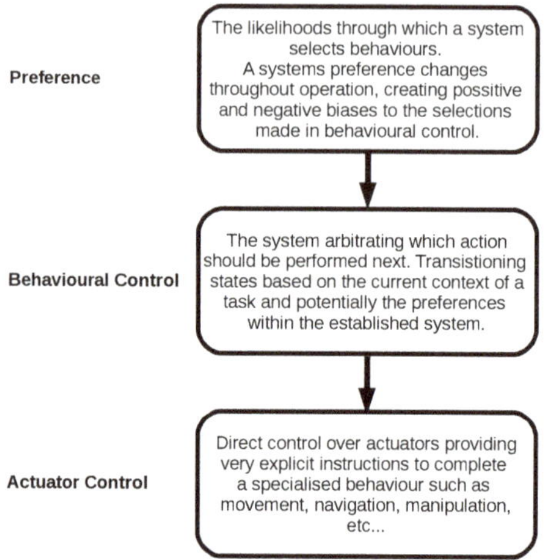

Figure 1. Behavioural hierarchy for those behaviours investigated within this paper.

Section 3 investigates virtual hormone driven motor control as a method to improve energy efficiency in the foraging swarm. This will focus on the need for adaptive motor speeds and their implementation.

Section 4 explores the compatibility between this new system and one governing sleep [5]. The potential energy efficiency benefits of combining a sleep system and a virtual hormone framework are examined.

Section 5, the swarm will be diversified, using the heterogeneous wheel types designed in [7], and a system capable of self analysis for task reallocation is combined with the previously established hormone speed and sleep regulation. Thus creating a system with 6 or more simultaneously acting virtual hormones in each member of the swarm, depending on the number of environments available to the swarm.

The implementation of this complex virtual hormone system will be effective for live adaptation and produce significant improvements to energy efficiency in foraging examples over individual hormone systems.

Finally, Section 7 gives a number of conclusions for the presented work and suggests future areas of investigation.

2. Background

Virtual hormones and hormone-inspired systems have previously been used to directly control the motor functions of a single robot. In [8] the authors presented a method that modelled a robot as two cells controlling the left and right motor of a puck robot, each motor was driven by their own hormones H_r and H_l with wheel speed changing proportionately with the magnitude of hormone value. The hormones for each cell were stimulated by a proximity sensor and were capable of diffusing between cells, acting as an inhibitor to the opposing hormone when present in the neighbouring cell.

With the hormone values corresponding to the wheel speeds on the respective sides of the robot, this produced an effective hormone controlled method for obstacle avoidance. The study found that this system could be successfully implemented in hardware and could be well studied with an exhaustive parameter sweep for 'reasonable computational cost'.

Similarly, work by Kernbach et al [9] produced a system which allowed hormones to regulate the movement of individual robots in a similar manner to [8]. This work added additional function to the virtual hormone, using the same hormone to regulate an additional behaviour state. In this new behaviour state the robots conjoined to produce a larger, specialised morphology. The hormone in this state was re-purposed to create a hormone gradient, regulating the size of the newly formed conjoined organism. This showed that, while explicit control over a robot is attainable with a virtual hormone system, virtual hormones can also be used to effectively arbitrating behaviour states.

Following this work additional hormone-inspired controllers have been successfully implemented to adapt swarm morphology, identifying context to environments via stimuli and then constructing appropriate formations [4,10]. These studies show that hormone-inspired systems can be engineered to provide an effective, computationally inexpensive method for robot control.

The need for mid-task adaptation for the energy efficient use of robot swarms has been highlighted in works such as [11]. In which the energy consumption of several bio-inspired robotic coordination procedures were investigated. This investigation found that energy consumption typically increased in line with parameters (e.g., swarm size, arena size, number of tasks). It is, therefore, important that such parameters are understood and controlled for before engaging in a task. This finding strengthens the demand for self allocating systems such as [5,7,12] that modulate the number of active robots performing a task. These self regulating systems reduce the need for a centralised decision on swarm size and means that swarms can instead perform multiple different tasks in series or explore different environments sequentially, without needing to return and redeploy.

In the previous implementations of hormone arbitration systems found in [5,7] the adaptive properties of hormone equations have been utilised for both task arbitration and robotic preference. By using hormone equations that provide a value that decays over time and increases as specified stimuli are encountered, environmental features can be extrapolated based on the current value provided by the equation. Through the comparison of hormone values receiving different stimuli or the comparison of hormone values present in different robots, hormone equations provide a powerful tool at a low computational cost for respectively regulating tasks or ranking the performance of robots within a swarm.

Using virtual hormones as a method for behavioural control, while providing a strong method for adaptation, does take an element of control away from the user. In traditional behavioural control where user defined thresholds or specific actions are used for behaviour transition, systems can be produced that behave consistently and repetitively in a manner that virtual hormones cannot. However, while this may be appropriate for individual robots whose performance and interactions can be predicted, in the context of swarms of robots exploring dynamic and volatile environments the level of on-line adaptation a virtual hormone system provides to the swarm will typically produce a better performance.

While the advantages of behavioural and preference control have be previously studied, there is very little literature on energy efficient speed control systems capable of adapting to demand. There are some examples of research investigating optimal speeds for energy efficiency [13,14]. However, this research only relates to rail vehicles, providing little relevance to puck robot vehicles. For this reason, the next section begins by obtaining data from real robots to obtain information that can be used to produce the motor driving hormone system before it is added to the other, previously developed hormone systems.

3. HIBAS Implementation for Control of a Foraging System with Deviating Motor Speeds

Hormone Inspired behavioural arbitration systems (HIBAS) have been studied using energy efficiency as the target output [5,7]. However, the speed at which robots move and the efficiency of their movement, vital to energy efficiency, have not been investigated. When simulating the energy consumption of robots it is typically assumed that robots in the swarm are either moving at a specific speed, stationary or consuming a fixed quantity of energy in a given behaviour state [5,12,15,16]. This section investigates the viability of virtual hormone implementation to directly control and adapt wheel speeds to achieve improved energy efficiency when foraging. A 'demand' concept will be present in the task that allows the user to specify, prior to or during use, the number of items to be gathered in a given time period. The purpose of this is to add an additional complexity for the swarm to overcome through adaptation.

3.1. Energy Characteristics of Psi Swarm Robot Hardware

In order to bring realism to the simulated experiments and reduce the reality gap data was taken from the PSI swarm robot platform [17] to obtain a power model similar to that produced in [18] for the MarXbot. This model would take an individual robots speed as an input and produce a realistic value for the power consumed at a given time step. As a result the total energy consumed by the swarm across an experiment could be recorded using reports of energy consumption from each swarm member, providing more meaningful data regarding changes of speed within the experiment.

To construct a power model, power consumption was measured using a Keysight N6705B power analyser [19]. Results for power consumption as speed increases were recorded through 10 repetitions and a quartic trend line was fit to the mean of these results, this is illustrated in Figure 2. The resultant equation for power consumption with speed as the input was:

$$P = 1.05 - 7.76 \times 10^{-3}s + 2.2 \times 10^{-3}s^2 - 8.89 \times 10^{-5}s^3 + 1.14 \times 10^{-6}s^4 \quad (1)$$

where P is energy consumption per second (Watts) and s is the current speed of the robot (cm/s). When implementing this equation in the robot swarm simulation, the offset of 1.05 was reduced to 0.05, as it was assumed that most of the offset was due to the base consumption of energy used by robot peripherals. The offset of 0.05 was left to ensure a negative power was never experienced during experiments. The equation was also scaled for the appropriate time frame, ensuring that the correct amount of power per wheel was collected per experiment tick. This equation was then used at each time step to calculate the current energy consumption based on the speed of each individual robot. Energy consumption could then be used to feed into the value of energy efficiency that would be used to measure the fitness of the systems tested in the experiments presented in this paper.

After implementing Equation (1) in the simulation, the analyses of energy efficiencies at different speeds were conducted. In these tests, 20 robots foraged in a simple environment for 500 simulated seconds or until 100 food items were gathered. The average final energy efficiency (food item per unit of energy consumed) from 50 trials at speeds ranging from 1 to 50 cm/s were then plotted (illustrated in Figure 3). Taking the peak value of energy efficiency for a given speed, a value was chosen to act as a baseline for the following experiments.

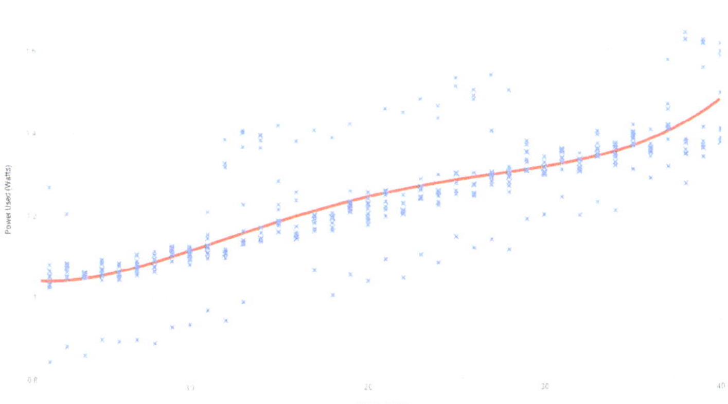

Figure 2. Graph displaying the results of the power consumption of a Psi Swarm robot increasing motor speed gradually. 10 repetitions were taken for these results and a trend line has been fit to the mean of these results and is shown in red.

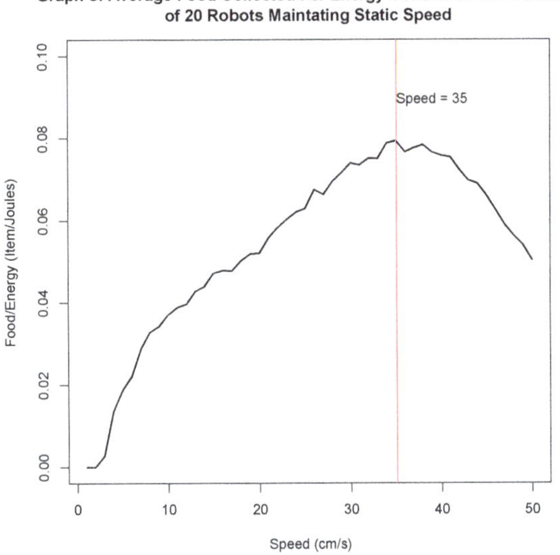

Figure 3. Graph showing average food gathered per energy unit consumed in a swarm of 20 foraging robots across 50 trials.

3.2. Hormone Interaction with Motor Speed

To produce a hormone equation that controlled motor speed in a direct manner and at appropriate speeds given context, it was decided that the two primary influencing factors should be: item demand and the evidence of negative performance.

The presence of frequent collisions and the decay present from failing to achieve task goals have been demonstrated as good indicators of negatives performance [5,7]. Collisions in these cases are identified by the detection of objects by short range proximity sensors, activating the avoidance behaviour, rather than a physical collision between a robot and another entity. These features were therefore used as the first step in the implementation of the new hormone system. The decay would reduce the hormone, and subsequently the speed, to an efficient settling point. Collisions would also reduce the hormone, thus inhibiting the speed of poorly performing robots and limiting their impact on energy consumption.

3.2.1. Demand

'Demand', as a new feature to the virtual hormone system, required the development of a novel formula accounting for: a target number of items to be collected (to be specified before deployment), the allotted time to collect said items, the current collection rate throughout the experiment. Following this, Equation (2) was created:

$$D(t) = \frac{I_T}{t_T} - \frac{I_c + 1}{t} \tag{2}$$

In this equation $D(t)$ represents the demand function, I_T is the total number of items desired by the end of the allotted time period, t_T is the end time for the allotted period, I_c is the current number of stored items and t is the current time step. Decentralisation is required to remain 'swarm like' during the experiment, therefore the 'demand value' is only accessible to individual robots in the nest. The value is updated as they leave and used as their stimuli throughout their next period of exploration.

Equation (2) models the demand value to fluctuate as items were collected without incurring an exponential increase near the end of the experiment should the swarm only be a few items away from the target collection. By setting the demand as the difference between the required average rate of collection and the current rate of collection, the hormone value and speed could increase with repeated failure to meet target collection rates. This meant that speed would only slightly deviate from the optimal speed of travel. Gradual incrementation in this manner prevented an inefficient burst of speed late in the experiment to compensate for a lack of items collected.

With a function for demand in place, the two Hormone equations were produced (Return Hormone and Speed Hormone, shown respectively in Equations (3) and (4)) to regulate the speeds and behaviours of each robot in the swarm. The hormones produced in these experiments were designed in the same format as [5,7] with λ representing decay and γ representing the coefficient of stimuli.

3.2.2. Return Hormone

The return hormone equation is as follows:

$$H_r(t) = \lambda_r H_r(t-1) + \gamma_r C \tag{3}$$

where t is the current time step H_r is the return hormone, λ_r is the decay for the system and γ_r is the stimuli weighting. The return hormone has a single stimulus, C, for collision detection. Although it does not regulate speed, it does feed into the speed hormone. The primary function of the Return Hormone is to identify the frequency of collisions detected by a robot, between walls or other robots. This information can then be used to decide if an individual robot should return to the nest having been unsuccessful, typically by exceeding either a fixed or similarly adaptive threshold. At this stage the threshold for returning was set to 50, with any value of H_r exceeding that resulting in a given robot changing behaviour state and travelling back to the nest site.

3.2.3. Speed Hormone

The speed hormone equation is as follows:

$$H_s(t) = \lambda_s H_s(t-1) + \gamma_{s1} D(t) - \gamma_{s2} H_r \tag{4}$$

where $H_s(t)$ is the Speed Hormone, λ_s is the decay rate for the system, γ_{s1} is the weighting for the stimuli and γ_{s2} is the weighting for the inhibitor. The speed hormone had two influencing factors.

A stimulus, $D(t)$ (Demand), and an inhibitor, H_r. With these features in place, higher demand would result in faster activity, consuming more energy but reducing the item demand. Conversely, the system would slow down robots in poor positions or in areas densely populated by other members of the swarm, consuming less energy while in a compromised position. It is worth noting that H_r was used in this case rather than C in order to smooth the response to collisions, rather than experiencing a sudden, large value inhibiting the system upon encountering a collision, H_r allows for the reduction to H_s to be smooth and gradual. This avoids the sudden loss of mobility in what could potentially be a one off collision.

While the speed of a robot does increase with the Speed Hormone, it doesn't have true direct control over the motor speed as has been seen in studies such as [8]. Instead, the Speed Hormone system allows the robot to operate at the optimal travelling speed for energy efficiency. To avoid deviation from this speed at low hormone levels, the speed hormone has no effect on speed until it exceeds the value of 10. Values below 10 in speed hormone would have very minimal effect on the actual speed of the robot while still reducing energy efficiency by deviating from the optimal speed. After the value of 10, the speed hormone effects the speed with the relationship shown in Equation (5), providing potential speeds ranging between 35, for H_s values below 10, and 50 when H_s is fully saturated.

$$S = 33.33 + \frac{H_s}{6} \tag{5}$$

3.2.4. Parameters

Parameter values for the hormone equations (shown in Table 1) were selected empirically using the context of the experiments to decide on appropriate time scales for decay, these time scales were then converted to decay values using Equation (6), taking values for H_{sat} (the numerical value at which the virtual hormone will saturate) and H_{fin} (the smallest value deemed relevant to the hormone system) as 100 and 1 respectively. The period of decay chosen for the sleep hormone was based on the amount of time it would take for an ideally operating robot to locate and retrieve two food items. i.e., the time it would take to reach the centre of available items and return twice, travelling in a straight line while operating at optimal speed. This meant that under ideal operation stimuli from the previous collection would still be present when returning for the second time, allowing the hormone value to build. The period for decay for the return hormone was calculated for only a single full collection and the collisions in a previous search period should have minimal bearing on that of the next.

$$\lambda = \sqrt[n]{\frac{H_{fin}}{H_{sat}}} \tag{6}$$

Stimuli coefficients were subsequently chosen to provide adequate response when interacting at expected minimum and maximum values of decay and rate of collision.

Table 1. Parameter values for the Return and Speed Hormones.

λ_r	γ_r	λ_s	γ_{s1}	γ_{s2}
0.9977	5	0.999	9	0.01

3.3. Comparison Systems

In order to test how effective the designed hormone systems were, two additional systems were produced for comparison. The first had no adaptive element, keeping all robots at optimal speed (35 cm/s) while foraging. This system was not influenced by 'demand' and should highlight the point at which speed adaptation is required to obtain remaining items required in the collection. In order to keep environmental awareness consistent across the three systems, the return hormone was implemented across all systems, allowing swarm members to return to the nest site should they encounter too many collisions.

The second comparison system featured an on-line adaptation method similar to reinforcement learning. This engineered adaptation was driven by the same function for demand as featured in the virtual hormone system. This style of online engineered adaptation has been used in the past to modify swarm traits, finding optimal partition lengths in [20] modifying travel distances based on success and failure of swarm individuals.

The adaptive system, designed for speed control, stepped the robot motor speeds up or down depending on the value of demand upon returning to the nest site. Positive demand values would increase speed, and negative values would decrease it. As with the hormone system, this would allow speed to be increased or decreased (and hence increase or decrease energy expenditure) in relation to collection requirements.

The increments and decrements made by the engineered system were influenced by demand, providing a variable adaptation to the system. A base change of 1 was applied based on the sign of the demand in addition to a change proportionate to the value of demand itself, increased by a coefficient of 20 to make suitable changes to the speed value. These values were tuned via iterative selection to produce strong rates of collection and energy efficiency across a wide variety of task demands.

The base change was used so that the swarm can catch up to the required collection rate even when demand is small. If this change was not implemented, increments based solely of demand would be too small to have a perceivable effect on robot speed. The same effect could not be achieved by increasing the coefficient of demand because the system could react too quickly to large disparities in current collection rate versus required rate and overcompensate by a large margin.

3.4. Analysis of Systems Highlighting the Need for Adaptation

After designing these systems, preliminary tests were conducted to demonstrate why adaptation is required for the foraging task. This section will elaborate on the environment in which the systems where tested, detail the key features of the simulations and discuss the results produced from the experiments.

3.4.1. Environments

The three systems discussed in this paper were tested in two environments. The first is a square environment measuring 15×15 m. The first 2 meters of the environment were assigned as the nest area, highlighted in grey as illustrated in Figure 4. This environment provided an arena for simple operation, identifying whether the system, under only the pressure of the specified demand could operate effectively.

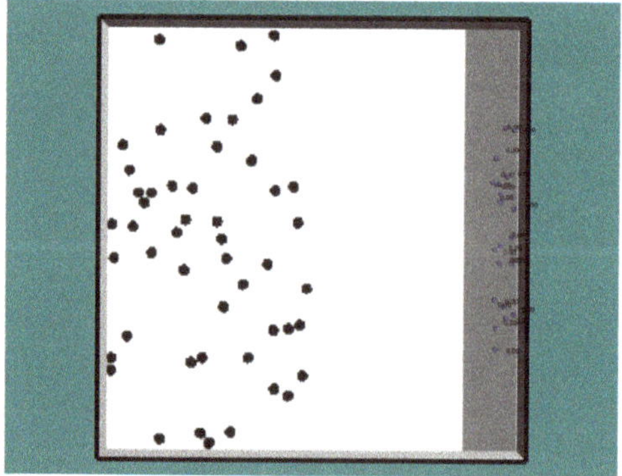

Figure 4. Screenshot of first simulated environment used. Food items are shown as black circles in the white environment, puck robots can be seen waiting in the nest area (light grey).

The second environment (illustrated in Figure 5) instead measured 20 × 10 m though retained a similar nest layout to the first. Four funnelled corridors were included in this environment to act as obstacles. These increase swarm density during exploration and provides additional difficulty to the tested systems, akin to that of a group of robots attempting to complete tasks in industrial settings such as mines, power plants or drainage systems, where space could be limited. This congestion will not only limit the success of the robots by slowing them down, but short range collision sensors will be triggered more frequently, meaning that the return hormone will potentially instruct robots to return home too early. This will heavily test the adaptability of the system, giving the combination of hormone systems a greater challenge, making the probability of one system disrupting the other in a negative fashion more likely.

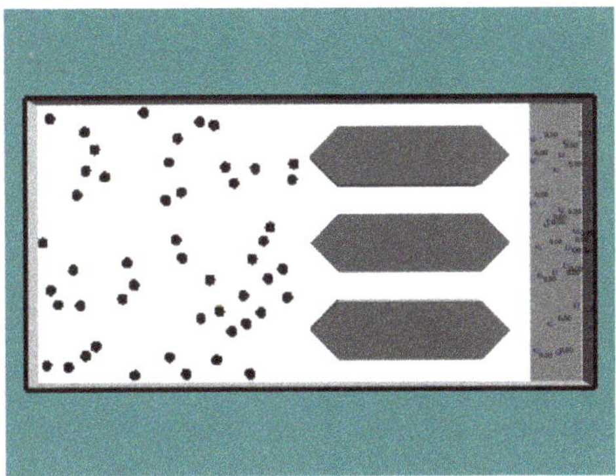

Figure 5. Screenshot of second simulated environment used. Food items are shown as black circles in the white environment, puck robots can be seen waiting in the nest area (light grey). Obstacles creating corridors are illustrated in dark grey.

3.4.2. Simulation

The experiments were performed in the ARGoS simulator [21] a multi robot simulator used to simulate large robot swarms. It was assumed that each of the robots was equipped with a food sensor, allowing them to identify food items within a 2m radius and each experiment featured a swarm of 20 robots.

Each test was executed for 500 simulated seconds (each simulation time step lasting 0.1 s) or until the target number of food items were collected.

The number of replicates required for consistent results were determined by performing cumulative mean tests as specified in [22]. This test indicated that the minimum number of trials required for consistency was 36. Therefore, 36 was the lowest number of replicates used when testing these systems.

3.5. Results

The results are illustrated in Figure 6 for environment 1 and Figure 7 for environment 2.

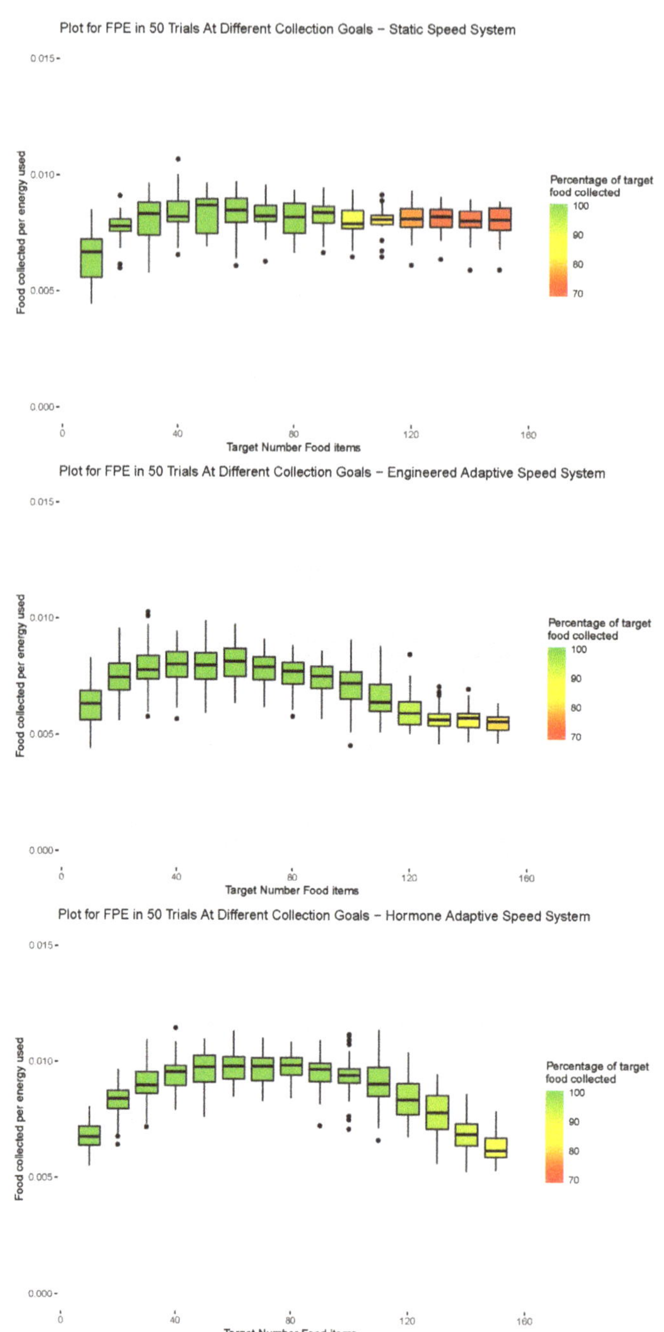

Figure 6. Target number for items collected ranged from 10 to 150 items of food. Percentage of the items requested versus those collected by the end of the simulation is indicated by colour (Green 100% and Red < 70%).

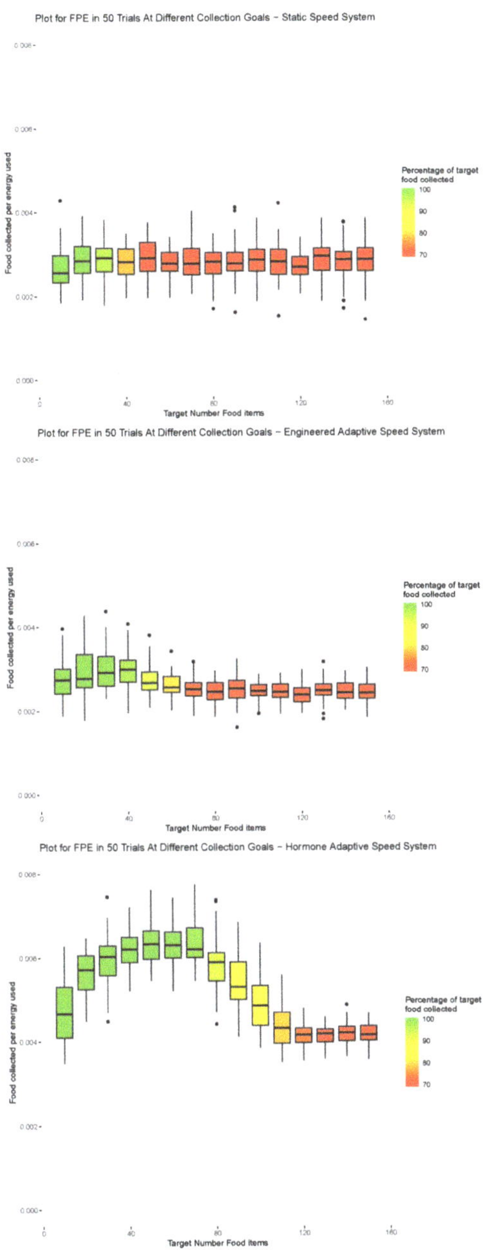

Figure 7. Three systems tested in in environment 2. Target number for items collected ranged from 10 to 150 items of food. Percentage of the items requested versus those collected by the end of the simulation is indicated by colour (Green 100% and Red < 70%).

3.5.1. Environment 1—Square Open Arena

Visual inspection of the first environment (Figure 6) shows that the static speed system has a fairly consistent level of food collected per energy unit used as the demand increases. This is expected due to the lack of change in speed, though the lowest target number for item collection does see a drop in energy efficiency when compared with the rest of the collection rates. This is because not all of the robots in the swarm will have returned to the nest by the time the experiment terminates having reached the target number of items. This will result in unnecessary energy consumption from the robots unable to return food items within the short period of the experiment.

The downside of this consistent energy consumption is the inability to reach greater item target numbers. This drawback can be seen in the discolouration of the box plots starting at 100 food items required and saturating to red, indicating a collection of less than 70% of the required items, by 130 required items.

Disregarding the lack of success in large item demand experiments, the results from the static speed system provide a strong baseline for energy efficiency. Giving a clear target for the other two more intelligent systems to aim for.

When inspecting the results of the two adaptive system it is immediately obvious that target collections are met more consistently with the demand function introduced to the system, with discolouration starting at 120 in the engineered system and 130 in the hormone system. In the engineered system the collection rate drops to approximately 80% by the 150 item goal while the hormone system still manages to collect upwards of 90%.

In terms of energy efficiency the engineered adaptive system follows a similar initial trend to the none adaptive system. The similarity is maintained until an item target of 50, at which point the engineered system becomes increasingly less efficient. Table 2 supports this, showing that there is no significant difference in the data sets of the Engineered and static systems until 70 target items. At this point the systems diverge as the engineered system consumes more energy.

Table 2. Environment 1: Wilcoxon rank sum tests comparing the three systems for the tested item collection targets between 10–150 in terms of energy efficiency. Significant differences (indicated by a p value of <0.05) are highlighted in **bold**.

System Type	Engineered vs. Static	Hormone vs. Static	Hormone vs. Engineered
Item Target Number			
10	0.8550	**0.0330**	0.0053
20	0.1648	$p < 0.0001$	$p < 0.0001$
30	0.1800	$p < 0.0001$	$p < 0.0001$
40	0.2626	$p < 0.0001$	$p < 0.0001$
50	0.0906	$p < 0.0001$	$p < 0.0001$
60	0.8227	$p < 0.0001$	$p < 0.0001$
70	**0.0068**	$p < 0.0001$	$p < 0.0001$
80	**0.0262**	$p < 0.0001$	$p < 0.0001$
90	$p < 0.0001$	$p < 0.0001$	$p < 0.0001$
100	$p < 0.0001$	$p < 0.0001$	$p < 0.0001$
110	$p < 0.0001$	$p < 0.0001$	$p < 0.0001$
120	$p < 0.0001$	**0.0199**	$p < 0.0001$
130	$p < 0.0001$	0.3984	$p < 0.0001$
140	$p < 0.0001$	$p < 0.0001$	$p < 0.0001$
150	$p < 0.0001$	$p < 0.0001$	$p < 0.0001$

These results also show that the hormone system managed to outperform both systems in regard to energy efficiency. With a significant difference versus the engineered adaptation and increased median result at every collection target excluding 10, the hormone system results can be seen arcing over those of the engineered system after starting at a similar point. Similarly, when compared to the static system, the hormone system shows significant increases to the food collected per energy used in all cases but targets of 10, 120 and 130 items. The similarity in energy efficiency of the hormone and speed systems at item targets of 120 and 130 can be explained by the speed increase of the hormone system in cases of very high item demand, actually reaching collection targets while the static system misses them by a large margin.

The efficiency of the hormone system over the static and engineered systems was explained by three factors:

Sensitivity: The hormone system is sensitive to collisions and capable of not only returning robots to the nest due to collisions, but also reducing speed due to the prolonged influence of collisions.

Dispersion: Rather than consistent speeds, or speeds of specific increments, the speeds of the hormone driven robots fluctuate during the search. Thus, dispersion is a by-product of efficiency as speed will be diverse amongst the swarm. This in turn will lead to less traffic and more energy efficient item collection.

Gradual variability: Speed can build over the duration of a search. This is contrary to the engineered system, which made relatively large (and potentially exaggerated) changes in speed on an individual's return to the nest.

3.5.2. Environment 2—Funnelled Corridor Arena

The results for the second environment, the increased length of environment and introduction of corridors, predictably show a notable decrease in percentage of target collection completed. The static system started to fail collection targets at 50 items and the engineered adaptive system starting to fail at 70. Compared with these, the change to collection rate in the hormone system is substantially less reduced. The results show the hormone system falling to a 70% collection rate at the 130 item target mark, showing a considerable increase in collection performance versus the two comparison systems.

In terms of energy efficiency there is again an expected drop in performance, when compared to the first environment, across all experiments due to the larger, more cluttered arena.

Analysing the systems tested in this environment, there is very little statistical similarity. Table 3 shows that almost all of the data sets at each item target number, with the exception of the first 5 item targets of the engineered versus static system, are all significantly different. The data produced from this environment does however follow very similar patterns those of the first environment. The static system maintains a consistent energy efficiency, though dipping slightly in the case of the smallest collection target. The Engineered system, while improving collection, does little to benefit energy consumption and lessens as target numbers increase. The hormone system, while exceeding the two comparison systems in both collection and energy efficiency, as it did in the first environment, does so in a much more exaggerated manner in the second environment.

Table 3. Environment 2: Wilcoxon rank sum tests comparing the three systems for the tested item collection targets between 10–150 in terms of energy efficiency. Significant differences (indicated by a p value of <0.05) are highlighted in **bold**.

System Type	Engineered vs. Static	Hormone vs. Static	Hormone vs. Engineered
Item Target Number			
10	0.2482	**$p < 0.0001$**	**$p < 0.0001$**
20	0.6918	**$p < 0.0001$**	**$p < 0.0001$**
30	0.3432	**$p < 0.0001$**	**$p < 0.0001$**
40	0.1010	**$p < 0.0001$**	**$p < 0.0001$**
50	0.0817	**$p < 0.0001$**	**$p < 0.0001$**
60	**0.0020**	**$p < 0.0001$**	**$p < 0.0001$**
70	**0.0002**	**$p < 0.0001$**	**$p < 0.0001$**
80	**$p < 0.0001$**	**$p < 0.0001$**	**$p < 0.0001$**
90	**$p < 0.0001$**	**$p < 0.0001$**	**$p < 0.0001$**
100	**$p < 0.0001$**	**$p < 0.0001$**	**$p < 0.0001$**
110	**$p < 0.0001$**	**$p < 0.0001$**	**$p < 0.0001$**
120	**$p < 0.0001$**	**$p < 0.0001$**	**$p < 0.0001$**
130	**$p < 0.0001$**	**$p < 0.0001$**	**$p < 0.0001$**
140	**$p < 0.0001$**	**$p < 0.0001$**	**$p < 0.0001$**
150	**$p < 0.0001$**	**$p < 0.0001$**	**$p < 0.0001$**

4. Introduction of the Sleep Hormone to a Foraging Swarm

The foundations of the introduced sleep hormone system are very similar to those presented in [5], following the same behaviour states as formerly designed. The system transitions through search, sleep, food collection and obstacle avoidance behaviours, directed by hunger, sleep and avoidance hormones. The hunger hormone was given an identical structure. However, due to the slight change in context to the foraging system, the stimulus to the sleep hormone in the system was edited from the original equation (first seen in [5]):

$$\text{Sleep Hormone (original): } H_\sigma(t) = \lambda_\sigma H_\sigma(t-1) + \gamma_\sigma H_A(t-1) \tag{7}$$

In these equations a sub index of 'σ' indicates relation to the sleep hormone and 'A' the relation to the avoidance hormone. Numbers ensuring these symbols indicate an additional coefficient relating to the parameter type i.e., decay or stimuli.

The additions to the original equation include both an α value and an inhibitor in the form of $\gamma_{\sigma 2}d$ (where $d(t)$ is the function of demand presented earlier in this paper) resulting in the new equation:

$$\text{New Sleep Hormone: } H_\sigma(t) = \alpha_\sigma + \lambda_\sigma H_\sigma(t-1) + \gamma_{\sigma 1} H_A(t-1) - \gamma_{\sigma 2} d(t) \tag{8}$$

The introduction of an α value offsets the settling point of the hormone. This allowed for the implementation of the demand based inhibitor ($\gamma_{Al2}d(t)$) and ensured that the hormone could fluctuate below the settling point without producing a negative value. The demand inhibitor itself created a larger decrease to the sleep hormone under high demand circumstances, assisting the decay already present in the hormone and reducing sleep times when the swarm's rate of collection was inadequate.

Meanwhile H_h (the sub index of h indicating a parameters relation to the hunger hormone) was kept in the same format, using the equation:

$$\text{Hunger Hormone: } H_h(t) = \alpha_h + \lambda_h H_h(t-1) + \gamma_h S \tag{9}$$

where S is a Boolean value representing whether the robot successfully returned a food item to the nest site or not.

The parameters used for the hunger and sleep hormones were calculated in a similar manner as Section 3.2.4, using the approximate time scale across which the hormones were expected to operate and thereafter tuning stimuli for the fitting reaction. The parameter values selected for the coming experiments are displayed in Table 4.

Table 4. Parameter values for the Return and Speed Hormones.

α_σ	λ_σ	$\gamma_{\sigma 1}$	$\gamma_{\sigma 2}$	α_h	λ_h	γ_h
0.01	0.999	0.01	0.06	0.0.015	0.999	10

4.1. *Preliminary Tests for Sleep Hormone in A Demand Lead Foraging Task*

The initial tests conducted on the new sleep hormone system used the same environments as the previous section and operated until a time limit of 500 seconds or until the target number of items was reached. A cumulative mean test indicated that a minimum of 14 trials were required. To ensure certainty, 20 trials were run.

4.1.1. Environment 1—Square Open Arena

Observing the results for the first environment (illustrated in Figure 8) the results appear very different to those of the previous three systems. The energy efficiency starts low, peaking momentarily and, after a dip to median performance, increases as the number of target items does. This pattern leads to an increased energy efficiency at all item targets compared to the previously tested static speed system and considerably better efficiency performance at item targets greater than 120 for the other two adaptive systems.

The initial spike in performance from this pattern is explained by the removal of poorly positioned robots at deployment. Those robots starting off in large groups will enter the sleep state either immediately or very soon after exploration. This initial state selection is then diluted as robots make more passes between the nest and the food area, seen as the Food Collected Per Energy used (FPE) reduces to a similar level as the non-adaptive system seen in Section 3.5. The gradual increase to FPE thereafter is due to the sleeping of poorly performing robots across greater periods of time, while robots with better positioning within the arena are able to collect food items more effectively.

While this system sees several increases to performance in terms of energy efficiency, it sacrifices this for poor performance in terms of item collection, with collection starting to drop at item targets of 90, lower than even the static system in the previous section. This is expected as the system actively impedes collection speed, with the sleep state removing swarm members for brief periods of time.

Though the collection percentage is lower in the sleep system than in the systems previously examined, this does show that it may be beneficial from the perspective of energy efficiency to combine the speed and sleep systems. With the intention of reducing the decrease to FPE seen in the adaptive speed systems as item target increases and using the speed system to compensate for the poor collection performance seen at targets above 90.

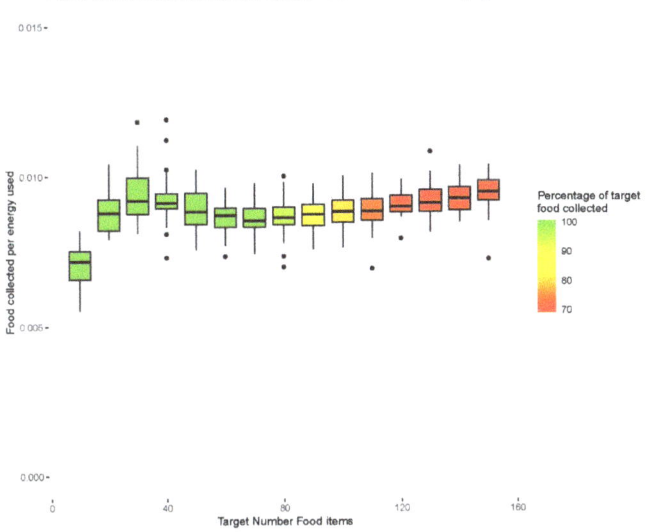

Figure 8. Hormone inspired sleep system tested in environment 1. Target number for items collected ranged from 10 to 150 items of food. Percentage of the items requested versus those collected by the end of the simulation is indicated by colour (Green 100% and Red < 70%).

4.1.2. Environment 2—Funnelled Corridor Arena

The benefit of this enhanced hormone system is further proven in the second environment. Following a similar pattern to the first environment, the energy efficiency increases with the target number (illustrated in Figure 9). In this environment, the sleep system is able to outperform the static and engineered system in terms of energy efficiency across all item target values. In addition to this, while not able to compete at lower item targets, after 80 items the sleep system largely outperforms the hormone speed system.

This increase to energy efficiency is due to the sleep system regulating the number of robots present in the corridors at any given moment, retaining poorly performing robots until demand is high and as a result increasing the productivity of the foraging swarm.

Again these results, while producing good values for energy efficiency, sacrifice collection rate. With collection similar to the static system, failing past 50 items.

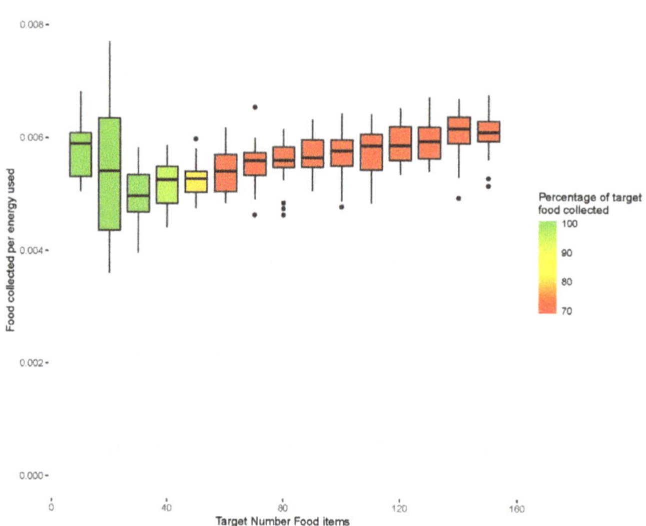

Figure 9. Hormone inspired sleep system tested in environment 2. Target number for items collected ranged from 10 to 150 items of food. Percentage of the items requested versus those collected by the end of the simulation is indicated by colour (Green 100% and Red < 70%).

4.2. Combining the Sleep Hormone with the Speed Deviating System

In an attempt to combine the benefits of both hormone systems and to identify the viability of combining existing hormone systems, the speed hormone was added to the already established sleep system. The parameter values established in prior testing were again used for the combined system. The speed hormone acted explicitly on motor speeds during exploration and the sleep hormone system regulated higher level behaviours.

The performance of this new combined system is illustrated in Figure 10. The first obvious improvement to the system can be seen in the results from the first environment, this set of data achieves the highest average collection rate of any system at a required collection of 150, obtaining an average of 92.3% of the needed items.

In addition to this the combined system achieves a significantly greater energy efficiency versus the sleep system at all item targets between 50 and 110 in the first environment and all item targets before 100 in environment 2 (p values for these tests can be found in Table 5). At higher value item targets the energy efficiency still crosses over, though the exceptional item collection rate more than compensates for this.

Relative to the adaptive speed system the combined hormone system obtains very similar results in the first environment at target item values below 70. Though there are large improvements to the energy efficiency at item targets larger than this. This increase to performance is mirrored in environment 2, though with a consistent increase at all item target values.

The substantial improvement in performance is proposed to be the mutually beneficial actions of the separate systems. It allows the system to avoid the circumstance in which positioned poorly robots in a high demand context might travel at high speeds that a cause large drain to power for poor returns.

It is clear from these results that these systems work better in combination than separately. This shows a strong symbiosis of already established hormone control, verifying the viability of combining hormone systems.

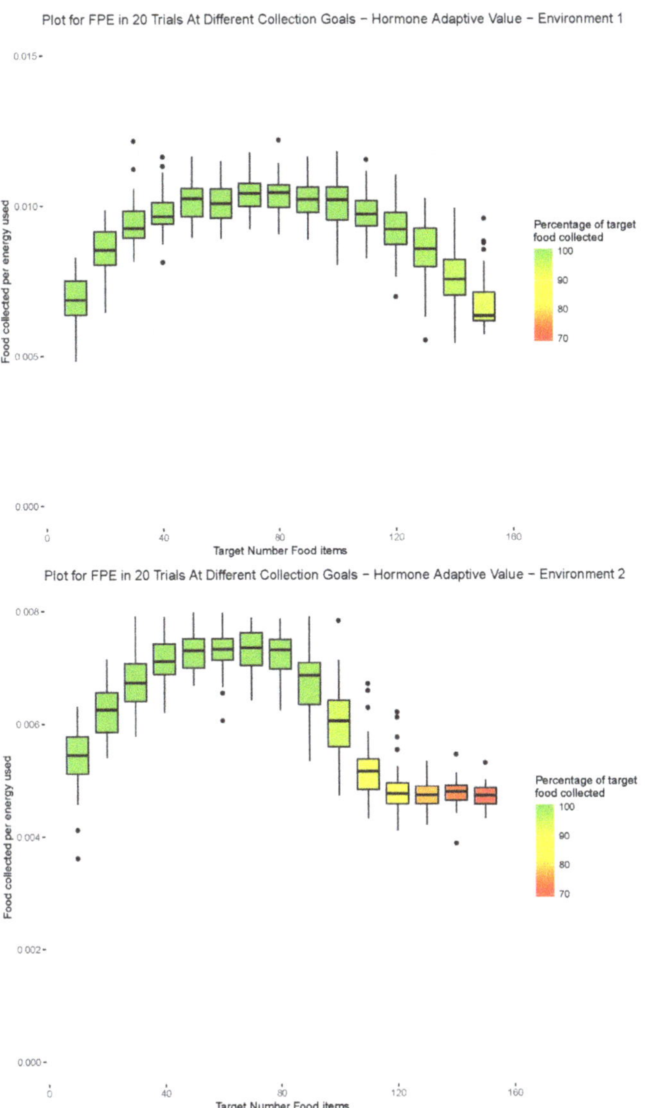

Figure 10. Combined hormone sleep and speed system tested in both environments. Target number for items collected ranged from 10 to 150 items of food. Percentage of the items requested versus those collected by the end of the simulation is indicated by colour (Green 100% and Red < 70%).

Table 5. Wilcoxon rank sum tests comparing the combined hormone system with both the speed hormone system and the sleep hormone system, in both of the previously established environments. Tests were conducted for item collection targets between 10–150 in terms of energy efficiency. Significant differences (indicated by a p value of <0.05) are highlighted in **bold**.

Item Target Number	Hormone Speed vs. Hormone Combination		Hormone Sleep vs. Hormone Combination	
	(Environment 1)	(Environment 2)	(Environment 1)	(Environment 2)
10	0.9680	**0.0047**	0.2315	**0.019**
20	0.8830	**0.0040**	0.1653	**0.0056**
30	0.5290	$p < 0.0001$	0.5831	$p < 0.0001$
40	0.6017	$p < 0.0001$	0.0810	$p < 0.0001$
50	0.5290	$p < 0.0001$	$p < 0.0001$	$p < 0.0001$
60	0.0809	$p < 0.0001$	$p < 0.0001$	$p < 0.0001$
70	**0.0283**	$p < 0.0001$	$p < 0.0001$	$p < 0.0001$
80	**0.0047**	$p < 0.0001$	$p < 0.0001$	$p < 0.0001$
90	0.0675	$p < 0.0001$	$p < 0.0001$	$p < 0.0001$
100	**0.0024**	$p < 0.0001$	$p < 0.0001$	**0.0430**
110	0.0910	$p < 0.0001$	**0.0024**	**0.0002**
120	0.0763	$p < 0.0001$	0.9042	$p < 0.0001$
130	0.1081	$p < 0.0001$	**0.0211**	$p < 0.0001$
140	**0.0227**	$p < 0.0001$	$p < 0.0001$	$p < 0.0001$
150	0.2648	$p < 0.0001$	$p < 0.0001$	$p < 0.0001$

5. Introduction of Environment Selection Hormones with Sleep and Speed Regulating Hormone Systems

With the viability of a larger hormone system confirmed, the next step taken was to combine the hormone combination system presented in the last section with yet more hormone arbitration. This was an important step because, while it has been shown that hormone systems can interact to produce satisfactory results, the current combination of hormone systems experience minimal detrimental interactions. In terms of behavioural arbitration, the speed and sleep hormone systems do not interfere with one another.

This section presents the amalgamation of the speed hormone, sleep hormone and a hormone system capable of monitoring the emergent success of the swarm under different conditions, implementing the designs shown in [7]. The monitoring of success and ensuing environmental preference, was driven by the speed at which items could be collected from the environments. Therefore, it is essential to investigate if the preference system will still be capable of categorising robots within a heterogeneous swarm effectively with an adaptive speed mechanism in place.

As such, the environment for these tests required a diverse terrain and multiple directional options alongside heterogeneity amongst the swarm.

5.1. Environmental Setup

The new environment used for testing in this scenario was identical to that used in [7] and can be seen in Figure 11. It features two different terrains designed to challenge robots with two specific wheel types, using 7 robots of each type for a total of 14 as previously studied.

Figure 11. Environment used for all tests requiring terrain preference and categorisation of heterogeneous robots within the swarm.

In order to incorporate heterogeneity into the swarm, while still using the energy characteristics presented in Section 3.1 to measure energy efficiency at different speeds, each wheel type was given a speed coefficient for respective terrains. These coefficients (displayed in Table 6) inhibited the speed properties of the wheels based on the ground a given robot was travelling on, these values are shown in Table 6. While not as realistic as the data used for wheel speeds in previous environment preference experiments, this allowed for the testing the combination of systems without extensive testing of robotic hardware.

Table 6. Speed Coefficients for Heterogeneous Robot Wheels on Different Terrains.

Terrain Type	Wood Suited Wheels	Grass Suited Wheels
Grass	0.6	0.7
Wood	1	0.8

5.2. *Effect of Demand on Environment Selection When the Speed Hormone Is Combined with Environmental Preference Hormone*

Before fully combining the systems, the speed hormone was added to the Environmental Preference System. The performance of the selection system was then measured by looking at the proportion of robots active in the environments they were best suited to as a percentage.

In order to incorporate the speed hormone to the directional preference hormone system, demand functions identical to that previously produced in Section 3.2.1 were created for both the north and south environments, taking only items collected in the respective environment into account when producing demand. Depending on the environmental preference when returning to the nest site, robots within the swarm would then update their demand stimuli with the corresponding demand value.

The full results of these tests are illustrated in Figure 12. Minimal differences were found in median categorisation across the range of item targets. Further, these were not found to differ from median categorisation found when the speed hormone was not included in the system. As the speed hormone did not appear to have a negative effect on the environmental preference hormones, it was deemed reasonable to further add the sleep hormone to the system.

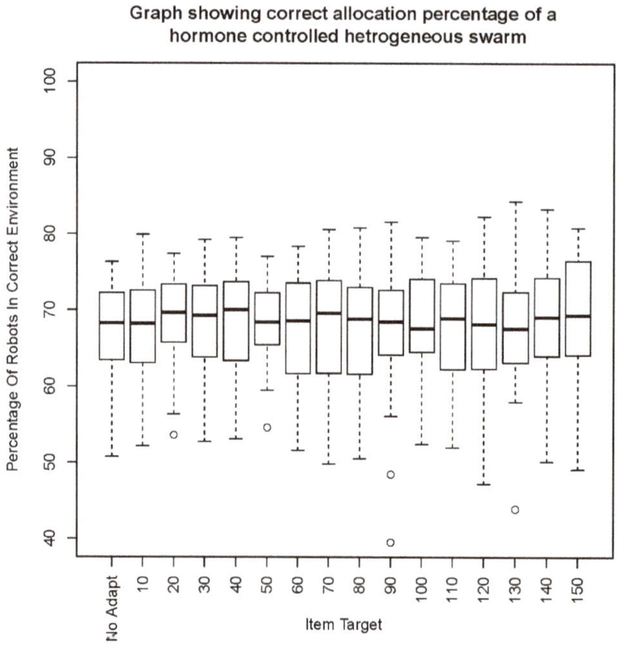

Graph showing correct allocation percentage of a hormone controlled hetrogeneous swarm

Figure 12. Effect of item target values driving different demands in the speed hormone on the percentage of robots taking preference to their optimal environment. The categorisation system running with no speed hormone present is marked as 'no adapt'.

With minimal negative interaction between the speed regulating and environmental preference hormones, it was deemed reasonable to continue with the implementation of the combined hormone system with the introduction of the hormone driven sleep system.

5.3. System Combining Sleep, Speed and Preference Hormones

To observe the performance of this new system, the various combinations of hormone systems were tested in combination in the new multi-terrain environment. First the preference system was tested on its own, the results from this are illustrated in Figure 13. These results would act as a baseline to the additional systems as results from this new environment, with new task complexities, would be incomparable with data from previous experiments.

When adding the sleep hormone to the system (results illustrated in Figure 14), results for energy efficiency are consistently raised past the first item targets of 5, as shown by median results increasing by approximately 25% for item targets past 70. While there is a large improvement in terms of energy efficiency, adding the sleep hormone only results in a slight increase to collection rate, with the cut off point for collection becoming poor (below 90% of the target item collection) shifted from 80 to 90.

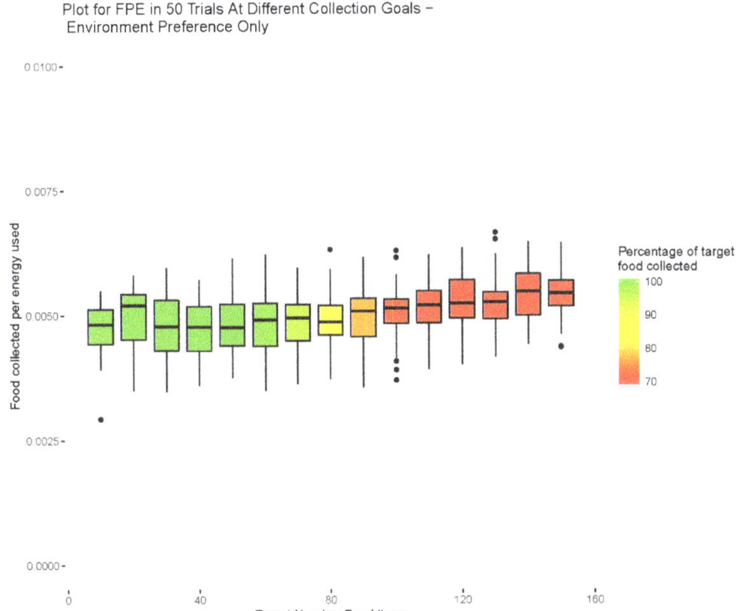

Figure 13. Hormone preference system tested in the environment containing two difference terrain types. Target number for items collected ranged from 10 to 150 items of food. Percentage of the items requested versus those collected by the end of the simulation is indicated by colour (Green 100% and Red < 70%).

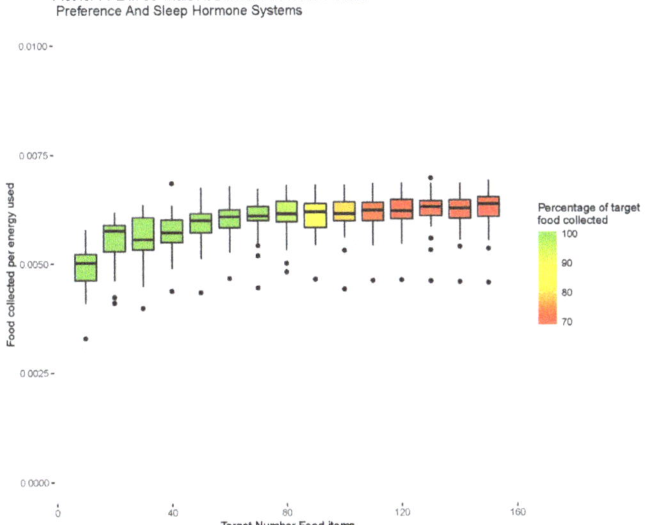

Figure 14. Hormone preference system, combined with the sleep hormone system tested in the environment containing two difference terrain types. Target number for items collected ranged from 10 to 150 items of food. Percentage of the items requested versus those collected by the end of the simulation is indicated by colour (Green 100% and Red < 70%).

When the speed hormone is added to the system in the absence of the sleep hormone energy efficiency suffers considerably. This is seen with the consistent drop in efficiency results illustrated in Figure 15 when compared to the baseline results. However, this drop in efficiency is traded for a substantial improvement to collection rate, moving the cut off point for poor collection to 120 target items. These results are expected with the additional speed fluctuation and if the results from previous hormone combinations hold consistent, the addition of the sleep system to the preference/speed hormone regulation should amend the poor energy efficiency while maintaining the item collection rate.

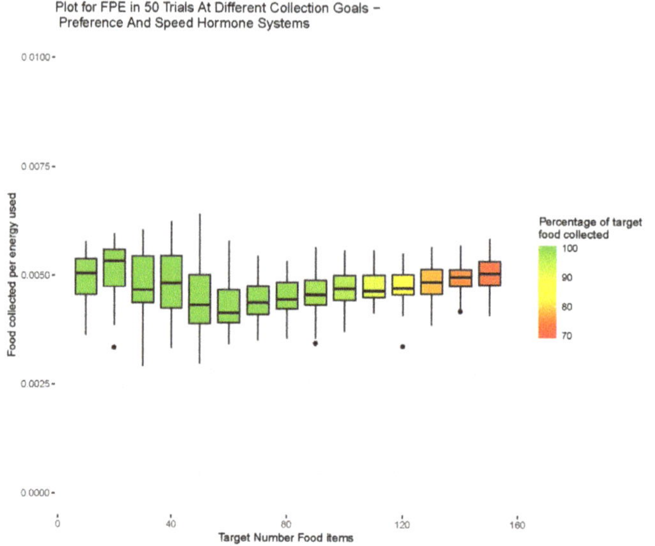

Figure 15. Hormone preference system, combined with the speed hormone system tested in the environment containing two difference terrain types. Target number for items collected ranged from 10 to 150 items of food. Percentage of the items requested versus those collected by the end of the simulation is indicated by colour (Green 100% and Red < 70%).

Combining all three systems (results illustrated in Figure 16) provides the best result in terms of item collection, maintaining adequate collection until the 130 target item trial. Simultaneously, the system that combines all three hormone types is capable of accomplishing competitive values for energy efficiency. These values show improvements across all item targets for the standard preference and combined speed hormone results. The three hormone system only marginally under performs in energy efficiency versus the combined preference and sleep system, although it shows much greater item collection percentages. These results suggest that the fully combined system as the strongest of the system permutations when considering both item collection and energy efficiency.

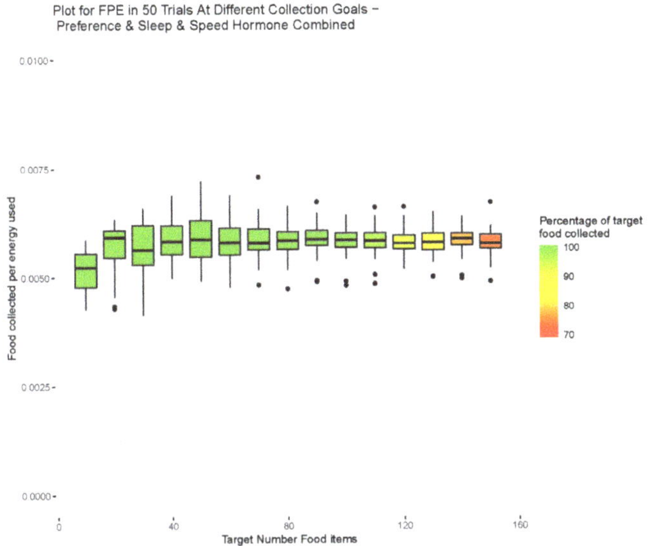

Figure 16. Hormone preference system combined with both the sleep and speed hormone system, tested in the environment containing two difference terrain types. Target number for items collected ranged from 10 to 150 items of food. Percentage of the items requested versus those collected by the end of the simulation is indicated by colour (Green 100% and Red < 70%).

6. Scalability of Final Amalgamated Hormone System

To introduce a key element of difficulty to the system presented in this paper a scalability test was conducted. With more robots in the swarm, the more difficult it will be to move effectively within the environment without slowing due to clutter. Along side this, with greater swarm density robots may receive over-stimulation from the transmitted hormones of the increased number of robots or there may be too much competition for food items, with multiple robots travelling to the same item simultaneously. With these additional negative features present it will be difficult for robots to form accurate preferences to terrain due to the fact that these negative features may have a greater effect on performance than the speed variance provided by the different wheel types.

The scalability tests were conducted by increasing the number of robots in each simulation by 6 for each set of trials, testing swarm sizes ranging between 12 and 60. In each experiment the target number of items was set to 100 and in every test all of these items were retrieved. The experiments conducted terminated after 500 simulated seconds or if the target number of items was reached. The item target of 100 was chosen due to the variability in performance at said target number across each of the previously tested systems. This indicated that this number of items is an area of interest, providing substantial challenge to some systems while still an achievable goal to others.

The results of the scalability test can be seen illustrated in Figure 17. It can be seen that energy efficiency decreases linearly with the increase in members of the swarm. This was expected with the increased difficulty to the task as, while the amalgamated system is able to augment performance with a given swarm size, additional or unneeded robots will still create detriment to performance. Through the linear nature of this performance degradation, a user can select a swarm size which is suitable for a given task, trading off energy efficiency for the speed at which items should be gathered.

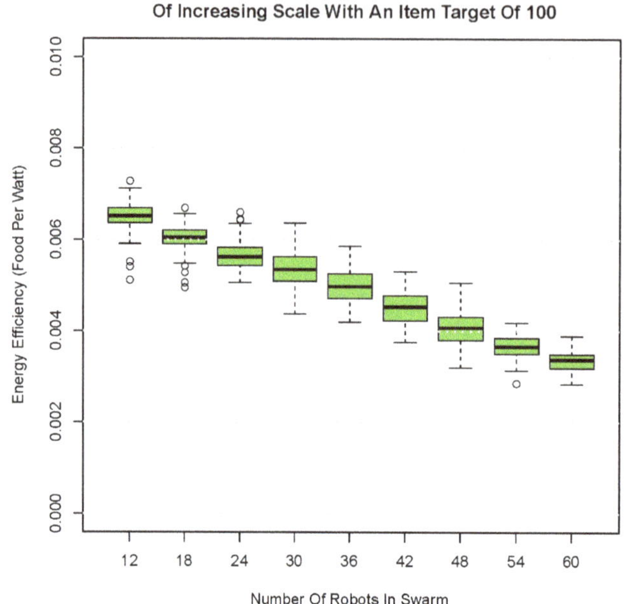

Figure 17. Results for energy efficiency as the number of robots in the swarm increases from 12 to 60.

7. Discussion

This paper has explored the viability of numerous simultaneously functioning hormone inspired systems. To address this, a speed controller for a foraging swarm was designed using a hormone inspired system and proven to be effective for energy efficient item collection at a number of different item targets. This system was then combined with a previously developed sleep system. The combination of these two systems addressed issues found amongst each of the individual systems, creating large increases to performance with minimal drawbacks. Based on this success a third hormone system was introduced, allowing members of a heterogeneous swarm to form a preference for environment, based on how successful individual robots assessed themselves to be in a given terrain. This new system tested with the speed adapting virtual hormone, identified as the system that would cause the most issue when attempting to categorise robots, was still able to effectively categorise robots, with limited change as demand increased.

Finally, the combination of all of the hormone systems was tested. While not producing the best energy efficiency of the tested systems, the amalgamated hormone system produced the best combination of collection rate and energy efficiency for the environment the system was tested in. Considering the total performance of the system should definitely take into account both energy efficiency and item collection, as the values they represent show task effectiveness and task completion respectively.

The results from the work in this paper have shown that a complex system controlled almost entirely by virtual hormones can be an effective adaptation system within a swarm robotic context.

Future work will involve robustness analysis of the systems, ensuring that performance is not drastically reduced by problems to be expected in real world scenarios such as motor wear or actuator failure. Along side this, additional testing can further close the reality gap, introducing gaussian noise to both the power model and wheel speeds. This will introduce an element of variability to both, as it is expected that a group of robots in reality would experience non-perfect energy consumption and navigational abilities. In addition to this tests using the PSI swarm robots (the robots approximated in simulation within this papers presented experiments) should take place to demonstrate the capabilities of a physically implemented system. These capabilities will provide evidence to suggest that swarms, equipped with complex hormone systems, would be capable of functioning well in real world applications that require on-line adaptation. These applications could involve disaster relief work, with systems investigating vast areas of volatile and changing environments associated with disaster aftermath, securing survivors or sustaining resources. Equally, hormone systems could be implemented to enable searching for valuable minerals or suitable areas for habitation on foreign planets with hostile and erratic weather.

Author Contributions: Conceptualization, J.W.; methodology, J.W.; software, J.W.; validation, J.W.; formal analysis, J.W.; investigation, J.W.; data curation, J.W.; Writing—Original draft preparation, J.W.; Writing—Review and editing, J.T. and A.T.; visualization, J.W.; supervision, J.T. and A.T.

Funding: This research received no external funding

Conflicts of Interest: The authors declare no conflict of interest.

References

1. Jaycox, E.R. Behavioral changes in worker honey bees (*Apis mellifera* L.) after injection with synthetic juvenile hormone (Hymenoptera: Apidae). *J. Kansas Entomol. Soc.* **1976**, *46*, 165–170.
2. Adkins-Regan, E. *Hormones and Animal Social Behavior*; Princeton University Press: Princeton, NJ, USA, 2005.
3. Baird, T.A. Lizards and other reptiles as model systems for the study of contest behaviour. In *Animal Contests*: Cambridge University Press: Cambridge, UK, 2013; pp. 258–286.
4. Kuyucu, T.; Tanev, I.; Shimohara, K. Hormone-inspired behaviour switching for the control of collective robotic organisms. *Robotics* **2013**, *2*, 165–184. [CrossRef]
5. Wilson, J.; Timmis, J.; Tyrrell, A. A Hormone Arbitration System for Energy Efficient Foraging in Robot Swarms. In *Annual Conference Towards Autonomous Robotic Systems*; Springer: Berlin, Germany, 2018; pp. 305–316.
6. Shen, W.; Will, P.; Galstyan, A.; Chuong, C. Hormone-inspired self-organization and distributed control of robotic swarms. *Auton. Robots* **2004**, *17*, 93–105. [CrossRef]
7. Wilson, J.; Timmis, J.; Tyrrell, A. A Hormone-Inspired Arbitration System For Self Identifying Abilities Amongst A Heterogeneous Robot Swarm. In Proceedings of the 2018 IEEE Symposium Series on Computational Intelligence (SSCI), Bangalore, India, 18–21 November 2018; pp. 843–850.
8. Stradner, J.; Hamann, H.; Schmickl, T.; Crailsheim, K. Analysis and implementation of an artificial homeostatic hormone system: A first case study in robotic hardware. In Proceedings of the 2009 IEEE/RSJ International Conference on Intelligent Robots and Systems, St. Louis, MO, USA, 10–15 October 2009; pp. 595–600.
9. Kernbach, S.; Meister, E.; Schlachter, F.; Jebens, K.; Szymanski, M.; Liedke, J.; Laneri, D.; Winkler, L.; Schmickl, T.; Thenius, R.; et al. Symbiotic robot organisms: REPLICATOR and SYMBRION projects. In Proceedings of the 8th workshop on performance metrics for intelligent systems, Gaithersburg, MD, USA, 19–21 August 2008; pp. 62–69.
10. Jin, Y.; Guo, H.; Meng, Y. Robustness analysis and failure recovery of a bio-inspired self-organizing multi-robot system. In Proceedings of the 2009 Third IEEE International Conference on Self-Adaptive and Self-Organizing Systems, San Francisco, CA, USA, 14–18 September 2009; pp. 154–164.
11. Palmieri, N.; Yang, X.S.; De Rango, F.; Marano, S. Comparison of bio-inspired algorithms applied to the coordination of mobile robots considering the energy consumption. *Neural Comput. Appl.* **2017**, *31*, 263–286. [CrossRef]
12. Liu, W.; Winfield, A.F.T.; Sa, J.; Chen, J.; Dou, L. Towards energy optimization: Emergent task allocation in a swarm of foraging robots. *Adapt. Behav.* **2007**, *15*, 289–305. [CrossRef]
13. De Martinisl, V.; Gallo, M.; D'Acierno, L. Estimating the benefits of energy-efficient train driving strategies: A model calibration with real data. *Urban Transp. XIX* **2013**, *130*, 201–211.
14. Miyatake, M.; Ko, H. Optimization of train speed profile for minimum energy consumption. *IEEJ Trans. Electr. Electron. Eng.* **2010**, *5*, 263–269. [CrossRef]

15. Lee, J.H.; Ahn, C.W. Improving energy efficiency in cooperative foraging swarm robots using behavioral model. In Proceedings of the 2011 IEEE Sixth International Conference on Bio-Inspired Computing: Theories and Applications, Penang, Malaysia, 27–29 September 2011; pp. 39–44.
16. Pang, B.; Zhang, C.; Song, Y.; Wang, H. Self-organized task allocation in swarm robotics foraging based on dynamical response threshold approach. In Proceedings of 2017 IEEE 18th International Conference on Advanced Robotics (ICAR), Hong Kong, China, 10–12 July 2017; pp. 256–261.
17. Hilder, J.; Horsfield, A.; Millard, A.G.; Timmis, J. The Psi Swarm: A Low-Cost Robotics Platform and Its Use in an Education Setting. In *Conference Towards Autonomous Robotic Systems*; Springer: Berlin, Germany, 2016; pp. 158–164.
18. Bonani, M.; Longchamp, V.; Magnenat, S.; Rétornaz, P.; Burnier, D.; Roulet, G.; Vaussard, F.; Bleuler, H.; Mondada, F. The marXbot, a miniature mobile robot opening new perspectives for the collective-robotic research. In Proceedings of the 2010 IEEE/RSJ International Conference on Intelligent Robots and Systems, Taipei, Taiwan, 18–22 October 2010; pp. 4187–4193.
19. Keysight Technologies. *N6700 Modular Power System Family*; Keysight Technologies: Santa Rosa, CA, USA, 2016.
20. Buchanan, E.; Pomfret, A.; Timmis, J. Dynamic Task Partitioning for Foraging Robot Swarms. In Prodeedings of the International Conference on Swarm Intelligence, Bali, Indonesia, 25–30 June 2016; Springer: Berlin, Germany, 2016; pp. 113–124.
21. Pinciroli, C.; Trianni, V.; O'Grady, R.; Pini, G.; Brutschy, A.; Brambilla, M.; Mathews, N.; Ferrante, E.; Di Caro, G.; Ducatelle, F.; et al. ARGoS: A Modular, Parallel, Multi-Engine Simulator for Multi-Robot Systems. *Swarm Intell.* **2012**, *6*, 271–295. [CrossRef]
22. Robinson, S. *Simulation: The Practice of Model Development and Use*; Wiley: Chichester, UK, 2004.

Article

Mixed Reality Simulation of High-Endurance Unmanned Aerial Vehicle with Dual-Head Electromagnetic Propulsion Devices for Earth and Other Planetary Explorations

Ashish Kumar [1,*], Sugjoon Yoon [1] and V.R.Sanal Kumar [2,3]

1 Department of Aerospace Engineering, Sejong University, Seoul 05006, Korea; sjyoon@sejong.edu
2 Indian Space Research Organisation, VSSC, Trivandrum 695022, India; vr_sanalkumar@yahoo.co.in
3 Kumaraguru College of Technology, Coimbatore 641049, India
* Correspondence: ashishkumar.aero@gmail.com

Received: 22 April 2020; Accepted: 23 May 2020; Published: 28 May 2020

Abstract: One of the major limitations of existing unmanned aerial vehicles is limited flight endurance. In this study, we designed an innovative uninterrupted electromagnetic propulsion device for high-endurance missions of a quadcopter drone for the lucrative exploration of earth and other planets with atmospheres. As an airborne platform, this device could achieve scientific objectives better than state-of-the-art revolving spacecraft and walking robots, without any terrain limitation. We developed a mixed reality simulation based on a quadcopter drone and an X-Plane flight simulator. A computer with the X-Plane flight simulator represented the virtual part, and a real quadcopter operating within an airfield represented the real part. In the first phase of our study, we developed a connection interface between the X-Plane flight simulator and the quadcopter ground control station in MATLAB. The experimental results generated from the Earth's atmosphere show that the flight data from the real and the virtual quadcopters are precise and very close to the prescribed target. The proof-of-concept of the mixed reality simulation of the quadcopter at the Earth atmosphere was verified and validated through several experimental flights of the F450 spider quadcopter with a Pixhawk flight controller with the restricted endurance at the airfield location of Hangang Drone Park in Seoul, South Korea. We concluded that the new generation drones integrated with lightweight electromagnetic propulsion devices are a viable option for achieving unrestricted flight endurance with improved payload capability for Earth and other planetary explorations with the aid of mixed reality simulation to meet the mission flight path demands. This study provides insight into mixed reality simulation aiming for Mars explorations and high-endurance missions in the Earth's atmosphere with credibility using quadcopter drones regulated by dual-head electromagnetic propulsion devices.

Keywords: mixed reality simulation; unmanned aerial vehicles; X-Plane flight simulator; visualized ground control station; dual-head electromagnetic propulsion device

1. Introduction

Multicopter rotorcraft unmanned aerial vehicles (UAVs) are less susceptible to turbulence as compared with similar-sized fixed wing aircraft. Among the rotorcraft, the quadcopter is the most commonly used UAV, because of its mechanical simplicity and performance [1–3]. Of late, the applications of quadcopter UAVs, i.e., drones are increasing dramatically due to less risk and more benefits to the operators and users in the Earth's atmosphere. A literature review revealed that the application of drones with improved payload capability for high-endurance planetary exploration has been emerging in aerospace industries worldwide owing to the fact that a drone can map a larger planet area than a rover at a resolution far better than the existing satellites or orbiters [4,5]. It is well known that airborne platforms cover much larger distances in a single mission than a rover and can transmit high-resolution images of very rocky or steep terrain better than state-of-the-art orbiting spacecraft. Orbiters extend the facility to map large areas for a longer period of time with restricted resolution. Landers can handle the planet's surface and atmospheric sampling but are limited to the close vicinity of the landing site. The mobility of the airborne platform is a major concern. To overcome these restrictions of orbiters and landers, we have proposed MR simulation of a high-endurance quadcopter UAV with dual-head electromagnetic propulsion (EMP) devices to maneuver to interesting sites lucratively for a longer duration through the prescribed trajectory.

A literature review revealed that there are eight listed planets and more than 160 moons known to us in the solar system. Among the planets, it has been reported that Venus, Earth, Mars, Jupiter, Saturn, Uranus, and Neptune have noteworthy atmospheres. The largest moon of Saturn, namely Titan, has been identified to have a dense atmosphere for facilitating the mobility of the airborne platforms. Over the decades, Mars has been one of the fascinating planets for scientific exploration. Flying a drone in the environment of Mars presents a major challenge, mainly because of the atmospheric characteristics of Mars. It has been reported that the density of the atmosphere of Mars is extremely low in the order of 1/70 as compared with that on Earth's surface [6], which demands the high-speed rotor to generate sufficient lift at a low Reynolds number. The speed of sound on Mars is approximately 20% less as compared with that on Earth [6], which creates the high Mach number flows. At this atmospheric condition, designing an airborne platform for planet exploration is a challenging task. In this paper, we proposed a viable option of a new generation quadcopter UAV integrated with lightweight feedback-controlled dual-head EMP devices which achieves the variable-speed spinning rotors to obtain a desirable lift with an efficient guidance, navigation, and control system, in accordance with local atmospheric properties, for high-endurance Earth and other planetary explorations with the aid of MR simulation to meet the flight path demands of the mission.

Although many studies have been carried out on the design and development of multicopter rotorcraft from a different perspective, a limited number of studies have been carried out on MR simulation based on a quadcopter UAV and an X-Plane flight simulator [7–9]. In quadcopters, two motors rotate in a clockwise (CW) direction, and the remaining two motors rotate in a counterclockwise (CCW) direction, which produce zero angular momentum. In order to control the yaw, where the copter turns left and right, either the CW or CCW propellers are required to speed up or slow down to cause angular momentum to turn the copter. Quadcopters are currently used in agricultural, scientific, and commercial fields, including the military. The quadcopter works according to the speed of each rotor. Flight controller hardware is the head of any UAV, including the quadcopter. Today, the Pixhawk controller is widely used for UAV applications due to its low cost and better performance [7–9] and it comes with autopilot open source software and firmware.

The necessity for UAV simulations is increasing largely because there is a great deal of research taking place on UAV exploration [9–11]. A literature review revealed that many undesirable accidents have occurred during UAV operations due to the lack of professional pilot training [12–14]. Therefore, it is necessary, rather desirable, and perhaps inevitable, to develop a training device, such as the UAV simulator for new UAV users, and also for further research and development. UAV simulations have shown many benefits, including better understanding of a system and experimental flight tests before a real UAV fight. In the industry, a considerable number of flight simulation softwares are available, including X-Plane [15–17], FlightGear [18], Gazebo [9], and MAV3DSim [19]. The three main types of existing simulations are virtual reality (VR) simulation [20], augmented reality (AR) simulation [21], and mixed reality (MR) simulation [22]. The virtual reality simulation is a fully immersive type of simulation; it is used in computer technology to create a simulated environment. The augmented reality simulation is overlaid digital information in the real world. The mixed reality simulation is a combination of the virtual part, together with the real part, and it interacts in real time. Various research methods have been conducted on these three simulations, including the UAV field. Gongjin Lan et al.

(2016) [23] developed UAV-based virtual reality systems and Shubo Wang et al. (2017) [24] constructed a virtual reality platform for UAV deep learning. Yuan Wang et al. and Li Yi-bo et al. introduced UAV with augmented reality technologies [25,26]. In MR simulation, Martin Selecký et al. [27] proposed a design for a communication architecture in unmanned systems. Fernando López Peña et al. (2017) [28] discussed an initial phase of the MR simulator for autonomous UAVs. Saimouli Katragadda et al. (2019) [29] developed a stereoscopic MR for UAV search and rescue.

Nowadays, UAVs are used for long-time missions for strategic defense, agricultural, surveillance, and rescue operations during a natural calamity. Additional surveillance applications include pipeline security, livestock monitoring, wildfire mapping, home security, road patrol, transportation, photography, and other entertainments. The main limitation of the long-time missions is the limited flight endurance due to the visibility problem of the UAV. It is well known that most UAVs have a live camera tracking system and global positioning system (GPS), but a user cannot see the overall view of an UAV during the use of these systems.

Although a large volume of simulation studies on UAVs are available in the open literature, there are no studies that address the overall view and performance of MR simulation of UAVs [27–29], which we have addressed, herein, along with the preliminary design of a quadcopter UAV governed by dual-head EMP devices for high endurance planetary explorations. More specifically, in this paper, we introduce a new MR simulation based on a quadcopter UAV and an X-Plane flight simulator (version 10.51). Herein, we present an overall view of the real-time performance of an UAV, which enables us to see the live performance of a real UAV quadcopter on a simulation platform. Using this MR simulation technique, we can solve the problem of limited flight endurance due to visibility problems during a long-time mission of any quadcopter.

2. Methodology

In this MR simulation, a computer with X-Plane software represented the simulation platform, and a real quadcopter within an airfield represented the real platform. In this paper, the interaction between the real and the virtual quadcopters in real time is called MR simulation. The data flow between the real and the virtual quadcopters are shown in Figure 1. More precisely, we developed a connection interface between the X-Plane flight simulator and the quadcopter ground control station (GCS) using the transmission control protocol/internet protocol (TCP/IP) [30] and the user datagram protocol (UDP) [31] in MATLAB. Both are the suite of communication protocols used for data transferring. For the simulation of the performance of the real quadcopter on the X-Plane platform, we needed the position (latitude, longitude, and altitude) and attitude (pitch, roll, and heading) data from the real quadcopter. The GCS receives the real-time flight data of the real quadcopter through a radio telemetry device. In our case, the developed connection interface needed two ways of communication; one to receive the real-time data from the GCS and the second to send the real-time position and attitude data from the received flight data to X-Plane. Therefore, here, we used, TCP/IP communication between the GCS and developed interface, and UDP communication between the developed interface and X-Plane. By using the developed connection interface, the GCS sends the real-time position and attitude data to X-Plane, and the virtual quadcopter in X-Plane follows the real quadcopter. Consequently, here, the virtual quadcopter interacted with the real quadcopter in real time (i.e., MR simulation).

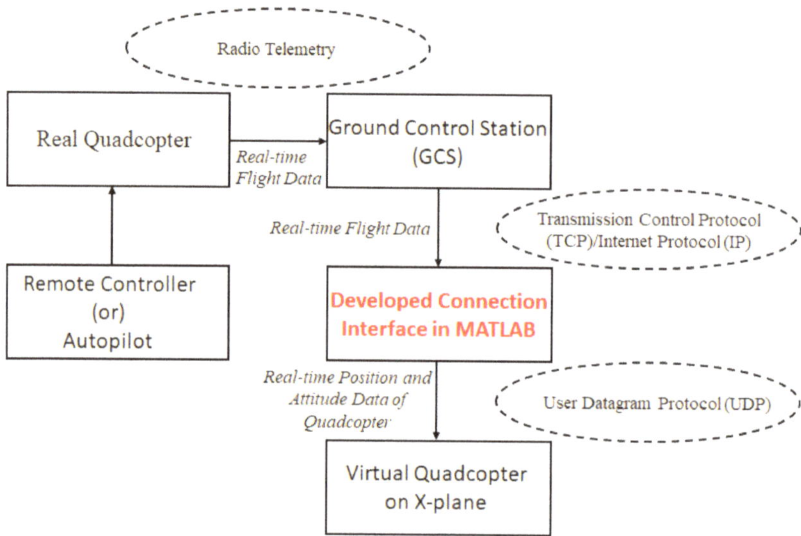

Figure 1. Data flow between the real and the virtual quadcopters (overview of mixed reality simulation).

3. Design of Quadcopter and Simulation Environment Setup

3.1. Design of Virtual Quadcopter

A real quadcopter and a virtual quadcopter are required to run a MR simulation. The real quadcopter performs within an airfield, and the virtual quadcopter interacts and follows the real quadcopter on the simulation platform. In this work, X-Plane is the simulation platform. X-Plane is a computer flight simulator software, produced by the Laminar Research Company of USA. In this study, an F450 spider quadcopter (see Figure 2) with a Pixhawk flight controller was made in house and used as the real quadcopter. The specifications of the F450 spider quadcopter are listed in Table 1. A virtual quadcopter with the parameters of the real quadcopter is needed to run the simulation platform in X-Plane. In this work, Plane Maker software (version 10.51) [32] was used to design a virtual quadcopter with the specifications of the F450 spider quadcopter.

Figure 2. F450 spider quadcopter.

Table 1. Specifications of the F450 spider quadcopter.

Quadcopter Specifications	Model/Values
Flight controller	Pixhawk 2.4.8
Total mass (including battery)	2 kg
Frame diagonal length	450 mm
Propellers dimension and pitch	10 × 3.8
No of blade	2
Motor (brushless)	2212–920 KV
Motor turn rate	4000–9000 RPM
Radio telemetry	433 MHZ

To make a virtual quadcopter in Plane Maker, we needed to design mainly five parts, i.e., the motors, propellers, fuselage, arms, and landing gears. Plane Maker consists of several sections which include fuselage, misc bodies, engine specs, landing gear, weight and balance, and visual texture regions. In this study, the fuselage section was used to design the fuselage of the quadcopter. The designed fuselage contained a central body part, rear tail portion, and camera holder, similar to a real quadcopter (F450 Spider). The arms of the quadcopter were designed using the "misc bodies" section. The section on "engine specs" was used to build the motors and propellers, using the real quadcopter's motor and propeller parameters including maximum turn rate; power set, pitch, root, and chord of the propeller; and propeller radius. We used the section on "landing gear" to design the four skid type landing gears similar to a real quadcopter. The total weight of the quadcopter was set the same as a real quadcopter using the "weight and balance" section. For more visibility in X-Plane, here, we used the color black (because the color of the real quadcopter was black) for the virtual quadcopter using the "visual texture regions" section. The fully designed virtual quadcopter in Plane Maker is shown in Figure 3. Note that it was essential to place the designed virtual quadcopter in the X-Plane's aircraft folder.

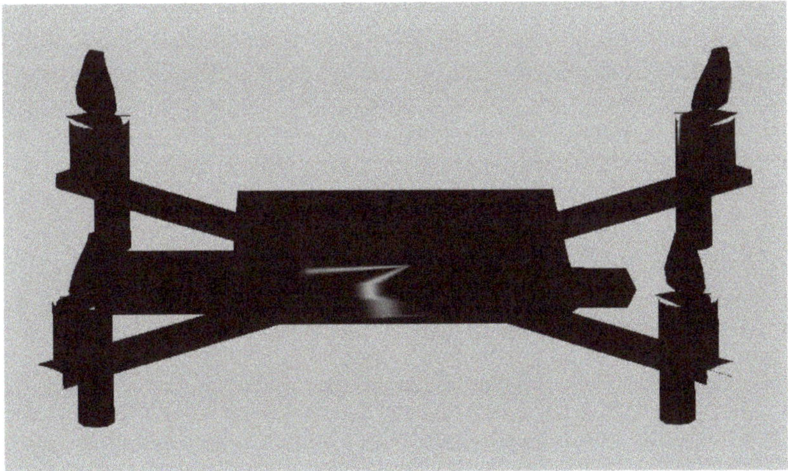

Figure 3. Fully designed virtual quadcopter.

3.2. Design of the Simulation Environment (Virtual Location)

For the MR simulation, it was essential to design a virtual location in X-Plane similar to the location of the real quadcopter. In this work, a drone airfield was chosen in the permissible east part of Seoul, South Korea. Figure 4 illustrates the chosen airfield location (Hangang Drone Park) in Seoul, South Korea. Note that X-Plane has only the sceneries of major airports, there is no scenery available for the chosen airfield location (Hangang Drone Park). To overcome this lacuna, a new scenery of the chosen airfield location was designed at the same geographic location using WorldEditor (version: 1.7.2). WorldEditor is a software, which is used to create and edit the scenery for X-Plane.

Figure 4. Airfield location (Hangang Drone Park) in Seoul, South Korea.

Note that the small runway in the Hangang Drone Park (see Figure 4) was taken as the initial reference point of the real quadcopter. Therefore, we created a virtual runway similar to the real runway in Hangang Drone Park. In WorldEditor, first, we chose the "create airport" option. After selecting the create airport option, we could see several design tools on WorldEditor, including runway, helipad, objects, forest, sealane, and facades. Here, we used the "runway" tool to build the runway. According to latitude, longitude, heading, and a few other known design parameters highlighted in Figure 5, the virtual location was built similar to the chosen airfield location. Note that the coordinates of the virtual and actual locations needed to be exactly the same, which we achieved; otherwise, an error would occur during the MR simulation.

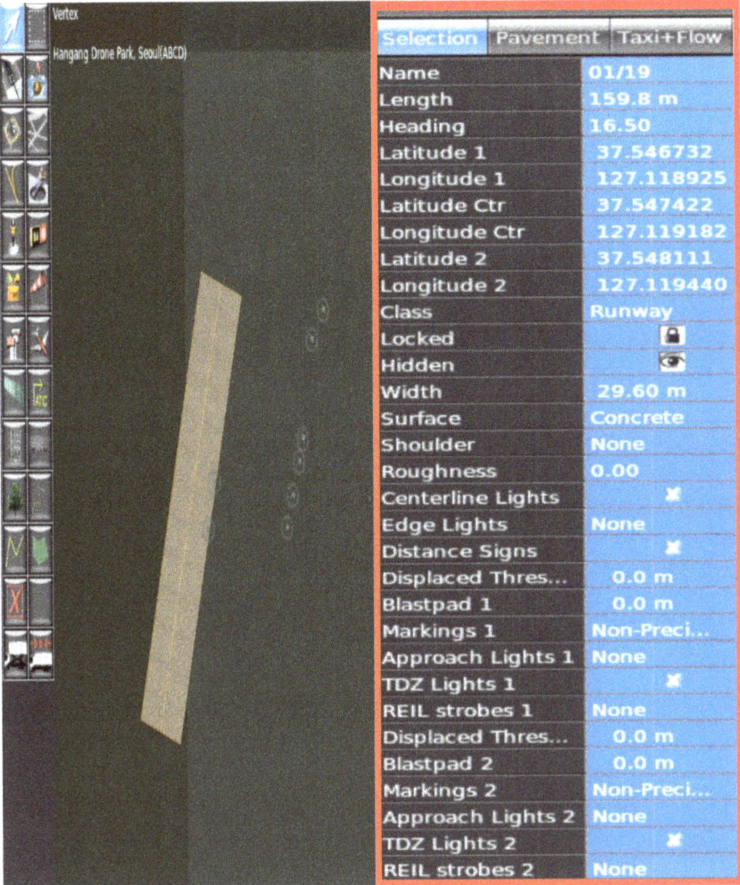

Figure 5. Design parameters of virtual location (on WorldEditor).

The designed virtual location was placed in the X-Plane environment (see Figure 6). Therefore, it was possible to fly the virtual quadcopter from the designed virtual location. Figure 7 illustrates the real and the virtual quadcopters in the real and the virtual environments, respectively (before MR simulation). In the following section, the interface between the real and the virtual quadcopters are presented.

Figure 6. Designed virtual location (Hangang Drone Park) in X-Plane.

Figure 7. The real quadcopter in the real environment (Hangang Drone Park) (**left**) and the virtual quadcopter in the X-Plane's virtual environment (designed Hangang Drone Park) (**right**).

4. Interface between the Real and Virtual Quadcopters (Mixed Reality Simulation Setup)

The MR simulation makes a connection between the real and virtual quadcopters. Herein, we connected our real quadcopter to the ground control station (GCS) through a radio telemetry device. The radio telemetry device contains two parts; one is a ground module which is connected to the computer with GCS, and the second is an air module which is connected to the real quadcopter. Here, we used Mission Planner (version: 1.3.66) [33] as a GCS. Mission Planner is a GCS software for a plane, copter, and rover developed by Michael Osborne. Through the radio telemetry device, the real quadcopter sends real-time data to the Mission Planner GCS. As we mentioned previously, X-Plane is our simulation platform, so in this work, we developed a connection interface between X-Plane and the Mission Planner GCS in MATLAB for mixed reality simulation.

Figure 8 shows the MR simulation setup. The outline of the developed connection interface is shown in Figure 9. The developed connection interface is the combination of three MATLAB program sets, viz., TCP/IP client program, data converter program, and UDP server program. First, we set up Mission Planner as a TCP/IP server and the developed connection interface as a TCP/IP client. Then, Mission Planner sends the real quadcopter's real-time flight data in the National Marine Electronics Association (NMEA) format.

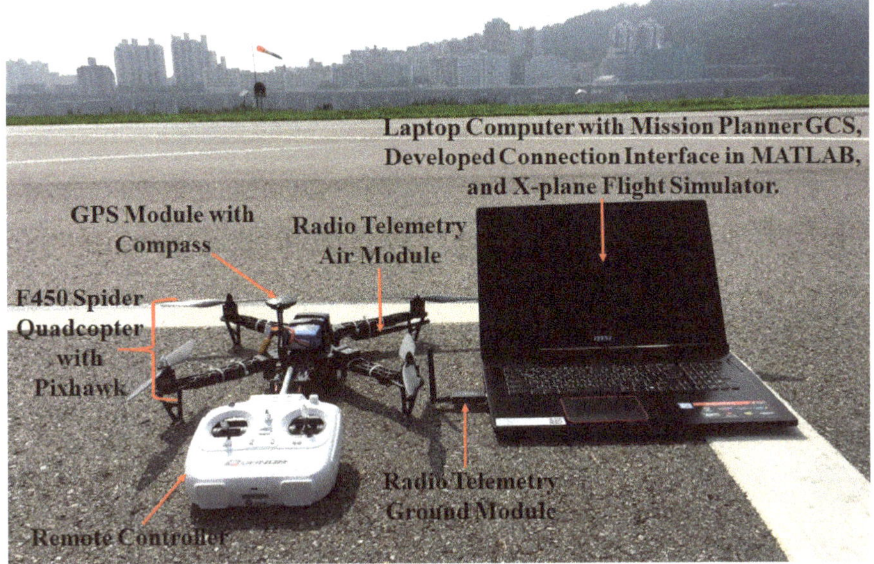

Figure 8. Mixed reality simulation setup.

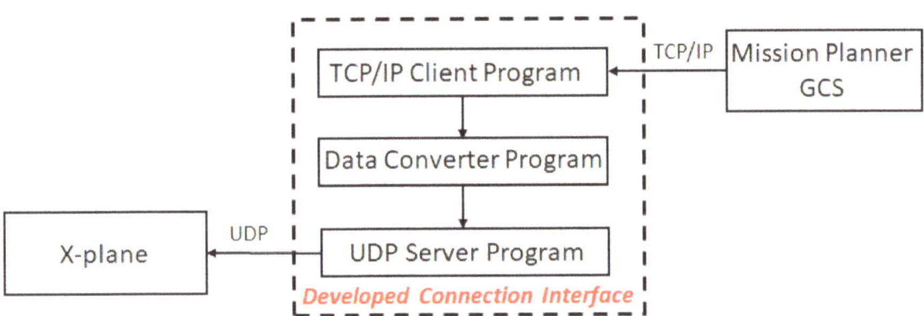

Figure 9. Outline of the developed connection interface.

Note that the NMEA is a standard data format supported by all GPS manufacturers [34]. The TCP/IP client program set in the developed connection interface collects the real-time NMEA formatted data from Mission Planner. The NMEA formatted data contains several interpreted sentences; the sentences contain the real-time flight data of the real quadcopter. The received NMEA formatted sentences are, GPGGA, GPGLL, GPHDG, GPVTG, GPRMC, and GPRPY. For MR simulation, we need only the real-time position and attitude data of the real quadcopter. Therefore, by using our data converter program (please see the link provided in the Supplementary Materials) set in the developed connection interface, we selected, split, and converted the GPGGA and GPRPY sentences from the NMEA formatted data, because the GPGGA sentence contains real-time latitude, longitude, and altitude data of the real quadcopter and the GPRPY sentence contains the real-time pitch, roll, and heading data of the real quadcopter. For the MR simulation, before running the program it is essential to set up X-Plane as the UDP client. Then, through the UDP server program set in the developed connection interface, the real-time latitude, longitude, altitude, pitch, roll, and heading data of the real quadcopter are sent to the X-Plane. Note that all the receiving, converting, and transmitting processes are in the real-time mode.

5. Visualized Ground Control Station for Quadcopter Using Mixed Reality Simulation

A ground control station (GCS) is an essential part of UAV flight, especially in the case of long-time missions. In this study, we developed a visualized GCS for a quadcopter UAV using a MATLAB/Simulink-based control system with our developed MR simulation technique. By using the visualized GCS, we controlled a quadcopter from a remote location, without a remote controller. Figure 10 illustrates the outline of the visualized GCS. In the first phase, we developed an open-loop control system in MATLAB/Simulink to control the quadcopter by a joystick. Note that for sending the control commands from the MATLAB/Simulink control system to the quadcopter, here, we attached a Raspberry Pi (Model 3 B+) single-board computer to the Pixhawk flight controller by a serial connection.

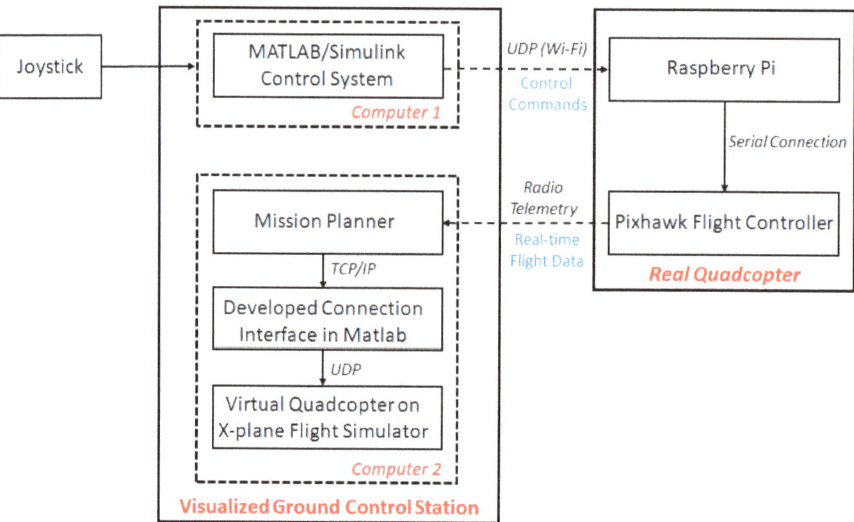

Figure 10. Overview of the visualized ground control station.

In the second phase, the MATLAB/Simulink control system sends the control commands (throttle, roll, pitch, and yaw command) to the Raspberry Pi. To do so, we developed a UDP interface in MATLAB/Simulink, which sent the control commands to Raspberry Pi via Wi-Fi. Figure 11 shows the control system with the UDP interface in MATLAB/Simulink. For communication between the Raspberry Pi and Pixhawk, here, we imported DroneKit-Python [35] in Raspberry Pi. DroneKit-Python is an open-source Python program package used to communicate with ArduPilot flight controllers, including Pixhawk. The DroneKit-Python coding in the Raspberry Pi sends the receiving control commands from the MATLAB/Simulink control system to Pixhawk.

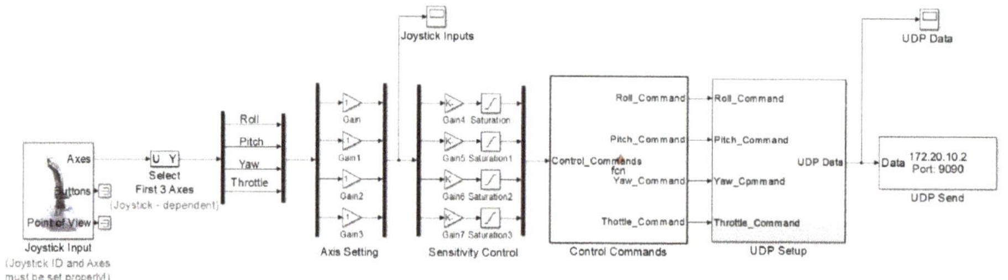

Figure 11. Control system with user datagram protocol (UDP) interface in MATLAB/Simulink.

In the third phase, we integrated our developed MR simulation technique with the control part for the visualization. The setup of the visualized GCS is shown in Figure 12. Note that the control part worked on a computer (Computer 1) with a joystick, and the visualization part ran on another computer (Computer 2) with the MR simulation setup. Therefore, instead of a remote controller, Computer 1 with the joystick controlled the quadcopter from a remote location using the visualization part on Computer 2.

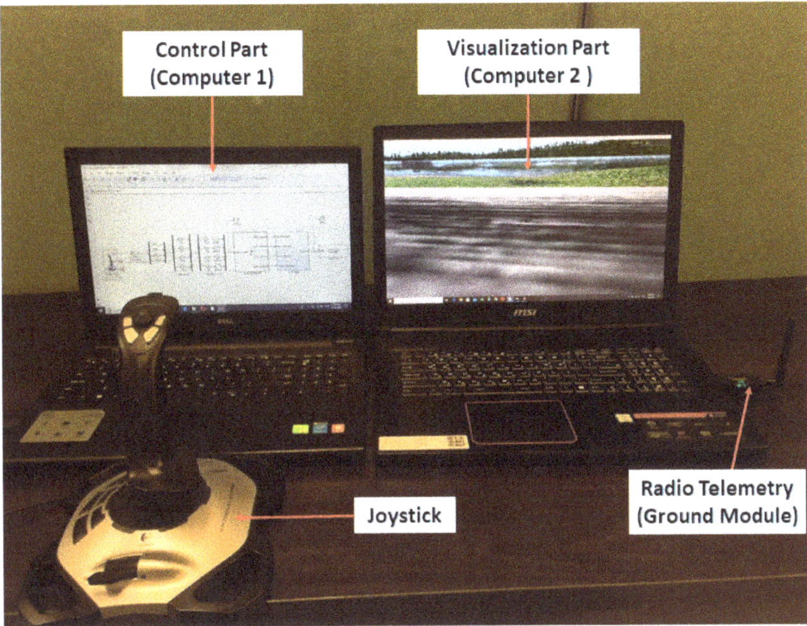

Figure 12. Visualized ground control station (GCS) setup.

6. Design of the Quadcopter with Dual-Head Electromagnetic Propulsion Devices

The origin of the science of electromagnetic propulsion (EMP) does not fall on any individual, group or institution, but many investigators have found enormous applications in multidisciplinary areas. The principle of EMP is well known, as it accelerates a body using a streaming electrical current, either to charge a field or oppose a magnetic field for the propulsion application. Recently, V.R.Sanal Kumar et al., [36–41] designed an innovative dual-head electromagnetic propulsion and energy conversion system for planet landers and other various industrial applications. The dual-head electromagnetic (DHEM) energy conversion system is found unique for the soft landing of landers on any planet with a variable density atmosphere [39]. The quadcopter design, presented in this paper, is an offshoot of the above-mentioned DHEM energy conversion system developed for planet landers [39–41]. In this study, we demonstrated the capability of a new generation quadcopter UAV integrated with four dual-head EMP devices to spin the rotors with variable speeds and generate the desired lift force in the desired direction in any atmosphere, and further continuously steer the drone for planet surveillance. The uninterrupted exploration of the drone is achieved using the reciprocating moment of a magnetic piston facilitated with each EMP device with a solar-powered polarity changer timing circuit (PCTC) along with a laser-based timing circuit (LBTC) for redundancy during the night zone [36]. A dual-head EMP device is capable of generating an uninterrupted propulsive force for spinning the UAV rotors using a connecting rod and crankshaft mechanism by creating a reciprocating moment of the magnetic piston in a vacuum cylinder by varying the polarity of magnets for attraction and repulsion. Figure 13a shows the experimental qualification test setup of a dual-head EMP device. Figure 13b shows the physical model of the quadcopter UAV with the dual-head EMP devices. Figure 13c shows the design details of the electromagnetic head (EMH).Figure 14 shows the idealized physical model of an EMP device for creating variable spinning speeds for the UAV rotors for flying in a variable density environment without any lift loss. Figure 15 shows the geometric layout of the pin location (A) of the magnetic piston, the crankpin location (B), and the crank center (C).

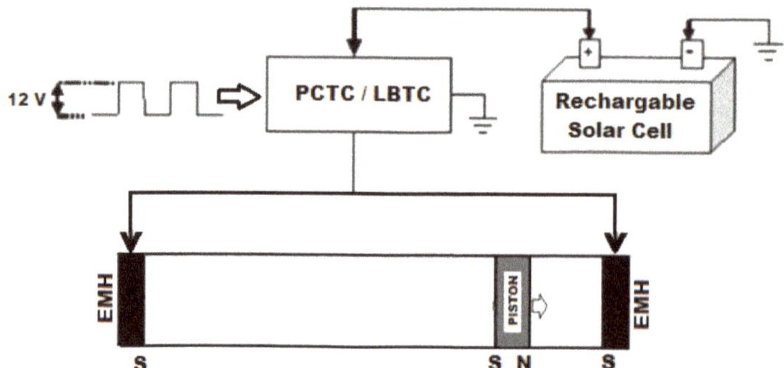

(a) Experimental qualification test setup of the dual-head electromagnetic propulsion (EMP) device

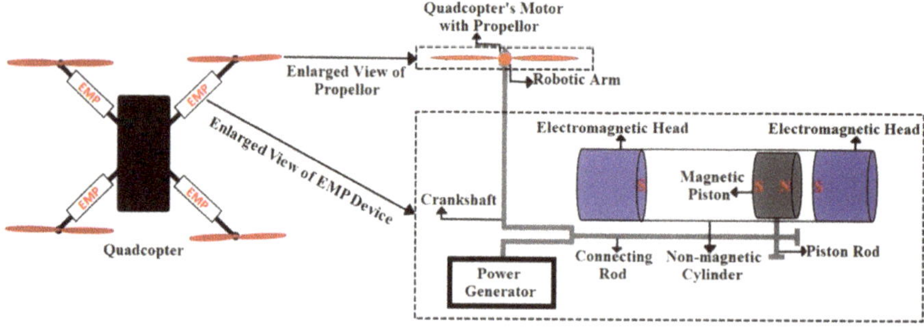

(b) Quadcopter UAV with EMP devices

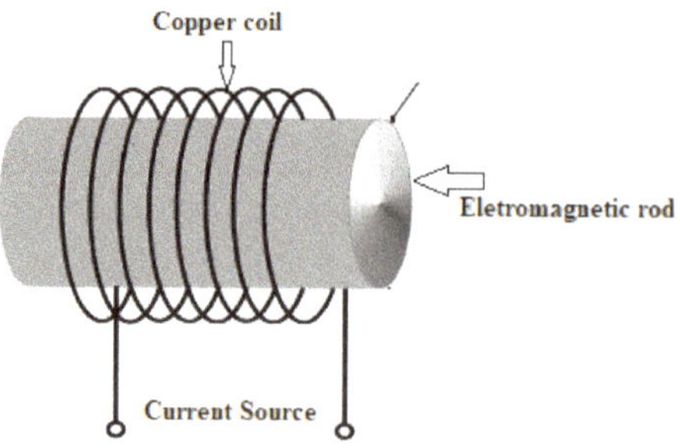

(c) Design details of the electromagnetic head (EMH)

Figure 13. (a–c) Ground testing and design details of the quadcopter unmanned aerial vehicle (UAV) with dual-head EMP devices.

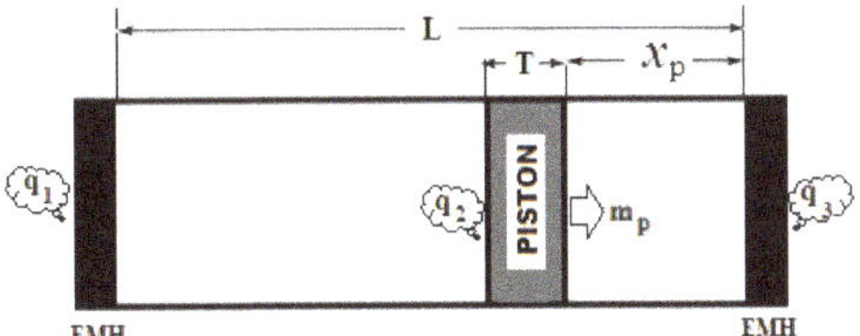

Figure 14. An idealized physical model of a dual-head EMP device for the UAV.

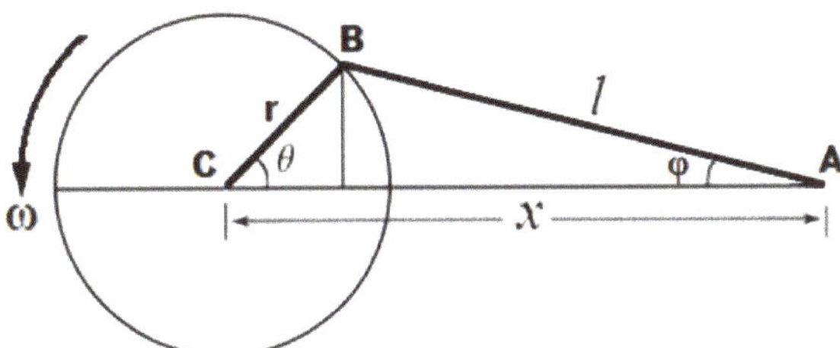

Figure 15. Geometric layout of the EMP device with a crankpin.

The basic equation of the reciprocating magnetic piston system considered in Figure 14 is obtained as,

$$m_p \frac{d^2x_p}{dt^2} - \frac{\mu q_1 q_2}{4\pi\, x^2{}_p} - \frac{\mu q_2 q_3}{4\pi \left(L - T - x_p\right)^2} = 0 \tag{1}$$

$$\frac{d^2x_p}{dt^2} = \frac{\mu}{4\pi\, m_p}\left[\frac{q_1 q_2}{x^2{}_p} + \frac{q_2 q_3}{\left(L - T - x_p\right)^2}\right] \tag{2}$$

where q_1, q_2, and q_3 are magnetic pole strength (see Figure 14), m_p is the mass of the magnetic piston, and μ is the permeability.

From Figure 15, the angular velocity of the crankshaft (ω) of the EMP device can be obtained using Equation (3) as follows:

$$\frac{dx_p}{dt} + \left[\sin\theta + \frac{\sin 2\theta}{2\sqrt{\left(\frac{l}{r}\right)^2 - sin^2\,\theta}}\right]\omega\, r = 0 \tag{3}$$

The acceleration of the magnetic piston (*dv/dt*) can be obtained from Equation (4), as given below:

$$\frac{dv}{dt} + r\,\omega^2\left[\cos\theta + \frac{\left(\frac{l}{r}\right)^2 \cos 2\theta + sin^4\,\theta}{\left(\left(\frac{l}{r}\right)^2 - sin^2\,\theta\right)^{3/2}}\right] = 0 \tag{4}$$

where r is the crank radius, *l* is the rod length, θ is the crank angle from the top dead center, x_p the axial position of the magnetic piston pin, t is the time, and v is the velocity of the magnetic piston (dx_p/dt), which is obtained from Equation (3).

Mixed Reality Simulation with Dual-Head EMP Devices to Control the Quadcopter in Space

A literature review revealed that an autonomous rotorcraft is suitable for planets with an atmosphere, including Mars and Venus [4,36–38]. In this paper, we present the MR simulation with dual-head EMP devices for controlling the quadcopter during a space mission. Figure 16 shows the overview of the MR simulation technique with an EMP device to control the quadcopter in space.

A GCS on Earth can communicate and control its device on any planet using the telemetry system. We developed an EMP device to control a quadcopter for a space mission based on the feedback system on atmospheric properties to the PCTC/LBTC controlled dual-head EMP devices. To control the EMP device from the GCS on Earth, we need a clear view and real-time performance of the space quadcopter. Therefore, here, we integrated our developed MR simulation technique with the GCS on Earth, then, using our MR simulation, we can see the clear view and real-time performance of the space quadcopter on our simulation platform (X-Plane). This preliminary study provides information for a real-time experiment with space agencies during the forthcoming planetary missions.

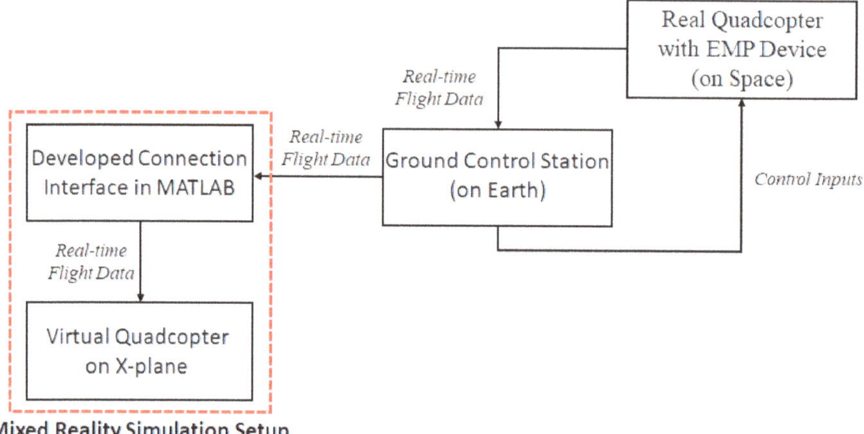

Figure 16. Overview of the mixed reality simulation for the quadcopter space activates using a dual-head EMP device.

7. Results and Discussion

7.1. Validation of Mixed Reality Simulation

In this paper, for the MR simulation of the quadcopter, three flight tasks are performed. Figures 17–19 represent the real-time flight trajectories of the real quadcopter in Mission Planner (the left side) and the real-time flight trajectories of the virtual quadcopter in X-Plane (the right side).

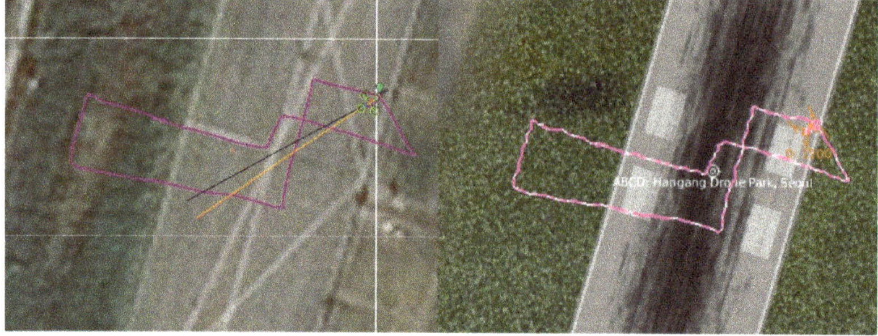

Figure 17. Flight Task 1, flight trajectory of the real (**left**) and the virtual quadcopters (**right**).

Figure 18. Flight Task 2, flight trajectory of the real (**left**) and the virtual quadcopters (**right**).

Figure 19. Flight Task 3, flight trajectory of the real (**left**) and the virtual quadcopters (**right**).

It can be seen from Figures 10–12 that the flight trajectories from Mission Planner (real quadcopter) and X-Plane (virtual quadcopter) are similar, which means that the virtual quadcopter followed the real quadcopter. Note that in the real and virtual trajectories, there are some micro differences, due to tiny differences in the real and the virtual locations and these kinds of micro differences can be neglect in the simulation field. To validate the simulation, it is necessary to be more precise, thus, one should go through the data analysis using the Pixhawk's data log file and the X-Plane's data file.

The Mission Planner and X-Plane softwares have a data acquisition system that can record a wide range of parameters. The latitude, longitude, altitude, pitch angle, roll angle, and heading data are considered for data comparison. The data comparison procedure is shown in Figure 20. From the X-Plane's data file, we can get flight data of the virtual quadcopter. In the case of the real quadcopter, we can get the flight data log file from Mission Planner or Pixhawk's micro SD card in the real quadcopter.

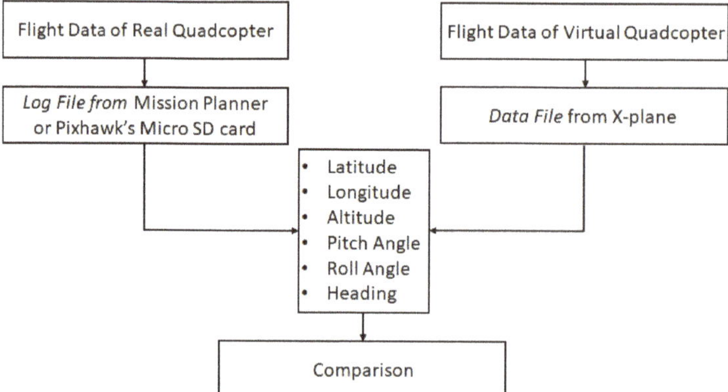

Figure 20. Data comparison procedure.

Figures 21–26 illustrate the flight data comparison of the real quadcopter and the virtual quadcopter. Here, we compared the flight data from Flight Task 1. The latitude, longitude, altitude, pitch angle, roll angle, and the heading comparison of the real and virtual quadcopters from Flight Task 1 are shown in Figures 21–26, respectively. In Figures 21–26, the continuous line indicates the real quadcopter's data, and the broken line represents the virtual quadcopter's data.

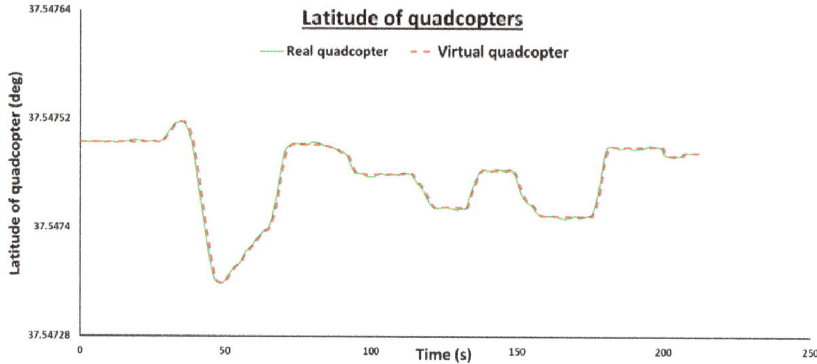

Figure 21. Latitude comparison of the real and the virtual quadcopters.

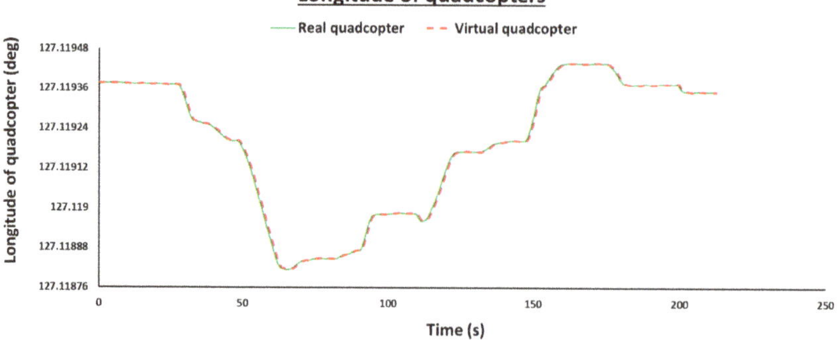

Figure 22. Longitude comparison of the real and the virtual quadcopters.

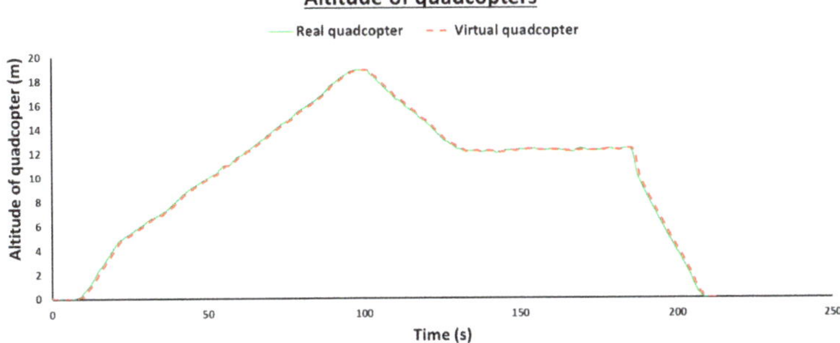

Figure 23. Altitude comparison of the real and the virtual quadcopters.

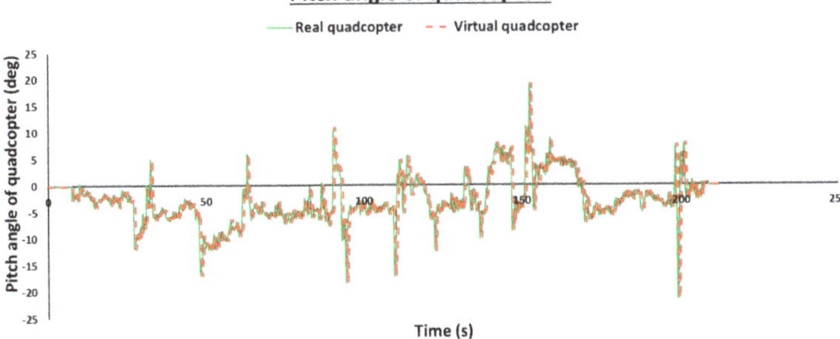

Figure 24. Pitch angle comparison of the real and the virtual quadcopters.

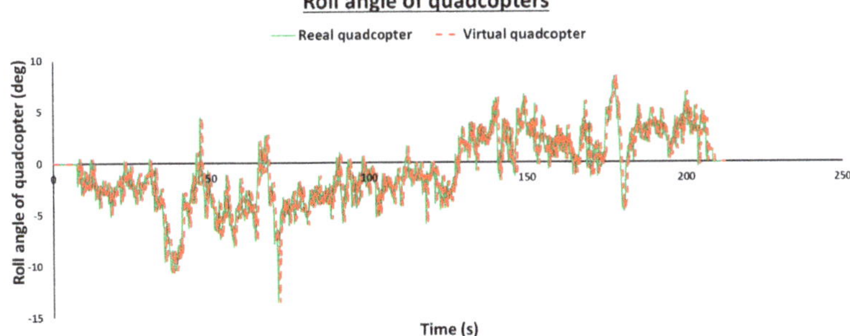

Figure 25. Roll angle comparison of the real and the virtual quadcopters.

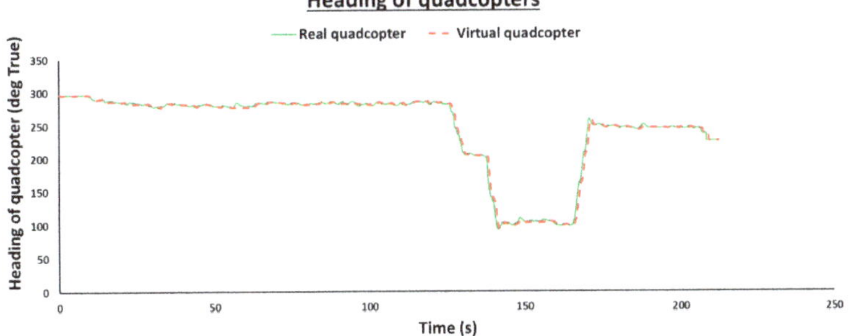

Figure 26. Heading comparison of the real and the virtual quadcopters.

It is crystal clear from Figures 21–26 that the real and the virtual quadcopters flight data are almost the same at every point of Flight Task 1. A comparison of the results shows that the latitude, longitude, and altitude of both quadcopters are nearly the same, which corroborates that the positions of both quadcopters are almost the same. The attitude comparison results (pitch angle, roll angle, and heading) of both quadcopters are also approximately equal. From the flight data comparison, we observed that on average a 400 ms time delay occurred in this MR simulation. Note that, if we observe the comparison results very precisely, we can see very tiny differences in the real and the virtual quadcopters' data which are due to the small time-delay. In the simulation field, we can neglect the small time-delay, and the tiny differences based on our allowable bandwidth.

Figure 27 shows the MR simulation of the quadcopter in real time (Flight Task 1). The stages 1 to 10 denoted (see Figure 27) that the full flight of the quadcopter (from take-off (1) to landing (10)) with precision close to a prescribed target was met. From Figure 27, we can see, by using our proposed architecture, that the virtual quadcopter in X-Plane (V series in the figure) followed the real quadcopter (R series in the figure) in real time. Therefore, here, the virtual quadcopter interacted with the real quadcopter in real time, and the MR simulation was carried out. In Figure 27, for the virtual part, we used Still Spot view in X-Plane, note that we could use different views on X-Plane including Chase, Panel, Circle, Free-Camera, and Linear Spot.

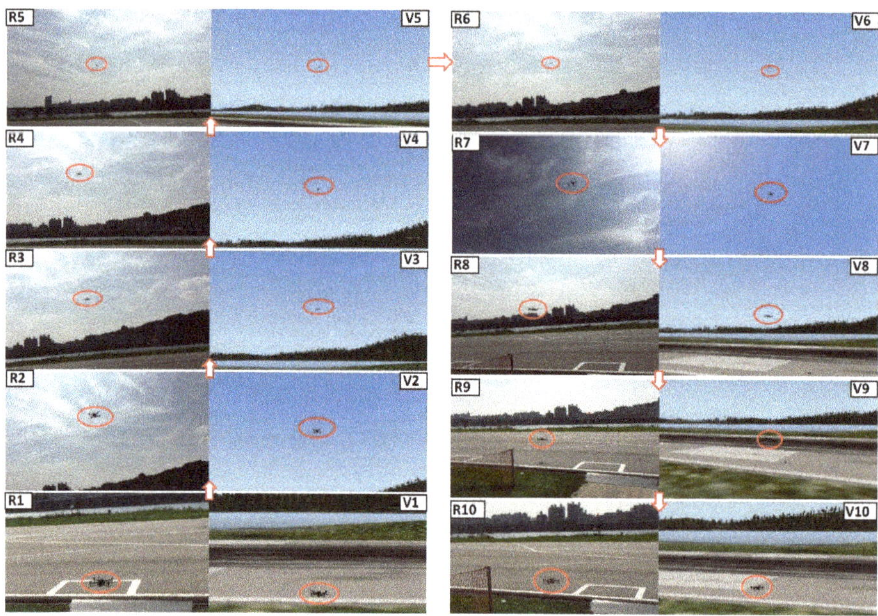

Figure 27. Mixed reality simulation of the quadcopter (R series = real part and V series = virtual part).

The comparison results from Figures 21–27 and the video in Supplementary Materials show that the performances of both quadcopters are almost identical in each and every point in the prescribed trajectory. Thereby, we could establish, herein, that using our developed architecture, the virtual quadcopter has interacted and followed the real quadcopter in real time. It proves conclusively that the MR simulation of a quadcopter is successfully achieved and validated.

By using our proposed MR simulation technique, we could see the clear view and the real-time performance of the real quadcopter on the simulation platform (X-Plane), which could possibly solve the visibility problems during a long-time mission.

7.2. Validation of Visualized Ground Control Station

For the validation and the stability check, we conducted a vertical takeoff and landing test of the quadcopter using visualized GCS. The latitude, longitude, altitude, heading, roll angle, and pitch angle of the quadcopter during the vertical takeoff and landing is illustrated in Figures 28–33, respectively.

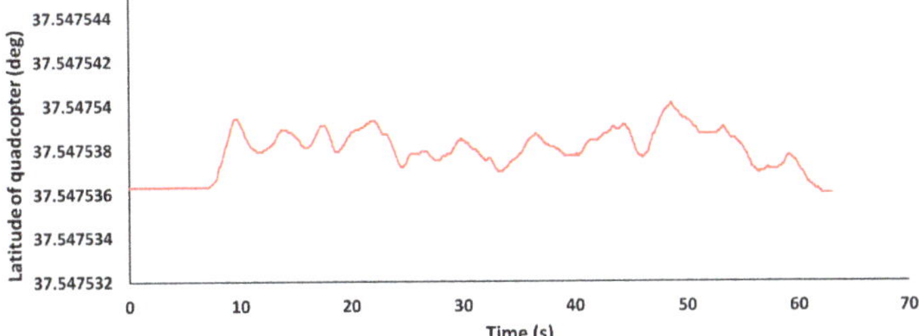

Figure 28. Latitude of the quadcopter during the vertical takeoff and landing.

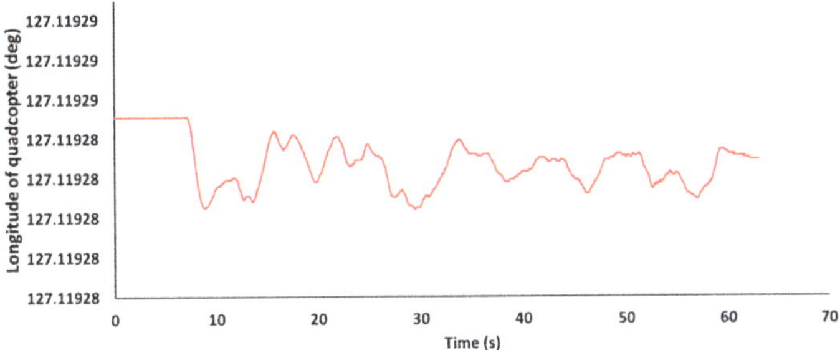

Figure 29. Longitude of the quadcopter during the vertical takeoff and landing.

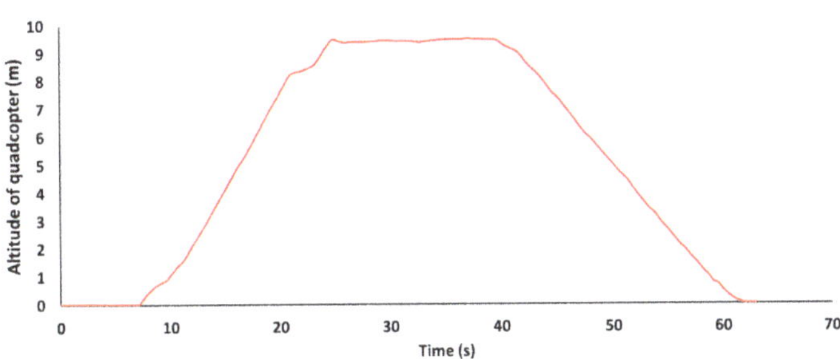

Figure 30. Altitude of the quadcopter during the vertical takeoff and landing.

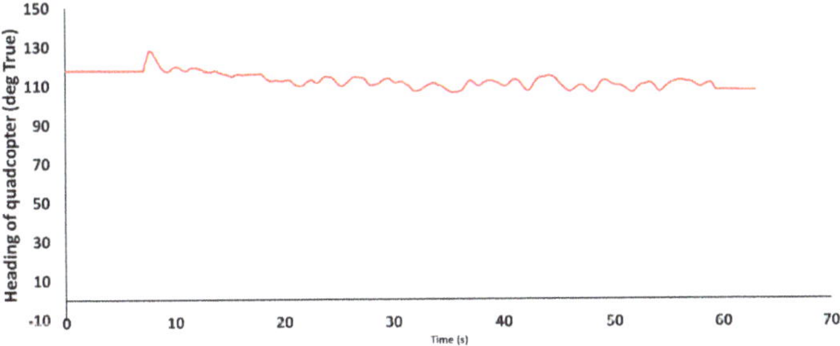

Figure 31. Heading of the quadcopter during the vertical takeoff and landing.

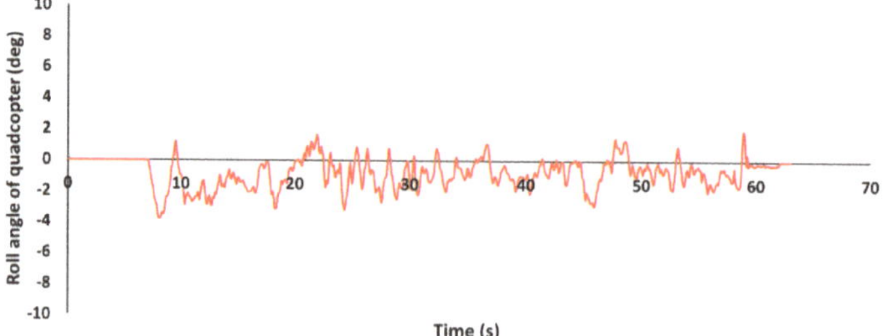

Figure 32. Roll angle of the quadcopter during the vertical takeoff and landing.

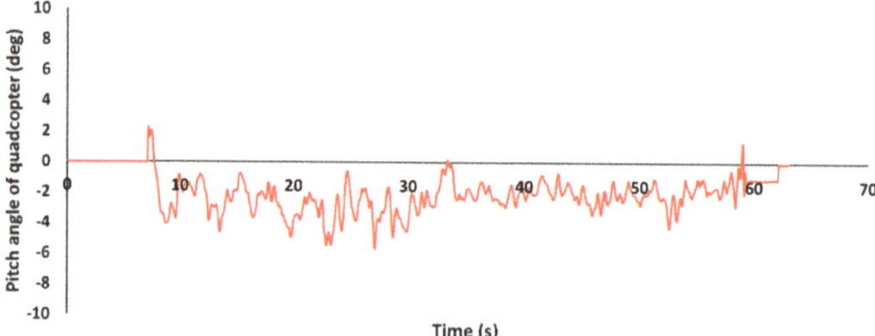

Figure 33. Pitch angle of the quadcopter during the vertical takeoff and landing.

From Figures 28 and 29, we can see that there is only a marginal difference in latitude and longitude position of the quadcopter from the initial stage to the final stage, during the vertical takeoff and landing. The heading changes that occurred in the quadcopter are also small (see Figure 31) and, both roll and pitch angle changes are between −4 to 4 degrees (see Figures 32 and 33). Note that the soft landing of any quadcopter flight is truly a difficult task, which we overcame in this study. Figure 30 shows that we landed our quadcopter safely using the visualized GCS. For the safe and soft landing, we took almost 22 s from around a 9 m height.

Figures 28–33 show that the performance of the quadcopter during the vertical takeoff and landing was stable using the visualized GCS. Figure 34 illustrates the comparison of the pictures of the real and visualization part during the vertical takeoff and landing of the quadcopter. Due to the distance limitation of the Wi-Fi network connectivity, the distance between the visualized GCS and the quadcopter was restricted herein to 20 m. If we use an interconnecting network system, we can use the developed visualized GCS for long-distance missions. The total time delay of the system, including the visualization part was approximately 420 ms.

Figure 34. Picture of the real (**left**) part and the visualization part (**right**) during the vertical take-off and landing of the quadcopter.

7.3. Validation of the Quadcopter with Electromagnetic Propulsion Devices

We have designed and laboratory tested a dual-head EMP device and qualified to integrate it with dual-head EMP devices for the next generation quadcopter UAV [40,41]. The flight experiment of a quadcopter UAV with four EMP devices and its MR simulation is beyond the scope of this paper.

8. Conclusions and Future Work

In this paper, we designed and laboratory tested a dual-head electromagnetic propulsion device for the new generation high-endurance quadcopter drone for lucrative earth and other planetary explorations. The beauty and novelty of this model comes from the fact that the proposed quadcopter can fly in an unknown environment without any lift loss for a longer duration than any existing design in the world. The fact is that a feedback control system is invoked for regulating the spinning speed of each rotor separately to retain the predetermined flight path and its hovering uninterruptedly in accordance with the density of the local atmosphere of the planet. In the case of any unexpected vacuum bubbles experienced during the surveillance, the feedback control system regulates the rotors to steer the drone in a favorable region through the polarity changer timing circuit and laser-based timing circuit. We conclude that the proposed new generation quadcopter drone integrated with lightweight electromagnetic propulsion devices is a viable option for achieving high-endurance with improved payload capability for earth and other planetary exploration with the aid of a mixed-reality simulation to meet the flight path demands of the mission. Using the base model of a drone, in this study, we developed a visualized ground control station using a Matlab/Simulink-based control system with mixed reality simulation. Additionally, we addressed mixed reality simulation based on a quadcopter and the X-Plane flight simulator. Through our comprehensive studies, mixed reality simulation is verified and validated. Herein, we connected the real quadcopter to the Mission Planner ground control station, through a radio telemetry device. Thereby, the real quadcopter could send real-time flight data to the Mission Planner. Note that we used X-Plane as the simulation platform. For mixed reality simulation, we developed a connection interface between Mission Planner and X-Plane in MATLAB. By using the developed connection interface, the virtual quadcopter in X-Plane could follow the real quadcopter in real time. The flight data from both quadcopters were gathered and compared. The comparison results show that the flight data from the real and the virtual quadcopters were almost the same at every point. In addition, we could see a similar flight trajectory from both quadcopters. Therefore, we concluded that the virtual quadcopter in X-Plane interacted and followed the real quadcopter in real time, which means that mixed reality simulation of the quadcopter UAV was executed and validated herein.

By using mixed reality simulation, we developed and tested a visualized ground control station that could control a quadcopter from a remote location, without a remote controller. Finally, in this phase, we introduced our mixed reality simulation technique with a dual-head electromagnetic propulsion device to control the quadcopter in any planet atmosphere. The real-time space flight experiment of a quadcopter drone with four electromagnetic propulsion devices and its mixed reality simulation is beyond the scope of this paper. However, it will be executed with the support of the space agencies worldwide, in the next phase of our work.

Note that, in this mixed reality simulation, we mainly focused on the interactions and performances of both quadcopters. In future work, we plan to consider more accurate scenery, weather conditions, and real-time movement of other objects, and also develop a ground control station for mixed reality simulation instead of the Mission Planner. We envision the advent of a new era of drones, a popular nickname for UAVs, that can autonomously fly in natural and man-made environments [41,42] with dual-head electromagnetic propulsion devices for a longer duration in any planet with high payload capability. Briefly, this paper provides insight for a credible mixed reality simulation for Mars explorations using quadcopter drones regulated by dual-head electromagnetic propulsion devices through multiple channel ultra-high speed wireless communication systems.

Supplementary Materials: The following are available online at http://www.mdpi.com/2076-3417/10/11/3736/s1, please see the video "MRS" provided herein for corroborating our claims on the mixed reality simulation of quadcopter UAV. Data converter program is available at https://github.com/ashishkumar2025/Data-converter-program-in-MATLAB.git.

Author Contributions: Conceptualization, A.K.; Methodology, A.K.; Experimental design, A.K.; Manuscript preparation, A.K.; Supervision, S.Y.; Review and editing, V.R.S.K. All authors have read and agreed to the published version of the manuscript.

Funding: This research was supported by the Agency for Defense Development, regarding the project of Design Study of Electronic Warfare Based on Modeling and Simulation

Acknowledgments: We thank our colleagues, Dong Cho Shin from the Agency for Defense Development and Ki Byung Jin from the LIG Nex1 Co., Ltd., who provided insight and expertise that greatly assisted the research, without prejudice to the final outcome of this manuscript.

Conflicts of Interest: The authors declare no conflict of interest.

References

1. Alkamachi, A.; Erçelebi, E. A proportional derivative sliding mode control for an overactuated quadcopter. *Proc. Inst. Mech. Eng. J. Part G Aerosp. Eng.* **2018**, *233*, 1354–1363. [CrossRef]
2. Kumar, A.; Yoon, S. Development of fast and soft landing system for quadcopter drone using fuzzy logic technology. *Int. J. Adv. Trends Comput. Sci. Eng.* **2019**, *9*, 624–629. [CrossRef]
3. Lee, K.; Kim, Y.; Hong, Y. Real-time swarm search method for real-world quadcopter drones. *Appl. Sci.* **2018**, *8*, 1169. [CrossRef]
4. Hassanalian, M.; Rice, D.; Abdelkefi, A. Evolution of space drones for planetary exploration: A review. *Prog. Aerosp. Sci.* **2018**, *97*, 61–105. [CrossRef]
5. Lemke, L.G.; Heldmann, J.L.; Young, L.A.; Gonzales, A.A.; Gulick, V.C.; Foch, R.E.; Marinova, M.M.; Gundlach, J.F. Vertical takeoff and landing UAVS for exploration of recurring hydrological events. In Proceedings of the Concepts and Approaches for Mars Exploration, Houston, TX, USA, 12–14 June 2012.
6. Colozza, A. Overview of innovative aircraft power and propulsion systems and their applications for planetary exploration. In Proceedings of the International Air and Space Symposium and Exposition cosponsored by the American Institute of Aeronautics and Astronautics, Dayton, OH, USA, 14–17 July 2003.
7. Rabah, M.; Rohan, A.; Talha, M.; Nam, K.; Kim, S.H. Autonomous vision-based target detection and safe landing for UAV. *Int. J. Control Autom. Syst.* **2018**, *16*, 3013–3025. [CrossRef]
8. Meier, L.; Tanskanen, P.; Fraundorfer, F.; Pollefeys, M. PIXHAWK: A system for autonomous flight using onboard computer vision. In Proceedings of the IEEE International Conference on Robotics and Automation, Shanghai, China, 9–13 May 2011; pp. 2992–2997.
9. Nguyen, K.D.; Ha, C. Development of hardware-in-the-loop simulation based on Gazebo and Pixhawk for unmanned aerial vehicles. *Int. J. Aeronaut. Space Sci.* **2018**, *19*, 238–249. [CrossRef]
10. Lyu, X.; Gu, H.; Zhou, J.; Li, Z.; Shen, S.; Zhang, F. Simulation and flight experiments of a quadrotor tail-sitter vertical take-off and landing unmanned aerial vehicle with wide flight envelope. *Int. J. Micro Air Veh.* **2018**, *10*, 303–317. [CrossRef]
11. Liu, D.; Hou, Z.; Gao, X. Flight modeling and simulation for dynamic soaring with small unmanned air vehicles. *Proc. Inst. Mech. Eng. J. Part G Aerosp. Eng.* **2016**, *231*, 589–605. [CrossRef]
12. Susini, A. A technocritical review of drones crash risk probabilistic consequences and its societal acceptance. In Proceedings of the Risk Information Management, Risk Models, and Applications (RIMMA) Conference, Berlin, Germay, 17–18 November 2014; pp. 27–38.
13. Asim, M.; Ehsan, D.R.N.; Rafique, K. Probable causal factors in UAV accidents based on human factor analysis and classification system. In Proceedings of the 27th Congress of the International Council of the Aeronautical Sciences, Nice, France, 19–24 September 2010; pp. 4881–4886.
14. Williams Kevin, W. *A Summary of Unmanned Aircraft Accident/Incident Data: Human Factors Implications*; Final Report; U.S. Department of Transportation (FAA): Washington, DC, USA, 2004.
15. Kumar, A.; Mondon, C.; Yoon, S. Development of mixed reality simulation based on an unmanned aerial vehicle. In Proceedings of the CHIRA 2019 3rd International Conference on Computer-Human Interaction Research and Applications, Vienna, Austria, 20–21 September 2019; pp. 89–96.

16. Kumar, A. Mixed Reality Simulation Based on an Unmanned Aerial Vehicle and X-Plane Flight Simulator. Korea Patent 1,615,010,612, 23 December 2019.
17. Garcia, R.; Barnes, L. Multi-UAV simulator utilizing X-plane. *J. Intell. Robot. Syst.* **2010**, *57*, 393–406. [CrossRef]
18. Qi, J.; Liu, J.; Zhao, B.; Mei, S.; Han, J.; Shang, H. Visual simulation system design of soft-wing UAV based on FlightGear. In Proceedings of the 2014 IEEE International Conference on Mechatronics and Automation, Tianjin, China, 3–6 August 2014; pp. 1188–1192.
19. Lugo-Cárdenas, I.; Flores, G.; Lozano, R. The MAV3DSim: A simulation platform for research, education and validation of UAV controllers. In Proceedings of the 19th International Federation of Automatic Control (IFAC) World Congress, Cape Town, South Africa, 24–29 August 2014; pp. 713–717.
20. Almousa, O.; Prates, J.; Yeslam, N.; Gregor, D.M.; Zhang, J.; Phan, V.; Nielsen, M.; Smith, R.; Qayumi, K. Virtual reality simulation technology for cardiopulmonary resuscitation training: An innovative hybrid system with haptic feedback. *Simul. Gaming* **2019**, *50*, 6–22. [CrossRef]
21. Hsu, K.; Wang, C.; Jiang, J.; Wei, H. Development of a real-time detection system for augmented reality driving. *Math. Probl. Eng.* **2015**, *2015*. [CrossRef]
22. Stevens, J.; Kincaid, P.; Sottilare, R. Visual modality research in virtual and mixed reality simulation. *J. Def. Model. Simul.* **2015**, *12*, 519–537. [CrossRef]
23. Lan, G.; Sun, J.; Li, C.; Ou, Z.; Luo, Z.; Liang, J.; Hao, Q. Development of UAV based virtual reality systems. In Proceedings of the 2016 IEEE International Conference on Multisensor Fusion and Integration for Intelligent Systems (MFI), Baden-Baden, Germany, 19–21 September 2016; pp. 481–486.
24. Wang, S.; Chen, J.; Zhang, Z.; Wang, G.; Tan, Y.; Zheng, Y. Construction of a virtual reality platform for UAV deep learning. In Proceedings of the 2017 Chinese Automation Congress (CAC), Jinan, China, 20–22 October 2017; pp. 3912–3916.
25. Wang, Y.; Huang, W.; Been-Lirn Duh, H. InspectAR: Unmanned aerial vehicle (UAV) with augmented reality (AR) technology. In Proceedings of the SA '16 SIGGRAPH ASIA 2016 Mobile Graphics and Interactive Applications, Macau, China, 5–8 November 2016. [CrossRef]
26. Li, Y.-B.; Kang, S.-P.; Qiao, Z.-H.; Zhu, Q. Development actuality and application of registration technology in augmented reality. In Proceedings of the 2008 International Symposium on Computational Intelligence and Design, Wuhan, China, 17–18 October 2008; pp. 69–74.
27. Selecký, M.; Jan, F.; Rollo, M. Communication architecture in mixed-reality simulations of unmanned systems. *Sensors* **2018**, *18*, 853. [CrossRef]
28. Peña, F.L.; Deibe, A.; Orjales, F. On the initiation phase of a mixed reality simulator for air pollution monitoring by autonomous UAVs. In Proceedings of the 2017 9th IEEE International Conference on Intelligent Data Acquisition and Advanced Computing Systems: Technology and Applications (IDAACS), Bucharest, Romania, 21–23 September 2017. [CrossRef]
29. Katragadda, S.; Benedict, A.M.; Deane, A. Stereoscopic mixed reality in unmanned aerial vehicle search and rescue. In Proceedings of the AIAA SciTech Forum, San Diego, CA, USA, 7–11 January 2019. [CrossRef]
30. Bai, H.; Atiquzzaman, M.; Ivancic, W. QoS support in ARINC 664 P8 data networks: ATN applications over TCP/IP ground-to-ground subnetworks. *J. Aerosp. Comput. Inf. Commun.* **2006**, *3*, 374–387. [CrossRef]
31. Wei, H.; Tung, Y.; Yu, C. Counteracting UDP flooding attacks in SDN. In Proceedings of the 2016 IEEE NetSoft Conference and Workshops (NetSoft), Seoul, Korea, 6–10 June 2016; pp. 367–371.
32. Bittar, A.; de Oliveira, N.M.F.; de Figueiredo, H.V. Hardware-in-the-loop simulation with X-Plane of attitude control of a SUAV exploring atmospheric conditions. *J. Intell. Robot. Syst.* **2014**, *73*, 271–287. [CrossRef]
33. Rahman, M.F.A.; Radzuan, S.M.; Hussain, Z.; Khyasudeen, M.F.; Ahmad, K.A.; Ahmad, F.; Ani, A.I.C. Performance of loiter and auto navigation for quadcopter in mission planning application using open source platform. In Proceedings of the 2017 7th IEEE International Conference on Control System, Computing and Engineering (ICCSCE), Penang, Malaysia, 24–26 November 2017; pp. 342–347.
34. Park, B.; Lee, J.; Kim, Y.; Yun, H.; Kee, C. DGPS enhancement to GPS NMEA output data: DGPS by correction projection to position-domain. *J. Navig.* **2013**, *66*, 249–264. [CrossRef]
35. Available online: https://dronekit-python.readthedocs.io/en/latest/about/index.html (accessed on 14 May 2020).

36. Kumar, V.R.S.; Mariappan, A.; Sukumaran, A.; Lal, V.K.V.; John, J.; Kumar, A.; Yoon, S.; Rajeev, J. Design of an Uninterrupted Propulsion System for Spinning Planet Landers for Soft Landing. In Proceedings of the AIAA Scitech 2019 Forum, San Diego, CA, USA, 7–11 January 2019. [CrossRef]
37. Mariappan, A.; Sukumaran, A.; Thianesh, U.K.; Sankar, P.G.; Kumar, A.; Yoon, S.; Kumar, V.R.S. Design of Planet Landers for Soft Landing with DHEM Propulsion System—Phase I. In Proceedings of the AIAA Propulsion and Energy 2019 Forum, Indianapolis, IN, USA, 19–22 August 2019. [CrossRef]
38. Kumar, A.; Yoon, S.; Amrith, M.; Thianesh, U.K.; Kumar, V.R.S. Mixed reality simulation of a quadcopter with a DHEM device for planet landers for soft landing—A review. *Int. J. Adv. Sci. Technol.* **2019**, *28*, 157–171.
39. Kumar, V.R.S.; Mariappan, A.; Thianesh, U.K.; Sukumaran, A.; Kumar, A.; Lal, V.K.V.; John, J. The Invention of an Electromagnetic Propulsion System for Drones and Plant Landers with an Unrestricted Flight Endurance. *Nature* **2020**, in press.
40. Kumar, V.R.S.; Lal, V.K.V.; Mariappan, A.; Sukumaran, A.; Kumar, A.; John, J.; Thianesh, U.K. Dual-Head Electromagnetic Propulsion and Energy Conversion System for Planet Landers and Other Various Industrial Application. India Patent 201,841,049,585, 28 December 2018.
41. Kumar, V.R.S. *Design and Testing of Dual-Head Electromagnetic Propulsion System for Spinning Venus Impact Probe*; ISRO Space Based Experiments to Study Venus—A Proposal; No. VRS/KCT/ISRO/VIP-P1; ISRO Space Science Programme Office, ISRO HQ, Antariksh Bhavan: Bangalore, India, 2017.
42. Floreano, D.; Robert, J.W. Science, technology and the future of small autonomous drones. *Nature* **2015**, *521*, 460–466. [CrossRef] [PubMed]

Article

Digital Twin and Virtual Reality Based Methodology for Multi-Robot Manufacturing Cell Commissioning

Luis Pérez [1,*], Silvia Rodríguez-Jiménez [1], Nuria Rodríguez [1], Rubén Usamentiaga [2] and Daniel F. García [2]

1 Fundación IDONIAL, Avda. Jardín Botánico 1345, 33203 Gijón (Asturias), Spain; silvia.rodriguez@idonial.com (S.R.-J.); nuria.rodriguez@idonial.com (N.R.)

2 Universidad de Oviedo, Campus de Viesques, 33204 Gijón (Asturias), Spain; rusamentiaga@uniovi.es (R.U.); dfgarcia@uniovi.es (D.F.G.)

* Correspondence: luis.perez@idonial.com; Tel.: +34-984-390-060

Received: 24 April 2020; Accepted: 20 May 2020; Published: 24 May 2020

Featured Application: The results of the work may find applications in process automation design, implementation, and commissioning.

Abstract: Intelligent automation, including robotics, is one of the current trends in the manufacturing industry in the context of "Industry 4.0", where cyber-physical systems control the production at automated or semi-automated factories. Robots are perfect substitutes for a skilled workforce for some repeatable, general, and strategically-important tasks. However, this transformation is not always feasible and immediate, since certain technologies do not provide the required degree of flexibility. The introduction of collaborative robots in the industry permits the combination of the advantages of manual and automated production. In some processes, it is necessary to incorporate robots from different manufacturers, thus the design of these multi-robot systems is crucial to guarantee the maximum quality and efficiency. In this context, this paper presents a novel methodology for process automation design, enhanced implementation, and real-time monitoring in operation based on creating a digital twin of the manufacturing process with an immersive virtual reality interface to be used as a virtual testbed before the physical implementation. Moreover, it can be efficiently used for operator training, real-time monitoring, and feasibility studies of future optimizations. It has been validated in a use case which provides a solution for an assembly manufacturing process.

Keywords: digital twin; virtual reality; collaborative robots; Industry 4.0; lean automation; multi-robot cell

---

1. Introduction

Lean manufacturing is a collection of synchronized methods and principles for organizing and controlling production sites in a technology-independent way to reach shortest lead time with minimum costs and the highest quality [1]. Due to the evolution of technology and its introduction in the factories, industry and manufacturing processes have changed and evolved throughout the industrial revolutions, from the introduction of mechanical production facilities powered by water and steam to the current cyber-physical production systems (CPPSs) and intelligent automation, keys of the Fourth Industrial Revolution, also known as "Industry 4.0" [2]. The integration of automation technologies and lean manufacturing is called "lean automation" [3]. These CPPSs monitor the physical processes, make decentralized decisions, and trigger actions, communicating and cooperating with each other and with humans in real time. Networked machines perform more efficiently, collaboratively, and resiliently [4].

The intelligent automation also requires the use of autonomous machines or robots, which are controlled by the CPPS and the humans. The main objective is to increase productivity and safety, which were traditionally limited by manual processes. The safety conditions have limited the use of robots in industrial environments, as they were isolated from people and some tasks were not affordable. The introduction of process automation and intelligent collaborative robots results in a rapid increase in productivity, major material and energy savings, and safer working conditions (repetitive and dangerous tasks can be done by robots). Robots provide versatility and flexibility; thus they are perfect substitutes for a skilled workforce for some repeatable, general, and strategically-important tasks [5]. Moreover, if robots' capacities are combined with humans' qualities, cost-effective productivity can be guaranteed [6].

Up until fairly recently, the information about any physical object or process was relatively inseparable from the physical object or process itself [7]. Digital data and artificial intelligence allow the dematerialization and the coexistence of the real factory with digital twins [8], where manufacturing processes are virtually simulated, monitored, and controlled. While at first, this digital twin was merely descriptive and static, in recent years it has become actionable and experimentable [9]. The digital information related to a physical system can be created as an entity itself. This means that there is a mirroring or twinning of systems between what exists in real space to what exists in virtual space and vice versa [7]. For this purpose, access to very realistic models of the current state of the process is necessary [10]. The use of sensors and 3D visualization technologies, such as virtual and augmented reality, makes this connection possible and facilitates the interaction with humans [11,12]. The digital twin is fed with the data from the sensors, PLCs, controllers, etc., and the 3D environment is visualized using the 3D glasses. A digital twin not only allows a static perspective at design stage, but also a real-time synchronization and optimization of the virtual object [13].

At this point, robotized processes can be mirrored using digital twin models during the design, implementation, and operation steps. Each robot manufacturer has its own simulation environment for cell design and program testing, but the challenge arises when the same cell contains robots from different manufacturers working collaboratively with humans. Multiple robots can perform tasks in parallel, speed up the execution time, and improve system performance [14]. Thus, this work presents a novel methodology for process automation design, enhanced implementation, and real-time monitoring in operation. The proposed approach is based on creating a digital twin of the manufacturing process with an immersive virtual reality interface to simulate and analyze the layout (the physical location of the elements) and to determine whether robots and other components are suitable. The results in this virtual testbed will easily permit modifications in the original design before the physical implementation, presenting a far more cost-efficient solution. In addition, once the new process has been implemented, the digital twin can be efficiently used for operator training, real-time process monitoring, and feasibility study of future optimizations, resulting in a novel and intelligent mirror with high potential benefits. As it is not a simple replica, it is able to process and understand data, and automatically react to changes according to them. The innovation of the proposed approach is that it combines design, feasibility studies, virtual and real commissioning, training, and monitoring of multi-robot cells in a unique application, which implies a cost-effective and affordable solution for manufacturing industries of all types. The theoretical outcome is the proposed methodology for robot-based automation, while the practical is the digital twin framework with an immersive virtual reality interface which is used as a testbed environment.

The proposed approach is validated in a real-life case study that provides a solution for the assembly of parts, demonstrating that the use of the digital twin-based methodology is feasible. Not only is the result of the proposed approach more visual, thanks to the integration of virtual reality, it also reduces costs and increases the productivity, as proven with real operations. Therefore, it is very likely to find potential applications in a number of different areas and multiple industries to create flexible and easily reconfigurable production lines.

The rest of the paper is organized as follows: Section 2 reviews the state of the art and previous work, Section 3 presents the proposed approach, Section 4 contains the use case, and main conclusions are found in Section 5.

2. Related Work

2.1. The Concept of the Digital Twin

The concept of the "digital twin" (DT) comes from NASA. In the early days of space exploration, they were pioneers in studying what was called "pairing technologies". Maintaining, repairing, and operating systems without physical access to them, were the challenges at that time. Indeed, the first twin was a hardware twin, and it consisted of two identical space vehicles, one in the space and the other on ground to enable engineers to better assist astronauts in orbit [15]. The use of digital twins is now very common at NASA, using a virtual environment to build and test their equipment, including spatial robots. Only after a total approval in the virtual environment, does the physical construction begin. The final result and the virtual twin are then linked through the sensors for a continuous improvement process. The general digital twin model of a product consists of the physical entities, the virtual models, and the connected data which tie physical and virtual worlds [16]. However, far away from this case, in general, current research on product lifecycle data mainly focuses on physical products rather than virtual models. The connection between physical and virtual product data is needed to support product design, manufacturing, and service [17].

Similarly to equipment or products, manufacturing systems are becoming more autonomous. They need access to realistic models and real-time information about the processes for smart production management and control [18]. The use of model-based simulation is necessary not only during design and planning, but also during the production phases for such purposes as diagnosis, control, and optimization [10]. Given the uncertainty involved during the process of machinery degradation, proper design and adaptability of a digital twin model remain challenges [19]. Digital twins can be applied from initial factory planning and design to commissioning and maintenance, giving them value throughout the production lifecycle [20,21]. The digital twin can be also used for risk prediction and prevention pertaining to operators in processing plants [22]. In this sense, robots are perfect candidates for digital twin applications.

2.2. Robot Simulators

The first developments were orientated towards the simulation of the real robot or the real cell environment mainly for robot programming and operator training, but these first virtual environments were disconnected from the real movements. They were isolated tools provided by robot manufacturers which enabled one to foresee the manipulator behavior in a simulated environment. Each manufacturer provided its own solution for its robots; thus it was not possible to combine robots from different manufacturers. Moreover, as robotic languages are dependent on each manipulator, the simulation faces the same complexity as the real programming—one of the major hurdles still preventing automation using industrial robots [23]. Benchmarking of multi-robot systems is crucial for comparing the different existing solutions. However, there are only few and limited tools to support it. For instance, Yan in [14] presented a simulation tool based on a *robot operating system* (ROS).

Recently, these robot simulators have evolved by including new technologies, such as virtual and augmented reality, or new features, such as human–robot collaboration. The collaboration between humans and robots is necessary to increase industrial competitiveness, and the application of virtual and augmented reality is essential to enable a smoother collaboration with 3D immersive visualization [24]. Virtual reality (VR) offers a way to simulate reality. Originally, it was mainly used for entertainment purposes, but nowadays the evolution of the technologies, the appearance of multiple applications, and the reduction of costs have extended it to the manufacturing industry for a safer human–machine interaction [25]. Nowadays, several commercial simulation tools with

VR visualization are available, such as *Visual Components* [26], *Robotics and Automation* [27], and *RoboDK* [28]. These tools are also used for the virtual commissioning of a robotic cell, which involves creating a digital twin and then testing and verifying the model in a simulated virtual environment [20]. ROS is also combined with virtual reality to create human interfaces [29].

2.3. Digital Twins and Robots

Several works relate the use of digital twins with robots. Kousi in [30] used the digital twin to adapt a robot's behavior in assembly tasks of the automobile industry. Malik in [31] presented a digital twin framework to support human–robot collaboration. Ma in [32] proposed a digital twin for enhanced human–machine interaction. Bilberg in [33] also combined the digital twin with human–robot collaboration but added a task allocation. Aivaliotis in [34] applied the digital twin of a robot for predictive maintenance. In the literature, the digital twin concept is applied not only to the single robot but also to the whole manufacturing cell [35,36].

In order to train people in virtual reality with systems that behave realistically, there is the interesting option of combining virtual reality and digital twin technologies [11]. Burghardt in [37] and Kuts in [38] present different methods for programming and controlling robots using virtual reality and digital twins, confirming that this combination facilitates human–robot interactions in terms of collaborative work, telecontrol, and programming.

3. Proposed Approach

Robots are perfect substitutes for a skilled workforce for some repeatable, general, and strategically-important tasks, but this substitution is not straightforward [5]. The automation of an industrial manufacturing process raises some previous issues:

1. What are the costs in terms of money, time, safety, etc., of the current manual process?
2. Is the use of robots technically feasible for the tasks?
3. Will the robot work isolated or collaboratively with humans?
4. What are the costs of the new automated or semi-automated process? Which costs are reduced and which ones are increased?
5. Is the new process cost-effective?
6. Does the automation reduce risks and enhance safety?

In order to answer the questions related to costs and safety, it is necessary to analyze and to compare the current situation (manual) with the new one using robots (total or partially automated). Thus, a sequential methodology with feedback loops has been defined to design, validate, implement, and operate the new robotized process. This methodology is based on using the digital twin of the new process as a virtual testbed to simulate and to analyze the layout and the suitability of the selected robots and the other components. An immersive VR-based interface permits a better visualization and understanding of the digital twin. This proposed approach permits the detection of design mistakes during the virtual commissioning before the real implementation, preventing costly and potentially unmanageable consequences. In addition, after the implementation, the digital twin can be used for operator training, thanks to the virtual reality interface; for real-time process monitoring, thanks to the real-time information received from sensors; and for testing future changes. All types of robots can be introduced in the digital twin framework, as it is independent of manufacturer.

Figure 1 illustrates the methodology to design the robotized process and its digital twin according to the proposed approach. As shown, this methodology is a sequential cascade process with feedback loops for redesign and verification. The overall process is detailed step by step:

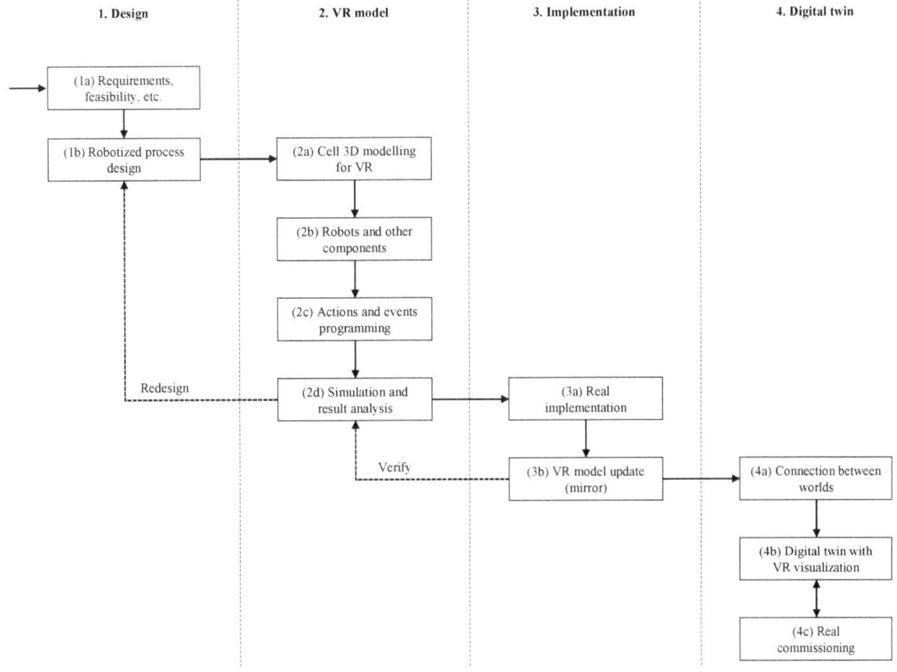

Figure 1. Proposed sequential methodology with feedback loops to create the digital twin.

1. Design:
   (a) Requirements, feasibility, etc. Analysis of the requirements of the new process and studying costs, technical solutions, number and type of robots, etc.
   (b) Robotized process design. Design and selection of the flowchart, the components, the layout, etc.
2. VR model:
   (a) Cell 3D modelling for VR. Environment 3D reconstruction or modelling in order to create a VR immersive experience.
   (b) Robots and other components. The elements of the cell (robots and others) are also included in the VR model.
   (c) Actions and events programming. For the immersive experience, actions and events should happen as in the real world.
   (d) Simulation and result analysis. The design process and cell are simulated and studied in the virtual environment to verify if the result fulfills the requirements. This is the virtual commissioning. At this point, if redesign is necessary, the process will go back to step (1b).
3. Implementation:
   (a) Real implementation. Once the process and the robot-based automation solution have been virtually tested, it is time for the real implementation.
   (b) VR model update (mirror). If during the real implementation there is any change from the original design, the virtual model should be updated in order to keep the mirror, and the simulation should be repeated by going back to (2d).
4. Digital twin:

(a) Connection between worlds. Sensor installation for real-time data communication between the real world and the virtual model to create the digital twin.
(b) Digital twin with VR visualization. Visualization of the real actions and events in the digital twin, and operator training.
(c) Real commissioning. Real functioning of the manufacturing process mirrored in its digital twin.

3.1. System Architecture

In general, a traditional robotic cell is composed by one or more robots, conveyor belts, the cell controller, physical safety systems, the human–machine interface (HMI), etc. If the cell is collaborative, more human factors need to be considered: safety, optimized task distribution, and human–robot interaction/adaptive control [39]. Here is where the digital twin becomes a key element for process automation design, enhanced implementation, and real-time monitoring in operation. As the digital twin mirrors real behavior, it should receive information about the movements of the robots and the other elements of the cell, including people. Therefore, additional sensors and a real-time connection between the real cell and the virtual one are necessary. Although the digital twin framework can be used afterwards to control the real manufacturing cell, it has not been considered in this work. It would only be necessary to add certain actuators in the cell which would receive the commands from the digital twin.

These components are structured and connected according to the architecture presented in Figure 2. The hardware is grouped in seven subsystems:

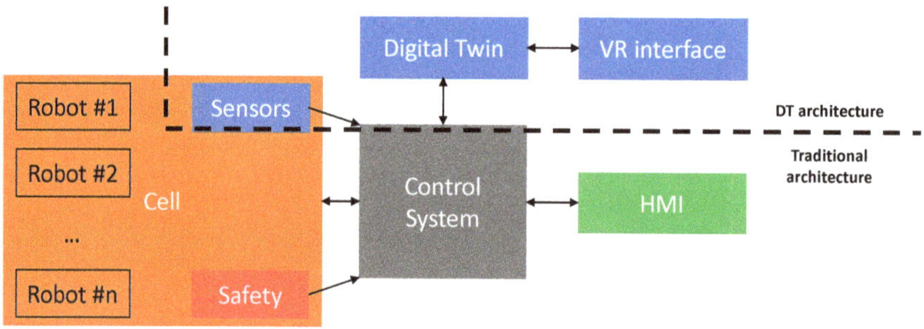

Figure 2. System architecture with the seven subsystems.

- Control system. The cell is managed by the control system with a graphical user interface.
- Robots and other cell components. The cell may be composed of one or several robots and other elements: conveyor belts, automatic tools, etc. As in a standard solution, the robots are controlled and commanded by their controllers.
- Safety system. As the operators can enter in the working area of the robots, the safety system is aimed at avoiding collisions between them. When there is any potential risk of collision, the robot controller gets the alert signal to reduce the speed, or to stop, depending on the distance between the human and the robot.
- HMI. During operation, the users will interact with the system using a HMI connected to the cell control system instead of dealing with the specific console of each robot.
- Digital twin. The real cell is mirrored in a virtual space.
- Sensors. In order to feed the digital twin with real-time information, additional sensors will be installed in cell.
- Virtual reality interface. The VR interface permits an immersive visualization of the digital twin.

The main advantage of this architecture is its modularity. If in a future application, for example, the VR visualization is not required, this module will not be necessary. If additional capabilities are necessary, new modules can be added to increase comprehension, intelligence, and services. Moreover, these subsystems can be developed independently and permit the integration of robots from different manufacturers in the living digital twin of the whole cell, which results in a novel and intelligent tool for design, simulation, real-time monitoring, training, and safe human–robot collaboration. These processes have been traditionally separated, which means that there was not a unique framework covering all the steps; thus different applications were necessary for each step and for each robot. The proposed approach covers all the steps with a unique application, increasing the efficiency of the automation of industrial manufacturing processes.

3.2. The Virtual Interface

Virtual reality replaces real sense perceptions by computer-generated ones describing a 3D scene and animations of objects within the scene. The user needs to feel a totally immersive and authentic experience in the VR application. This is achieved by a realistic behavior of the elements, avoiding the latencies between actions and feedback, and creating a high quality 3D reconstruction to transmit to the user the sense of presence [25]. In order to avoid latencies and permit real-time interaction, the VR system requires a powerful dedicated computer [40] (this computer contains the digital twin of the cell also). Moreover, computer graphics and algorithms can be used to improve the rendering process. The high quality 3D reconstruction for the virtual interface can be achieved following the procedure described in [25]:

1. Scan the real scenario. Firstly, using a 3D reconstruction scanner, a dense 3D point cloud is obtained. The scanner of the mentioned reference can be used, but it is possible to use others.
2. Process the resulting 3D point cloud and apply filters. The point cloud is processed and filtered with the software of the 3D scanner in order to reduce noise and the number of points, guaranteeing a continuous density of points to facilitate the next step.
3. Model the point cloud to render the virtual environment. The point cloud is modeled to create the 3D reconstruction with the real dimensions of the real environment and the immersive effect for the user.
4. Implement the elements' behavior and the human interaction. Finally, the virtual scene is completed with the configuration of the physical behavior of the elements (animations, movements, events, actions, etc., so that virtual elements act similarly to the real ones), a set of virtual buttons, floating text charts, etc., to permit the immersive user interaction and the data visualization.

4. Use Case

A good assembly process plan can increase efficiency and quality, and decrease cost and time of the whole manufacturing process [41]; thus the design step is critical to achieving a successful implementation. For this reason, the proposed approach is applied to the design, implementation, and operation of an assembly manufacturing process, where humans and robots work collaboratively. This is a very representative case for different manufacturing industries wherein the final product is the result of the integration of several parts, such as aerospace, automotive, pharmaceutical, food and beverage, and electronics industries. Thus, it is very likely to find potential applications to create flexible and easily reconfigurable production lines.

4.1. Design

The aim of this process is to classify the different manufactured parts (Figure 3), to assemble the parts and the covers, and to put them on a tray for inspection and delivery. For this purpose, the parts and the covers are spread in different containers. The operator will prepare the batches according to the manufacturing orders, extracting the parts from the containers, placing them on the trays, and assembling the covers.

Figure 3. Design of the manufactured parts and the covers for the use case.

Productivity and safety were limited by manual processes in the traditional industry. The introduction of automation and intelligent collaborative robots in the industrial manufacturing processes is resulting in a rapid increase in productivity, major material and energy savings, and safer working conditions [42]. Thus, in this use case robots will assist the operator during the tasks, and inspection systems will verify whether parts, covers, and batches are correct.

The overall assembly process is composed of the following steps (Figure 4 shows the flowchart of the process):

1. Batch preparation and individual inspection:
   (a) Once the operator and the robot are on the assembly table, the system indicates to the operator the batch and the first part to take.
   (b) Following the instructions, the operator puts the part in the buffer.
   (c) The robot verifies whether the part is correct with an on-board camera [43]. If the part is correct (type, dimensions, and color), it is picked and placed in the tray. If not, the robot puts away the wrong part.
   (d) The process is repeated for all the parts of the batch. Related to the wrong part, the program has a list of pending parts; thus as long as a part is still on this list, it is requested again. It will only be deleted if it is seen correctly in the buffer.
2. Covers assembly and batch inspection:
   (a) When all the positions of the template of the tray are completed, the robot verifies again that all the parts are the required ones and that they are in the right position.
   (b) The operator puts the covers inside the holes.
   (c) The robot verifies whether all the covers have been placed. If not, it notifies to the operator that the covers are not ok.
3. Batch distribution:
   (a) The robot takes the tray and puts it on the conveyor belt for delivery.
   (b) On the other side of the conveyor, another robot receives the tray with assembled parts.

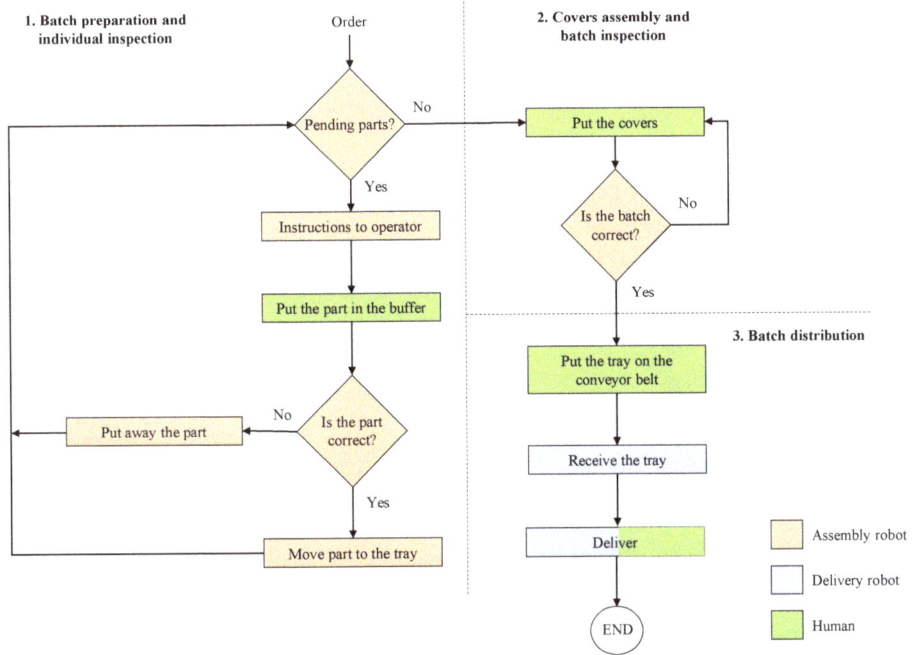

Figure 4. Flowchart of the use case process with the tasks of each key player.

To carry out this process, two different collaborative robots (cobots) and a conveyor belt have been selected instead of an autonomous mobile platform due to the spatial limitations and in order to save costs. Thus, the following components are necessary:

- A table for the containers of parts and trays.
- The assembly robot.
- An assembly table with the containers of covers and discarded parts as well as a buffer where the operator will put the parts which the robot will move to the tray.
- A vision system for part location and inspection.
- A conveyor belt where the robot will place the tray once completed.
- The delivery robot which is waiting the tray at the other side of the conveyor.

Figure 5 shows the particularized architecture for the use case process:

- Control system. The unit controller is a *Raspberry Pi 3 B+*.
- Robots and other cell components. Two light weigh collaborative robots with grippers have been selected, the *Omron TM5-900* and the *Universal Robots UR5e*. The first one is used in the assembly operation as it has an on-board vision system which is used for the inspection. The tool of this robot is the *Robotiq 2F-140* gripper. The second robot is used for the delivery of the trays with the *Robotiq Hand-e* gripper as tool. The conveyor belt is own-manufacture and it is controlled by the *Siemens S7-1200* PLC. Moreover, it contains presence sensors in order to control the position of the trays.
- Safety system. The movements of the operators inside the working area of the robots are monitored by the *Sick microScan 3 Core* scanner. If the operator is in the collaborative area, the robot controller gets the alert signal to reduce the speed, and if he/she is too close to the robot, the robot controller gets the alert signal to stop.

- HMI. A touch screen is connected to the control system to permit the human interaction so that the operator can know what operations and actions must perform, and get feedback of his/her performance from the system.
- Digital twin. The PC which hosts the virtual mirror of the real cell is connected to the cell network to receive the real-time information.
- Sensors. The real-time information about movements, events, etc. is provided by the safety system, the presence sensors, the robots, the inspection systems, etc.
- Virtual reality interface. Among the different available commercial hardware, *HTC Vive* glasses were selected. *Unity3D* is used as the VR engine.

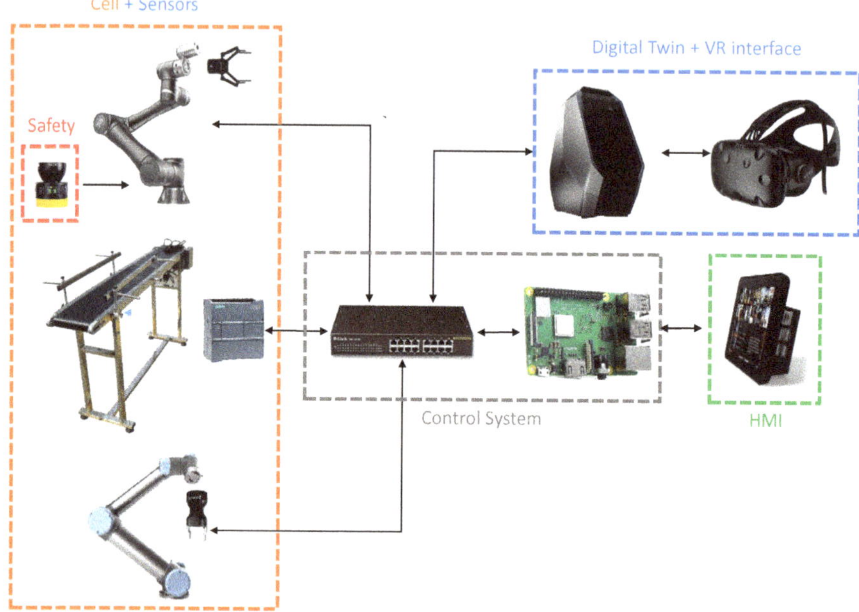

Figure 5. System architecture particularized to the use case: subsystems and components.

Figure 6 shows the layout that has been designed according to the previous steps.

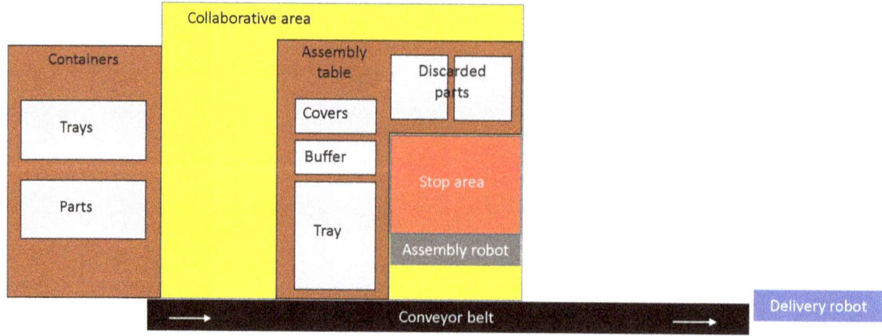

Figure 6. Layout for the physical location of the use case components.

4.2. VR Model

According to the proposed approach, the real scenario is created in virtual reality to visualize the designed process and to evaluate this layout, for instance, in terms of the robots working range, among others. Figure 7 shows the creation of the virtual reality environment: Figure 7a is the original area of the facilities, Figure 7b is the 3D point cloud resulting from scanning this area with *FARO Focus3D X130 HDR*, and Figure 7c is the *Blender*-resulting virtual environment ready for its use in virtual reality. The VR model is completed adding the virtual model of all the components of the cell (robots, conveyor belt, tables, parts, etc.). Finally, the physical behavior of the elements and the human interaction is implemented using *Unity3D*. The final result is shown in Figures 8 and 9, where the high quality of the 3D reconstruction to create the immersive effect in the VR interface can be noticed.

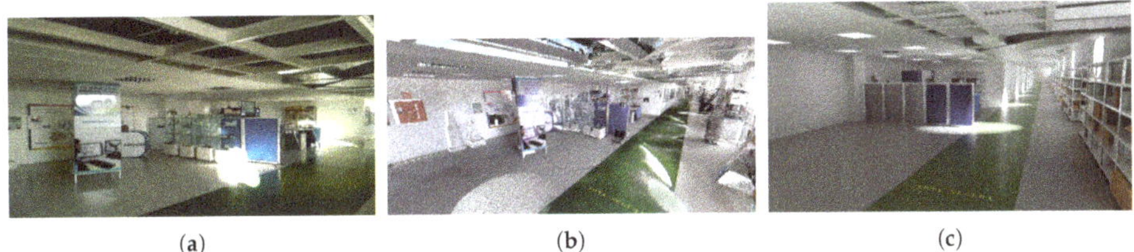

(a) (b) (c)

Figure 7. From the real to the virtual environment: (**a**) real environment, (**b**) 3D point cloud, and (**c**) virtual environment.

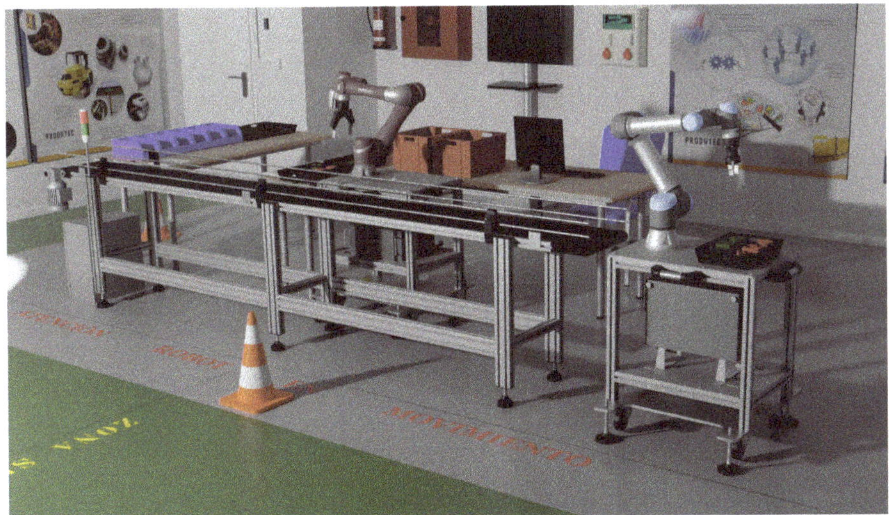

Figure 8. General view of the virtual cell.

(a) (b)

Figure 9. Virtual cell: (**a**) containers with the parts and assembly robot, and (**b**) delivery robot.

Using this virtual scenario, the process has been simulated and evaluated for the virtual commissioning via applying lean automation concepts and verifying that the design fulfills the requirements. Different work models were defined, assuming a distribution of tasks between the operator and the collaborative robots. Each of these models was mainly characterized by the layout, the logic of behavior, and the parameters associated with production management. The manual process without robots was also evaluated. After analyzing and comparing the models in terms of (1) efficiency and optimization, (2) reduction of movements and transportation, and (3) possibility of future extensions, these are the main conclusions which validate the designed process (including elements, layout, flowchart, etc.):

1. Efficiency and optimization:
   - The parts are inspected previously to the assembly of the covers, avoiding spending time and wasting materials on wrong parts.
   - The operator receives instructions to continue his/her tasks in parallel with robot's tasks. It is not necessary to wait.
   - The buffer avoids bottlenecks.
   - The robot discards automatically the wrong parts. The operator does not have to wait for the inspection result.
   - If a part is pending, it is requested at the end of the batch, increasing process flexibility and avoiding confusions to the operator.
   - If all the tasks are done manually without any automation, the operator will spend time in repetitive tasks, the inspection will be subjective, and materials will be wasted, among other inefficiencies. Thus, it will not be the optimal situation.
2. Reduction of movements and transportation:
   - Containers and trays are close to the operator in order to reduce movements.
   - The conveyor belt permits the transportation of the completed batches from the assembly area to the delivery area, avoiding the use of a mobile platform or an automatic guided vehicle (AGV).
3. Possibility of future extensions:
   - New workstations can be added.
   - Augmented reality can be used as HMI instead of the current screen. For this purpose, it is necessary to locate QR-codes in the real scenario. Their possible location has been studied in the virtual scenario as it can be appreciated in Figure 9a.

4.3. Implementation

Once the process and the layout have been validated in the virtual space, they are physically implemented in the real scenario, as shown in Figure 10. Comparing Figures 8 and 10, the high accurate 3D reconstruction to create the immersive effect in the VR interface can be noticed. The virtual environment is totally accurate to the real one, including the minimum details (real dimensions, textures, colors, lighting effects, etc.) to transmit to the user the sense of presence. Furthermore, Figures 11 and 12 show the elements in detail.

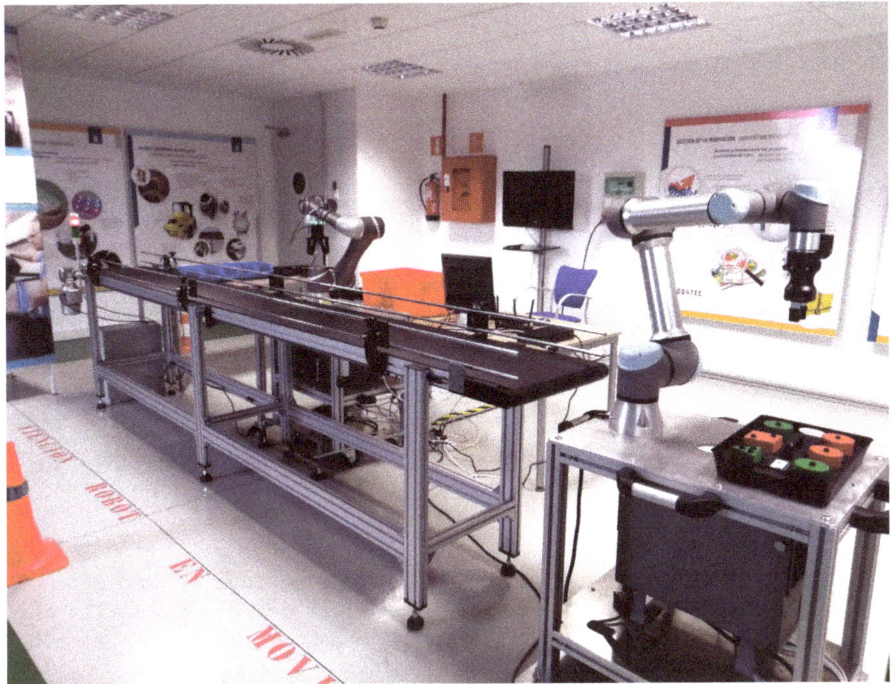

Figure 10. General view of the real scenario.

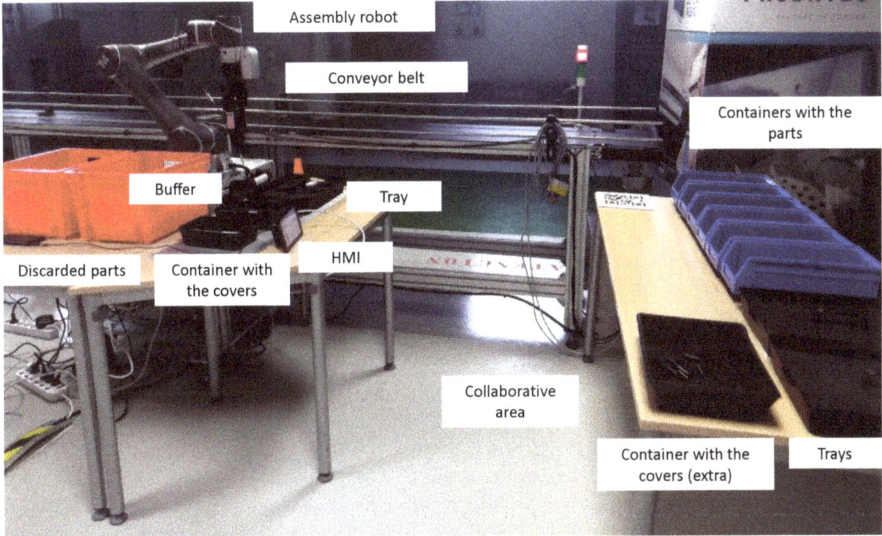

Figure 11. Detailed view of the collaborative area (real scenario).

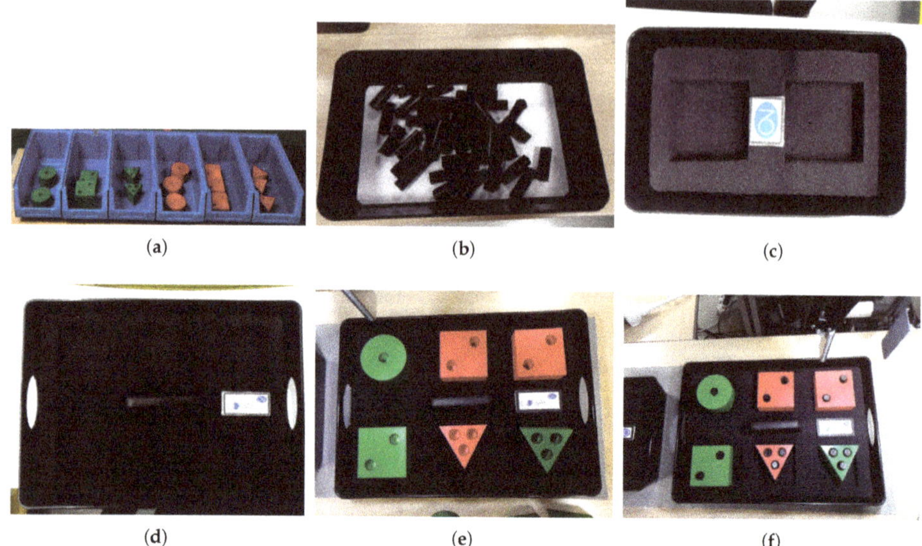

Figure 12. Detailed view of the use case components: (**a**) parts in the containers, (**b**) covers, (**c**) buffer, (**d**) empty tray, (**e**) parts in a tray, and (**f**) completed batch.

4.4. Digital Twin

Finally, it is necessary to connect the real process with the virtual one in order to create the digital twin and feed it with real-time information. For this purpose, the virtual reality controller communicates with the cell control system to receive information related to the movements of the humans, the robots, the parts, the trays, etc.:

- Manufacturing orders. They are directly transmitted to the digital twin at the beginning of the assembly process.
- Operators. Their positions are controlled by the safety system, thus they are transmitted to the digital twin.
- Robots. The control system is executing predetermined (previously programmed) movements which are communicated to the digital twin.
- Inspection results. The control system receives the results of the part and batch inspections; thus they can be communicated and replicated in the digital twin.
- Conveyor belt. The PLC which controls the movement of the conveyor is connected to the cell control system. It captures the data from the presence sensors, the speed, etc. This information is transmitted to the digital twin.

5. Results and Discussion

This work presents a novel methodology for multi-robot manufacturing cell design and operation, combining digital twin and virtual reality. The proposed framework and the modular architecture permit simulation and real-time monitorization. The fulfillment of the requirements is verified in a digital twin framework based on virtual reality, which permits the immersive visualization of the design and the simulation of the possible modifications to find the optimal solution during the virtual commissioning. Once the automated process is implemented in the real world, it is mirrored and linked to its digital twin in the virtual world, which permits real-time monitoring and continuous training and improvement. Thus, this work implies a theoretical outcome, which is the proposed methodology for robot-based automation, and a practical one, which is the digital twin framework with VR visualization used as testbed environment. Results show that the proposed methodology permits the efficient design and real commissioning of multi-robot manufacturing processes, including human–robot collaborative cells, which implies an intelligent, efficient, and unique work environment with high potential applications for process design, implementation, and control. Moreover, digital twins with VR visualization allow humans the possibility to work in totally safe environments with robots.

Table 1 shows the comparison between simulation tools from robot manufacturers, commercial simulation tools with virtual reality, and the digital twin based on virtual reality in terms of low acquisition costs (labeled as "Low investment" in the table), integration of robots from different manufacturers ("Multi-robot"), orientation to human–robot collaboration ("Human-robot collab."), immersive effect and virtual reality ("Immersive"), environment customization ("Customization"), usability for training ("Training"), and versatility to include new functionalities ("Versatility"). The scale 1–3 represents a relative comparison between the tools, where "1" means the worst, or not supported, and "3" means the best. Many companies cannot purchase a specific simulation software for each type of robot when they are studying the introduction of robots in their manufacturing processes. The proposed methodology based on the digital twin is totally affordable as it only requires the VR system as an additional component, which is a mass consumer product. Although the methodology can be extended to the automation of other manufacturing processes, the disadvantage is that it requires an expert developer for the creation of the customized digital twin model and the immersive virtual environment. However, this fact provides great versatility to add new features and functionalities according to the needs of the company.

Table 1. Comparative between simulation tools and the proposed approach.

	Robot Manufacturers	Commercial Sim. Tools with VR	DT Based on VR
Low investment	2	1	3
Multi-robot	1	3	3
Human-robot collab.	1	1	3
Immersive	1	3	3
Customization	1	2	3
Training	1	2	3
Versatility	1	2	3

The proposed approach has been validated in the real commissioning of a representative use case of an assembly manufacturing process, where humans and robots from different manufacturers work collaboratively in classification, assembly, inspection, and delivery of batches of parts. Results show that the presented combination of the digital twin concept with virtual reality permits the design, simulation, training, and real-time monitoring of the manufacturing process. The digital twin of the robotic cell permits an efficient and optimized design, evaluating different options for the layout, the use and the number of robots, and other parameters to find the best solution according to lean automation concepts. All of them are validated in the virtual commissioning before the physical implementation. After the implementation, the cell is mirrored in the digital twin, monitoring productivity and safety for the real commissioning—key issues for industrial leadership.

Figure 13 shows the tests in the virtual and real scenarios. Testers point out that the proposed methodology increases the efficiency as the same tool includes all the necessary steps for the real commissioning of the cell, integrating all types of robots and collaborative applications. The intermediate virtual commissioning includes all the minimum details, and the immersive VR visualization gives a sense of total realism (sense of presence). Moreover, this solution is safe, dynamic, and cost-effective. Potential applications can be found in different industries. Thus, a very likely outcome is extending these results to introduce robots in the manufacturing processes of multiple industries and to increase their efficiency. In this sense, the next steps of this work will focus on more complex manufacturing processes, and extend the capabilities to conduct data analysis.

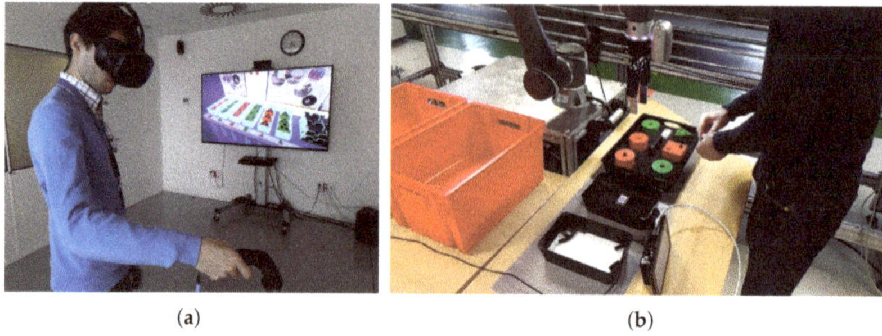

(a) (b)

Figure 13. Tests and validation: (**a**) tester in the virtual scenario, and (**b**) tester in the real scenario.

In the current Industry 4.0 revolution, where manufacturing technologies are continuously changing in order to achieve personalized products, in contrast with the traditional serial production, intelligent automation is a core element to increase the productivity and to improve the competitiveness of the industry. Robots are the future of the industry, and thus the design, the commissioning, and the operation of the robotized cells are critical to achieving success. The proposed approach demonstrates that the synergies between Industry 4.0 technologies, such as digital twins, virtual reality, and collaborative robotics, enable working at new levels and parallel environments which have not been accomplished yet. The future of manufacturing requires the interaction between humans and multiple types of robots, and between physical and virtual scenarios. Each manufacturing process will have its digital twin not only for visualizing or controlling, but also for continuous improving.

Author Contributions: Conceptualization, L.P.; methodology, L.P.; validation, L.P., S.R.-J., and N.R.; writing—original draft, L.P.; writing—review and editing, L.P., R.U., and D.F.G. All authors have read and agreed to the published version of the manuscript.

Funding: This research was funded by Gobierno del Principado de Asturias Programa Asturias grant number IDI/2018/00063, Robots 4.0.

Acknowledgments: The authors would like to thank Eduardo Diez and Paulino San Miguel for their programming support.

Conflicts of Interest: The authors declare no conflict of interest.

Abbreviations

The following abbreviations are used in this manuscript:

3D	Three-Dimensional
AGV	Automatic Guided Vehicle
CPPS	Cyber-Physical Production Systems
DT	Digital Twin
HMI	Human-Machine Interface
PLC	Programmable Logic Controller
QR	Quick Response
ROS	Robot Operating System
VR	Virtual Reality

References

1. Ohno, T. *Toyota Production System: Beyond Large-Scale Production*; CRC Press: Boca Raton, FL, USA, 1988.
2. Henning, K. *Recommendations for Implementing the Strategic Initiative INDUSTRIE 4.0*; National Academy of Science and Engineering: Munich, Germany, 2013.
3. Kolberg, D.; Zühlke, D. Lean automation enabled by industry 4.0 technologies. *IFAC-PapersOnLine* **2015**, *48*, 1870–1875. [CrossRef]

4. Lee, J.; Bagheri, B.; Kao, H.A. A cyber-physical systems architecture for industry 4.0-based manufacturing systems. *Manuf. Lett.* **2015**, *3*, 18–23. [CrossRef]
5. Pérez, L.; Rodríguez-Jiménez, S.; Rodríguez, N.; Usamentiaga, R.; García, D.F.; Wang, L. Symbiotic human–robot collaborative approach for increased productivity and enhanced safety in the aerospace manufacturing industry. *Int. J. Adv. Manuf. Technol.* **2020**, *106*, 851–863. [CrossRef]
6. Vanderborght, B. *Unlocking the Potential of Industrial Human–Robot Collaboration*; Technical Report; Publications Office of the European Union: Brussels, Belgium, 2019.
7. Grieves, M.; Vickers, J. Digital twin: Mitigating unpredictable, undesirable emergent behavior in complex systems. In *Transdisciplinary Perspectives on Complex Systems*; Springer: Berlin/Heidelberg, Germany, 2017; pp. 85–113.
8. Boschert, S.; Rosen, R. Digital twin—The simulation aspect. In *Mechatronic Futures*; Springer: Berlin/Heidelberg, Germany, 2016; pp. 59–74.
9. Schluse, M.; Rossmann, J. From simulation to experimentable digital twins: Simulation-based development and operation of complex technical systems. In Proceedings of the 2016 IEEE International Symposium on Systems Engineering (ISSE), Edinburgh, UK, 3–5 October 2016; pp. 1–6.
10. Rosen, R.; Von Wichert, G.; Lo, G.; Bettenhausen, K.D. About the importance of autonomy and digital twins for the future of manufacturing. *IFAC-PapersOnLine* **2015**, *48*, 567–572. [CrossRef]
11. Havard, V.; Jeanne, B.; Lacomblez, M.; Baudry, D. Digital twin and virtual reality: A co-simulation environment for design and assessment of industrial workstations. *Prod. Manuf. Res.* **2019**, *7*, 472–489. [CrossRef]
12. Zhu, Z.; Liu, C.; Xu, X. Visualisation of the Digital Twin data in manufacturing by using Augmented Reality. *Procedia CIRP* **2019**, *81*, 898–903. [CrossRef]
13. Cimino, C.; Negri, E.; Fumagalli, L. Review of digital twin applications in manufacturing. *Comput. Ind.* **2019**, *113*, 103130. [CrossRef]
14. Yan, Z.; Fabresse, L.; Laval, J.; Bouraqadi, N. Building a ros-based testbed for realistic multi-robot simulation: Taking the exploration as an example. *Robotics* **2017**, *6*, 21. [CrossRef]
15. Schleich, B.; Anwer, N.; Mathieu, L.; Wartzack, S. Shaping the digital twin for design and production engineering. *CIRP Ann.* **2017**, *66*, 141–144. [CrossRef]
16. Tao, F.; Sui, F.; Liu, A.; Qi, Q.; Zhang, M.; Song, B.; Guo, Z.; Lu, S.C.Y.; Nee, A. Digital twin-driven product design framework. *Int. J. Prod. Res.* **2019**, *57*, 3935–3953. [CrossRef]
17. Tao, F.; Cheng, J.; Qi, Q.; Zhang, M.; Zhang, H.; Sui, F. Digital twin-driven product design, manufacturing and service with big data. *Int. J. Adv. Manuf. Technol.* **2018**, *94*, 3563–3576. [CrossRef]
18. Zhuang, C.; Liu, J.; Xiong, H. Digital twin-based smart production management and control framework for the complex product assembly shop-floor. *Int. J. Adv. Manuf. Technol.* **2018**, *96*, 1149–1163. [CrossRef]
19. Wang, J.; Ye, L.; Gao, R.X.; Li, C.; Zhang, L. Digital Twin for rotating machinery fault diagnosis in smart manufacturing. *Int. J. Prod. Res.* **2019**, *57*, 3920–3934. [CrossRef]
20. Association, R.I. Digital Twins and Virtual Commissioning in Industry 4.0. Available online: https://www.robotics.org/content-detail.cfm/Industrial-Robotics-Tech-Papers/Digital-Twins-and-Virtual-Commissioning-in-Industry-4-0/content_id/8173 (accessed on 3 April 2020).
21. Rocca, R.; Rosa, P.; Sassanelli, C.; Fumagalli, L.; Terzi, S. Integrating Virtual Reality and Digital Twin in Circular Economy Practices: A Laboratory Application Case. *Sustainability* **2020**, *12*, 2286. [CrossRef]
22. Bevilacqua, M.; Bottani, E.; Ciarapica, F.E.; Costantino, F.; Di Donato, L.; Ferraro, A.; Mazzuto, G.; Monteriù, A.; Nardini, G.; Ortenzi, M.; et al. Digital Twin Reference Model Development to Prevent Operators' Risk in Process Plants. *Sustainability* **2020**, *12*, 1088. [CrossRef]
23. Pan, Z.; Polden, J.; Larkin, N.; Van Duin, S.; Norrish, J. Recent progress on programming methods for industrial robots. In Proceedings of the ISR 2010 (41st International Symposium on Robotics) and ROBOTIK 2010 (6th German Conference on Robotics), Munich, Germany, 7–9 June 2010; pp. 1–8.
24. Penttilä, S.; Lund, H.; Ratava, J.; Lohtander, M.; Kah, P.; Varis, J. Virtual Reality Enabled Manufacturing of Challenging Workpieces. *Procedia Manuf.* **2019**, *39*, 22–31. [CrossRef]
25. Pérez, L.; Diez, E.; Usamentiaga, R.; García, D.F. Industrial robot control and operator training using virtual reality interfaces. *Comput. Ind.* **2019**, *109*, 114–120. [CrossRef]

26. Components, V. Visual Components: 3D Manufacturing Simulation and Visualization Software—Design the Factories of the Future. Available online: https://www.visualcomponents.com (accessed on 10 April 2020).
27. Siemens. Engineer Automated Production Systems Using Robotics and Automation Simulation. Available online: https://www.plm.automation.siemens.com/global/es/products/manufacturing-planning/robotics-automation-simulation.html (accessed on 10 April 2020).
28. RoboDK. Simulate Robot Applications. Available online: https://robodk.com (accessed on 8 May 2020)
29. Roldán, J.J.; Peña-Tapia, E.; Garzón-Ramos, D.; de León, J.; Garzón, M.; del Cerro, J.; Barrientos, A. Multi-Robot Systems, Virtual Reality and ROS: Developing a new generation of operator interfaces. In *Robot Operating System (ROS)*; Springer: Berlin/Heidelberg, Germany, 2019; pp. 29–64.
30. Kousi, N.; Gkournelos, C.; Aivaliotis, S.; Giannoulis, C.; Michalos, G.; Makris, S. Digital twin for adaptation of robots' behavior in flexible robotic assembly lines. *Procedia Manuf.* **2019**, *28*, 121–126. [CrossRef]
31. Malik, A.A.; Bilberg, A. Digital twins of human robot collaboration in a production setting. *Procedia Manuf.* **2018**, *17*, 278–285. [CrossRef]
32. Ma, X.; Tao, F.; Zhang, M.; Wang, T.; Zuo, Y. Digital twin enhanced human-machine interaction in product lifecycle. *Procedia CIRP* **2019**, *83*, 789–793. [CrossRef]
33. Bilberg, A.; Malik, A.A. Digital twin driven human–robot collaborative assembly. *CIRP Ann.* **2019**, *68*, 499–502. [CrossRef]
34. Aivaliotis, P.; Georgoulias, K.; Arkouli, Z.; Makris, S. Methodology for enabling digital twin using advanced physics-based modelling in predictive maintenance. *Procedia CIRP* **2019**, *81*, 417–422. [CrossRef]
35. Vachálek, J.; Bartalský, L.; Rovný, O.; Šišmišová, D.; Morháč, M.; Lokšík, M. The digital twin of an industrial production line within the industry 4.0 concept. In Proceedings of the 2017 21st International Conference on Process Control (PC), Strbske Pleso, Slovakia, 6–9 June 2017; pp. 258–262.
36. Zhang, C.; Zhou, G.; He, J.; Li, Z.; Cheng, W. A data-and knowledge-driven framework for digital twin manufacturing cell. *Procedia CIRP* **2019**, *83*, 345–350. [CrossRef]
37. Burghardt, A.; Szybicki, D.; Gierlak, P.; Kurc, K.; Pietruś, P.; Cygan, R. Programming of Industrial Robots Using Virtual Reality and Digital Twins. *Appl. Sci.* **2020**, *10*, 486. [CrossRef]
38. Kuts, V.; Otto, T.; Tähemaa, T.; Bondarenko, Y. Digital Twin based synchronised control and simulation of the industrial robotic cell using Virtual Reality. *J. Mach. Eng.* **2019**, *19*. [CrossRef]
39. Zhang, J.; Fang, X. Challenges and key technologies in robotic cell layout design and optimization. *Proc. Inst. Mech. Eng. Part C J. Mech. Eng. Sci.* **2017**, *231*, 2912–2924. [CrossRef]
40. Burdea, G.C.; Coiffet, P. *Virtual Reality Technology*; John Wiley & Sons: Hoboken, NJ, USA, 2003.
41. Wang, L.; Keshavarzmanesh, S.; Feng, H.Y.; Buchal, R.O. Assembly process planning and its future in collaborative manufacturing: A review. *Int. J. Adv. Manuf. Technol.* **2009**, *41*, 132. [CrossRef]
42. European Factories of the Future Research Association. *Factories of the Future: Multi-Annual Roadmap for the Contractual PPP under Horizon 2020*; Publications Office of the European Union: Brussels, Belgium, 2013.
43. Pérez, L.; Rodríguez, Í.; Rodríguez, N.; Usamentiaga, R.; García, D.F. Robot guidance using machine vision techniques in industrial environments: A comparative review. *Sensors* **2016**, *16*, 335. [CrossRef]

Article

Mapless Collaborative Navigation for a Multi-Robot System Based on the Deep Reinforcement Learning

Wenzhou Chen [1,†], Shizheng Zhou [1,†], Zaisheng Pan [1,2], Huixian Zheng [2] and Yong Liu [1,*]

1 Institute of Cyber-Systems and Control, Zhejiang University, Hangzhou 310027, China; wenzhouchen@zju.edu.cn (W.C.); 21632051@zju.edu.cn (S.Z.); panzs@zju.edu.cn (Z.P.)
2 Zhejiang Chipkong Technology Co., Ltd., Hangzhou 310000, China; zhenghuixian@nz-ic.com
* Correspondence: yongliu@iipc.zju.edu.cn
† These authors contributed equally to this work.

Received: 25 August 2019; Accepted: 24 September 2019; Published: 9 October 2019

Abstract: Compared with the single robot system, a multi-robot system has higher efficiency and fault tolerance. The multi-robot system has great potential in some application scenarios, such as the robot search, rescue and escort tasks, and so on. Deep reinforcement learning provides a potential framework for multi-robot formation and collaborative navigation. This paper mainly studies the collaborative formation and navigation of multi-robots by using the deep reinforcement learning algorithm. The proposed method improves the classical Deep Deterministic Policy Gradient (DDPG) to address the single robot mapless navigation task. We also extend the single-robot Deep Deterministic Policy Gradient algorithm to the multi-robot system, and obtain the Parallel Deep Deterministic Policy Gradient (PDDPG). By utilizing the 2D lidar sensor, the group of robots can accomplish the formation construction task and the collaborative formation navigation task. The experiment results in a Gazebo simulation platform illustrates that our method is capable of guiding mobile robots to construct the formation and keep the formation during group navigation, directly through raw lidar data inputs.

Keywords: multi-robot; collaborative navigation; deep reinforcement learning

1. Introduction

Autonomous navigation for mobile robots is one of the most practical and essential challenges in robotics. The navigation systems for mobile robots mainly rely on the localization in a given map and the decision-making system. The relative technique for localization is called Simultaneous Localization and Mapping (SLAM) [1,2], which can obtain the map of the environment and get the robot poses simultaneously. In addition, the corresponding decision-making system which consists of planning [3–7] and control [8–10] would generate a safety trajectory and control the mobile robot to follow it until reaching the goal. The decision-making system plays an important role to connect the preceding localization stage and the following manipulation stage. This paper provides a decision-making method to address the core problem of robot navigation. We particularly focus on dealing with the motion planning and control problems with a navigation method based on the deep reinforcement learning.

The traditional decision-making system of the mobile robot can be hierarchically structured into four components, the route planning, the behavioral decision-making, the motion planning and the robot control [11]. For indoor navigation tasks, the decision-making system can be simplified into the motion planning part and the navigation control part. The mobile robot is supposed to plan a trajectory from its current position to the target destination with the specific indoor environment. Then, the motion controller will guide the mobile robot precisely follow the trajectories to the target position. The required efficient motion planning strategies and stable motion controllers are mainly based on the mathematical computations and geometric relationships. Although useful in many situations, the applicability of traditional decision-making systems is limited by their flexibility and versatility. The complexity of the environment and the dynamic obstacles can both increase the computational efficiency and decrease the navigation performance. In addition, the errors of each step will be accumulated to the end. Furthermore, most of the traditional methods can't address the mapless navigation tasks in the complex environment. However, there are many practical situations in which the mobile robots can't obtain the accurate map of the environment and only have the relative position relationship between the mobile robots and the targets. Thus, we proposed an end-to-end policy module with the raw sensor inputs to address these issues in mapless navigation tasks.

The deep reinforcement learning techniques have been greatly developed and widely applied in various fields of study. This kind of technical training the agent is conducted through the interaction trajectories between the agent and the environment. Intuitively, the character of interactive learning is quite similar to humans. When we consider the mapless navigation tasks, with a given target value and the current position, one person is likely to find the target position heuristically with his intuition. If one person is informed about the relative position values or the polar relationship to the target, he can easily find the target position based on the prior knowledge and the basic navigation strategies in an indoor environment. Compared with the traditional navigation methods, one human has a more efficient way to navigate without computing the precise mathematical module of the unknown environment. In addition, the deep reinforcement learning technique paves the way to accomplish the robot mapless navigation tasks like humans. With the aim of teaching the robot to navigate like humans, this work presents an approach to training the robot to navigate with the human intuition. By utilizing the proposed deep reinforcement learning method, the policy module of the mobile robot can make decisions through the raw 2D lidar sensor data and navigate to the target position in an indoor environment.

When we extend the navigation tasks to the multi-robot system, the coordination between a group of mobile robots becomes more complex. Similarly, we analyze the human intuition to get the inspiration. If there are a group of humans navigating in the unknown indoor environment, they would communicate and collaborate with each other to attain the goal. There are several forms of communications such as sharing the experience and observations. For a multi-robot navigation system, the robots can share the sensor observations, the training data and the relative state parameters. Inspired by the analysis of human intuition, our work provides a new insight into the multi-robot collaborative navigation task with deep reinforcement learning. By sharing the training experience and observing the states of other robots, the multi-robot system can keep the formation during group navigation.

In our proposed method, we assume that the relative position values of the targets are easily acquirable for the robots via cheap localization solutions such as WiFi [12] and QR code [13]. By improving the classical deep reinforcement learning methods, the proposed algorithm can accomplish the single robot mapless navigation task with human intuition. We also extend the method into the multi-robot navigation cases with some useful training strategies, and then evaluate the performance in the simulation platform. Particularly, the contributions of this paper can be summarized as follows:

1. We improve the classical Deep Deterministic Policy Gradient (DDPG) to address the end-to-end single robot mapless navigation problems directly through the raw lidar data inputs.
2. The Parallel Deep Deterministic Policy Gradient (PDDPG) is proposed to accomplish the multi-robot formation construct and the formation keeping during collaborative navigation. We also evaluate the proposed methods in the Gazebo simulation platform.

This paper is organized as follows: Section 2 describes the deep reinforcement learning algorithms and their applications in robot navigation tasks. In Section 3, an overview of the proposed algorithm is presented. The methods for different situations and the training details are described separately.

Section 4 presents the evaluations of the proposed methods in the simulated environment and analyzes the experiment results. Section 5 concludes this work.

2. Related Works

Benefiting from the development of the deep neural networks, the deep reinforcement learning techniques show great potential for solving the decision-making tasks. By deploying deep neural networks as powerful nonlinear function approximators, the deep reinforcement learning algorithms can handle the complex decision-making problems with high-dimensional state and action spaces.

2.1. Deep Reinforcement Learning

With the aim of maximizing the expected return of behaviors, deep reinforcement learning considers the agent learning a good policy by interacting with its environment. Mnih et al. [14] first kickstarted the revolution in deep reinforcement learning. They proposed the deep Q network (DQN) that could learn to play Atari 2600 games at a superhuman level only based on the image inputs. This work convincingly demonstrates that deep reinforcement learning agents can be trained with high-dimensional observations. The later research [15–20] updated these kinds of methods and promoted the performance in the subproblems of the Atari games.

The second standout success of the deep reinforcement learning was the AlphaGo [21]. The AlphaGo merged the supervised learning and the Monte Carlo Tree Search (MCTS) [22] techniques into the deep reinforcement learning framework, and defeated the world champion human in Go by learning from human knowledge. After the AlphaGo received widespread attention, the AlphaGo Zero [23] defeated the AlphaGo completely. Unlike utilizing the human knowledge in the training stage, the AlphaGo Zero mastered the game of Go through self-play. Furthermore, the researchers extended the AlphaGo Zero to the other board games and proposed the AlphaZero [24].

2.2. Deep Reinforcement Learning for Navigation

Besides the well-known works in Games, the DRL algorithms also have been successfully applied to a wide range of problems, such as manipulators [25] and mobile robots [26,27]. In this section, we discuss the research about the mobile robot navigation tasks.

With the progress of the deep learning techniques, the powerful representation capabilities shed a new light on learning control policies directly from raw sensor inputs with the reinforcement learning framework. In recent years, there have been lots of proposed methods to tackle the autonomous navigation tasks with deep reinforcement learning algorithms. To address the navigation problems in reinforcement learning ways, these methods formulate the navigation process as the Markov decision process(MDP) 111 or partially observable Markov decision process(POMDP), and stack the observations from sensor readings as the states. Then, the methods will find the optimal policy module that is capable of guiding the robot to the target position.

Kretzschmar et al. [28] and Pfeiffer et al. [29] used the maximum entropy inverse reinforcement learning (IRL) methods to learn interaction models for pedestrians from demonstration in occupied environments. Zhu et al. [30] used the image of the target object and the current observation as the input to the Siamese actor–critic model. Then, they formulated their task as a target-driven navigation problem, and evaluated the performance in an indoor simulator [31]. Zhang et al. [32] proposed a deep reinforcement learning algorithm based on the successor features, which can transfer knowledge from previous navigation problems to new situations. By using additional supervision signals from auxiliary tasks, Mirowski et al. [33] greatly boosted the training and improved the task performance of their DRL algorithm in 3D mazes navigation tasks. Unlike addressing the navigation tasks in the static environment, Chen et al. [34] developed a time-efficient navigation method in dynamic environments with pedestrians. Moreover, Long et al. [35,36] extended the robot navigation task to the multi-robot case, and focused on addressing the collision avoidance problem.

2.3. Multi-Agent Deep Reinforcement Learning

If we ignore the the kinematics and only consider the behavior strategies of the group of robots, there have been plenty of novel works called multi-agent reinforcement learning (MARL) in recent years. Raileanu et al. [37] proposed a new approach for learning in self other-modeling. This method used its own policy to estimate the other agent's actions and updated its belief of the hidden state. Then, the estimations were used to choose new actions. Yang et al. [38] simplified the communication of agents into an average effect. By introducing the mean-field theory, they mainly studied one agent with the average effect of the others. Wang et al. [39] estimated an opponent's future behavior by utilizing the history information. Tacchetti et al. [40] proposed the relational forward models to address the MARL tasks. They added the relational graph model in the action making stage, and used the recurrent neural network (RNN) to model the relationships between agents. However, with the ideal assumptions, there is still a lot of work to do if we want to apply these methods to the multi-robot navigation tasks.

3. Methods

In this section, we first formulate the problem for Multi-Robot collaborative navigation. Then, we describe our Parallel Deep Deterministic Policy Gradient (PDDPG) method for this task.

3.1. Problem Formulation

This paper aims to provide an end-to-end navigation method for multi-robot systems. We try to find such a learnable policy module π:

$$a_t^i = \pi(x_t^i, \varphi_t^i, a_{t-1}^i), \tag{1}$$

where x_t^i is the observation from the raw lidar sensor information of robot i at timestep t, φ_t^i is the relative parameters about the target, and a_{t-1}^i is the control action in the last timestep. In the multi-robot reinforcement learning system, these inputs can be regarded as the state of whole system $s_t = (x_t, \varphi_t)$. For a single robot at each timestep t, the robot makes observations $s_t \in \mathcal{S}$ and selects actions $a_t \in \mathcal{A}$ with respect to its policy $\pi\colon \mathcal{S} \to \mathcal{P}(\mathcal{A})$, which maps states to a probability distribution over the actions. Then, the robot can receive the reward $r(s_t, a_t)$ and arrive at the next state. The state-action value function describes the expected return of a state-action trajectory according to π. It is commonly used to evaluate the quality of a policy as defined in Equation (2):

$$Q_\pi(s,a) = \mathbb{E}\left[\sum_{t=0}^{\infty} \gamma^t r\left(s_t, a_t\right)\right] \quad where \;\; s_t \sim p\left(\cdot|s_{t-1}, a_{t-1}\right), a_t = \pi\left(s_t\right). \tag{2}$$

The recursive form of the state-action value function, known as the Bellman equation, can be defined as Equation (3). The policy module π directly maps the state perception to the control law of robots with the collaborative consideration. E represents the environment. In addition, the notation $\sim$ means the former variable follows the distribution of the latter:

$$Q^\pi\left(s_t, a_t\right) = \mathbb{E}_{r_t, s_{t+1} \sim E}\left[r\left(s_t, a_t\right) + \gamma \mathbb{E}_{a_{t+1} \sim \pi}\left[Q\left(s_{t+1}, a_{t+1}\right)\right]\right]. \tag{3}$$

In order to get a good policy module π, the multi-robot system has to overcome several challenges. First of all, it is difficult to extract valuable information from the raw sensor data and merge it properly with the relative parameters of robot state. Then, based on the processed information, a policy module needs to learn a stable transition rule between the perceptions and the decision to make. Additionally, the robots need to avoid the collision and consider the collaboration in some situation.

3.2. Single-Robot End-to-End Navigation

To accomplish the final goal, we start at the single robot navigation case in this section. In the end-to-end mapless navigation tasks, the relationship between the perception inputs and the output control law can be very complex. In our work, we consider a number of modifications to the Deep Deterministic Policy Gradient (DDPG) and use it as the basic framework.

3.2.1. Basic Reinforcement Learning Framework

Classical Deep Deterministic Policy Gradient (DDPG) [41] is a kind of remarkable reinforcement learning algorithm to address the continuous control problem. As we all know, the trade-off between the exploration and the exploitation is one of the most important problems in the reinforcement learning field. To increase the sample efficiency, the DDPG uses the deterministic target policy $\mu : \mathcal{S} \leftarrow \mathcal{A}$ rather than the stochastic policy π, and the corresponding exploration decrease is made up of the off-policy technique. The Bellman equation can be rewritten as Equation (4), where the γ term represents the discount factor of the equation:

$$Q^{\mu}\left(s_t, a_t\right) = \mathbb{E}_{r_t, s_{t+1} \sim E}\left[r\left(s_t, a_t\right) + \gamma Q^{\mu}\left(s_{t+1}, \mu\left(s_{t+1}\right)\right)\right]. \tag{4}$$

At each timestep, the actor networks and the critic networks are updated by sampling a minibatch uniformly from the memory buffer. The algorithm also creates a copy of the actor and critic networks, $Q'\left(s, a|\theta^{Q'}\right)$ and $\mu'\left(s|\theta^{\mu'}\right)$, respectively; these target networks are then updated softly by the learned networks: $\theta' \leftarrow \tau\theta + (1-\tau)\theta'$ with $\tau \ll 1$. This trick can greatly improve the stability of learning.

The loss function for the critic networks can be formulated as Equation (5):

$$L = \frac{1}{N}\sum_i \left(y_i - Q\left(s_i, a_i|\theta^Q\right)\right)^2, \ where\ y_i = r_i + \gamma Q'\left(s_{i+1}, \mu'\left(s_{i+1}|\theta^{\mu'}\right)|\theta^{Q'}\right). \tag{5}$$

In addition, the actor networks can be updated by using the sampled policy gradient as shown in Equation (6):

$$\nabla_{\theta^{\mu}} J \approx \frac{1}{N}\sum_i \nabla_a Q\left(s, a|\theta^Q\right)\Big|_{s=s_i, a=\mu(s_i)} \nabla_{\theta^{\mu}} \mu\left(s|\theta^{\mu}\right)\Big|_{s_i}. \tag{6}$$

3.2.2. Network Structure

Owing to the structure of the 2D lidar data, we add one-dimensional, convolutional neural network (CNN) layers to the feature extraction part. To enrich the perception information of the policy module, some relative parameters about the target are added to our framework. It enables the policy module to simultaneously consider the sensor observation and the robot state information. Both of them are essential for efficient navigation.

As shown in Figure 1, the raw lidar sensor data are filtered into 180 dimensions as inputs. As mentioned before, there are two kinds of neural networks in our module. One is the actor network and the other is the critic network. Both of them use three one-dimensional, CNN layers for feature extraction. After the data feature extraction, the networks use residual building blocks with shortcut connections [42] to reduce the training complexity. Then, we pack the target distance, the target angle and the actions at the last timestep as the state parameter set. The state parameter set and the lidar data feature extraction will be fused together and fed to the actor and critic networks at the same time. For the actor network, the fusion data will pass three fully connected layers and output the linear velocity and angular velocity through different activation functions. Besides controlling the robot navigation, the output of the actor network will be transmitted to the critic network. The output action, the state parameter set and the feature extraction of the critic network will be fused into one data stream. After passing through four fully connected layers, the data stream will finally become a Q value to evaluate the policy and train the networks.

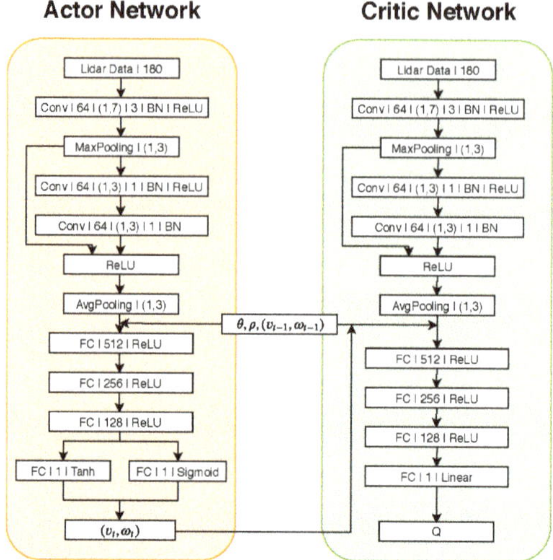

Figure 1. The structure of the modified Deep Deterministic Policy Gradient for single robot mapless navigation tasks.

3.2.3. Reward Shaping

The reward function plays an important role in reinforcement learning: it greatly influences the navigation of the robot system and serves as the truth holder of the learning process. For the single robot mapless navigation task, the reward function shown in Equation (7) consists of three different parts: the arrival award, the collision penalty and the approaching award. If the distance value between the robot and the target point is less than the arrival threshold, the robot can get a positive reward r_{arrive} if the robot collides with the obstacles during navigation. In other words, one of the lidar distance observations is less than the safety threshold. The robot will get a negative reward $r_{collision}$. To enable the intuition of approaching the target, we add the approaching award to the robot, where the reward is noted as $k(\rho_{t-1} - \rho_t)$. The k term is a constant to adjust the amplitude of the approaching reward. The ρ term represents the distance between the target and the mobile robot at timestep t. To unify the range of different parameters, they will be normalized before utilizing:

$$r(s,a) = \begin{cases} r_{arrive} & (\rho < d_{goal}) \\ r_{collision} & (\min(d_1, d_2, ..., d_{180}) < d_{collision}) \\ k(\rho_{t-1} - \rho_t) & (At\ each\ time\ step\ t). \end{cases} \quad (7)$$

3.2.4. The Switchable Controller

Even if we formulate the reward function to encourage the robot to navigate efficiently and safely, the state space of the robot navigation is still large. Utilizing the switchable controller is an intuitive way inspired by imitation learning. For example, if the robot is close to the obstacle, the basic navigation rules will give a large angle velocity to control the robot in order to avoid crashing.

When the robot starts training, the switchable controller can guide the robot to follow some basic navigation rules, rather than randomly exploring in the unknown environment. In addition, the probability of choosing the basic controller will decrease slowly. This means that the robot mainly learns the navigation policy from the basic controller at the beginning, and the basic controller plays an important role as an expert. The trajectories saved in the experience memory will teach the basic navigation rules to the robot. After learning the basic navigation rules, the robot will gradually increase the probability of choosing the learned navigation policy. With that practical framework, the distribution of the policy module can converge to the appropriate shape faster than random exploration.

3.3. Multi-Robot Collaborative Navigation

In this section, we would extend the single navigation algorithm to the multi-robot situation. In the multi-robot system, the observations of each robot are contained by the other robots, the others can be regarded as the dynamic obstacles. If the group of robots don't know how to collaborate with others, they will interfere during navigation.

3.3.1. Parallel Deep Deterministic Policy Gradient

To address the aforementioned issues, we proposed a parallel deep deterministic policy gradient (PDDPG). Compared with the improved DDPG for a single robot mapless navigation task, the PDDPG algorithm trains the robots in parallel. The algorithm attains the target of a whole multi-robot system by sharing the experience memory data and the navigation policy. The robots of the system use the same policy module to make decisions, and the trajectories of robots will be saved in the shared experience replay buffer. The basic network structure unit of the PDDPG algorithm is the same as improved DDPG. We only modify the input dimension of lidar observations and the robot states.

3.3.2. Curriculum Design for Reward Shaping

In the last section, we modified the network structure and proposed the PDDPG to address multi-robot navigation tasks. Besides the good network structure like PDDPG, proper reward formulation is one of the most essential parts of enabling a collaborative capability for a multi-robot navigation system.

When the robots navigate in a mapless environment, the reward would be very sparse. PDDPG would took a long time to converge to a satisfying policy module. A direct training with PDDPG on the collaborative navigation task with eight robots does not yield acceptable performance. To address this issue, we use the curriculum learning [43] which trains the robot with a sequence of progressively more difficult tasks. It paves the way to train the agent to accomplish the difficult tasks. In our proposed method, we only divide the collaborative navigation task into two stages. First, the group of robots needs to construct the formation. In addition, then the robots would be trained to keep the initial formation during navigation.

For the formation construction task, we adjust the reward function of the single robot navigation task to the multi-robot version. The approaching reward of each robot would be summed up as a group approaching reward, as shown in Equation (8). When any of the robots collide, the environment will be reset and the episode will be ended. In the view of training a good reinforcement learning agent, terminating the episode at an appropriate step can efficiently accelerate the agent training. After that, we train the robot to accomplish the collaborative navigation task based on the formation construction policy module. Besides the aforementioned reward, we add the formation constraints in the group reward term $r_{formation}$. This term represents some deformation penalties in our system. Specifically, the algorithm calculates the distances between the robots and save their ratios at the first timestep. When the group of robots are navigating, the algorithm would check the distance ratios and give relative rewards. Moreover, if the speed of one robot is much higher than the mean, the formation keeping constraints would penalize the robot:

$$r(s,a) = \begin{cases} r_{arrive} & (\rho < d_{goal}) \\ r_{collision} & (\min(d_1, d_2, ..., d_{180}) < d_{collision}) \\ \Sigma_1^n k(\rho_{t-1}^i - \rho_t^i) & (At\ each\ time\ step\ t) \\ r_{formation} & (Only\ at\ the\ formation\ keeping\ stage). \end{cases} \tag{8}$$

4. Experiments

To evaluate the performance of our PDDPG algorithm, sufficient experiments are illustrated in the simulation platform. To construct the whole navigation environment of the mobile robots, the simulation is built by a robot simulator named Gazebo, which is used for fast moduling and validating the proposed PDDPG algorithm. The proposed method is implemented on a PC with 15.6 G memory, i7-7700 CPU and Geforce GTX1080Ti on an Ubuntu16.04 operating system.

4.1. Mobile Robot Construction

Robot Operating System (ROS) is a robotics middleware; it provides various tools and services designed for robot research, such as package management, hardware abstraction and the low-level device control. This work utilizes the Gazebo simulation platform to construct the simulation environment. Gazebo is a well-designed simulator with a robust physics engine and ROS support. It offers the ability to rapidly design robots and test algorithms. By using the gym-Gazebo toolkit [44], the deep reinforcement learning agent could be trained on the Gazebo platform efficiently.

As shown in Figure 2a, we build differential driven robot modules in simulation, and add the 2D lidar sensor to perceive the environment. The perception angle of 2D lidar is 270° with 0.25° resolution. The measurement distance of the lidar is from 0.1 m to 30 m, and the frequency of it is 40 Hz. As shown in Figure 2b, the blue lines represent the lasers. When the lasers are blocked by the barriers, each laser will return a distance measurement to the lidar sensor. If the distance measurement is longer than 30 m, the return value will be restricted at 30 m.

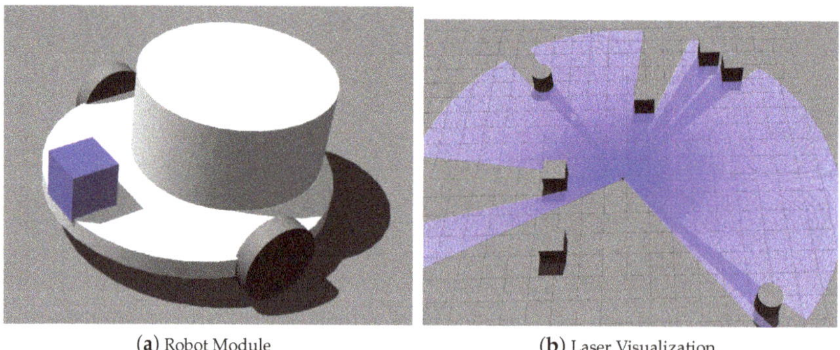

(**a**) Robot Module (**b**) Laser Visualization

Figure 2. (**a**) the module of mobile robots in our Gazebo simulation platform; (**b**) the visualization of the lasers.

4.2. Single-Robot Experiments

In this section, we discuss the single robot mapless navigation task. The training procedure of our robot is implemented in the indoor scene simulated by Gazebo. We train our robot in several indoor scenes with different obstacle placement. We randomly set the robot at the start position with different initial directions. To guarantee the variety of the navigation trajectories and overcoming the overfitting, the targets are randomly chosen outside the restricted zone within the indoor scenes. It can reduce the gap between the training and testing of the policy module.

In the simulation environment, the training procedure can be sped up 10 times. With the simulation acceleration, the control frequency of the mobile robot can reach 100 Hz. The safety threshold of the mobile robot is defined as 0.3 m. In addition, the arrival threshold is also defined as 0.3 m. In other words, if the distance between the mobile robot and the target is less than 0.3 m, this episode will be terminated. Other parameters of the single robot experiments are illustrated in Table 1.

Table 1. Hyperparameters of the networks on a single-robot navigation task.

Parameter	Batch Size	Max Step	LR (Actor)	LR (Critic)	Discount	Buffer Size	Explore Decay
Value	128	800	0.001	0.0001	0.99	100,000	0.998

To compare the performance between the improved DDPG and the classical DDPG, we trained both of them with the same hyperparameters. As shown in Figure 3, the negative Q value of the improved DDPG illustrated by the orange line has a steady decline, while the negative Q value of the classical DDPG illustrated by the blue line increases slightly. Since the Q value evaluates the performance of the policy module, this means the improvement of DDPG becomes better during training, and the classical DDPG falls into local optimum. By analyzing the experiment results, we can infer that the modification in the improved DDPG can address the single robot mapless navigation tasks properly.

Figure 3. The negative Q value comparison between the improved deep deterministic policy gradient (DDPG) and the classical deep deterministic policy gradient(DDPG) and the classical DDPG of the single robot navigation experiments.

For the navigation task that has a high-dimensional decision space, the improved DDPG with the switchable controller and the prioritized experience replay [17] technique can quickly constrain the searching field of policy module. Then, the policy module will be improved gradually. In the simulation environment, the robot with classical DDPG has the possibility to go around in circles. In addition, the robot with improved DDPG can arrive to the target position in most cases. We tested the improved DDPG policy module in the training environments 80 times, and its arrival rate can even reach 95%. When we move it to the unseen environment, the arrival rate declines to 87.5%. Figure 4 shows the single robot navigation experiments with the improved DDPG. The left part shows the simulation environment, and the right part illustrates the navigation trajectories of the mobile robot. The trajectories are separately visualized by the Rviz toolkit. The robot starts to navigate from the star and ordinally traverses through target 1 to target 6. When the robot arrives at target 6, it will set target 1 as the next destination and repeat this cycle.

4.3. Multi-Robot Experiments

Based on the single robot navigation results, we discuss the experiments of the multi-robot navigation tasks in this section. The proposed Parallel Deep Deterministic Policy Gradient (PDDPG) algorithm will be evaluated and analyzed carefully. With the curriculum learning setup, the experiments would be illustrated in a two-stage arrangement: the formation construction and the collaborative navigation.

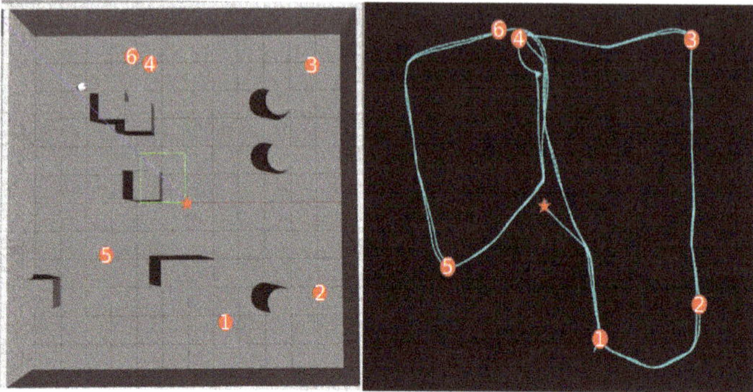

Figure 4. The trajectory of the mobile robot in the single robot mapless navigation tasks. The left part is the navigation environment in Gazebo, the right part is the trajectory of the mobile robot.

4.3.1. Formation Construction

In the formation construction experiments, eight mobile robots are deployed in the indoor scene. The robots are initialized at different positions with various environmental characteristics. We set the target formation of the robots as a rectangle; then, the eight robots will be trained together with the proposed Parallel Deep Deterministic Policy Gradient (PDDPG) algorithm. The hyperparameters of the PDDPG are listed in Table 2. In the training stage, we add the random bias to the target position values in each episode. It can enrich the formation data and reduce the overfitting problem of multi-robot training. Figure 5a,b illustrate one of the formation construction experiments in the testing stage. Figure 5a shows the initialization of the multi robot system; the robots are deployed in the indoor scene with various poses. Figure 5b shows the final results of the formation construction task. The curves with different colors represent the navigation trajectories of different mobile robots. As illustrated in the figures, all the mobile robots arrive at the target and construct the formation properly.

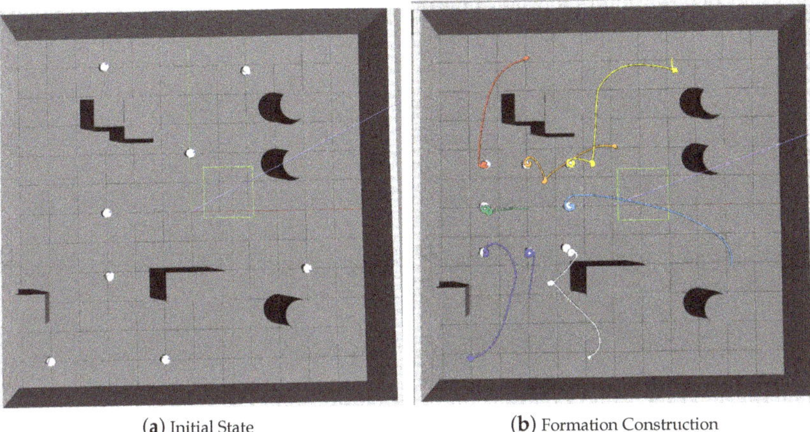

(a) Initial State (b) Formation Construction

Figure 5. (a) the initial state of the group of mobile robots; (b) the terminal state of the group of mobile robots after accomplishing the formation construction navigation.

Table 2. Hyperparameters of the networks on multi-robot navigation tasks.

Parameter	Batch Size	Max Step	LR (Actor)	LR (Critic)	Discount	Buffer Size	Explore Decay
Value	256	2000	0.0001	0.0001	0.99	500,000	0.998

4.3.2. Collaborative Navigation

After obtaining a good policy module in the formation construction task, we use it as the pretrained module in this section. The collaborative reward functions and the constraints are added in these experiments. As shown in Figure 6, the group of mobile robots can keep the rectangle formation during navigation. The trajectories of different mobile robots are illustrated by the colorful curves. In addition, the mobile robots at several timesteps are merged into one figure after getting the trajectory.

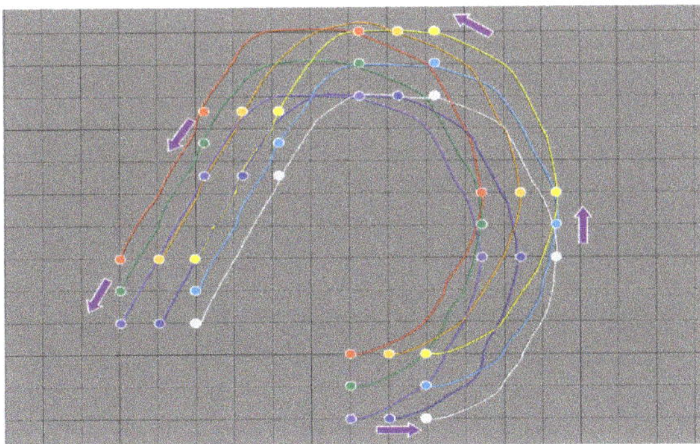

Figure 6. The trajectory of the multi-robot collaborative navigation.

To evaluate the versatility of the multi-robot navigation module, we also evaluate the module with the arrow formation. Figure 7 consists of four different figures, and they are recorded at the different timesteps. The intersection point of three colorful lines represents the origin of the environment. The group of robots navigated from the bottom left to the top right. This experiment illustrates that the mobile robots can keep the arrow formation during navigation.

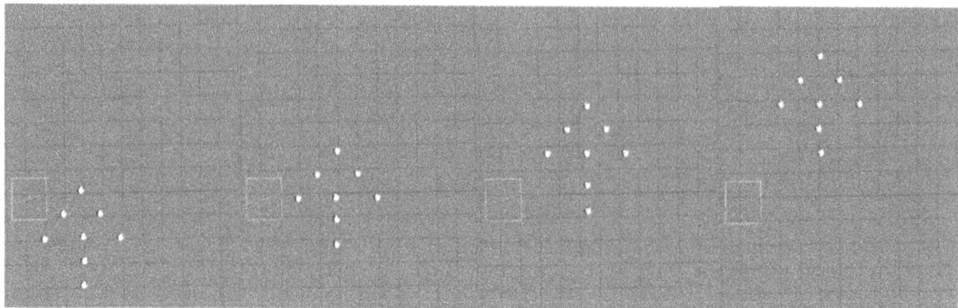

Figure 7. The arrow formation of the multi-robot navigation system.

As illustrated in the collaborative navigation experiments, we mainly evaluate the formation keeping navigation tasks in the simulation environment without obstacles, since the distance of the obstacles can influence the performance of formation keeping. Generally, the multi-robot collaborative navigation task with formation keeping has two levels: one is the formation keeping navigation; the other is the multi-robot obstacle avoidance in a complex environment with constant formation. In this paper, we mainly focus on the former. In our work, the policy module of a collaborative navigation task utilizes the pre-trained policy module in the formation keeping task. There is lots of prior knowledge about obstacle avoidance that has been learned in the pre-trained module. Thus, the collaborative navigation module is sensitive about the obstacles. If there are obstacles in the navigation path, the formation of mobile robots can't keep the neat shape. The size of the obstacles can also influence the performance of formation keeping. In this paper, we mainly illustrate the formation keeping navigation tasks in the simulation environment without obstacles. In addition, we will address the multi-robot obstacle avoidance in a complex environment with constant formation keeping in future work.

To discuss the execution time of our method, we add the experiments to compute the execution time for the proposed Parallel Deep Deterministic Policy Gradient (PDDPG) algorithm. Since our end-to-end policy module can directly obtain the linear velocity and angular velocity through the raw sensor data, we can infer the execution time by computing the navigation velocity of the multi-robot system. There are two kinds of times in the Gazebo simulation platform: the simulation time and real time. The platform can accelerate the simulation or slow it down to fit specific tasks. To speed up the experimental efficiency, our work accelerates the simulation during training. Thus, we convert the simulation time to real time and compute the velocity of our multi-robot navigation system. As shown in Table 3, the group of robots navigates in four different kinds of trajectories with rectangular formation. The navigation distances, the time durations and the execution velocities are listed as follows. According to the results, we can infer that our method has an acceptable execution time.

Table 3. Experiments related to the navigation velocities of the multi-robot system.

Trajectory Type	Straight Line	L Shape	Rectangular Shape	S Shape
Distance (m)	12.73	18.50	35.50	35.50
Time (s)	5.01	8.45	18.49	18.98
Velocity (m/s)	2.54	2.19	1.92	1.87

5. Conclusions

This work mainly studied the collaborative formation and navigation of multi-robot system by using the deep reinforcement learning algorithm. By taking the raw 2D lidar sensor data and the relative target positions as inputs, the proposed Parallel Deep Deterministic Policy Gradient (PDDPG) algorithm could directly control the group of mobile robots to construct the formation and maintain it during navigation. The end-to-end policy module for mapless navigation was evaluated both in the single-robot situation and the multi-robot situation. Our experimental results demonstrated that the proposed method could teach the multi-robot system to learn human intuition and accomplish the collaborative navigation tasks with a high arrival rate.

Author Contributions: Conceptualization, W.C. and S.Z.; Data curation, S.Z.; Formal analysis, S.Z.; Funding acquisition, Y.L.; Investigation, S.Z. and H.Z.; Methodology, W.C. and S.Z.; Project administration, W.C., S.Z. and Y.L.; Resources, Z.P. and H.Z.; Software, S.Z.; Supervision, Z.P. and Y.L.; Validation, W.C. and S.Z.; Visualization, W.C.; Writing—original draft, W.C.; Writing—review and editing, W.C.

Funding: This work is supported by the National Natural Science Foundation of China under Grant U1509210, 61836015.

Conflicts of Interest: The authors declare no conflict of interest.

References

1. Cadena, C.; Carlone, L.; Carrillo, H.; Latif, Y.; Scaramuzza, D.; Neira, J.; Reid, I.; Leonard, J.J. Past, present, and future of simultaneous localization and mapping: Toward the robust-perception age. *IEEE Trans. Robot.* **2016**, *32*, 1309–1332. [CrossRef]
2. Bresson, G.; Alsayed, Z.; Yu, L.; Glaser, S. Simultaneous localization and mapping: A survey of current trends in autonomous driving. *IEEE Trans. Intell. Veh.* **2017**, *2*, 194–220. [CrossRef]
3. Karaman, S.; Frazzoli, E. Sampling-based algorithms for optimal motion planning. *Int. J. Robot. Res.* **2011**, *30*, 846–894. [CrossRef]
4. Schmerling, E.; Janson, L.; Pavone, M. Optimal sampling-based motion planning under differential constraints: The driftless case. In Proceedings of the 2015 IEEE International Conference on Robotics and Automation (ICRA), Seattle, WA, USA, 26–30 May 2015; pp. 2368–2375.
5. Li, Y.; Littlefield, Z.; Bekris, K.E. Sparse methods for efficient asymptotically optimal kinodynamic planning. In *Algorithmic Foundations of Robotics XI*; Springer: New York, NY, USA, 2015; pp. 263–282.
6. Janson, L.; Schmerling, E.; Clark, A.; Pavone, M. Fast marching tree: A fast marching sampling-based method for optimal motion planning in many dimensions. *Int. J. Robot. Res.* **2015**, *34*, 883–921. [CrossRef]
7. Webb, D.J.; Van Den Berg, J. Kinodynamic RRT*: Asymptotically optimal motion planning for robots with linear dynamics. In Proceedings of the 2013 IEEE International Conference on Robotics and Automation, Karlsruhe, Germany, 6–10 May 2013; pp. 5054–5061.
8. Kim, E.; Kim, J.; Sunwoo, M. Model predictive control strategy for smooth path tracking of autonomous vehicles with steering actuator dynamics. *Int. J. Automot. Technol.* **2014**, *15*, 1155–1164. [CrossRef]
9. Gáspár, P.; Szabó, Z.; Bokor, J. LPV design of fault-tolerant control for road vehicles. *Int. J. Appl. Math. Comput. Sci.* **2012**, *22*, 173–182. [CrossRef]
10. Doumiati, M.; Sename, O.; Dugard, L.; Martinez-Molina, J.J.; Gaspar, P.; Szabo, Z. Integrated vehicle dynamics control via coordination of active front steering and rear braking. *Eur. J. Control* **2013**, *19*, 121–143. [CrossRef]
11. Paden, B.; Čáp, M.; Yong, S.Z.; Yershov, D.; Frazzoli, E. A survey of motion planning and control techniques for self-driving urban vehicles. *IEEE Trans. Intell. Veh.* **2016**, *1*, 33–55. [CrossRef]
12. Biswas, J.; Veloso, M. Wifi localization and navigation for autonomous indoor mobile robots. In Proceedings of the 2010 IEEE International Conference on Robotics and Automation, Anchorage, AK, USA, 3–7 May 2010; pp. 4379–4384.
13. Nazemzadeh, P.; Fontanelli, D.; Macii, D.; Palopoli, L. Indoor localization of mobile robots through QR code detection and dead reckoning data fusion. *IEEE/ASME Trans. Mechatron.* **2017**, *22*, 2588–2599. [CrossRef]
14. Mnih, V.; Kavukcuoglu, K.; Silver, D.; Rusu, A.A.; Veness, J.; Bellemare, M.G.; Graves, A.; Riedmiller, M.; Fidjeland, A.K.; Ostrovski, G. Human-level control through deep reinforcement learning. *Nature* **2015**, *518*, 529. [CrossRef]
15. Van Hasselt, H.; Guez, A.; Silver, D. Deep reinforcement learning with double q-learning. In Proceedings of the Thirtieth AAAI Conference on Artificial Intelligence, Phoenix, AZ, USA, 12–17 February 2016.
16. Wang, Z.; Schaul, T.; Hessel, M.; Van Hasselt, H.; Lanctot, M.; De Freitas, N. Dueling network architectures for deep reinforcement learning. *arXiv* **2015**, arXiv:1511.06581.
17. Schaul, T.; Quan, J.; Antonoglou, I.; Silver, D. Prioritized experience replay. *arXiv* **2015**, arXiv:1511.05952.

18. Fortunato, M.; Azar, M.G.; Piot, B.; Menick, J.; Osband, I.; Graves, A.; Mnih, V.; Munos, R.; Hassabis, D.; Pietquin, O.; et al. Noisy networks for exploration. *arXiv* **2017**, arXiv:1706.10295.
19. Dabney, W.; Rowland, M.; Bellemare, M.G.; Munos, R. Distributional reinforcement learning with quantile regression. In Proceedings of the Thirty-Second AAAI Conference on Artificial Intelligence, New Orleans, LA, USA, 2–7 February 2018.
20. Hessel, M.; Modayil, J.; Van Hasselt, H.; Schaul, T.; Ostrovski, G.; Dabney, W.; Horgan, D.; Piot, B.; Azar, M.; Silver, D. Rainbow: Combining improvements in deep reinforcement learning. In Proceedings of the Thirty-Second AAAI Conference on Artificial Intelligence, New Orleans, LA, USA, 2–7 February 2018.
21. Silver, D.; Huang, A.; Maddison, C.J.; Guez, A.; Sifre, L.; Driessche, G.V.D.; Schrittwieser, J.; Antonoglou, I.; Panneershelvam, V.; Lanctot, M. Mastering the game of Go with deep neural networks and tree search. *Nature* **2016**, *529*, 484. [CrossRef]
22. Browne, C.B.; Powley, E.; Whitehouse, D.; Lucas, S.M.; Cowling, P.I.; Rohlfshagen, P.; Tavener, S.; Perez, D.; Samothrakis, S.; Colton, S. A survey of monte carlo tree search methods. *IEEE Trans. Comput. Intell. AI games* **2012**, *4*, 1–43. [CrossRef]
23. Silver, D.; Schrittwieser, J.; Simonyan, K.; Antonoglou, I.; Huang, A.; Guez, A.; Hubert, T.; Baker, L.; Lai, M.; Bolton, A.; et al. Mastering the game of Go without human knowledge. *Nature* **2017**, *550*, 354. [CrossRef] [PubMed]
24. Silver, D.; Hubert, T.; Schrittwieser, J.; Antonoglou, I.; Lai, M.; Guez, A.; Lanctot, M.; Sifre, L.; Kumaran, D.; Graepel, T.; et al. Mastering chess and shogi by self-play with a general reinforcement learning algorithm. *arXiv* **2017**, arXiv:1712.01815.
25. Kalashnikov, D.; Irpan, A.; Pastor, P.; Ibarz, J.; Herzog, A.; Jang, E.; Quillen, D.; Holly, E.; Kalakrishnan, M.; Vanhoucke, V.; et al. Qt-opt: Scalable deep reinforcement learning for vision-based robotic manipulation. *arXiv* **2018**, arXiv:1806.10293.
26. Kober, J.; Bagnell, J.A.; Peters, J. Reinforcement learning in robotics: A survey. *Int. J. Robot. Res.* **2013**, *32*, 1238–1274. [CrossRef]
27. Tai, L.; Zhang, J.; Liu, M.; Boedecker, J.; Burgard, W. A survey of deep network solutions for learning control in robotics: From reinforcement to imitation. *arXiv* **2016**, arXiv:1612.07139.
28. Kretzschmar, H.; Spies, M.; Sprunk, C.; Burgard, W. Socially compliant mobile robot navigation via inverse reinforcement learning. *Int. J. Robot. Res.* **2016**, *35*, 1289–1307. [CrossRef]
29. Pfeiffer, M.; Schwesinger, U.; Sommer, H.; Galceran, E.; Siegwart, R. Predicting actions to act predictably: Cooperative partial motion planning with maximum entropy models. In Proceedings of the 2016 IEEE/RSJ International Conference on Intelligent Robots and Systems (IROS), Daejeon, Korea, 9–14 October 2016; pp. 2096–2101.
30. Zhu, Y.; Mottaghi, R.; Kolve, E.; Lim, J.J.; Gupta, A.; Fei-Fei, L.; Farhadi, A. Target-driven visual navigation in indoor scenes using deep reinforcement learning. In Proceedings of the 2017 IEEE International Conference on Robotics and Automation (ICRA), Singapore, 29 May–3 June 2017; pp. 3357–3364.
31. Kolve, E.; Mottaghi, R.; Gordon, D.; Zhu, Y.; Gupta, A.; Farhadi, A. Ai2-thor: An interactive 3d environment for visual ai. *arXiv* **2017**, arXiv:1712.05474.
32. Zhang, J.; Springenberg, J.T.; Boedecker, J.; Burgard, W. Deep reinforcement learning with successor features for navigation across similar environments. In Proceedings of the 2017 IEEE/RSJ International Conference on Intelligent Robots and Systems (IROS), Vancouver, BC, Canada, 24–28 September 2017; pp. 2371–2378.
33. Mirowski, P.; Pascanu, R.; Viola, F.; Soyer, H.; Ballard, A.J.; Banino, A.; Denil, M.; Goroshin, R.; Sifre, L.; Kavukcuoglu, K.; et al. Learning to navigate in complex environments. *arXiv* **2016**, arXiv:1611.03673.
34. Chen, Y.F.; Everett, M.; Liu, M.; How, J.P. Socially aware motion planning with deep reinforcement learning. In Proceedings of the 2017 IEEE/RSJ International Conference on Intelligent Robots and Systems (IROS), Vancouver, BC, Canada, 24–28 September 2017; pp. 1343–1350.
35. Long, P.; Liu, W.; Pan, J. Deep-learned collision avoidance policy for distributed multiagent navigation. *IEEE Robot. Automat. Lett.* **2017**, *2*, 656–663. [CrossRef]
36. Long, P.; Fanl, T.; Liao, X.; Liu, W.; Zhang, H.; Pan, J. Towards optimally decentralized multi-robot collision avoidance via deep reinforcement learning. In Proceedings of the 2018 IEEE International Conference on Robotics and Automation (ICRA), Brisbane, Australia, 21–25 May 2018; pp. 6252–6259.
37. Raileanu, R.; Denton, E.; Szlam, A.; Fergus, R. Modeling others using oneself in multi-agent reinforcement learning. *arXiv* **2018**, arXiv:1802.09640.

38. Yang, Y.; Luo, R.; Li, M.; Zhou, M.; Zhang, W.; Wang, J. Mean field multi-agent reinforcement learning. *arXiv* **2018**, arXiv:1802.05438.
39. Wen, Y.; Yang, Y.; Luo, R.; Wang, J.; Pan, W. Probabilistic recursive reasoning for multi-agent reinforcement learning. *arXiv* **2019**, arXiv:1901.09207.
40. Tacchetti, A.; Song, H.F.; Mediano, P.A.; Zambaldi, V.; Rabinowitz, N.C.; Graepel, T.; Botvinick, M.; Battaglia, P.W. Relational forward models for multi-agent learning. *arXiv* **2018**, arXiv:1809.11044.
41. Lillicrap, T.P.; Hunt, J.J.; Pritzel, A.; Heess, N.; Erez, T.; Tassa, Y.; Silver, D.; Wierstra, D. Continuous control with deep reinforcement learning. *arXiv* **2015**, arXiv:1509.02971.
42. He, K.; Zhang, X.; Ren, S.; Sun, J. Deep residual learning for image recognition. In Proceedings of the IEEE Conference on Computer Vision and Pattern Recognition, Las Vegas, NV, USA, 27–30 June 2016; pp. 770–778.
43. Bengio, Y.; Louradour, J.; Collobert, R.; Weston, J. Curriculum learning. In Proceedings of the 26th Annual International Conference on Machine Learning, Montreal, QC, Canada, 14–18 June 2009; pp. 41–48.
44. Zamora, I.; Lopez, N.G.; Vilches, V.M.; Cordero, A.H. Extending the openai gym for robotics: A toolkit for reinforcement learning using ros and gazebo. *arXiv* **2016**, arXiv:1608.05742.

Article

A Survey on Robotic Technologies for Forest Firefighting: Applying Drone Swarms to Improve Firefighters' Efficiency and Safety

Juan Jesús Roldán-Gómez [1,*], Eduardo González-Gironda [2] and Antonio Barrientos [2]

[1] Departamento de Ingeniería Informática, Escuela Politécnica Superior, Universidad Autónoma de Madrid, Francisco Tomás y Valiente, 11, 28049 Madrid, Spain

[2] Centro de Automática y Robótica (UPM-CSIC), Universidad Politécnica de Madrid, José Gutiérrez Abascal, 2, 28006 Madrid, Spain; eduardo.gonzalez.gironda@alumnos.upm.es (E.G.-G.); antonio.barrientos@upm.es (A.B.)

* Correspondence: juan.roldan@uam.es

Abstract: Forest firefighting missions encompass multiple tasks related to prevention, surveillance, and extinguishing. This work presents a complete survey of firefighters on the current problems in their work and the potential technological solutions. Additionally, it reviews the efforts performed by the academy and industry to apply different types of robots in the context of firefighting missions. Finally, all this information is used to propose a concept of operation for the comprehensive application of drone swarms in firefighting. The proposed system is a fleet of quadcopters that individually are only able to visit waypoints and use payloads, but collectively can perform tasks of surveillance, mapping, monitoring, etc. Three operator roles are defined, each one with different access to information and functions in the mission: mission commander, team leaders, and team members. These operators take advantage of virtual and augmented reality interfaces to intuitively get the information of the scenario and, in the case of the mission commander, control the drone swarm.

Keywords: robotics; multi-robot systems; swarms; drones; firefighting

Citation: Roldán-Gómez, J.J.; González-Gironda, E.; Barrientos, A. A Survey on Robotic Technologies for Forest Firefighting: Applying Drone Swarms to Improve Firefighters' Efficiency and Safety. *Appl. Sci.* **2021**, *11*, 363. https://doi.org/10.3390/app11010363

Received: 29 November 2020
Accepted: 25 December 2020
Published: 1 January 2021

1. Introduction

Forest fires are one of the most common and, at the same time, serious emergencies facing humanity. They threaten not only natural areas, where they cause important losses of plant and animal diversity, but also urban areas, where they can cause dramatic human and material losses. Furthermore, forest fires cause significant emissions of greenhouse gases and consequently are contributing to global warming. For these and other reasons, states develop policies for fire prevention, early detection, and rapid intervention.

Quantifying the fires and their consequences along the world is not a trivial task. According to the World Fire Statistics [1], a report published by the International Association of Fire and Rescue Services that collects data from multiple governments, there were 4.5 million fires and 30,800 deaths in countries with 2700 million inhabitants in 2018, which means 1.7 fires per 1000 and 1.1 deaths per 100,000 inhabitants that year. Although these figures do not take into account the whole world, they allow us to quantify the magnitude of the problem. The information provided by several space agencies supports this thesis: the European Space Agency (ESA) publishes the World Fire Atlas with the information collected by ATSR-2 [2] and Sentinel-3 [3]), whereas the National Aeronautics and Space Administration (NASA) does the same with the Global Fire Atlas [4].

Current forest firefighting missions consider prevention, surveillance, and extinguishing tasks. The first ones seek to prevent the occurrence of fires and limit their consequences, the second ones look for detecting fires early, and the third ones search to put them out quickly and safely. Firefighters reveal the lack of human and material means and the degraded information of the scenario as the main problems in these tasks. They routinely use multiple types of vehicles and machinery to improve the performance and safety of these operations. However, the use of robots and especially drones is not common, although these autonomous systems could solve some of the current challenges.

This paper aims at analyzing the current problems in forest firefighting missions and the potential of robotic technologies to solve them. Therefore, we pose the following two research questions:

1. What are the main problems in current forest firefighting missions?
2. How can robotic technologies contribute to solving them?

For this purpose, the paper analyzes the data provided by governments, the results of two original surveys on firefighters, and the literature on robotics applied to forest firefighting.

Finally, the paper proposes a concept of operation for the application of drone swarms to fire prevention, surveillance, and extinguishing tasks.

The remainder of the paper is organized as follows: Section 2 addresses the current situation of firefighting, analyzing the public statistics provided by multiple countries and presenting the results of our surveys and interviews to professionals. Section 3 collects various works developed in the context of academy and industry that apply robots to firefighting tasks. The concept of operations using drone swarms to support firefighters in all these tasks is presented in Section 4. Finally, the main conclusions of the work are summarized in Section 5.

2. Firefighting State

This section analyzes the current state of firefighting. For this purpose, it describes current operations of fire prevention, surveillance, and extinguishing, collects relevant statistics to identify main problems, and presents the opinions of professionals through two surveys. Note that most of the information presented in this section is from Spain, but can be generalized to at least European and Mediterranean countries.

In 2019, 10,883 fires burned 83,963 ha in Spain: 7290 of these fires affected less than 1 ha, whereas 3593 affected more than 1 ha [5]. As reported, these figures were similar to the previous years, having an average of 12,182 fires and 99,082 ha per year between 2009 and 2018. In other words, every year, 0.356% of the forest surface of Spain suffers the consequences of fires.

Most of these fires occur in spring and summer, especially in March, July, August and September, whereas the worse consequences occur in July and August when more surface burns than the rest of the year [6]. In the case of summer, this behavior can be explained by the high temperatures, which favor the appearance of fires and their spread throughout the territory. In the case of March, most of these fires occur in the north-west of the country and are caused by an accidental, negligent, or intentional use of fire. However, firefighting is performed throughout the year, since it involves not only extinguishing fires but also preventing them.

Fire prevention involves a set of activities that seek to reduce the probability of fire occurrence, as well as to limit their effects if they occur [7]. More than half of forest fires in Spain between 2006 and 2015 were caused intentionally, whereas 28% were caused by accidents or negligent behaviors, 7% had natural origins, and 12% still have unknown causes. Therefore, there are two main groups of activities: prevention on causes and prevention on combustibles. The first category groups those activities that seek to reduce the risks and usually present a social character, such as the awareness campaigns to avoid the use of fire in the primary sector and negligent behaviors in natural environments. The second one covers the actions performed on land uses and vegetation distribution, which seek to generate discontinuities to prevent the expansion of potential fires.

Fire surveillance involves the activities performed to detect fires as early as possible. The damages caused by forest fires highly depend on detection and response times. The information of Spain in 2019 is clear: the average burned surface when response time was shorter than 1 h was 7.10 ha, whereas the one when response time was longer than 1 h was 30.66 ha [5]. For this reason, minimizing detection and response times is key for firefighting.

Currently, detection times are addressed with a watchmen network distributed throughout the land and, to a lesser extent, ground and aerial mobile surveillance. In Spain, 60% of fires are detected thanks to citizen collaboration, 27% by static watchmen, 1.6% by mobile watchmen, and 0.5% by aerial means [6]. Meanwhile, response times are addressed by the effective coordination of the teams and the use of helicopters to deploy firefighters in the affected area.

Fire extinguishing involves not only the actions performed to put out the flames but also some activities that support these actions, such as creating firewalls, routes for entry and exit of vehicles, runways, heliports, etc. In Spain between 2006 and 2015, these activities involved the participation of humans (100% of fires), ground vehicles and machinery (94.8%), and aerial means (23.5%) [6]. Extinguishing operations are dangerous because any accident can cause injuries or even deaths among the professionals. The government of Spain reports 24 accidents between 2006 and 2015 with 37 deaths, including only firefighting professionals [6]. According to this study, the causes of these deaths were air accidents (43%), entrapments (30%), medical problems (8%), falls (8%), accidents with vehicles (5%), and accidents with machinery (5%). Therefore, it would be good if technological solutions could reduce both accident rate and mortality in the cases of entrapments and falls, which can be caused by the lack of information about fire evolution and terrain features.

We performed a set of surveys and interviews with firefighting professionals to check and broaden this information. The surveys allowed us to involve a high number of professionals and distinguish collective consensus from individual opinions. Meanwhile, the interviews were done before and after the surveys: the first ones allowed us to prepare the questions, whereas the second ones provided us with more details about the answers. These activities aimed to collect information about current problems of firefighting and opinions about potential technological solutions.

Two surveys were carried out with firefighting professionals: one focused on their problems at work (see Section 2.1),and another on their opinion about multiple technologies (see Section 2.2). Both surveys had between 10 and 20 questions and required fewer than 5 min to maximize the answer ratio. The separation of problem and technology surveys prevented the influence of the questions of one on the responses of the other.

The dissemination of the surveys sought to reach professionals who perform all the firefighting roles in most of the regions of Spain. For this purpose, we sent the surveys by email to fire stations and firefighter unions, as well as share them in firefighting groups on various social networks. In this way, we avoided getting a sample biased towards a specific firefighting role or geographic area.

2.1. Problem Survey

Our first survey was focused on the problems on current forest firefighting missions. We performed this survey to obtain more information about the first research question. Although the official data previously analyzed are useful to answer this question, the opinions of the professionals involved in these activities are also relevant.

In this survey, we pose the following questions:

- Importance of prevention tasks: As previously mentioned, there are two prevention strategies: those focused on causes and those centered on combustibles. This question seeks the importance that professionals give to each one of these strategies.
- Problems in prevention tasks: This question seeks to find the most relevant problems in current prevention activities, according to the opinions of firefighters.
- Importance of surveillance means: As previously pointed out, forest fires can be detected by citizen collaboration, ground watchmen, and aerial means. This question seeks the importance that professionals give to each one of these means.
- Problems in surveillance tasks: This question seeks to find the most relevant problems in current surveillance activities, according to the opinions of firefighters.

- Problems in extinguishing tasks: This question seeks to find the most relevant problems in current extinguishing activities, according to the opinions of firefighters.

We had the support of several firefighting professionals in the writing of the questions and their possible answers. In this way, we could check that our surveys were sound and easy to understand by our target public. Furthermore, we sent the questionnaires to a sample of twenty professionals before their dissemination to check if they could understand them adequately. Finally, we allowed open answers to some questions and shared our contact data to receive doubts.

This survey was sent via email to fire stations, unions, and associations, as well as shared with firefighters' communities on several social networks. A total of 140 professionals from different regions of Spain took part in that survey in three weeks (note that this survey is still open to new responses (Forest fires in Spain: Problem survey (https://forms.gle/e4327HBxqqWVMUbY7) [in Spanish])). A summary of the results is shown in Figure 1.

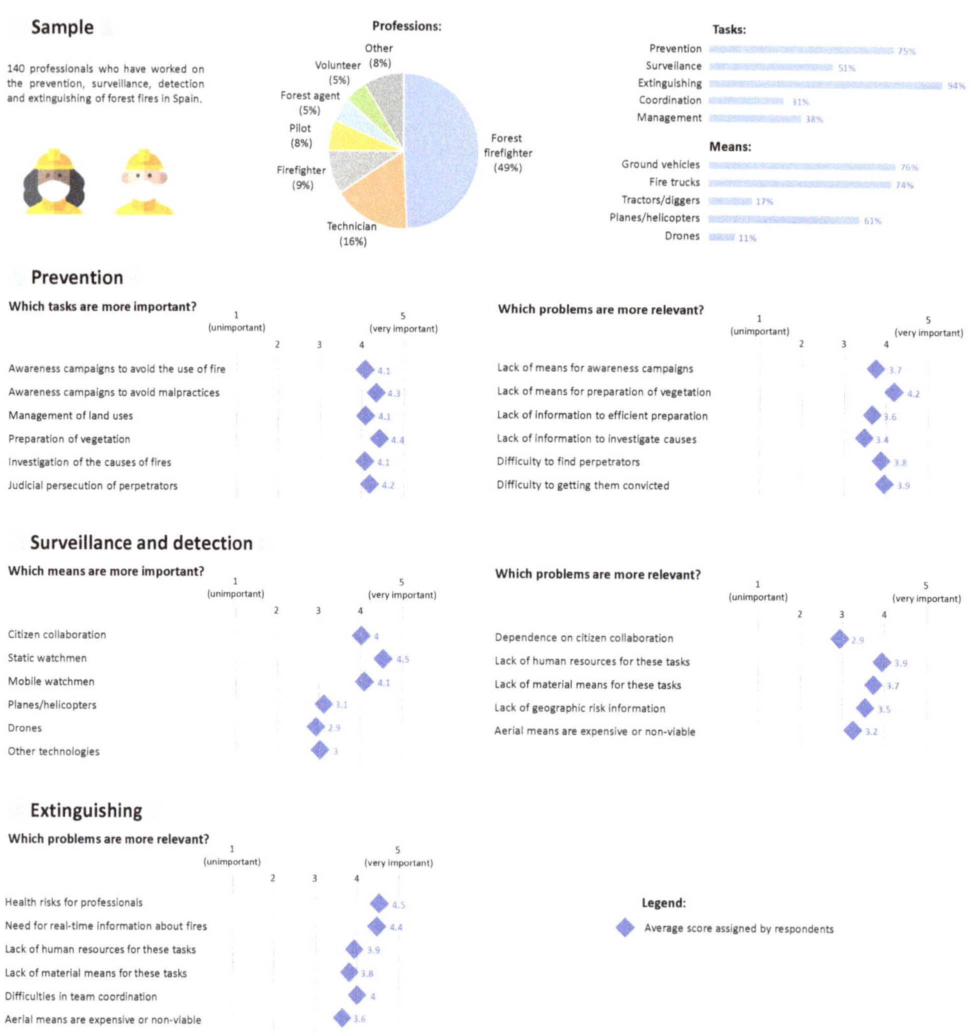

Figure 1. Results of the problem survey about forest fires in Spain.

The survey sample is representative of the professionals involved in forest firefighting in Spain, given that it includes not only the main roles (forest firefighters, technicians, firefighters, forest agents, and volunteers) but also other less common roles (pilots, army forces, researchers, support staff, meteorologists...), in comparison to the official reports [5]. The vast majority of them take part in extinguishing tasks (94%), whereas three quarters have experience in prevention and a half in surveillance tasks. Besides these fundamental tasks, around one-third of the respondents have carried out the coordination of operations (31%) and management of human and material resources (38%). Regarding the means used to perform these tasks, three quarters used ground vehicles and fire trucks, sixty percent aerial vehicles (this includes pilots and airborne firefighters), and fewer tractors (17%) and drones (11%).

Spanish firefighters often say that "summer fires should be extinguished in winter", remarking on the importance of prevention activities in firefighting. In this sense, the professionals surveyed assign very close scores to all the prevention tasks. Mainly, they give a slightly higher score to the preparation of vegetation (4.4 in a scale from 1 to 5), and prioritize awareness campaigns against malpractices (e.g., barbecue, smoke or throwing glass bottles in the bush) over campaigns against the use of fire in agricultural and livestock activities (4.3 vs. 4.1, respectively). There are more differences between the main problems faced in prevention tasks. They highlight the lack of means for preparing the vegetation (4.2), together with the difficulty to find perpetrators (3.8) and convict them (3.9).

Regarding surveillance and detection tasks, the participants highlight static watchmen (4.5), mobile watchmen (4.1), and citizen collaboration (4). Note that this evaluation does not coincide with the actual situation, given that sixty percent of fires are detected by citizen collaboration, whereas only twenty-six percent are detected by static watchmen and less than two percent by mobile watchmen. Aerial means are considered less important for fire surveillance and detection: planes and helicopters receive 3.1 points and drones 2.9 points. According to their opinions, the main problem in these tasks is the lack of human resources (3.9), followed by the lack of material means (3.7), and the lack of risk information, which would allow reinforcing surveillance in areas with a higher risk of fire.

Finally, the professionals surveyed consider health risks and the need for real-time information as the main problems in extinguishing tasks with 4.5 and 4.4 points, respectively. Both problems are closely related, considering that most accidents are caused by the lack of information about the fire evolution, such as entrapments and falls. Other relevant problems are the lack of human and material resources (3.9 and 3.8, respectively), and the difficulties to coordinate the teams on the ground (4).

2.2. Technology Survey

Our second survey was focused on some technologies that can contribute to solving the reported problems. In this case, we performed this survey to obtain more information about the second research question. The objective was to collect opinions from professionals to estimate their predisposition to use these technologies.

For this purpose, the survey included the target technologies of this study (drone swarms and immersive interfaces), together with some control technologies. These technologies were chosen after a review of research and the commercial literature and served as a reference in the evaluation of target technologies. Incentive systems

- Prevention: The survey considers a solution of prevention on causes (incentive systems for farmers/ranchers to prevent their use of fire) and two solutions of prevention on combustibles (drone and satellite images to support the preparation of vegetation). In this way, two comparisons can be performed: one among the two strategies for prevention, and another between the two technologies that support the vegetation preparation.
- Surveillance: The survey considers two detection systems: one with drones and another with fixed cameras. In this way, the target technology can be compared with a well-known and widely-used surveillance system. Additionally, it includes the use

of artificial intelligence to predict the risk of fire, which allows performing this task over specific areas.

- Extinguishing: The survey asks about the application of drones to monitor the evolution of fires. In addition, it considers three alternatives to receive the information during field operations: an immersive interface, a mobile device, and a voice assistant. In this way, the target technology can be compared to two common methods to receive information.

We took the same measures as in the previous survey to ensure that questions and possible answers were understandable.

The participants of the first survey who gave their emails were invited to fill the second survey. In this case, a total of 70 professionals submitted their responses in the first three weeks (again, the survey is still open to new responses (Forest fires in Spain: Technology survey (https://forms.gle/XV4ScxL9jyCgr4fJ7) [in Spanish])). A summary of the results is shown in Figure 2.

Sample

70 professionals who have worked on the prevention, surveillance, detection and extinguishing of forest fires in Spain.

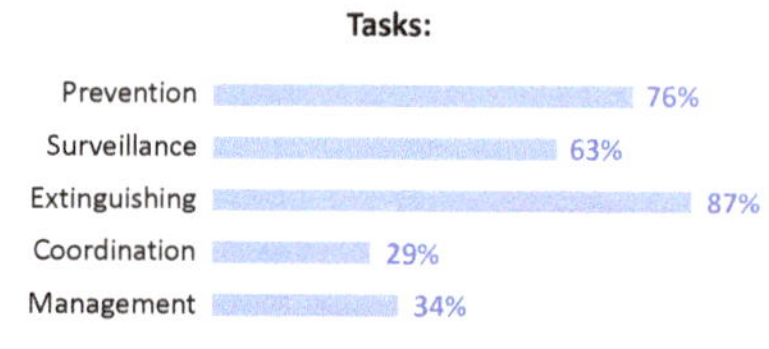

Technology

What is your opinion on the following technologies that can help in fighting fires?

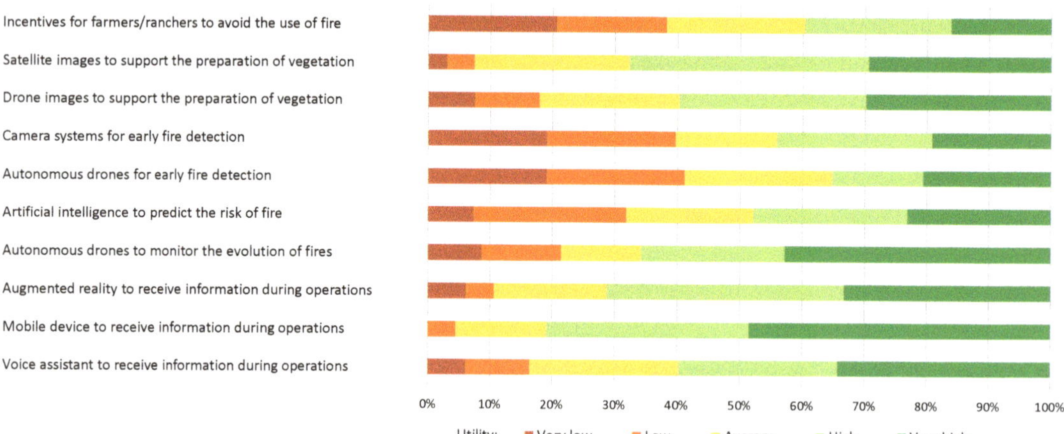

Figure 2. Results of the technology survey about forest fires in Spain.

As seen in Figures 1 and 2, the survey samples are very similar, only presenting small variations in the professions and tasks. The professionals could evaluate the utility of the different technologies with five ratings: "very high", "high", "average", "low", and "very low". However, we analyze the results considering three evaluations: positive (including "very high" and "high" answers), neutral (equivalent to "average" answer), and negative (including "low" and "very low" answers).

Three technological solutions for supporting prevention were evaluated: a system with incentives to avoid the use of fire in primary sector activities, satellite images to support the preparation of vegetation, and drone images for the same purpose. In this case, the professionals surveyed evaluate more positively the use of satellite and drone images (approximately, 60–70% positive, 20% neutral, and 10–20% negative).

Another three technological solutions were presented for surveillance and detection tasks: a system with cameras to monitor large and/or remote areas, autonomous drones to cover hard-to-reach areas, and the use of artificial intelligence to predict the risk of fire in every location. This last system received the best rating with 48% positive, 20% neutral, and 32% negative, whereas the other two received ratings of 35–44% positive, 16–24% neutral, and 40–41% negative.

A combination of two technologies was considered for extinguishing tasks: a first one for collecting real-time data from the fire, and a second one to display this data to the firefighter teams. For the first, we presented a fleet of autonomous drones that can monitor fire evolution in real-time. This solution is one of the best valued and the best that uses drones, having 66% positive, 12% neutral, and 22% negative evaluations. For the second, we posed three alternatives: augmented reality headset, mobile device, and voice assistant. The three alternatives are evaluated positively, but the mobile device receives the best score: 81% positive, 15% neutral, and 4% negative. Augmented reality receives 71% positive, 18% neutral, and 11% negative, whereas voice assistant achieves 60% positive, 23% neutral, and 17% negative. These results reveal that participants prefer visual over aural feedback and well-known over new devices, as well as they do not matter using their hands to manage these devices during operations.

A certain bias was detected in the evaluations of technologies by professionals, which positively affects those that are presented as a support for their current operations and negatively those that can change those operations or even threaten their jobs. This can be seen in detection, where they think that artificial intelligence can improve their effectiveness in surveillance, while they feel more threatened by autonomous systems with drones or cameras. Furthermore, the preference for well-known systems for receiving information (mobile devices) over more innovative ones (augmented reality) also reveals this conservative bias.

3. Firefighting Robots

Once we have analyzed the current state of firefighting and the opinions of professionals, we must address a new question: "can technology help to solve any of the presented problems?" This section collects the most relevant works that apply robotic and automation technologies to firefighting activities. Our analysis focuses on multi-robot systems and aerial robots used for the prevention, surveillance, and extinguishing of forest fires. However, relevant works that propose other types of robots and consider urban or indoor scenarios are also featured.

As previously occurred with industry, agriculture, and services, robots are being applied to intervene in emergencies and, more specifically, to fight against fires. According to our survey, firefighters are receptive to these technologies when they support their work and do not change its conditions. A previous study with fire chiefs of New Jersey (United States of America) supports these conclusions: they are willing to use drones in firefighting operations, but they point out budget, manpower, and regulation issues [8]. The public opinion about the use of drones for cargo, passenger, and commercial transportation is analyzed by [9], including explicitly firefighting in this last group of applications. The participants of this study support the use of drones for cargo and commercial applications, but they prefer piloted aircrafts for passenger transportation. Finally, a comprehensive survey on the public opinion about drones considering multiple applications and risks can be found in [10].

As most of the relevant articles focus on one or a few specific tasks, we have classified them into prevention (Section 3.1), surveillance (Section 3.2), and extinguishing

(Section 3.3). This classification is supported by several papers in the literature: for instance, Ref. [11] distinguishes between activities before fire (vegetation mapping, surveillance, and risk estimation), during fire (detection and extinguishing), and after fire (ember search and damage assessment).

3.1. Prevention

As already explained, prevention is considered the first step of firefighting and encompasses two classes of actions. The first ones involve social activities that seek to prevent fires from occurring, usually developing awareness campaigns targeting key groups as farmers or tourists. The second ones group multiple works on vegetation to reduce the risk of fire and generate discontinuities to difficult their propagation. Logically, the potential of robots to improve current results is higher in these latter operations. The preparation of vegetation is an activity that requires remarkable efforts, where a lack of human and material resources is perceived. Robotic technologies can make this activity more efficient in two ways.

On the one hand, drones can take aerial images that can be used to plan these tasks: detecting the most problematic areas, selecting the vegetation to remove, planning routes for its extraction, etc. Some techniques developed for precision agriculture can be applied in this context [12], such as the detection and identification of plants and trees in high-resolution images [13], three-dimensional LIDAR scans [14], and multispectral images [15] acquired by drones. In all these tasks, drones have been revealed as a suitable alternative to satellites, since they offer greater availability at a lower cost, as well as they are less dependent on the weather conditions in the area of interest [16].

On the other hand, ground robots can support the activities aimed at remove vegetation in forests, playing an intermediate role between the manual labor of firefighters and the heavy machinery used by them. These robots can reach a compromise between the flexibility and precision of firefighters and the quickness and performance of machinery. Forestry and agricultural robots share some challenges and requirements [17], such as the locomotion in rough terrains, localization and mapping in unstructured environments, and planning under uncertainty [18].

A comprehensive fire prevention solution is being developed in the SEMFIRE Project [19], which proposes a multi-robot system to reduce the fuel accumulation in forests and assist in landscaping maintenance. This system consists of small flying robots for vegetation mapping and large-sized tracked mobile robots for forestry mulching.

3.2. Surveillance

Fire surveillance is the most covered activity in the literature about robotics for firefighting. Most of the proposals involve the use of different kinds of aerial robots (fixed-wing and multi-rotor drones) equipped with various types of cameras (RGB, infrared, multispectral...) to watch over the forests from above. Fire surveillance tasks may have up to four objectives: search of potential fires, detection to alert firefighters, diagnosis to get relevant data about the fire, and prognosis to predict fire propagation [20]. The early detection of fire is as important as the complete analysis of it, given that firefighting teams need information such as the ignition and danger potential to organize their operations [21].

Unmanned aerial vehicles (UAVs) with on-board vision systems have considerable potential in the detection and monitoring of forest fires, since they offer high maneuverability, flexible perspective and resolution, and limited risks to people [22]. For this purpose, surveillance systems should integrate six elements: a fleet of UAVs with payloads, sensor fusion and image processing methods, guidance, navigation and control (GNC) algorithms, coordination and cooperation strategies, path planning algorithms, and ground control stations (GCS) [23]. The selected UAVs shall meet a set of requirements, such as long flight time, accurate localization with the data obtained by the Inertial Measurement Unit (IMU) and Global Navigation Satellite System (GNSS), stable and robust flight, and good image quality [24].

There are multiple approaches to develop vision systems to detect fires. The work in [25] comprehensively analyzes the potential sensors and methods for terrestrial, aerial, and satellite-based fire detection systems. Regarding the hardware, they use visible [26,27], thermal [28,29], multispectral [30,31] and infrared cameras [20,32], as well as environmental sensors (mostly used in indoor scenarios [33], but also proposed for forests [21]). Regarding the software, traditional computer vision algorithms [22,34] compete with recent artificial intelligence solutions [35,36]. The most common features used to recognize fires in aerial images are color, geometry, and movement. Color and geometry allow detecting potential fires in isolated frames, whereas movement is relevant to check these detections with the whole sequence of frames [22]. A challenge for these algorithms is adapting to different types of fires and scenarios: for instance, subterranean fires show up as columns of smoke, in contrast to common surface forest fires [37].

Heterogeneous multi-robot systems are also considered for fire surveillance. The work presented in [38] proposes an air-ground robotic team, where the Unmanned Ground Vehicles (UGVs) compensate for the weaknesses of UAVs, such as their limitations in autonomy (flight time) and payload (weight capacity). This work proposes the use of UGVs to transport UAVs to the fire scenario, where UAVs can take off, perform their tasks, and land again. Additionally, UGVs are used as base stations for UAVs, centralizing the communications between the fleet, processing the data collected by them, and coordinate their tasks in the scenario. Moreover, the work published in [35] proposes the use of two different types of drones: fixed-wing UAVs for medium-altitude flights searching fires and rotary-wing UAVs for low-altitude flights checking detections. The need for checking detection to avoid false alarms is also expressed in [39], which suggests the use of multiple drones to collect simultaneous information of every area, as well as the use of various features to detect fires in the provided images (e.g., color and movement).

3.3. *Extinguishing*

In general terms, fire extinguishing is the last task of firefighting after having detected and checked the fire. Currently, this task is mostly performed with ground and aerial means that require human intervention. Sometimes, the presence of humans results in risky situations for their lives due to the virulence of fires. Although this is a good reason to try to use robots in these tasks, these autonomous systems are only used experimentally. The literature considers two main approaches: one for aerial extinguishing and another for supporting ground operations.

The main idea of the firefighting drones is to attack the fire when it is in its first stages, trying to avoid the spread of it. An extinguishing quadcopter equipped with a bucket to capture and release water is presented in [40]. Although this design is similar to those used in current firefighting helicopters, the limitation in the payload capacity of the quadcopter reduces its performance.

An aerial hose-type robot that can fly directly into the fire source by a water jet is presented in [41]. This robot receives a continuous intake of water for fighting the fires and controlling its stability. In this way, it solves the limited payload issues of conventional drones, but it requires a water source close to the fire scenario. Another alternative is the use of gases instead of water. A quadcopter that carries a balloon filled with helium is proposed by [42]. This inert gas is used because it can reduce the amount of oxygen of the flames, as well as it is light enough to be transported by a quadcopter. The scalability of this system to forest fires must be validated, including the mechanism to release the balloons on the exact points.

A common idea for putting out fires is the utilization of extinguishing balls [43]. These elements burst when they come into contact with high temperatures, releasing some chemical components that put out the fire. Ref. [44] proposes a quadcopter that can launch an extinguishing ball to the flames of urban and wildfires. Following the same approach, Refs [45,46] propose some alternatives for the release mechanism to allow throwing multiple balls and keep the stability of the drone. Finally, ref. [47] poses a swarm of UAVs that can perform monitoring and extinguishing tasks, demonstrating the scalability of fire extinguishing systems based on drones that release balls.

Different types of multi-robot systems are proposed for fire extinguishing missions. There is a trend in the literature to apply multiple light robots instead of developing drones with the capabilities of planes and helicopters. For instance, a drone fleet is proposed in [48] and a drone swarm in [49]. When multiple drones work in the same scenario, the coordination of the fleet becomes relevant. The literature contains various proposals of algorithms to allocate targets among the drones, seeking to minimize traveling distance for every drone. Some examples are [11], which proposes that the team shares all the information of the mission and runs an auction-based mechanism to distribute the tasks, and [50], which describes a deep learning method to allocate tasks, overcoming the sensing, communication, and motion limitations of drones.

In addition to fire extinguishing tasks, robots can be used to monitor fires and provide information to firefighters. Ref. [51] describes a novel algorithm for safe human-robot coordination in wildfires. The drones track the evolution of fires, which can be stationary, moving, and moving/spreading, and a human safety module detects if there are humans close to fire spots. Moreover, ref [52] three types of drones to perform patrolling, confirmation, and monitoring tasks, as well as a fire-spreading model to use the information collected from the fires to predict their behavior.

4. System Overview

After analyzing the current state of firefighting operations and proposals of robotic systems to perform them, we present a comprehensive concept of operation to apply drone swarms in firefighting missions. This concept of operation is shown in Figure 3 and described in the following subsections: mission in Section 4.1, drone swarm in Section 4.2, team in Section 4.3, and required infrastructure in Section 4.4.

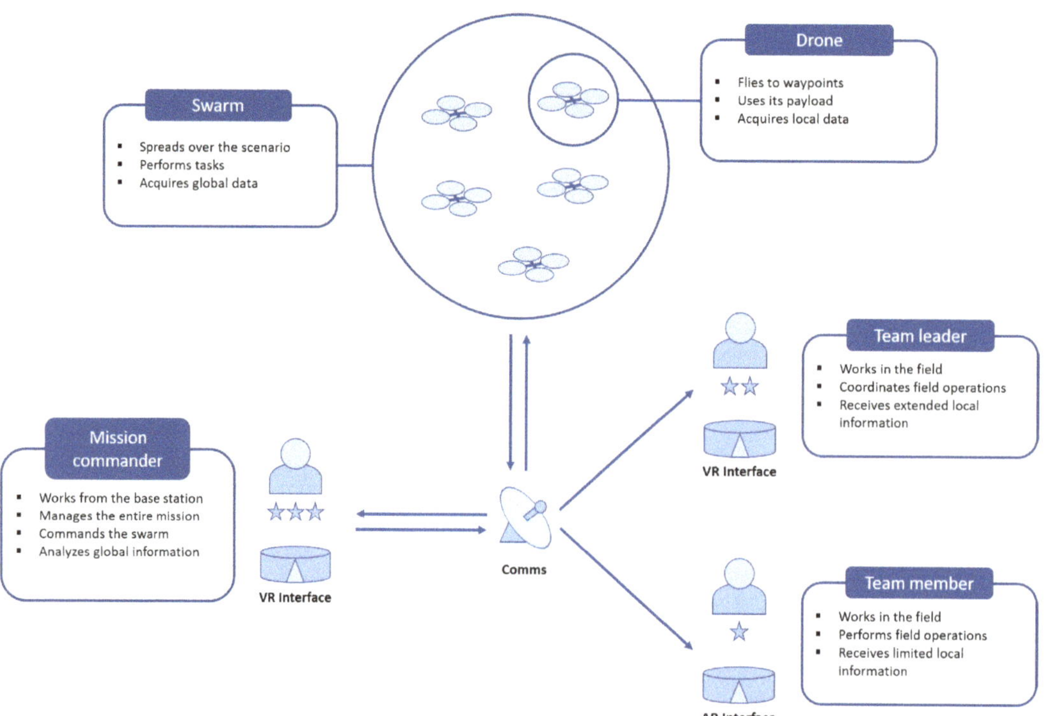

Figure 3. System overview.

4.1. Mission

The mission has been designed based on current firefighting operations and including research contributions addressed in Sections 2 and 3, respectively. It considers the tasks that could require the participation of the drone swarm, but excludes aerial extinguishing because it would need other types of drones currently in development.

- Prevention: This phase groups the tasks that seek to avoid fires from occurring and control their spread.

  Vegetation mapping: In this task, the drones fly over an area of interest to take ground pictures and build a vegetation map. The number of drones, flight pattern and altitude, and other variables can be tuned to efficiently cover the area and obtain high-quality images. The drones must integrate conventional and multispectral cameras to perform this task. The base station processes images, build a mosaic, detect trees and plants, and recommend actions to the firefighters.

  Fire investigation: This task is developed after the fire is detected. The objective is to find evidence to identify and pursue the perpetrators of the fire. For this purpose, the drones must search around the fire to detect suspicious people, objects, and situations, monitoring static targets, and tracking mobile targets. Although this task is performed after the fire has occurred and the drones have detected it, it is considered a prevention task because it can prevent the occurrence of more outbreaks of the fire. In practice, few drones can perform this task while the rest are carrying out extinguishing tasks.

- Surveillance: This phase considers the tasks that seek to detect fires and alarm firefighting teams early.

  Risk mapping: This task is very similar to vegetation mapping, but creating a map with the risk of fire. This map is useful to know in which areas there is more probability of fire and reinforce surveillance over them. The drones must be equipped with conventional and thermal cameras to perform this task.

  Fire surveillance: In this task, the drones fly over an area of interest looking for potential fires. When one of the drones detects a possible fire, this or another drone must fly closer to check it. For this purpose, the drones must integrate conventional and thermal cameras, as well as environmental sensors: temperature, humidity, and concentrations of combustion gases.

- Extinguishing: This phase groups the task aimed at extinguishing fires and supporting firefighters.

  Fire monitoring: This task is performed to collect information about the fire while the teams on the ground extinguish it. Spatial and temporal information is useful to know the outline of the fire, locate new sources, and predict its evolution. For this purpose, the drones must fly around the fire to incorporate new information from the periphery while keeping updated information from the center. This task needs the same equipment in the drones as risk mapping and fire surveillance.

  Firefighter support: This task aims at supporting the firefighters that are working on the ground to extinguish the fire. For this purpose, the drones must fly around the firefighting teams to collect data about their surroundings and recommend them safe paths and effective actions. Additionally, the drones can transport light resources to firefighters, such as communication devices and protection equipment.

4.2. Drone Swarm

The mission described in the previous section can be addressed by several types of aerial robot systems. The first approach is using a heterogeneous drone fleet, so different types of drones can adapt to different types of tasks, increasing the efficiency of the whole mission. For instance, fixed-wing drones can do the tasks that require covering large areas as surveillance and mapping, whereas rotary-wing drones can do the tasks that require stationary flights as monitoring and support. However, our proposal involves the use of a homogeneous drone swarm to solve this mission. This system relies on the cooperation between drones to accomplish the tasks and not on the adaptation of them to specific tasks. In this case, the same type of drones can perform surveillance and monitoring, but in a different number.

Both systems have advantages and disadvantages in the defined scenario. As already mentioned, heterogeneous fleets can optimize the missions by allocating their different resources to different tasks. Additionally, these systems are easier to control because the drones have more capabilities and need less coordination. On the other hand, drone swarms are more scalable and have more flexibility to adapt to the changes in the scenario. Besides, these systems have better fault tolerance because they can recover from losing one or more members.

We contemplate the definition of robot swarm drawn from [53]: a robot swarm is a group of simple robots, which individually can only perform rudimentary actions, but collectively form an intelligent system and can perform complex tasks. Therefore, we consider a quadcopter fleet as a robot swarm when the fleet consists of a dozen robots, single robots cannot cover the target scenarios, and individual robots are not able to perform the considered tasks. Firefighting missions involve large and complex terrains, where single quadcopters can only collect local information and perform simple actions.

The quadcopters considered for this application shall have the following features:

- Size and weight: No more than 1600 × 1600 × 800 mm unfolded and 15 kg including drone and payload.
- Autonomy: A minimum of 30 min of flight.
- Navigation: Fusion of IMU measurements, visual odometry and GPS/GLONASS/GALILEO signal.
- Control: Capability of reaching and hovering on waypoints.
- Communications: Telemetry and video links in a range of 5 km.
- Payload: Conventional, thermal, and multispectral cameras, as well as temperature, humidity, and gas sensors.

The size and weight were established looking for a compromise between versatility and load capacity. On the one hand, the drones must be light enough to be transported to the fire area in a vehicle and deployed in the field by a person. On the other hand, they must carry up to three cameras, environmental sensors, and communication devices. Finally, we have taken into account the impact of these parameters on flight range and maneuverability. Furthermore, autonomy is an essential aspect of the system: practically, the longer the flight time of the drones, the better the viability of the system in real missions. Current high-performance commercial drones offer around 30 min of continuous flight, but this figure may increase in the following years.

The navigation capabilities of the drones are another relevant aspect of the operation of the system. We have chosen to combine multiple sources to get high accuracy and fault tolerance. Specifically, we consider a high-performance IMU to provide linear acceleration, rotation speed, and orientation, as well as a GNSS receiver to obtain the position, velocity, and time with high frequency. Additionally, on-board cameras can get terrain features, which allow estimating drone motion. Multiple models can integrate the data provided by these sources to obtain the accurate location of the drone, such as Kalman [54] and particle filters [55]. In this way, the drones can preserve enough autonomy to perform their tasks even in GNSS denied or limited environments.

Finally, communications are often a challenge to apply drones in large and distant scenarios. In fire fighting missions, there must be a continuous exchange of information between the different agents: data from the drones to the base station, commands from the base station to the drones, information from the base station to the firefighters, etc. Our proposal to maintain these communications during the missions is to use the vehicles and robots involved in them as communications relays. However, we estimate that the drones must have a communication range of 5 km to enable this system in the considered scenarios.

As shown in Figure 3, each quadcopter is only able to fly to waypoints and use its payload, whereas the whole fleet can spread over the scenario and perform the required tasks. For instance, a quadcopter can move through a list of waypoints taking images of the terrain, whereas the fleet can cover the whole area monitoring the evolution of the fire. It is made possible thanks the control and coordination algorithms executed by the drones, which allow them to make individual decisions based on local data that produce collective behaviors to perform global tasks. The most representative are behavior-based algorithms, whose efficiency has been validated for surveillance, search, and monitoring tasks in previous works [53,56,57].

Behavior-based algorithms usually consist of multiple behaviors, which process the information and generate possible actions following different patterns, and decision-making module, which fuses the outputs of them and computes the final action. Some common behaviors are inspired in nature, such as "keep distance" and "keep velocity", which are followed by birds' flocks and fishes' shoals. However, some others are devoted to solving specific robot tasks, such as search and surveillance. In both cases, the behaviors have multiple parameters that can be tuned to adapt them to different scenarios.

The drone swarm shall perform the following generic tasks partially drawn from [58]:

- Search: This task involves flying over an area of interest to find some targets, covering every point in that area at least once.
- Surveillance: This task involves flying over an area of interest to find some targets, covering every point multiple times to get updated data.
- Reconnaissance: This task involves flying to a list of points of interest to acquire data.
- Mapping: This task involves flying over an area of interest to build a map, covering every point once to acquire images or data.
- Monitoring: This task involves flying over an event of interest to acquire data.
- Support: This task involves flying over teams that work on the ground to provide them with information about their environment.
- Tracking: This task involves following a mobile target to acquire information or control it.
- Transport: This task involves taking a load from one point to another.

These generic tasks can be used individually or in combination to represent the specific tasks of firefighting missions described above. For instance, fire surveillance can be represented as a combination of surveillance and reconnaissance having fires as targets. The specific tasks and their corresponding generic tasks are collected in Table 1.

Table 1. List of tasks considered for firefighting missions.

Missions	Specific Tasks	Generic Tasks
Prevention	Vegetation mapping	Mapping
	Fire investigation	Search, Monitoring, Tracking
Surveillance	Risk mapping	Mapping
	Fire surveillance	Surveillance, Reconnaissance
Extinguishing	Fire monitoring	Monitoring, Search
	Firefighter support	Support, Transport

4.3. Team

Regarding the crew, we consider three principal roles: mission commander, team leaders, and team members. There can be other roles according to the mission and scenario, such as analysts, maintenance workers, communications technicians, etc. As shown in Figure 3 and described below, each role entails different functions, access to information, workplace, and available actions.

- Mission commander: They monitor and controls the mission from the base station, which does not have to be in the fire scenario. All the data collected by the drones is received in the base station and processed to obtain valuable information. Therefore, the mission commander has access to full information on the mission, including the telemetries of drones and measurements on the fire. They must use this information to manage the mission, coordinating the teams on the ground and commanding the swarm. The drone swarm is controlled through high-level commands (e.g., defining areas of interest, variables that must be measured, and required tasks) instead of through low-level orders (e.g., sending specific waypoints and actions to specific drones). This feature is one of the most remarkable strengths of robot swarms, which can self configure to accomplish tasks most accurately, efficiently, and safely. Finally, the mission commander communicates with the team leaders to deliver high-level orders for their teams, establishing the areas where they must work, the tasks that they must perform, and the resources that they can use.
- Team leader: They work in the fire scenario, preferably in a facility or vehicle to ensure communications with the base station. The task of a team leader is to coordinate the field operations of a firefighting team. For this purpose, they receive high-level orders from the mission commander (e.g., area of work and tasks to be performed) and sends low-level commands to the team (e.g., move along a path and attack some flames). In this role, local information is managed both geographically and functionally, that is, the events that happen in the work area and affect the performed tasks.
- Team members: They work in the fire scenario, executing prevention, surveillance, and extinguishing tasks. For this purpose, they can exercise their workforce or use different types of vehicles and machinery. They have access to limited local information, mainly related to the paths that must follow and the actions that must perform. The amount of information should be limited to avoid distractions, but should be enough to ensure their safety.

4.4. *Infrastructure*

A minimal infrastructure is required for the operation of the system. This infrastructure consists of multiple elements that sustain the autonomy of the swarm, enable the communications among the agents, and allow the human-swarm interaction.

As mentioned above, autonomy is a major challenge for applying drone swarms to firefighting missions. Some of the tasks imply continuous flights over target areas, such as fire surveillance and monitoring, whereas some others require a rapid deployment there, such as fire investigation and firefighter support. Therefore, the drones must be able to charge their batteries in the scenario to increase their availability during the missions. For this purpose, charging stations can be distributed throughout the scenario, even using the ground vehicles involved in the mission.

Adaptive and immersive interfaces can improve the situational awareness and reduce the workload of operators in the considered mission. These results have been validated in similar missions, such as the control of multiple robots to perform complex missions [59] and the analysis of the information collected by a drone swarm from a smart city [57].

These interfaces adapt their displays to the mission state and operator preferences, in order to reduce the amount of information and the workload of operator. For this purpose, they can integrate mission and operator models. The first ones allow following the state of the mission and selecting the relevant information according to it, whereas the second ones allow adapting the interface to the operator preferences. The adaptation can be performed through artificial intelligence models like neural networks.

These interfaces apply immersive technologies like virtual reality (VR), augmented reality (AR), and mixed reality (MR) to introduce the operator in the scenario, improving their perception of the environment where the robots are working. VR reproduces virtual environments and allows interacting with their elements; AR enhances real environments with virtual elements with which the operator can interact, and MR combines real and virtual elements and allows interacting with them [60].

In this work, we consider VR interfaces for the mission commander and AR interfaces for team leaders and members. The mission commander works away from the scenario, so they can focus on the information from the mission. A VR interface can reproduce the scenario, incorporating the real-time information of the swarm and its environment, allowing the operator to move around the scene searching for the best point of view. Meanwhile, team leaders and members work in the scenario, so they must pay most of their attention to the mission. In this case, an AR interface can provide them with relevant information about the mission while keeping their attention in their environment.

5. Conclusions

This paper analyzes the current state of firefighting missions and potential technologies that can be applied in the future. To this end, we have conducted two surveys of firefighters to know the main problems they face in their work and their point of view on possible technological solutions. According to the results, the most common problems are the lack of human and material resources for all the activities and the need for real-time information about the evolution of fires during extinguishing tasks. The proposed technologies are positively evaluated when they support their tasks and do not threaten their jobs. Specifically, firefighters support the use of drones as a tool to collect relevant information for prevention, surveillance, and extinguishing activities. In the cases of prevention and surveillance, they approve the generation of maps that help to organize the tasks for preparing vegetation and detect the areas with the highest risk of fires, respectively. In the case of extinguishing, they consider that drones can provide them with real-time information about fires to make their actions safer and more effective.

A review of the literature has been developed to find proposals of robotic systems applied to firefighting tasks. In the case of prevention, there are no proposals for robotizing the vegetation preparation tasks, but there are some developments in the context of forestry and agriculture applicable to them. Conversely, there are multiple proposals for robotizing surveillance tasks, including homogeneous and heterogeneous fleets of drones equipped with conventional, multispectral, and thermal cameras. Finally, in the case of extinguishing, there are multiple approaches to put out fires using autonomous drones, but less to use them to support the firefighters working on the ground.

This paper proposes a concept of operation to apply drone swarms to support fire prevention, surveillance, and extinguishing activities. It considers a fleet of homogeneous quadcopters that individually are only able to visit waypoints and use payloads, but collectively can perform tasks of search, surveillance, reconnaissance, mapping, monitoring, support, tracking, and transport. Three operator roles are defined: mission commander, who commands the swarm and coordinates the mission; team leaders, who coordinate a team on the ground; and team members, who perform the tasks in different areas. These operators have access to different levels of information on the mission through virtual and augmented reality interfaces. On the one hand, this system addresses some of the problems of current operations reported by the firefighters in our survey. It provides the professionals with enhanced information of the scenarios, having an impact on the efficiency of some tasks (e.g., vegetation preparation and fire surveillance) and the safety of some others (e.g., fire extinguishing). On the other hand, some challenges must be overcome, such as the scalability of the system, the training of operators, and the current limitations in the autonomy and communications of drones.

In future works, we are going to develop a complete simulation prototype of the system, as well as a minimum viable product (MVP) with real drones, in order to design, develop and validate the required algorithms.

Author Contributions: J.J.R.-G.: Conceptualization, Data curation, Formal analysis, Investigation, Methodology, Project administration, Writing—original draft, Writing—review & editing; A.B.: Funding acquisition, Writing—original draft, Writing—review & editing; E.G.-G.: Writing—original draft, Investigation, Writing—review & editing. All authors have read and agreed to the published version of the manuscript.

Funding: This research received no external funding.

Institutional Review Board Statement: Not applicable.

Informed Consent Statement: Not applicable.

Data Availability Statement: Statistical results are contained within the article. Raw data are not publicly available due to data protection regulation.

Acknowledgments: This work has been partially developed within the "UAM Emprende" initiative of the "Universidad Autónoma de Madrid". The authors would like to thank the collaboration of the people involved in this initiative and especially Carlos Romero Moreno. Last but not least, the authors would like to express their gratitude to all the professionals that have taken part in the surveys and interviews. This paper would not have been possible without their knowledge and commitment.

Conflicts of Interest: The authors declare no conflict of interest.

Abbreviations

The following abbreviations are used in this manuscript:

AR	Augmented Reality
ESA	European Space Agency
GCS	Ground Control Station
GLONASS	Global'naya Navigatsionnaya Sputnikovaya Sistema
GNC	Guidance, Navigation and Control
GNSS	Global Navigation Satellite System
GPS	Global Positioning System
IMU	Inertial Measurement Unit
LIDAR	Light Detection and Ranging
MR	Mixed Reality
MRS	Multi-Robot System
MVP	Minimum Viable Product
NASA	National Aeronautics and Space Administration
UAV	Unmanned Aerial Vehicle
UGV	Unmanned Ground Vehicle
VR	Virtual Reality

References

1. Brushlinsky, N.; Ahrens, M.; Sokolov, S.; Wagner, P. *World Fire Statistics*; Technical Report 23; Center of Fire Statistics: Moscow, Russia, 2018.
2. Mota, B.; Pereira, J.; Oom, D.; Vasconcelos, M.; Schultz, M. Screening the ESA ATSR-2 World Fire Atlas (1997-2002). *Atmos. Chem. Phys.* **2006**. [CrossRef]
3. Xu, W.; Wooster, M.J.; He, J.; Zhang, T. First study of Sentinel-3 SLSTR active fire detection and FRP retrieval: Night-time algorithm enhancements and global intercomparison to MODIS and VIIRS AF products. *Remote Sens. Environ.* **2020**, *248*, 111947. [CrossRef]
4. Andela, N.; Morton, D.C.; Giglio, L.; Paugam, R.; Chen, Y.; Hantson, S.; Van Der Werf, G.R.; Randerson, J.T. The Global Fire Atlas of individual fire size, duration, speed and direction. *Earth Syst. Sci. Data* **2019**, *11*, 529–552. [CrossRef]
5. Los Incendios Forestales en España: Avance Informativo 2019. Technical report, Área de Defensa contra Incendios Forestales, Ministerio para la Transición Ecológica y el Reto Demográfico, Gobierno de España. 2020. Available Online: https://www.mapa.gob.es/es/desarrollo-rural/estadisticas/avance_1_enero_31_diciembre_2019_tcm30-537398.pdf (accessed on 31 december 2020).
6. Los Incendios Forestales en España: Decenio 2006–2015. Technical report, Área de Defensa contra Incendios Forestales, Ministerio de Agricultura, Pesca y Alimentación, Gobierno de España. 2019. Available Online: https://www.mapa.gob.es/es/desarrollo-rural/estadisticas/incendios-decenio-2006-2015_tcm30-511095.pdf (accessed on 31 december 2020).
7. Muñoz, R.V. *La Defensa Contra Incendios Forestales: Fundamentos y Experiencias*; McGraw-Hill Interamericana de España: Madrid, Spain, 2009.
8. Russell, M.T. *Fire Chief Perception of Unmanned Aircraft Systems: A Diffusion Study*; New Jersey City University: Jersey City, NJ, USA, 2016.

9. MacSween, S. A public opinion survey-unmanned aerial vehicles for cargo, commercial, and passenger transportation. In Proceedings of the 2nd AIAA" Unmanned Unlimited" Conference and Workshop & Exhibit, San Diego, CA, USA, 15–18 September 2003; p. 6519.
10. Aydin, B. Public acceptance of drones: Knowledge, attitudes, and practice. *Technol. Soc.* **2019**, *59*, 101180. [CrossRef]
11. Ghamry, K.A.; Kamel, M.A.; Zhang, Y. Multiple UAVs in forest fire fighting mission using particle swarm optimization. In Proceedings of the 2017 International Conference on Unmanned Aircraft Systems (ICUAS), Miami, FL, USA, 13–16 June 2017; pp. 1404–1409.
12. Roldán, J.J.; del Cerro, J.; Garzón-Ramos, D.; Garcia-Aunon, P.; Garzón, M.; de León, J.; Barrientos, A. Robots in agriculture: State of art and practical experiences. *Serv. Robot.* **2018**. [CrossRef]
13. Valente, J.; Sari, B.; Kooistra, L.; Kramer, H.; Mücher, S. Automated crop plant counting from very high-resolution aerial imagery. *Precis. Agric.* **2020**, *21*, 1366–1384. [CrossRef]
14. Anders, N.; Valente, J.; Masselink, R.; Keesstra, S. Comparing Filtering Techniques for Removing Vegetation from UAV-Based Photogrammetric Point Clouds. *Drones* **2019**, *3*, 61. [CrossRef]
15. Zhang, C.; Valente, J.; Kooistra, L.; Guo, L.; Wang, W. Opportunities of uavs in orchard management. *Int. Arch. Photogramm. Remote Sens. Spat. Inf. Sci.* **2019**. [CrossRef]
16. Barrientos, A.; Colorado, J.; Cerro, J.d.; Martinez, A.; Rossi, C.; Sanz, D.; Valente, J. Aerial remote sensing in agriculture: A practical approach to area coverage and path planning for fleets of mini aerial robots. *J. Field Robot.* **2011**, *28*, 667–689. [CrossRef]
17. Bergerman, M.; Billingsley, J.; Reid, J.; van Henten, E. Robotics in agriculture and forestry. In *Springer Handbook of Robotics*; Springer: Berlin/Heidelberg, Germany, 2016; pp. 1463–1492.
18. Aguiar, A.S.; dos Santos, F.N.; Cunha, J.B.; Sobreira, H.; Sousa, A.J. Localization and Mapping for Robots in Agriculture and Forestry: A Survey. *Robotics* **2020**, *9*, 97. [CrossRef]
19. Couceiro, M.S.; Portugal, D.; Ferreira, J.F.; Rocha, R.P. SEMFIRE: Towards a new generation of forestry maintenance multi-robot systems. In Proceedings of the 2019 IEEE/SICE International Symposium on System Integration (SII), Paris, France, 14–16 January 2019; pp. 270–276.
20. Yuan, C.; Liu, Z.; Zhang, Y. Fire detection using infrared images for UAV-based forest fire surveillance. In Proceedings of the 2017 International Conference on Unmanned Aircraft Systems (ICUAS), Miami, FL, USA, 13–16 June 2017; pp. 567–572.
21. Bouabdellah, K.; Noureddine, H.; Larbi, S. Using wireless sensor networks for reliable forest fires detection. *Procedia Comput. Sci.* **2013**, *19*, 794–801. [CrossRef]
22. Yuan, C.; Liu, Z.; Zhang, Y. Vision-based forest fire detection in aerial images for firefighting using UAVs. In Proceedings of the 2016 International Conference on Unmanned Aircraft Systems (ICUAS), Arlington, VA, USA, 7–10 June 2016; pp. 1200–1205.
23. Yuan, C.; Liu, Z.; Zhang, Y. UAV-based forest fire detection and tracking using image processing techniques. In Proceedings of the 2015 International Conference on Unmanned Aircraft Systems (ICUAS), Denver, CO, USA, 9–12 June 2015; pp. 639–643.
24. Allauddin, M.S.; Kiran, G.S.; Kiran, G.R.; Srinivas, G.; Mouli, G.U.R.; Prasad, P.V. Development of a Surveillance System for Forest Fire Detection and Monitoring using Drones. In Proceedings of the IGARSS 2019—2019 IEEE International Geoscience and Remote Sensing Symposium, Yokohama, Japan, 28 July–2 August 2019; pp. 9361–9363.
25. Barmpoutis, P.; Papaioannou, P.; Dimitropoulos, K.; Grammalidis, N. A Review on Early Forest Fire Detection Systems Using Optical Remote Sensing. *Sensors* **2020**, *20*, 6442. [CrossRef] [PubMed]
26. Dang-Ngoc, H.; Nguyen-Trung, H. Aerial Forest Fire Surveillance-Evaluation of Forest Fire Detection Model using Aerial Videos. In Proceedings of the 2019 International Conference on Advanced Technologies for Communications (ATC), Hanoi, Vietnam, 17–19 October 2019; pp. 142–148.
27. Barmpoutis, P.; Stathaki, T.; Dimitropoulos, K.; Grammalidis, N. Early fire detection based on aerial 360-degree sensors, deep convolution neural networks and exploitation of fire dynamic textures. *Remote Sens.* **2020**, *12*, 3177. [CrossRef]
28. Valero, M.; Rios, O.; Pastor, E.; Planas, E. Automated location of active fire perimeters in aerial infrared imaging using unsupervised edge detectors. *Int. J. Wildland Fire* **2018**, *27*, 241–256. [CrossRef]
29. Bosch, I.; Serrano, A.; Vergara, L. Multisensor network system for wildfire detection using infrared image processing. *Sci. World J.* **2013**, *2013*, 402196. [CrossRef] [PubMed]
30. Veraverbeke, S.; Dennison, P.; Gitas, I.; Hulley, G.; Kalashnikova, O.; Katagis, T.; Kuai, L.; Meng, R.; Roberts, D.; Stavros, N. Hyperspectral remote sensing of fire: State-of-the-art and future perspectives. *Remote Sens. Environ.* **2018**, *216*, 105–121. [CrossRef]
31. Carvajal-Ramírez, F.; Marques da Silva, J.R.; Agüera-Vega, F.; Martínez-Carricondo, P.; Serrano, J.; Moral, F.J. Evaluation of fire severity indices based on pre-and post-fire multispectral imagery sensed from UAV. *Remote Sens.* **2019**, *11*, 993. [CrossRef]
32. Martínez-de Dios, J.R.; Merino, L.; Ollero, A. Fire detection using autonomous aerial vehicles with infrared and visual cameras. *IFAC Proc. Vol.* **2005**, *38*, 660–665. [CrossRef]
33. Fonollosa, J.; Solórzano, A.; Marco, S. Chemical sensor systems and associated algorithms for fire detection: A review. *Sensors* **2018**, *18*, 553. [CrossRef]
34. Zhang, L.; Wang, B.; Peng, W.; Li, C.; Lu, Z.; Guo, Y. Forest fire detection solution based on UAV aerial data. *Int. J. Smart Home* **2015**, *9*, 239–250. [CrossRef]
35. Kinaneva, D.; Hristov, G.; Raychev, J.; Zahariev, P. Early forest fire detection using drones and artificial intelligence. In Proceedings of the 2019 42nd International Convention on Information and Communication Technology, Electronics and Microelectronics (MIPRO), Opatija, Croatia, 20–24 May 2019; pp. 1060–1065.

36. Jiao, Z.; Zhang, Y.; Xin, J.; Mu, L.; Yi, Y.; Liu, H.; Liu, D. A Deep Learning Based Forest Fire Detection Approach Using UAV and YOLOv3. In Proceedings of the 2019 1st International Conference on Industrial Artificial Intelligence (IAI), Shenyang, China, 23–27 July 2019; pp. 1–5.
37. Burke, C.; Wich, S.; Kusin, K.; McAree, O.; Harrison, M.E.; Ripoll, B.; Ermiasi, Y.; Mulero-Pázmány, M.; Longmore, S. Thermal-Drones as a Safe and Reliable Method for Detecting Subterranean Peat Fires. *Drones* **2019**, *3*, 23. [CrossRef]
38. Ghamry, K.A.; Kamel, M.A.; Zhang, Y. Cooperative forest monitoring and fire detection using a team of UAVs-UGVs. In Proceedings of the 2016 International Conference on Unmanned Aircraft Systems (ICUAS), Arlington, VA, USA, 7–10 June 2016; pp. 1206–1211.
39. Sudhakar, S.; Vijayakumar, V.; Kumar, C.S.; Priya, V.; Ravi, L.; Subramaniyaswamy, V. Unmanned Aerial Vehicle (UAV) based Forest Fire Detection and monitoring for reducing false alarms in forest-fires. *Comput. Commun.* **2020**, *149*, 1–16. [CrossRef]
40. Qin, H.; Cui, J.Q.; Li, J.; Bi, Y.; Lan, M.; Shan, M.; Liu, W.; Wang, K.; Lin, F.; Zhang, Y.; et al. Design and implementation of an unmanned aerial vehicle for autonomous firefighting missions. In Proceedings of the 2016 12th IEEE International Conference on Control and Automation (ICCA), Kathmandu, Nepal, 1–3 June 2016; pp. 62–67.
41. Ando, H.; Ambe, Y.; Ishii, A.; Konyo, M.; Tadakuma, K.; Maruyama, S.; Tadokoro, S. Aerial hose type robot by water jet for fire fighting. *IEEE Robot. Autom. Lett.* **2018**, *3*, 1128–1135. [CrossRef]
42. Ogawa, S.; Kudo, S.; Koide, M.; Torikai, H.; Iwatani, Y. Development and control of an aerial extinguisher with an inert gas capsule. In Proceedings of the 2014 IEEE International Conference on Robotics and Biomimetics (ROBIO 2014), Bali, Indonesia, 5–10 December 2014; pp. 1320–1325.
43. Aydin, B.; Selvi, E.; Tao, J.; Starek, M.J. Use of fire-extinguishing balls for a conceptual system of drone-assisted wildfire fighting. *Drones* **2019**, *3*, 17. [CrossRef]
44. Alshbatat, A.I.N. Fire extinguishing system for high-rise buildings and rugged mountainous terrains utilizing quadrotor unmanned aerial vehicle. *Int. J. Image, Graph. Signal Process.* **2018**, *12*, 23. [CrossRef]
45. Soliman, A.M.S.; Cagan, S.C.; Buldum, B.B. The design of a rotary-wing unmanned aerial vehicles–payload drop mechanism for fire-fighting services using fire-extinguishing balls. *SN Appl. Sci.* **2019**, *1*, 1259. [CrossRef]
46. Manimaraboopathy, M.; Christopher, H.; Vignesh, S.; others. Unmanned Fire Extinguisher Using Quadcopter. *Int. J. Smart Sens. Intell. Syst.* **2017**, *10*, 471–481. [CrossRef]
47. Sherstjuk, V.; Zharikova, M.; Sokol, I. Forest Fire Fighting Using Heterogeneous Ensemble of Unmanned Aerial Vehicles. In Proceedings of the 2019 IEEE 5th International Conference Actual Problems of Unmanned Aerial Vehicles Developments (APUAVD), Kiev, Ukraine, 22–24 October 2019; pp. 218–223.
48. Islam, S.; Razi, A. A path planning algorithm for collective monitoring using autonomous drones. In Proceedings of the 2019 53rd Annual Conference on Information Sciences and Systems (CISS), Baltimore, MD, USA, 20–22 March 2019; pp. 1–6.
49. Innocente, M.S.; Grasso, P. Self-organising swarms of firefighting drones: Harnessing the power of collective intelligence in decentralised multi-robot systems. *J. Comput. Sci.* **2019**, *34*, 80–101. [CrossRef]
50. Haksar, R.N.; Schwager, M. Distributed deep reinforcement learning for fighting forest fires with a network of aerial robots. In Proceedings of the 2018 IEEE/RSJ International Conference on Intelligent Robots and Systems (IROS), Madrid, Spain, 1–5 October 2018; pp. 1067–1074.
51. Seraj, E.; Silva, A.; Gombolay, M. Safe Coordination of Human-Robot Firefighting Teams. *arXiv* **2019**, arXiv:1903.06847.
52. Sherstjuk, V.; Zharikova, M.; Sokol, I. Forest fire-fighting monitoring system based on uav team and remote sensing. In Proceedings of the 2018 IEEE 38th International Conference on Electronics and Nanotechnology (ELNANO), Kiev, Ukraine, 24–26 April 2018; pp. 663–668.
53. Garcia-Aunon, P.; Roldán, J.J.; Barrientos, A. Monitoring traffic in future cities with aerial swarms: Developing and optimizing a behavior-based surveillance algorithm. *Cogn. Syst. Res.* **2019**, *54*, 273–286. [CrossRef]
54. Al-Turjman, F. A novel approach for drones positioning in mission critical applications. *Trans. Emerg. Telecommun. Technol.* **2019**, e3603. [CrossRef]
55. Wang, L.; Sanchez-Matilla, R.; Cavallaro, A. Tracking a moving sound source from a multi-rotor drone. In Proceedings of the 2018 IEEE/RSJ International Conference on Intelligent Robots and Systems (IROS), Madrid, Spain, 1–5 October 2018; pp. 2511–2516.
56. Garcia-Aunon, P.; Barrientos Cruz, A. Comparison of heuristic algorithms in discrete search and surveillance tasks using aerial swarms. *Appl. Sci.* **2018**, *8*, 711. [CrossRef]
57. Roldán-Gómez, J.J.; Garcia-Aunon, P.; Mazariegos, P.; Barrientos, A. SwarmCity project: Monitoring traffic, pedestrians, climate, and pollution with an aerial robotic swarm. *Pers. Ubiquitous Comput.* **2020**, 1–17. [CrossRef]
58. Roldán, J.J.; del Cerro, J.; Barrientos, A. A proposal of methodology for multi-UAV mission modeling. In Proceedings of the 2015 23rd Mediterranean Conference on Control and Automation (MED), Torremolinos, Spain, 16–19 June 2015; pp. 1–7.
59. Roldán, J.J.; Peña-Tapia, E.; Auñón, P.G.; Del Cerro, J.; Barrientos, A. Bringing Adaptive and Immersive Interfaces to Real-World Multi-Robot Scenarios: Application to Surveillance and Intervention in Infrastructures. *IEEE Access* **2019**, *7*, 86319–86335. [CrossRef]
60. Martín-Barrio, A.; Roldán, J.J.; Terrile, S.; del Cerro, J.; Barrientos, A. Application of immersive technologies and natural language to hyper-redundant robot teleoperation. *Virtual Real.* **2019**, *24*, 541–555. [CrossRef]

Article

Director Tools for Autonomous Media Production with a Team of Drones

Ángel Montes-Romero [1], Arturo Torres-González [1], Jesús Capitán [1,*], Maurizio Montagnuolo [2], Sabino Metta [2], Fulvio Negro [2], Alberto Messina [2] and Aníbal Ollero [1]

1 GRVC Robotics Lab, University of Seville, 41092 Seville, Spain; amromero@us.es (Á.M.-R.); arturotorres@us.es (A.T.-G.); aollero@us.es (A.O.)

2 Centre for Research and Technological Innovation, Radiotelevisione Italiana (RAI), 10138 Turin, Italy; maurizio.montagnuolo@rai.it (M.M.); sabino.metta@rai.it (S.M.); fulvio.negro@rai.it (F.N.); alberto.messina@rai.it (A.M.)

* Correspondence: jcapitan@us.es

Received: 6 February 2020; Accepted: 13 February 2020; Published: 21 February 2020

Featured Application: This work can be applied for media production with aerial cameras. The system supports media crew to film outdoor events with an autonomous fleet of drones.

Abstract: This paper proposes a set of director tools for autonomous media production with a team of drones. There is a clear trend toward using drones for media production, and the *director* is the person in charge of the whole system from a production perspective. Many applications, mainly outdoors, can benefit from the use of multiple drones to achieve multi-view or concurrent shots. However, there is a burden associated with managing all aspects in the system, such as ensuring safety, accounting for drone battery levels, navigating drones, etc. Even though there exist methods for autonomous mission planning with teams of drones, a media director is not necessarily familiar with them and their language. We contribute to close this gap between media crew and autonomous multi-drone systems, allowing the director to focus on the artistic part. In particular, we propose a novel language for cinematography mission description and a procedure to translate those missions into plans that can be executed by autonomous drones. We also present our director's *Dashboard*, a graphical tool allowing the director to describe missions for media production easily. Our tools have been integrated into a real team of drones for media production and we show results of example missions.

Keywords: autonomous media production; drone cinematography; multimedia tools

---

1. Introduction

There is a clear trend toward using drones for aerial cinematography and media production in general. The main reasons for the emergence of this technology are twofold: First, small drones can be equipped with high-quality cameras, but, at the same time, they are not too expensive, which makes them appealing for amateur and professional users. Second, due to their maneuverability, they broaden the aesthetic possibilities for media production, as they can create novel and unique shots. Besides, media production applications to cover outdoor events can benefit from the use of teams of drones, mainly to produce multi-view shots and film multiple action points concurrently. From a logistic perspective, a multi-drone system could be deployed easily (e.g., instead of camera cranes) to operate in such large-scale, outdoor scenarios without requiring the pre-existence of complex infrastructure.

The main issue with a multi-drone system for media production is its complexity of operation. Currently, two operators per drone are usually required: one to pilot the drone and another to handle camera movement. The media *director* is the person in charge of the whole system from the production point of view. However, there are additional aspects that need to be accounted for: ensuring safety in operations avoiding collisions and no-fly zones, deciding which cameras allocate to each shot, considering battery levels of drones, etc. For that, enhancing the system with autonomous capabilities to plan and execute shots is rather helpful to alleviate the director's burden and allow her/him to focus on the artistic part.

There exist methods for autonomous mission planning with teams of multiple drones. Indeed, general purpose methods for multi-robot task allocation and scheduling could be adapted to tackle this problem. However, a media director is not necessarily familiar with these kinds of algorithms and their robotics language. Therefore, our objective is to close this gap between a media crew and these autonomous, intelligent systems, so that a director can use an autonomous fleet of drones for media production. In particular, we think that there is a need for a standard language and tools for cinematography mission description. With such tools, a director could talk to her/his autonomous cinematographers in a transparent manner, abstracting herself/himself from the planning and execution procedures behind.

In this paper, we propose a set of tools for mission description in media production. First, a novel language for mission description is presented. This language is used by the media director to specify the desired shots when filming an event. The output is an XML-based file that contains details for all shots specified by the director at each action point. Then, this *shooting mission* is interpreted by the system to be translated into a list of tasks that can be understood and executed by the drones. The proposed language acts as a scripting system for director story-telling, who can hence focus on the artistic part without specific knowledge in multi-robot autonomous planning. Last, we also propose a graphical tool, the so-called director's *Dashboard*, to create and manage XML-based shooting missions. This Dashboard allows the director to interact with the fleet of drones by sending/canceling missions and monitoring them during execution.

This paper continues our prior work [1], where we proposed a taxonomy of cinematography shots to implement and a preliminary version of our Dashboard. Here, we add the language to specify autonomous cinematography missions, and a complete description of the final version of the Dashboard, enhanced with new functionalities. Besides, we showcase the use of our tools in example scenarios to film sport and outdoor events, like a rowing or cycling race. We integrated our system with a real team of drones within the framework of the EU-funded project MULTIDRONE (https://multidrone.eu/), which aims at building a team of several drones for autonomous media production. We describe the whole process to write, translate, and execute example shooting missions, as well as details of the actual aerial platforms developed for media production.

The remainder of this paper is as follows. Section 2 reviews related work. Section 3 gives an overview of the system, Section 4 presents our language for shooting mission description. Section 5 describes our director's Dashboard. Section 6 depicts some experimental results showcasing the use of our tools, and Section 7 gives conclusions and future work.

2. Related Work

Media production is adopting drones as replacements for dollies (static cranes) and helicopters, as their deployment is easier and has less costs associated. Thus, the use of drones in cinematography has increased recently, in parallel with the improvement on their technology. For instance, the New York City Drone Film Festival (https://www.nycdronefilmfestival.com.) was the world's first film festival dedicated to drones in cinematography. Besides, drones are quite attractive for live coverage of outdoor events, as they can provide novel angles and visual effects. In October 2016, drones were relevant for news agencies in the coverage of the massive destruction caused by Hurricane Matthew on Haiti [2]. They have also been used in major international sport events, such as the Olympics [3,4].

In the current state-of-the-art solutions for media production with drones, the director usually specify targets subjects or points of interest to be filmed in pre-production, together with a temporarily ordered script, the camera motion types, etc. Then, the drone pilot and the cameraman must execute the plan manually in a coordinated fashion. There are also commercial drones, such as DJI [5], AirDog [6], 3DR SOLO [7], or Yuneec Typhoon [8], that can implement certain cinematographic functionalities autonomously. However, they are typically prepared to track a target visually or with a GNSS and keep it on the image frame (*follow-me* mode), not focusing on high-level cinematography principles for shot performance.

Besides, there are some end-to-end solutions for semi-autonomous aerial cinematographers [9,10]. In these works, a director specifies high-level commands such as shot types and positions for single-drone shooting missions. Then, the drone is able to compute autonomously navigation and camera movement commands to execute the desired shots, but the focus is on static scenes. In [9], an outdoor application to film people is proposed, and different types of shots from the cinematography literature are introduced (e.g., close-up, external, over-the-shoulder, etc.). Timing for the shots is considered by means of an easing curve that drives the drone along the planned trajectory (i.e., this curve can modify its velocity profile). In [10], an iterative quadratic optimization problem is formulated to obtain smooth trajectories for the camera and the look-at point (i.e., the place where the camera is pointing at). No time constraints or moving targets are included. In general, these approaches use algorithms to compute smooth camera trajectories fulfilling aesthetic and cinematographic constraints, which can be formulated as an optimization problem [11–13].

In a multi-drone context, there is little work on autonomous systems for media production. In [14], the optimal number of drones to cover all available targets without occlusion is computed. However, cameras must always be facing targets and smooth transitions are not considered. A more advanced method is presented in [15], where they propose an online planning algorithm to compute optimal trajectories for several drones filming dynamic, indoor scenes. User-defined aesthetic objectives and cinematographic rules are combined with collision and field-of-view avoidance constraints, but they do not address interfacing with the media director.

Regarding user interfaces, there exist several drone *apps* to support photographers, video makers, photogrammetrists, and other professional profiles during their activities. As relevant features, they usually include drone mapping, geo-fencing, and flight logging. In particular, based on regulatory information from each country, some apps can help users indicating no-flight zones such as airports, control zones (CTZ), etc. Local meteorological information (e.g., wind speed, direction, temperature, etc.) is also used in some apps to inform the pilot about flight safety. Moreover, these apps can support the creation of accurate, high-resolution maps, 3D models, and real-time 2D live maps. More specifically on media production, some apps offer autonomous tracking features for shooting video, see for example in [16]. This app supports the execution of several shooting modes, such as *Orbit me* and *Follow me*, as well as the planning in advance of specific flights and shots using a *Mission Hub* on a computer. Thus, the user can pre-program paths that will be later followed by the drone [17,18]. This is done specifying a set of temporally ordered, desired *key-frames* on a 3D reconstruction of the scene, as a rudimentary cinematography plan. Nevertheless, these apps are thought and implemented for single-drone flight and shooting. From a media perspective, the functionality of using more than one drone within the same shooting mission is not properly addressed.

3. System Overview

We present in this paper a set of tools so that a director can interface with an autonomous fleet of drone cinematographers and govern the system from an editorial point of view. The main contributions are a novel language for mission description in media production and a graphical tool to define those missions and interact with the autonomous system. The general architecture of the system is depicted in Figure 1.

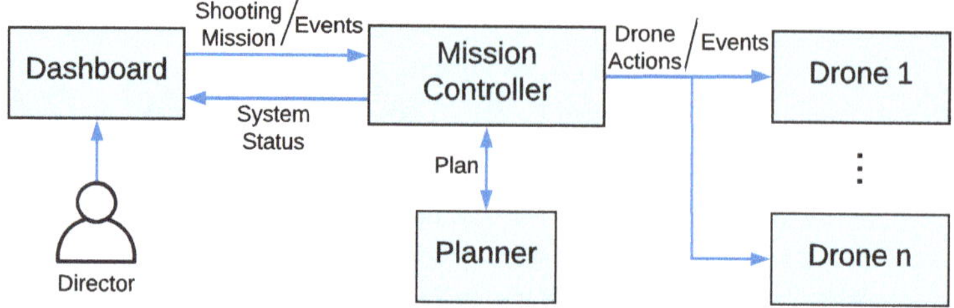

Figure 1. General scheme of the system architecture. The director defines shooting missions with the Dashboard and sends them to the Mission Controller. This Mission Controller uses a Planner to compute plans that are then sent to the drones. During the execution of the plan, the Mission Controller sends events to the drones to trigger actions and reports periodically to the Dashboard about the execution/system status.

The director and the editorial team can specify a set of artistic shots and associate them with different events happening in time (e.g., a race start or racers reaching a relevant point). This is done through the *Dashboard*, which is a web-based graphical tool that allows her/him to interact with the rest of the system. The whole set of director shots, together with information of the events with which they are associated, constitutes the so-called *shooting mission*, which is saved in a database during the editing phase. After finishing this editorial phase, the shooting mission is encoded in an XML-based language and sent to the *Mission Controller*, which is the central module managing autonomous planning.

The Mission Controller can understand the director's dialect and is in charge of interpreting the shooting mission, as sequential or parallel *shooting actions* that will be assigned to the available drones in the team. A list of *tasks* to be executed by the drone fleet is compiled, where each task has a starting position and a duration, extracted from the shooting actions' descriptions. Then, the Mission Controller can use a *Planner* to obtain a feasible plan to accomplish that shooting mission.

The Planner should take into account spatial and temporal constraints of the tasks, as well as drone battery levels and starting positions, in order to assign tasks to drones efficiently. The plan consists of a list of *drone actions* for each drone, specifying where the drone should go at each instant and what to do. Basically, each drone gets assigned a sequence of shooting actions, so its list of actions is made up of single-drone shooting actions plus the required navigation actions in between. In general, this Planner module could be implemented by any standard planner for task allocation with temporal and resource constraints and it is not the focus of this paper. Some preliminary ideas can be seen in [19].

Once the plan is computed, it is sent to the drones, which are able to execute it autonomously. During execution, the Mission Controller provides some feedback to the director through the Dashboard, reporting on the status of each drone and the mission itself.

Section 4 describes the XML-based language to describe shooting missions and the process to translate them into plans made up of drone actions. Section 5 describes the design and functionalities of the Dashboard.

4. Language for Cinematography Mission Description

There are professional drones for filming in the market, and many of them include autonomous capabilities to implement certain specific shots, for example, a panoramic view or tracking a moving target. However, there is no general framework to specify complete missions for autonomous cinematography. In this section, we present a novel language to describe aerial cinematography missions for multiple drones.

We use a vocabulary that the editorial team can understand and define an XML-based dialect to write shooting missions. In the following, we explain the structure of the XML files and how they can be converted into lists of tasks for autonomous drones. The complete XML schema is also publicly available (https://grvc.us.es/downloads/docs/multidrone_schema.xsd). Our XML schema to describe shooting missions has the following main information entities:

- ***Event***: This is the central entity of information. An event represents a real-world occurrence with a spatial and temporal localization that is going to be filmed. For instance, a sport event like a rowing/cycling race. An event can have child events, i.e., events that occur relatively to the timeline of their parent. For example, different stages of a cycling race or different parts of a particular stage. Thus, there can be a tree-like structure of events where each of them can have a *planned start time* and *duration*.
- ***Mission***: This describes the aerial shots designed by the editorial team for a given event. Missions can only be associated with leaf events, i.e., events with no children. A leaf event may contain an arbitrary number of missions, each of which has a specific *role*. This is because the director could design different missions for the occurrence of the same event, depending on contextual conditions. For example, the director might plan to have two distinct (and mutually exclusive) missions, in case of sunny or cloudy weather, respectively.

  Each mission can have associated a specific *drone team*, made up of several drones. This choice has been made to foresee settings in which more than one multi-drone team are available, so the corresponding team for each mission needs to be specified. Moreover, a mission can have associated an *editorial team*, i.e., a group of Dashboard users, each with their own role, who will be granted permission to manage and modify the mission data.
- ***Shooting Action Sequence***: A mission consists of one or more shooting action sequences (SASs), which can also have different *roles*. Thus, the director can plan different actions within the mission depending on contextual circumstances. For example, if the associated event is a race finish line, the director may decide to have two possible shooting action sequences: one in case of arrival in the sprint, another in case of a solo attack. All shooting action sequences with the same role within a mission will happen concurrently as the associated event occurs.
- ***Shooting Action***: A shooting action sequence is made up of an ordered sequence of shooting actions (SAs). Therefore, shooting action sequences with the same role within a mission are aimed to run in parallel, whereas the shooting actions within each of them occur sequentially. A shooting action represents an aerial visual shot to be performed with one or several drones. A *duration* or *length* can be specified to indicate action endurance in terms of time or space (distance), respectively. A shooting action has a *reference target* (RT) that is tracked to move the drone formation alongside (it could be virtual just to define a desired formation movement). An *RT trajectory*, the *origin of the RT*, and *origin of the formation* need to be specified in global coordinates. Thus, it is indicated the origin of the center of the drone formation, which should then move along the RT trajectory as the shooting takes place. The *formation speed* is optional to indicate how fast the formation should move along the RT trajectory (mainly when RT is virtual).
- ***Shooting Role***: In a shooting action, there can be one or several shooting roles, one per drone involved. This is helpful to indicate the role of each drone in shots with multiple drones. It has a *shooting type* defining the type of movement of the camera, e.g., static, lateral, orbit, etc. Each shooting role has some *shooting parameters* defining the geometry of the flying formation. These parameters vary depending on the specific type of shot, for instance, to indicate the lateral distance of the drone in a lateral shot or the radius and angle rate in an orbit. A shooting role has also a *framing type*: long shot, medium shot, close-up, etc. Even though all shooting roles in a shooting action have the same RT to move in a linked way, each shooting role could have a different *shooting target* to point the camera, which is specified as an identifier. This identifier may match a certain GNSS transmitter or visual tracker.

In summary, the relation of the main components of the XML schema is as follows.

```
<event> Main event: e.g.,~a race to film
<event> Leaf event 1: e.g.,~start line
<mission>
<shootingActionSequence>
<shootingAction>
<shootingRole>
<event> Leaf event 2: e.g.,~finish line
<mission>
<shootingActionSequence>
<shootingAction>
<shootingRole>
```

In a pre-production stage, the director and the editorial team will manage a database with all the above entities through the Dashboard. Then, an XML file with all events and missions associated is generated, i.e., a shooting mission. The Mission Controller receives that file and computes plans for all possible combinations of mission roles and shooting action sequence roles. Before starting the execution of the mission, the director will specify the final selected role for the missions and for the shooting action sequences, which will determine the actual plan to be executed. Note that multiple plans for the different roles are presented to the director, who must choose unique roles for the missions and the SASs. Once the plan is selected, the Mission Controller sends their corresponding actions to the drones and waits for events. Anytime a leaf event occurs, this should be notified to the drones, so that they can trigger the associated shooting actions. The occurrence of these leaf events is either indicated manually by the director (e.g., start of a race) or detected automatically by the Mission Controller (e.g., target reaching a relevant position).

To compute a plan, the Mission Controller extracts the relevant information from the XML file and creates a list of data objects of type `SHOOTING ACTION`, as indicated in Table 1. Most of the fields of the `SHOOTING ACTION` data structure come directly from the `<shootingAction>` XML element. However, some of them come from data in the `<shootingActionSequence>`, `<mission>`, and `<event>` elements, for instance, *Start event*, *Mission ID*, or *Action sequence ID*. Other fields are calculated from the received data, such as the *RT displacement*, which is the difference between the `<originOfFormation` and the `<originOfRT>`.

Table 1. Structure for the data type `SHOOTING ACTION`.

SHOOTING ACTION		
Field Name	**Data Type**	**Comment**
Start event	String	Identifies the leaf event that triggers this shooting action
Planned start time	Time in seconds	Leaf event time with respect to the parent event
Mission ID	String	Identifies the associated mission
Action sequence ID	String	Identifies the associated shooting action sequence
Action ID	String	Identifies this shooting action
Duration	Time in seconds	Duration of the shot
Length	Distance in meters	Expected length of the RT trajectory
RT trajectory	List of global positions	Estimated path of the RT
RT displacement	Vector of floats	Displacement between the origin of the drone formation and the RT trajectory.
Formation speed	Float	Speed for drone formation along RT trajectory
Shooting roles	List of `SHOOTING ROLE`	Look at Table 2

Table 2. Structure for the data type SHOOTING ROLE.

SHOOTING ROLE		
Field Name	**Data Type**	**Comment**
Shooting type	Discrete value	Lateral, chase, static, orbit, fly-through, etc.
Framing type	Discrete value	Long shot, medium shot, close up, etc.
Shooting parameters	Set of parameters	e.g., distance to the target in a lateral, angular velocity in an orbit, etc.
Shooting target	Natural number	Identifies the shooting target to film

The Planner would receive a list of these SHOOTING ACTION objects, each with one or multiple SHOOTING ROLE objects included. Then, each individual SHOOTING ROLE represents a task, and the Planner should solve the problem of allocating tasks to drones holding with constraints such as tasks start time and duration, drones' remaining battery, etc. For instance, if a shooting action implements a shot with several drones involved, this is translated into several tasks with the same starting time. Once that assignment is done, the plan for each drone is produced. This plan consists of a list of DRONE ACTION objects, some of them implementing navigation actions and other actual shots, which is specified by an *action type* field. If the action to perform is a shot, all the related information is included a SHOOTING ACTION object with a single SHOOTING ROLE for that drone. Table 3 shows the data structure for the DRONE ACTION objects. Last, the Mission Controller sends the corresponding list of DRONE ACTION objects to each drone and the mission execution can start.

Table 3. Structure for the data type DRONE ACTION.

DRONE ACTION		
Field Name	**Data Type**	**Comment**
Action type	Discrete value	Takeoff, land, go-to-waypoint or shooting
Action ID	String	Identifies the corresponding shooting action
Start event	String	Identifies the event that triggers this drone action
Shooting action	SHOOTING ACTION	The shooting action if type is "shooting"
Path	List of waypoints	List of waypoints to follow if type is "go to waypoint"

5. Director's Dashboard

The director's Dashboard is a Graphical User Interface (GUI) designed specifically to allow the editorial team to define shooting missions in an easy manner during a pre-production phase and interact with the autonomous system during mission execution. We designed this tool for media production in several steps. First, we defined the high-level interactions between the Dashboard and its users, to come up with an UML diagram depicting use cases. Second, we created a database to store all information entities considered in our XML-based cinematography language. Third, we used UML activity diagrams to define the workflows that shall be supported by the Dashboard software, and built upon them the GUI appearance and behavior.

5.1. Use Case Analysis

Figure 2 shows the use case breakdown of the Dashboard. The roles involved in the process include both human actors and machine system components (shown, respectively, as man icons and gray rectangles in Figure 2). The director is responsible for defining, leading, and coordinating the editorial shooting missions. The role of the cameramen is to operate the cameras on board the drones, thus replacing their autonomous operation when necessary due to production requirements. The editorial staff deals with the data entry of events, possibly organizing them hierarchically in sub-events. The system administrator is responsible for effective provisioning, configuration, operation, and maintenance of the Dashboard.

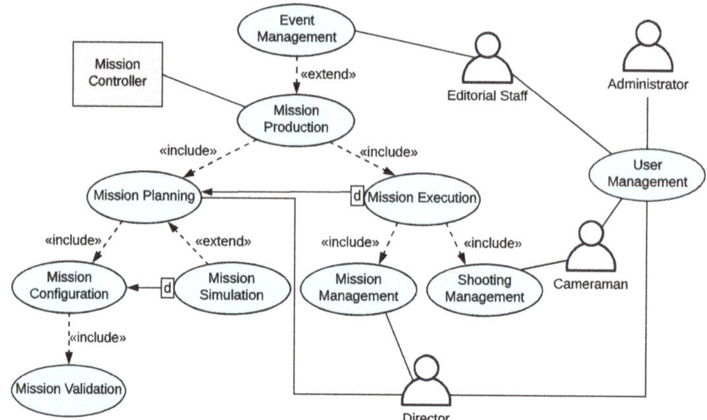

Figure 2. High-level use case diagram of the Dashboard.

The Mission Controller is the autonomous module in charge of controlling the mission production workflow. It generates signals upon the occurrence of leaf events in to trigger associated shooting actions on board the drones. Some leaf events may be manually triggered by the director, e.g., to start a race. Others may be generated automatically, e.g., after detecting the appearance of a point of interest. The Mission Controller is also in charge of managing semantic maps, i.e., Keyhole Markup Language (KML) files [20] representing a set of geolocalized features such as landing/take-off spots, no-fly zones, or points of interest. These maps are first created offline during the pre-production phase and displayed on the Dashboard. Then, during mission execution, they might be automatically updated if new features were detected, e.g., obstacle or crowd regions detected by the drone cameras.

The human and machine actors defined above interact with each other in order to create and manage an editorial shooting mission (see Figure 2). There is a higher-level process preceding any shooting mission, which is that of event management, i.e., the process by which events are organized hierarchically, and by which specific events are associated with missions planned to be executed in reaction to them.

The process of mission production is split into mission planning and mission execution. Mission planning is the process in which all aspects related to each mission are defined. These include, e.g., expected starting time and duration, shot types, shooting targets like the leader of a race or a geographic points of interest within the race route, etc. The mission planning process can be further broken down into a mission configuration process, optionally supported by a mission simulation process, which depends on the former. A fundamental subprocess of mission planning is mission validation, i.e., the process by which the flight plan originated by a shooting mission is checked and validated in terms of safety and security.

Mission execution is the process during execution, and it always follows mission planning, i.e., a shooting mission cannot be started if it has not been previously planned and validated. Mission execution is broken down into two distinct subprocesses, namely, mission management and shooting management. Mission management is the process by which the director finally takes decisions about which among the several missions associated with an event will be actually executed, i.e., specific mission and shooting action sequence roles are selected. Shooting management is the process that allows the cameraman to watch the live streams coming from the available video sources (drone cameras) and to change some camera parameters such as zoom, translation and rotation.

5.2. Dashboard Implementation

The Dashboard is implemented as a 3-tier Java EE7 application, consisting of, from bottom to top, data access layer, business logic layer, and presentation layer.

The data access layer communicates with a MySQL database, translating data from/to database format to/from domain format. The database was designed and built based on the information modeling described in Section 4. Java entity beans and data access objects are used to map database entries to Java objects and vice versa. The value of this approach is that if something changes in the database structure, changes are automatically reflected on the application data model.

The business logic layer manages user authentication/authorization, and it performs actions based on user inputs and expected outputs. It provides a set of RESTful HTTP endpoints used to drive most of the functionality of the Dashboard, including administration facilities, as well as CRUD (create, read, update, and delete) operations on the database data.

The presentation layer handles user requests and displays information to users. The GUI consists of a web application based on HTML5, Bootstrap [21], and Angular [22]. The GUI is designed to be responsive, meaning that one may access the platform through any kind of device from smartphones to desktop computers. The main functions of the GUI are described in the following.

5.2.1. Director Home Page

Once the user is logged into the system as a director, the homepage with the list of created missions is presented (see Figure 3), together with the related events, role of the mission, and date of the event. The status of each mission is also shown, so that the director may know the next steps of the planning or execution workflows that can be taken for every mission. For example, if a mission is in "Pending" status, it means that the director requested to validate the mission, and the Mission Controller is performing the required safety and security checks. Several actions can be performed from this page, as editing a mission or deleting it. The deletion of a mission implies the deletion of all the content of the mission itself (i.e., any related shooting action sequences and shooting actions). On the left side of the homepage, a menu allows the director to be redirected to the event management section, a specific page of the Dashboard used to manage already created events. Through this menu, the director has also access to the page for creating new events, and to the page for executing validated missions.

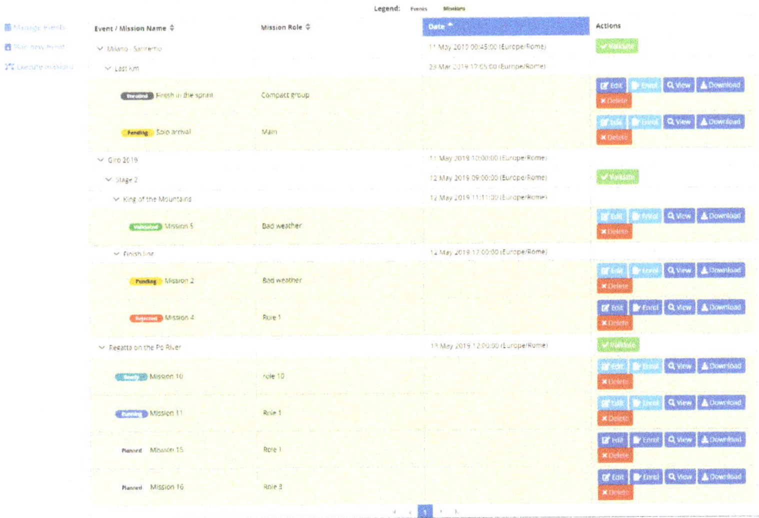

Figure 3. Example screenshot of the director's home page.

5.2.2. Event Management

The event management page shows both information on already created events, such as their planned date and time, and available actions to be performed on single events (i.e., edit and delete). An example is shown in Figure 4. If the director or the editorial staff need to create new events, they can select the "Plan new event" item on the left menu of the director home page. This action will redirect to the event configuration page, where a form containing information about the new event can be filled in and then saved on the database. For each event, users can create new child events or new missions. Figure 5 illustrates an example.

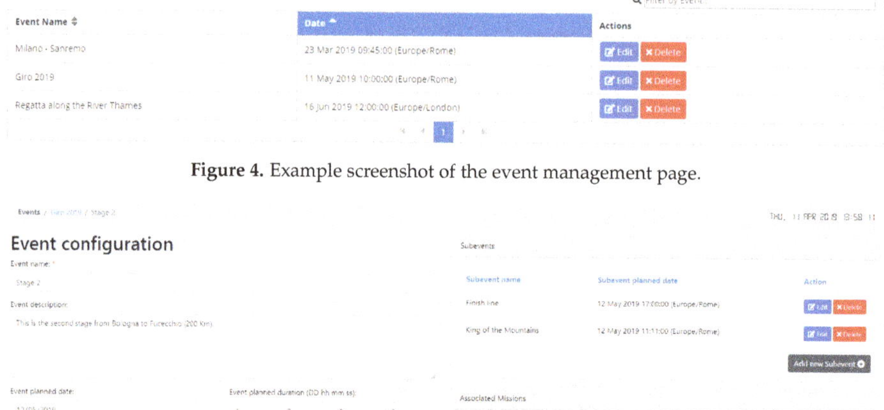

Figure 4. Example screenshot of the event management page.

Figure 5. Example screenshot of the event configuration page, used to create a new event.

5.2.3. Mission Planning

Once the hierarchy of events has been created, missions for leaf events (i.e., events with no more children) can be configured. Figure 6 shows an example. The event is the shooting of the last kilometer of the "Milano-Sanremo" cycling race. The mission is called "Finish in the sprint" and it will be executed in case of a sprint finish of the race. The crew enrolled in the mission includes four people: one director, two cameramen, and one production technician.

The mission configuration page also shows, if present, a list of SASs related to the mission. The user can either add new SASs or update existing ones by simply filling in a short web-form. For each SAS, the user can add one or more shooting actions, detailing through appropriate web forms what the drone formation will do during the live event and what each single drone will do as well. Three tabs, as shown in Figure 7, compose the shooting action configuration page. The first tab describes the main metadata of the shooting action, like the duration. The second tab allows the user to fill in information about the drone formation. This includes, for example, the RT trajectory on a map and the speed of the formation. The third tab collects specific information about each shooting role in the drone formation, such as the shooting parameters (e.g., drones' displacements), framing types (e.g., long shot, medium shot, close-up shot, etc.), and shooting target.

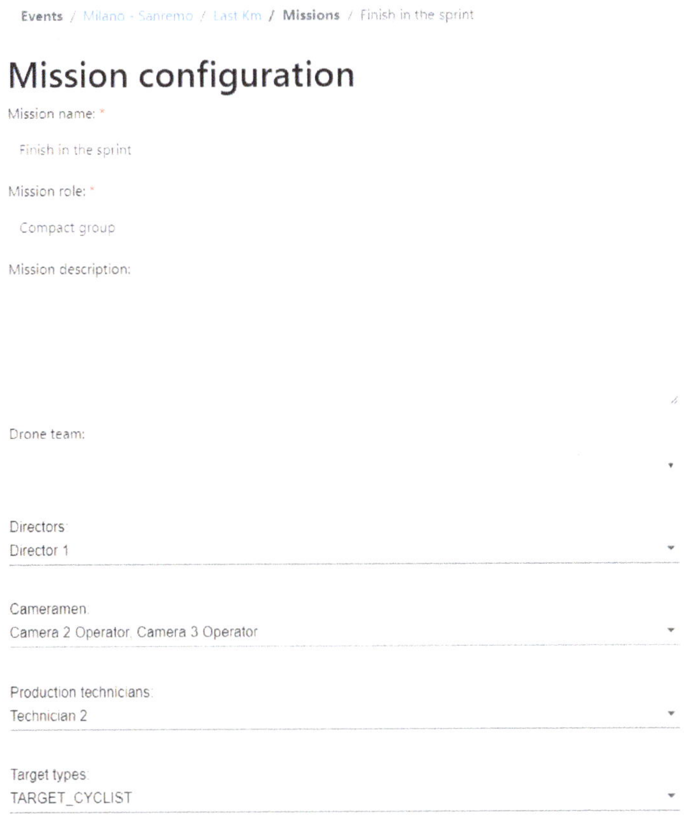

Events / Milano - Sanremo / Last Km / Missions / Finish in the sprint

Mission configuration

Mission name: *

Finish in the sprint

Mission role: *

Compact group

Mission description:

Drone team:

Directors:
Director 1

Cameramen:
Camera 2 Operator, Camera 3 Operator

Production technicians:
Technician 2

Target types:
TARGET_CYCLIST

Figure 6. Example screenshot of the mission configuration page. The people and drone team involved can be specified.

5.2.4. Mission Execution

Once the plans for a shooting mission have been computed and validated, the director can select any valid combination of mission and SAS roles to execute. This is done by triggering a "run" command from the Dashboard, which will notify the Mission Controller to start the selected missions/SASs. The execution of a selected SAS can be monitored in real-time through a dedicated page of the Dashboard, as illustrated in Figure 8. The page shows information about the corresponding mission, including mission metadata (e.g., name, role, date, and a countdown timer), and associated SASs. Selecting a SAS will show the corresponding timelines with the shooting actions involved, as well as the video streams coming from drones cameras, when available. Furthermore, for each drone, a console to manually adjust camera parameters (i.e., gimbal orientation, zoom level, and focus) is shown.

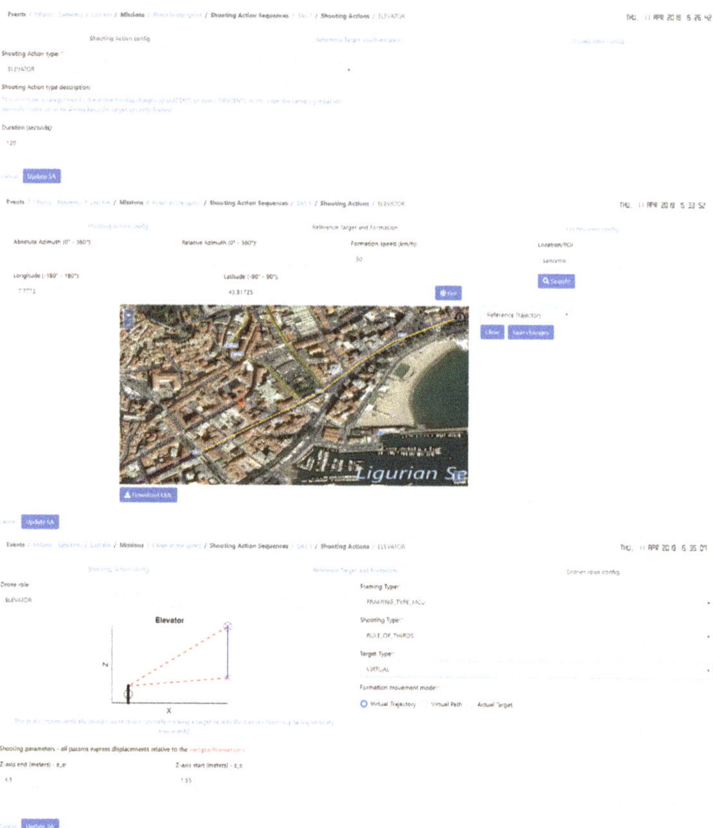

Figure 7. Example screenshots of the three tabs of the shooting action configuration page. From top to bottom: main data, drone formation data, and shooting role data.

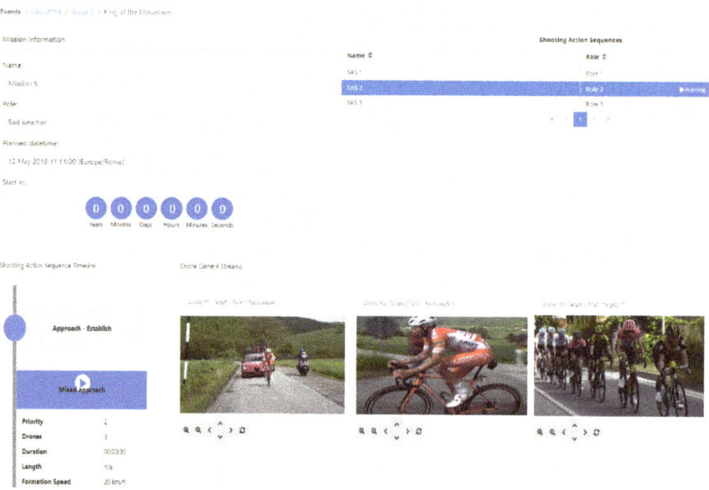

Figure 8. Example screenshot of the mission execution page.

6. Experiments on Shooting Mission Planning

In this section, we showcase the use of our tools in example scenarios to film outdoor events. Within the framework of MULTIDRONE project, we integrated our tools for autonomous cinematography into a real system with multiple drones. In the following, we first describe an example mission to illustrate the whole process to write, translate, and execute missions. Then, we show the execution of that example shooting mission in a simulator developed in MULTIDRONE. Last, we show some results of a real experimental media production performed in north Germany shooting a rowing race.

6.1. Example of a Shooting Mission

This section details an example of a shooting mission to cover an outdoor event with a single moving target to film. The objective is to illustrate the whole process to write the shooting mission through our language in Section 4, how to translate it into a list of tasks to feed a Planner, and a possible output plan computed for each drone.

6.1.1. Shooting Mission XML

The example shooting mission consists of a parent event that is a race with a single target to be filmed by a drone team. There could be several leaf events to associate missions, as for example, the start of the race or the finish line. For simplicity, our parent event contains a single leaf event named "START_RACE" with a duration of 20 s. Associated with the start of the race there is a mission with role "main" described by two shooting action sequences with the same role "main", i.e., to be executed in parallel. One of the shooting action sequences has two consecutive shooting actions with a single shooting role each, i.e., single-drone shooting actions. The other shooting action sequence has a shooting action with two shooting roles, i.e., a shooting action to be performed by two drones simultaneously.

Figure 9 shows an example of the XML code. For the sake of simplicity, only the main elements of the XML are included. Tags relative to target information and positioning data (e.g., RT trajectory, origin of formation, shooting target, etc.) have been omitted. A complete version of the XML file of the example (also including alternative missions and shooting action sequences with other roles) can be seen online (https://grvc.us.es/downloads/docs/example0.xml).

6.1.2. List of Shooting Actions

Once the Mission Controller receives the XML shown in the previous section, this is parsed to extract the relevant information and build a list of objects of type `SHOOTING ACTION` and `SHOOTING ROLE`. The next code block shows the resulting list with most relevant information, where `SHOOTING ACTION` data type is denoted as `SA` whereas `SHOOTING ROLE` data type as `SR`.

The idea in this example mission is to take a lateral shot with medium-shot framing for 10 s, followed by a static shot with close-up framing. In parallel, an orbital shot with two drones (one with medium-shot framing and another with long-shot framing) should be taken for 20 s. The shooting parameters depend on the type of shot; in this example, they indicate relative positioning of the drone formation with respect to the RT (Cartesian coordinates for the lateral and static shots and polar coordinates for the orbital shot).

In the XML, duration was not specified for shooting actions SA2 and SA3 (values set to 0). Therefore, they are estimated as the difference between the event duration (20 s in the XML) and the duration of previous shooting actions, if any (10 s for SA1). Thus, SA2 will last 10 s, whereas SA3 20 s. When the duration of shooting actions is specified in time, as in this example, the XML tag "length" is set to 0.

```
- <event>
    <description>Example 0, car race.</description>
    <name>Car race</name>
    <plannedStartTime>2019-08-01T09:25:00Z</plannedStartTime>
  - <childEvents>
    - <event>
        <uuid>START_RACE</uuid>
        <plannedDuration>20</plannedDuration>
      - <missions>
        - <mission>
            <uuid>M1</uuid>
            <role>main</role>
          - <shootingActionSequences>
            - <shootingActionSequence>
                <role>main</role>
                <uuid>S1</uuid>
              - <shootingActions>
                - <shootingAction>
                    <uuid>A1</uuid>
                    <duration>10</duration>
                    <length>0</length>
                    <name>SA1</name>
                  - <shootingRoles>
                    - <shootingRole>
                        <shootingTypeName>SHOOT_TYPE_LATERAL</shootingTypeName>
                        <framingTypeName>FRAMING_TYPE_MS</framingTypeName>
                        <params>{"y":-7,"z":8}</params>
                      </shootingRole>
                    </shootingRoles>
                  </shootingAction>
                - <shootingAction>
                    <uuid>A2</uuid>
                    <duration>0</duration>
                    <length>0</length>
                    <name>SA2</name>
                  - <shootingRoles>
                    - <shootingRole>
                        <shootingTypeName>SHOOT_TYPE_STATIC</shootingTypeName>
                        <framingTypeName>FRAMING_TYPE_CU</framingTypeName>
                        <params>{"x":5,"y":-2,"z":10}</params>
                      </shootingRole>
                    </shootingRoles>
                  </shootingAction>
                </shootingActions>
              </shootingActionSequence>
            - <shootingActionSequence>
                <role>main</role>
                <uuid>S2</uuid>
              - <shootingActions>
                - <shootingAction>
                    <uuid>A3</uuid>
                    <duration>0</duration>
                    <length>0</length>
                    <name>SA3</name>
                  - <shootingRoles>
                    - <shootingRole>
                        <shootingTypeName>SHOOT_TYPE_ORBIT</shootingTypeName>
                        <framingTypeName>FRAMING_TYPE_LS</framingTypeName>
                      - <params>
                          { "radius":3, "initial azimuth":0, "z":3, "relative angular speed":0.52 }
                        </params>
                      </shootingRole>
                    - <shootingRole>
                        <shootingTypeName>SHOOT_TYPE_ORBIT</shootingTypeName>
                        <framingTypeName>FRAMING_TYPE_MS</framingTypeName>
                      - <params>
                          { "radius":3, "initial azimuth":180, "z":3, "relative angular speed":0.52 }
                        </params>
                      </shootingRole>
                    </shootingRoles>
                  </shootingAction>
                </shootingActions>
              </shootingActionSequence>
            </shootingActionSequences>
          </mission>
        </missions>
      </event>
    </childEvents>
  </event>
```


Figure 9. Example XML describing our race shooting mission.

```
list<SA>:
SA1:
start_event: START_RACE
planned_start_time: 0
mission_ID: M1
sequence_ID: S1
action_ID: A1
duration: 10
shooting_roles:
SR1:
shooting_type: SHOOT_TYPE_LATERAL
framing_type: FRAMING_TYPE_MS
shooting_parameters:
{``y'':-7,``z'':8}
SA2:
start_event: -
planned_start_time: 10
mission_ID: M1
sequence_ID: S1
action_ID: A2
duration: 10
shooting_roles:
SR1:
shooting_type: SHOOT_TYPE_STATIC
framing_type: FRAMING_TYPE_CU
shooting_parameters:
{``x'':5,``y'':-2,``z'':10}
SA3:
start_event: START_RACE
planned_start_time: 0
mission_ID: M1
sequence_ID: S2
action_ID: A3
duration: 20
shooting_roles:
SR1:
shooting_type: SHOOT_TYPE_ORBIT
framing_type: FRAMING_TYPE_MS
shooting_parameters:
{``initial_azimuth'':0,
``angular_speed'':0.52,
``radius'':3,``z'':3}
SR2:
shooting_type: SHOOT_TYPE_ORBIT
framing_type: FRAMING_TYPE_LS
shooting_parameters:
{``initial_azimuth'':180,
``angular_speed'':0.52,
``radius'':3,``z'':3}
```

6.1.3. Generated Plan Per Drone

Once the list from previous section is sent to the Planner, this is in charge of solving the task assignment problem computing a plan fulfilling with time and battery constraints. Shooting actions with several shooting roles, i.e., multi-drone shots, are decomposed into several tasks with the same starting time, one per drone. With the location and timing information from all tasks, as well as the position and battery for the available drones, the Planner computes a schedule solving the task allocation problem. Different planners could be used and its implementation is not the focus of this paper (see [19] for more information). In our example, at least three drones are necessary to implement the shooting mission. One drone could perform two consecutive tasks (corresponding to SA1 and SA2), while the other two drones perform one single task each (corresponding to each of the shooting roles in SA3). The final plan could consist of the following list of `DRONE ACTION` objects per drone.

```
Drone 1:
DA1-1:
action_type: TAKE-OFF
action_id: -
start_event: GET_READY
shooting_action: -
path: -
DA1-2:
action_type: GOTOWAYPOINT
action_id: A1
start_event: -
shooting_action: -
path: (planned path to SA1
start position)
DA1-3:
action_type: SHOOTING
action_id: A1
start_event: START_RACE
shooting_action: SA1
path: -
DA1-4:
action_type: GOTOWAYPOINT
action_id: A2
start_event: -
shooting_action: -
path: (planned path to SA2
start position)
DA1-5:
action_type: SHOOTING
action_id: A2
start_event: -
shooting_action: SA2
path: -
DA1-6:
action_type: GOTOWAYPOINT
action_id: -
start_event: -
shooting_action: -
```

```
path: (planned path to
landing station)
DA1-7:
action_type: LAND
action_id: -
start_event: -
shooting_action: -
path: -
Drone 2:
DA2-1:
action_type: TAKE-OFF
action_id: -
start_event: GET_READY
shooting_action: -
path: -
DA2-2:
action_type: GOTOWAYPOINT
action_id: A3
start_event: -
shooting_action: -
path: (planned path to SA3
start position)
DA2-3:
action_type: SHOOTING
action_id: A3
start_event: START_RACE
shooting_action: SA3' (as SA3 but
only with SR1)
path: -
DA2-4:
action_type: GOTOWAYPOINT
action_id: -
start_event: -
shooting_action: -
path: (planned path to landing
station)
DA2-5:
action_type: LAND
action_id: -
start_event: -
shooting_action: -
path: -
Drone 3:
DA3-1:
action_type: TAKE-OFF
action_id: -
start_event: GET_READY
shooting_action: -
path: -
DA3-2:
```

```
action_type: GOTOWAYPOINT
action_id: A3
start_event: -
shooting_action: -
path: (planned path to SA3
start position)
DA3-3:
action_type: SHOOTING
action_id: A3
start_event: START_RACE
shooting_action: SA3'' (as SA3 but
only with SR2)
path: -
DA3-4:
action_type: GOTOWAYPOINT
action_id: -
start_event: -
shooting_action: -
path: (planned path to landing
station)
DA3-5:
action_type: LAND
action_id: -
start_event: -
shooting_action: -
path: -
```

These lists of `DRONE ACTION` objects are sent to the respective drones. Shooting actions are interleaved with navigation actions. A `GET_READY` event has been included. This is generated manually by the director at the beginning of the mission to get the system started and take off the drones. Thus, each drone takes off and navigates to its starting position in its associated shooting action. Then, it waits there for the start event of that shooting action, to trigger the shooting action itself. This process is repeated sequentially until the drone is commanded to land. In other missions, there may be landing operations in between shooting actions to recharge batteries, depending on the drones maximum flight time.

6.2. Simulation of Example Mission

This section shows a simulation of the example shooting mission described in the previous section. A simulator developed in MULTIDRONE project is used. It is based on GAZEBO [23] and the PX4 [24] SITL (Software In The Loop) functionality. GAZEBO is a well-known open source multi-robot simulator. It allows for fast robot prototyping and creation of 3D scenarios. PX4 is an open-source autopilot software, implementing a set of functionalities to fly a drone both manually and autonomously. The SITL is a functionality to use PX4 with GAZEBO, simulating drone's sensors such as Global Navigation Satellite System (GNSS) and Inertial Measurement Unit (IMU). Our simulator also uses *UAV Abstraction Layer* (UAL) [25], an open source library to interact with autonomous drones. With this simulator and our tools for media production, it is easy to run and test different examples without flying real drones but using the very same software that would run in the real system.

Recall that the mission consists of two shooting action sequences, one with a drone performing first a lateral shot (SA1) and then a static shot (SA2), and another with two drones performing an orbital shot together (SA3). In the simulation, the shooting target is simulated with a car following a straight road that the drones must track and film. Figure 10 depicts the timeline of the executed mission. Everything starts with the `GET_READY` event, which makes all drones take off and go to their corresponding waypoints. Then, the `START_RACE` event triggers both sequences. Finally, after 20 s, the drones come back to the home position and land.

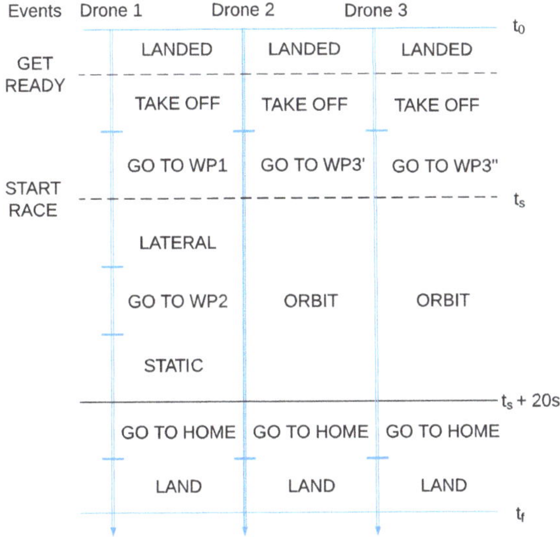

Figure 10. Execution of the example mission. WP1, WP2, WP3′, and WP3″ are drone start positions for SA1, SA2, SA3′, and SA3″, respectively.

Figure 11 shows *Drone 1* doing the lateral shot while *Drone 2* and *Drone 3* are orbiting around the target. Figure 12 shows *Drone 1* doing the static shot while *Drone 2* and *Drone 3* are still orbiting (The complete simulation can be seen in: https://youtu.be/qRPXTid9dFI.).

Figure 11. Screenshots of drones executing SA1 and SA3 in our simulation.

Figure 12. Screenshots of drones executing SA2 and SA3 in our simulation.

6.3. Integration with Real Drones

After validating the system in simulation, we integrated our tools with custom-built drones to run media production on real outdoor events. Figure 13 shows one of our drones, based on an hexarotor platform (120 cm of diagonal distance) with a Pixhawk 2.1 autopilot running PX4 [24]. The payload consists of an Intel NUC computer, a Here + RTK GNSS, a communication board, and a gimbal with a Blackmagic Micro Cinema camera.

Figure 13. One of the drones used for autonomous media production.

The MULTIDRONE consortium ran some experimental media production in a countryside area near Bothkamp, in the north of Germany. We tested our system in mock-up, outdoor scenarios, implementing missions (e.g., shooting cyclists and rowers) with one to three drones and different shot types. The objective was to demonstrate the integration of the whole system and to explore the artistic possibilities of our tools for media production.

In particular, Figure 14 depicts some snapshots from an experiment where two drones have to shoot a rowing race in a lake. In that experiment, the original shooting mission consists of a lateral shot that runs concurrently with a fly-through followed by a static panoramic. The first shot is performed by one of the drones, which flies parallel to the boats over the yard. The other two shots are carried out by the other drone, which goes over the lake coming through the trees, and then keeps static following the rowers with the camera. In this case, one of the boats carries a GNSS receiver sending their position to the computers on board the drones, so that the gimbal cameras can track them all time as shooting targets. A video showing the whole shooting mission is available online (https://www.youtube.com/watch?v=COay0hZsMzk&feature=youtu.be).

Figure 14. Experimental media production of a rowing race with two drones. The images are taken from the cameras on board the drones: the drone taking the lateral shot from the yard (**left**), and the drone taking the static shot after the fly-through (**right**).

7. Conclusions and Future Work

This paper described a set of tools to support autonomous media production with multiple drones. We contributed with an XML-based language for shooting mission description. The language builds upon new concepts, such as events, shooting actions, or shooting roles, to create media production missions. High-level cinematography principles and different shot types can be implemented. Our language has proved to be flexible and adaptable enough to describe current cinematography operations, as new shots and parameters can easily be incorporated. Moreover, the system is able to bridge the gap between common vocabulary from editorial teams and robotics planning tools. We also contributed with our Dashboard, a graphical tool to support the editorial crew creating and executing shooting missions. The Dashboard has been designed to be intuitive and simple, and users can define complex shooting missions by navigating through a few configuration pages, describe shot geometry by clicking on aerial maps, and validate and execute plans for the missions. This tool generates XML files that can be interpreted by our Mission Controller, to compute mission plans and execute them.

Our system has been integrated and tested in a realistic simulator and during experimental media production with actual drones. As future work, we plan to explore the combination of more complex multi-camera shots to assess the capabilities of multi-drone teams for media production from an artistic and editorial point of view.

Author Contributions: Conceptualization, A.T.-G., J.C., M.M., F.N., and A.M.; software and validation, Á.M.-R., A.T.-G., M.M., and S.M.; writing and editing, A.T.-G., J.C., M.M., and S.M.; supervision, F.N., A.M., and A.O.; funding acquisition, A.M. and A.O. All authors have read and agreed to the published version of the manuscript.

Funding: This work has received funding from the European Union's Horizon 2020 research and innovation programme under grant agreement No 731667 (MULTIDRONE). This publication reflects the authors' views only. The European Commission is not responsible for any use that may be made of the information it contains.

Conflicts of Interest: The authors declare no conflicts of interest.

Abbreviations

The following abbreviations are used in this manuscript:

CRUD	Create, Read, Update and Delete
CTZ	Control Zone
GNSS	Global Navigation Satellite System
GUI	Graphical User Interface
HTML	HyperText Markup Language
HTTP	HyperText Transfer Protocol
IMU	Inertial Measurement Unit
KML	Keyhole Markup Language
RT	Reference Target
RTK	Real Time Kinematic
SA	Shooting Action
SAS	Shooting Action Sequence
SITL	Software In The Loop
SR	Shooting Role
UAV	Unmanned Aerial Vehicle
UML	Unified Modeling Language
XML	Extensible Markup Language

References

1. Mademlis, I.; Mygdalis, V.; Nikolaidis, N.; Montagnuolo, M.; Negro, F.; Messina, A.; Pitas, I. High-Level Multiple-UAV Cinematography Tools for Covering Outdoor Events. *IEEE Trans. Broadcast.* **2019**, *65*, 627–635. [CrossRef]
2. NBC News. New Drone Video Captures Scale of Haiti Hurricane Damage. 2016. Available online: http://www.nbcnews.com/video/new-drone-video-captures-scale-of-haiti-hurricane-damage-784114243853 (accessed on 22 February 2020).
3. Charlton, A. Sochi Drone Shooting Olympic TV, Not Terrorists. 2014. Available online: http://wintergames.ap.org/article/sochi-drone-shooting-olympic-tv-not-terrorists (accessed on 22 February 2020).
4. Gallagher, K. How Drones Powered Rio's Olympic Coverage. 2016. Available online: http://www.simulyze.com/blog/how-drones-powered-rios-olympic-coverage (accessed on 22 February 2020).
5. DJI. 2019. Available online: https://www.dji.com (accessed on 22 February 2020).
6. Airdog. 2019. Available online: https://www.airdog.com (accessed on 22 February 2020).
7. 3DR. 3DR SOLO. 2019. Available online: https://3dr.com/solo-drone/ (accessed on 22 February 2020).
8. Yuneec. Yuneec Typhoon 4K. 2019. Available online: https://us.yuneec.com/typhoon-4k-overview (accessed on 22 February 2020).
9. Joubert, N.; Goldman, D.B.; Berthouzoz, F.; Roberts, M.; Landay, J.A.; Hanrahan, P. Towards a Drone Cinematographer: Guiding Quadrotor Cameras using Visual Composition Principles. *arXiv* **2016**, arXiv:cs.GR/1610.01691.
10. Gebhardt, C.; Hepp, B.; Nägeli, T.; Stevšić, S.; Hilliges, O. Airways: Optimization-Based Planning of Quadrotor Trajectories according to High-Level User Goals. In Proceedings of the Conference on Human Factors in Computing Systems (CHI), San Jose, CA, USA, 7–12 May 2016; pp. 2508–2519. [CrossRef]
11. Joubert, N.; Roberts, M.; Truong, A.; Berthouzoz, F.; Hanrahan, P. An interactive tool for designing quadrotor camera shots. *ACM Trans. Graph.* **2015**, *34*, 1–11. [CrossRef]
12. Lino, C.; Christie, M. Intuitive and efficient camera control with the toric space. *ACM Trans. Graph.* **2015**, *34*, 82:1–82:12. [CrossRef]
13. Galvane, Q.; Fleureau, J.; Tariolle, F.L.; Guillotel, P. Automated Cinematography with Unmanned Aerial Vehicles. In *Proceedings of the Eurographics Workshop on Intelligent Cinematography and Editing*; Eurographics Association: Goslar, Germany, May 2016. [CrossRef]
14. Saeed, A.; Abdelkader, A.; Khan, M.; Neishaboori, A.; Harras, K.A.; Mohamed, A. On Realistic Target Coverage by Autonomous Drones. *arXiv* **2017**, arXiv:1702.03456.
15. Nägeli, T.; Meier, L.; Domahidi, A.; Alonso-Mora, J.; Hilliges, O. Real-time planning for automated multi-view drone cinematography. *ACM Trans. Graph.* **2017**, *36*, 1–10. [CrossRef]
16. VC Technology. Litchi. 2019. Available online: https://flylitchi.com/ (accessed on 22 February 2020).
17. Garage, A. Skywand. 2016. Available online: https://skywand.com/ (accessed on 22 February 2020).
18. FreeSkies. FreeSkies CoPilot. 2016. Available online: http://freeskies.co/ (accessed on 22 February 2020).
19. Torres-González, A.; Capitán, J.; Cunha, R.; Ollero, A.; Mademlis, I. A multidrone approach for autonomous cinematography planning. In Proceedings of the Iberian Robotics Conference (ROBOT'), Seville, Spain, 22–24 November 2017.
20. Open Geospatial Consortium. OGC KML 2.3. 2015. Available online: http://docs.opengeospatial.org/is/12-007r2/12-007r2.html (accessed on 22 February 2020).
21. Otto, M.; Thornton, J. Bootstrap. 2018. Available online: https://getbootstrap.com/ (accessed on 22 February 2020).
22. Google. Angular. 2018. Available online: https://angular.io/ (accessed on 22 February 2020).
23. Koenig, N.; Howard, A. Design and Use Paradigms for Gazebo, An Open-Source Multi-Robot Simulator. In Proceedings of the IEEE/RSJ International Conference on Intelligent Robots and Systems, Sendai, Japan, 28 September–2 October 2004; pp. 2149–2154.

24. Meier, L.; Gubler, T.; Oes, J.; Sidrane, D.; Agar, D.; Küng, B.; Babushkin, A.; px4dev; Charlebois, M.; Bapst, R.; et al. PX4/Firmware: v1.7.3 Stable Release. 2018. Available online: https://doi.org/10.5281/zenodo.1136171 (accessed on 22 February 2020).
25. Real, F.; Torres-González, A.; Ramón-Soria, P.; Capitán, J.; Ollero, A. UAL: An Abstraction Layer for Unmanned Aerial Vehicles. In Proceedings of the 2nd International Symposium on Aerial Robotics, Philadelphia, PA, USA, 11–12 June 2018.

Article

Development of an Effective Information Media Using Two Android Robots

Toshiaki Nishio [1,*], Yuichiro Yoshikawa [1], Kohei Ogawa [1] and Hiroshi Ishiguro [1,2]

1 Graduate school of Engineering Science, Osaka University, Osaka 560-8531, Japan
2 Intelligent Robotics and Communication Laboratories, ATR, Kyoto 619-0237, Japan
* Correspondence: nishio.toshiaki@irl.sys.es.osaka-u.ac.jp

Received: 28 July 2019; Accepted: 18 August 2019; Published: 21 August 2019

Abstract: Conversational robots have been used to convey information to people in the real world. Android robots, which have a human-like appearance, are expected to be able to convey not only objective information but also subjective information, such as a robot's feelings. Meanwhile, as an approach to realize attractive conversation, multi-party conversation by multiple robots was the focus of this study. By collaborating among several robots, the robots provide information while maintaining the naturalness of conversation. However, the effectiveness of interaction with people has not been surveyed using this method. In this paper, to develop more efficient media to convey information, we propose a scenario-based, semi-passive conversation system using two androids. To verify its effectiveness, we conducted a subjective experiment comparing it to a system that does not include any interaction with people, and we investigated how much information the proposed system successfully conveys by using a recall test and a questionnaire about the conversation and androids. The experimental results showed that participants who engaged with the proposed system recalled more content from the conversation and felt more empathic concern for androids.

Keywords: human robot interaction; android robot; passive social conversation; multiple conversation robots

1. Introduction

In recent years, communication robots have become popular in the real world [1]. To promote their use in the current society, the capability of communication in robots should be improved so that users can interact with robots more easily. One approach to realizing easy communication with a user is using humanoid robots. For example, robots have been used as communication media that provide information autonomously to visitors in public spaces such as a science museum [2] and a shopping mall [3]. Humanoid robots that resemble the appearance of human beings are called android robots. Owing to their appearance, an observer who sees android robots considers android robots as human for a few seconds [4]. Android robots can present a stronger sense of existence than other media such as video, or a speakerphone [5]. Android robots can effectively represent human-like mental states, including subtle emotion, by using nonverbal communication [6]. Therefore, it is argued that android robots can become influential conversation media that can effectively convey information [7].

In previous work on the conversation of agents, voice has been widely adopted as one of the most accessible and most universal modalities in many applications, such as an application for weather forecasts [8], for the guiding of buses [9], and for providing information to visitors in a science museum [10]. However, voice recognition in the real world has still been a formidable challenge [11]. Furthermore, especially in robot applications, it becomes more difficult because a human would often speak in less formally correct ways, such as in dialects, toward the human-like agent. Adopting a very human-like appearance in the android robot also causes a risk called the adaptation gap [12]: humans tend to be easily disappointed by the robot's utterance because they have heightened expectations for a human-like, contextually natural response due to the appearance of human-like organs corresponding to voice production, i.e., mouth and throat.

To adopt verbal communication in robot applications without suffering from potential failure and disappointments, a specific form of communication called multi-party conversation has been developed in the field of human–robot interaction [13]. There is a study of robots as a passive social medium, where robots provide information for people with dialogue between two robots while not interacting with people directly [14]. It was reported that passive social robots succeeded in getting more attention from observers than a single robot receives alone. People who observe a scene in which two robots communicate with each other are more likely to treat those robots as if they are human and evaluate their dialogue as more understandable than they would for robots without communication between them [15]. However, without listening to a human response, it is difficult for robots to keep a human paying attention to the conversation for an extended period. Arimoto et al. showed that inter-robot turn-taking triggered by a human response contributes to maintaining the sense of conversation, even though the robots produced pre-scripted utterances and ignored the content of the human response [13]. However, the extent to which their conversation was successfully conveyed to humans has not been examined, especially in the case of affective content. Therefore, we constructed and examined a scenario-based, semi-passive conversation system using two androids that basically talked to each other and sometimes actively listened for human answers to their questions, including affective ones. To validate the usefulness of semi-passive social androids as a medium to provide information for people, a comparative experiment with passive social androids was conducted. Specifically, we evaluated the feasibility of semi-passive social androids to transmit the feelings of androids (subjective information) and the contents of the conversation (objective information) to people. To confirm the functionality of communicating subjective information, we prepared a script that had some statements for making participants feel empathic concern with androids and asked for their impressions of the androids through a questionnaire. Additionally, to confirm the functionality of communicating objective information, we added general information about androids in the script and measured how much information the participants could remember with a recall test. The between-subjects design experiment was conducted, where the subjects talked with either semi-passive social androids or passive social androids for ten minutes. After the talk, participants evaluated the androids and the conversations with a recall test and a questionnaire. From the result, it was verified that the semi-passive social condition was assessed higher than the passive social condition based on the information memorized by the subjects and the emotions conveyed and evoked by the androids.

2. Materials and Methods

This section describes details of the semi-passive dialogue system using two androids, as well as the experiment to validate its effect on the conveyance of the objective and subjective information.

2.1. System Overview

The dialogue system developed in this research was a scenario-based verbal communication system. The system supposed multi-party conversation among two androids and a person. Two androids normally led the conversation by talking with each other and sometimes engaged the person in the conversation by asking his/her opinion or experience. After receiving an answer from the person, the androids recognized it and responded to it. Androids spoke and moved according to directions written in a prepared script, which included the contents of speech and behavior for each android. Some templates for responding to the person were also written in the script, and the system selected a suitable template as a response.

Figure 1 shows an overview of the system. The system was comprised of four modules: the speech recognition module, the text-to-speech module, the androids' behavior control module, and the scenario manager module. Androids could vocally communicate with people by integrating these modules together. The function of each module was as follows:

- Speech recognition module: This module's function was to capture the voice data of a person and convert the voice data to text.
- Text-to-speech module: This module's function was to convert the text received from the scenario manager module into voice data and playing the resulting voice data.
- Androids' behavior control module: This module controlled the body movement of the androids, such as the mouth movement for utterance, head or eye motion for looking at people or the other robot, breathing motion, and emotional expression.
- Scenario manager module: This module was used for reading the script and controlling other modules by sending order sequences.

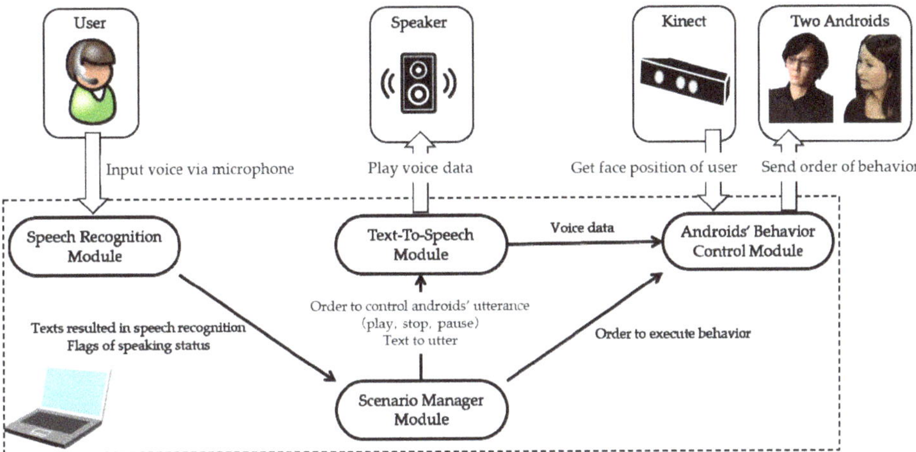

Figure 1. System diagram.

2.1.1. Android: The Robot Having Human-Like Appearance

An android robot is a humanoid robot that has a very human-like appearance. For the conversation system, we employed a female android called Geminoid F (Figure 2a) and a male android called Geminoid HI4 (Figure 2b). Both androids were developed in a collaborative project between Osaka University and Advanced Telecommunications Research Institute International (ATR). These androids have skin made of silicon rubber, and the shape of their skin was molded from an actual human. Because the tactile sensation and texture of the surfaces are very close to that of a human, they can convey a human-like presence. In addition, these androids have many degrees of freedom (DOFs) in their faces (Geminoid F: 7 DOFs, Geminoid HI4: 8 DOFs), with enough DOFs on their face allowing them to convey a natural human-like presence. Since the androids are driven by air compression actuators, they can move softly and show their behavior without producing loud noises. Due to these factors, we concluded that Geminoid F and Gemenoid HI4 were suitable for showing human-like communication with people. On the other hand, neither F nor HI4 have any sensor to get environmental information inside their body, and therefore, additional external sensors were needed to obtain the voices and positions of people. Each robot had a speaker inside its body; however, a sound played from the speaker was not clear. Thus, we designed external speakers for each android.

(a)

(b)

Figure 2. Appearance of Geminoids: (**a**) Geminoid F: female android; (**b**) Geminoid HI4: male android.

2.1.2. Speech Recognition Module

The speech recognition module converts voice data from people into text and sends the recognized text to the scenario manager module. A non-directional close-talking microphone is used to capture voices, and Dragon Speech 11 (Nuance Communication Inc., Burlington, MA, United States) is used for speech recognition. The module also detects utterance sections using a threshold of sound pressure. Specifically, the module measures an average of sound pressure sampling of 44.1 kHz for every millisecond. When an average pressure over the specified threshold has been detected for more than 500 milliseconds continuously, the module determines that people are in talking status. When an average pressure under the threshold has been detected for more than 1500 milliseconds in the talking status, the module determines that the person has finished speaking. The module also sends flags to represent the talking and talk-end status to the scenario management module. To avoid detecting noises, the module sends text recognized by Dragon Speech only during the talking status.

2.1.3. Text-To-Speech Module

The text-to-speech module converts text received from the scenario manager module to sound data and plays it. Using the AITalk Custom Voice (AI Inc., Tokyo, Japan) as a voice synthesizer, the module can do real-time synthesis, in which the synthesizing of voice data and playback on a device are executed in parallel. The module receives three types of orders with the text from the scenario manager module: start playing, stop playing, and pause playing. The module can not only stop/pause playing voice data immediately but also optionally stop/pause it at the end of a current phrase. The module has variable parameters for the start playing order, including volume, speed, pitch, range, length of short pauses in the sentence, length of long pauses in the sentence, and the length of the end pause of the sentence.

2.1.4. Androids' Behavior Control Module

The Androids' Behavior Control Module controls the actuators of androids to express various behavior such as emotional expression, mouth movement with synchronized voice, and motion to make an android look at people or the other android. The behavior of the android is described in a motion file, which contains the positions of each joint at 50 milliseconds intervals called frames. The position of a joint is set using a value from 0 to 255. Androids execute idling behaviors, which are minimal behaviors like blinking, as well as breathing, which is expressed by slight movements of the shoulders, waist, and neck. Moreover, an android can express these behaviors at any time by executing prepared motion files. Some motions for which preparation is difficult, like looking at the person's face, are realized using sensors. When looking at a face, the face position is detected in three-dimensional space using Kinect for Windows v2 by Microsoft, and is associated with the three axes of the neck of the android. Values of these three axes are transformed from the face position using a projection matrix calculated via calibration using the least squares method with 16 pairs of position of the axes of the neck and the position of the detected face. Regarding lip motion synchronized with voice, the motion is generated using formant information included in the voice [16].

2.1.5. Scenario Manager Module

The scenario manager module controls the conversation and behavior of two androids by following a prepared script. In the script, the details of a dialogue, such as speech contents, the behavior of the two androids, and the timings of speech recognition, are specified, and the module sends orders to the other modules sequentially. Moreover, the script also has templates to be used as a response of an android to utterances from people at every scene, for which an android asks people something and the person may utter. Androids encourage the person to utter by asking him/her a yes-or-no question. When the person utters something, the module identifies the utterance as positive or negative, and the androids generate a response using a template for each category. If the person has not spoken anything or the speech recognition module fails to recognize speech, the module selects a template for an ambiguous response.

The module has a classifier to categorize the utterance of the person and learns it from training data, which is constructed using pairs of a recognized text of the utterance of people and a label of it. After morphological analysis of training data, vectors of bag of words (BOW) are created. Incidentally, stop words included in SlothLib [17], which is a programming library, are removed from the data. Also, high-frequency words included in the top 0.9% of the data and low-frequency words appearing only two times are removed. Additionally, to reduce the amount of sampling used for categorization, the dimension of BOW vectors is diminished until reaching the number of classes by using Latent Semantic Indexing (LSI). The module learns support vector machine (SVM) using BOW vectors and labels.

Morphological analysis is performed using JUMAN [18] which is a Japanese morpheme analysis program. For SVM, Scikit-Learn [19] is used, which is an open source machine learning library. Gemsim [20] is used for creating vectors of bag of words, and is a topic model library in Python.

After selecting a template, the module creates a text response by inputting words included in the person's utterance. What is extracted from the person's utterance are words with polarity. To extract polarity words, Japanese Sentiment Polarity Dictionaries (Volume of Verbs and Adjectives) ver. 1.0, created by Inui-Okazaki laboratory at Tohoku University, is used.

2.2. *Experiment*

This section describes a subject experiment between participants conducted to reveal the effectiveness of semi-passive social androids as communication media to convey objective information and subjective information. The semi-passive social androids tried to engage the participants by repeatedly giving directions to them in the conversation. This was expected to help the participants to remain engaged in and concentrating on the conversation. This engagement and concentration were expected to help the participants to recall the message in the conversation. Meanwhile, it was clear the messages when the messages were being directed to the participants because the android uttered toward the participant, and the participants were expected to feel the stronger will of the androids conveying them, and as a result, the messages were expected to be recognized as stronger. Accordingly, with respect to these expected effects of the semi-passive social androids, we examined three hypotheses in this experiment: (i) Participants will recall more objective information given in the semi-passive social conversation of two androids than in the passive social one. (ii) Participants will feel the subjective messages as being stronger in the semi-passive social conversation than in the passive social conversation. (iii) Participants will be moved to follow the subjective messages in the semi-passive social conversation more than in the passive social conversation. In addition to the three hypotheses, two other points, namely the degree of participant engagement with the conversation and the difficulty of the conversation, were surveyed to confirm that the system was able to run as intended and the experiment was conducted as expected.

2.2.1. Participants

The current study was approved by The ethics committee for research involving human subjects at the Graduate School of Engineering Science, Osaka University. The participants were 28 university students (14 male and 14 female students), whose ages ranged from 19 to 22 years old. They were randomly assigned to two conditions. In the passive condition, seven males and seven females participated in conversation with two androids that talked and directed their gaze to each other but did not do so to the participant. In the semi-passive one, the remaining seven males and seven females participated in the conversation with two androids that talked and directed their gaze not only to each other but also to the participant. No participants had any experience directly interacting with the androids before the experiment.

2.2.2. Apparatus

In the experimental room, the participants faced two androids across a table. The distances between the two androids and the participant formed an equilateral triangle with 1.2-m sides (Figure 3). Figure 4 shows a scene of the experiment.

The topic of the conversation was related to android robots and the Intelligent Robotics Laboratory where they were developed. We designed an approximately ten-minute scenario for the conversation to introduce both the subjective and objective information. To verify how efficiently the subjective information was conveyed, we included utterances by which the androids explained their situation and how they felt. For example, "Visitors touch my body as they like. I would like them to refrain from it a little more." and "I am seldom quickly repaired (by the laboratory members) when I am broken. Don't you think it's terrible treatment?" Note that, in the semi-passive social condition, when they mentioned "you," they looked at the participant. In contrast, in the passive one, they did not look at him or her. Meanwhile, to verify how efficiently the objective information was conveyed, we included utterances with numbers or proper nouns to introduce facts about the laboratory and the androids. For example, "In fact, our bodies are driven by air pressure at about 0.7 MPa." Figure 5 shows a part of the conversation used for the experiment. In addition to the script of the passive social condition, nine utterances toward the participant were included in the conversation used for the semi-passive condition. Note that, to avoid providing further information by these nine utterances, the androids were limited to utterances asking about the participant's agreement or opinions in relation to the androids' conversation.

The responses from the participants were classified into the four categories: no answer, positive answer, negative answer, and other. This allowed the androids to choose an appropriate but brief response without providing any further information. To categorize the participant's answer as one of the four types, the system was trained before the experiment. For this, we constructed another system that could select a response for the androids using the Wizard of Oz method and collected morphemes as training data via conversations between eight males and the system.

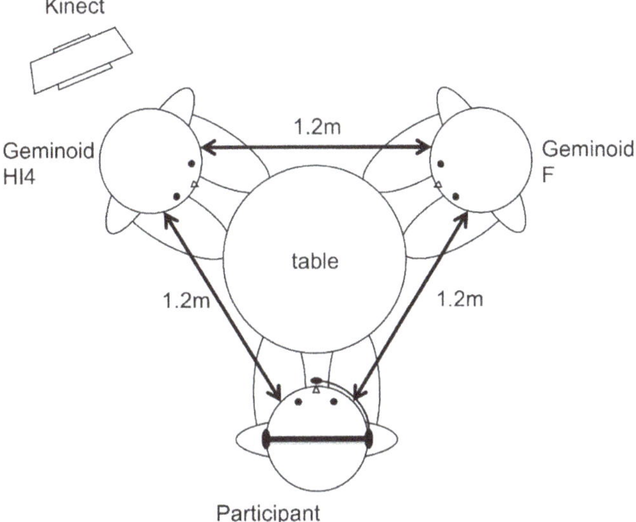

Figure 3. The environment of the experiment.

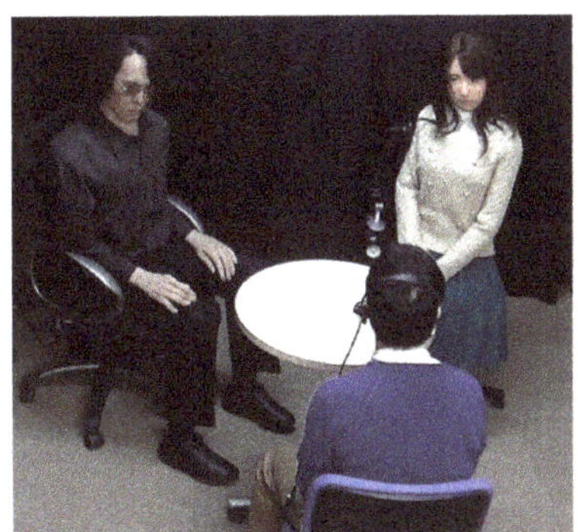

Figure 4. A scene of the experiment.

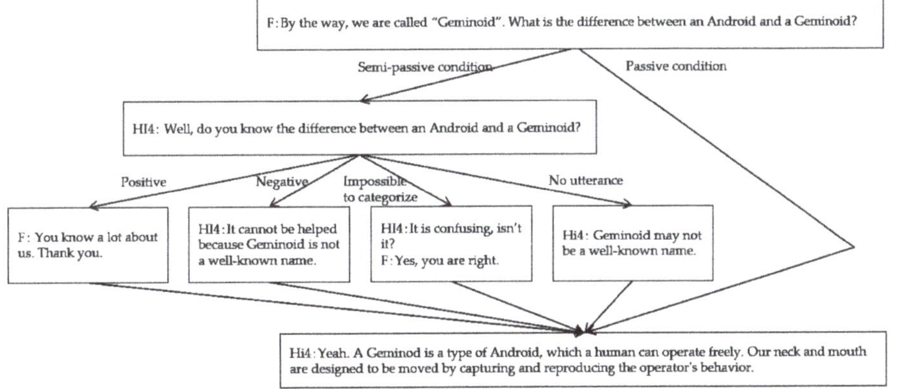

Figure 5. Example of the conversation.

2.2.3. Procedure

The experiment was conducted using the following procedure, one participant at a time. First, a participant waited in an anteroom until the start of the experiment. At a specified time, an experimenter explained an overview of the experiment in the anteroom and got the participant's consent regarding participation. After that, the experimenter provided instruction regarding the procedure of the experiment. In the instruction, the participant was requested to listen to the conversation between the androids. However, if the androids asked them questions during the conversation, the participant was permitted to answer it. Next, the participant put on a headset and configured the microphone using a function of Dragon Speech, which is an automatic speech recognition software. The software can adjust the volume level and check the quality of voice input by having the participant read a few prepared sentences aloud. Then, the participant moved to an experimental room containing the two androids. The participant sat across the table from the two androids. When the participant took their seat, the two androids started to talk each other. The conversation lasted for around 10 min, and when the conversation finished, the experimenter called out to the participant, directing them to go back to the anteroom. The participant removed the headset and answered a questionnaire and a recall test about the conversation. Lastly, the experimenter interviewed the participant.

2.2.4. Measurements

First, we checked whether the participants were adequately stimulated from the following two points of view. Namely, we checked whether the contents of the conversation presented in the experiment were too difficult for the participants to understand and whether the participants in the semi-passive condition were able to engage in the conversation. Regarding the difficulty of the conversation, it was checked whether the contents of conversation made sense (D-1) to the participants. Meanwhile, it was supposed to not be too difficult for participants (D-2) to avoid a floor effect in evaluating how much objective information was conveyed to participants.

D-1. Were you able to understand the conversation entirely?
D-2. Were you able to consent the conversation entirely?

Regarding the degree of engagement to the conversation, it is important for the system to correctly handle the replies from the participants so as to prevent them from feeling ignored to provide the experience of semi-passive conversation. If the system failed to do that, it was considered that the androids' attitude toward engagement in the conversation with the subjects, i.e., the feeling of being given attentive (E-1) and interested (E-4), as well as the participants' attitude to engage in the conversation with the androids, i.e., the motivation to speak (E-3) and to join the conversation (E-2), would not be recognized. Note that only the participants in the semi-passive condition were evaluated for this as a comparison to the passive condition in this regard was considered to be unfair because the androids never paid their attention to the participants in the passive one. The degree of engagement with the conversation was surveyed using the following questions:

E-1. Did you feel that the female android and the male android paid attention to you?
E-2. Did you feel that you were able to join the conversation?
E-3. Did you feel that you could give your opinion to the female android and the male android?
E-4. Did you feel that the female android and the male android were interested in your opinion?

To verify the first hypothesis regarding objective information, we conducted a recall test to confirm how much the participants memorized the content of the conversation. We created ten questions to ask about the contents that the androids conveyed in the conversation. For example, the participants were asked: "How many androids existed in this laboratory?", "To which country did the female android want to go?", and so on. We counted how many questions the participants could correctly answer. To easily judge whether the answers were correct, questions were chosen so that the correct answers could be described as a single word.

The second and third hypotheses are related to the effective conveyance of the subjective information. The second one concerns whether participants regarded messages from the androids more strongly in the semi-passive condition than in the passive one. To verify this, the strength of messages from the androids about the subject's experience in the conversation was evaluated by using the below questionnaire:

2-1. Androids requested you to help them leave the bad situation where they got a rough deal in the laboratory. Did you think of telling the laboratory members the request from the android?
2-2. Did you feel that the female android and the male android had firm beliefs?
2-3. Did you feel that you were urged strongly by the female android and the male android?
2-4. Did you feel that the statements of the female android and the male android were persuasive?

Furthermore, as predicted in hypothesis (iii), the participants were expected to be moved to follow the subjective messages in the semi-passive social conversation than in the passive social conversation. To verify this, we measured how much empathic concern the androids elicited from the subjects by using a questionnaire. In this experiment, we focused on empathic concern as a typical example of emotional response because the androids talked about their story to effect empathic concern in the conversation. Specifically, the following two questions were used:

3-1. Androids said in the conversation that they were often told that they were scary or spooky. Did you feel that it was cruel?
3-2. Androids said in the conversation that they were carried as baggage. Did you feel that it was cruel?

A seven-point Likert scale was adopted for questions about the strength of the message, empathic concern, the difficulty of contents, and engagement, which ranged from 1: Strongly disagree, 2: Disagree, 3: More or less disagree, 4: Undecided, 5: More or less agree, 6: Agree, to 7: Strongly agree.

3. Results

3.1. Manipulation Check

Regarding the difficulty of the conversation, no participant in either condition marked under 4 for the two questions (QD-1 and QD-2), which was interpreted as indicating that the contents of the conversation were appropriately prepared so as not to be too difficult. Regarding the degree of engagement to the conversation, the median scores obtained from the participants attending the semi-passive conversation for four questions (QE-1, QE-2, QE-3, and QE-4) were not less than intermediate points, which was interpreted as indicating that the system could provide the experience of conversation with a moderate level of engagement.

3.2. Recall Test for Hypothesis (I)

We evaluated the difference between the semi-passive condition and the passive condition using a recall test and the questionnaire. Figure 6a shows the average number of questions that the participants correctly answered. Mann–Whitney's U test was used to compare the scores of the two conditions. The result shows that the median of the number of correct answers of the semi-passive condition (Mdn = 8) is significantly higher than that of the passive condition (Mdn = 7) (U = 54.5, $p < 0.05$).

3.3. Evaluation of Questionnaire for Hypotheses (II) and (III)

Figure 6b,c show the questionnaire results. Medians of each question between the two conditions were compared using Mann–Whitney's U test. Figure 6b shows scores of four questions about the strength of subjective message (Q2-1, Q2-2, Q2-3, and Q2-4), and no significant difference was suggested. Figure 6c shows scores from the two questions about empathic concern from participants toward the androids (Q3-1 and Q3-2). The result of Q3-1 shows that the score of the semi-passive condition (Mdn = 5.5) was higher than that of the passive condition (Mdn = 4.5) with a marginal difference

(U = 57.5, $p < 0.1$). The result of Q3-2 shows that the score of the semi-passive condition (Mdn = 4.5) was significantly higher than that of the passive condition (Mdn = 3) (U = 48.5, $p < 0.05$).

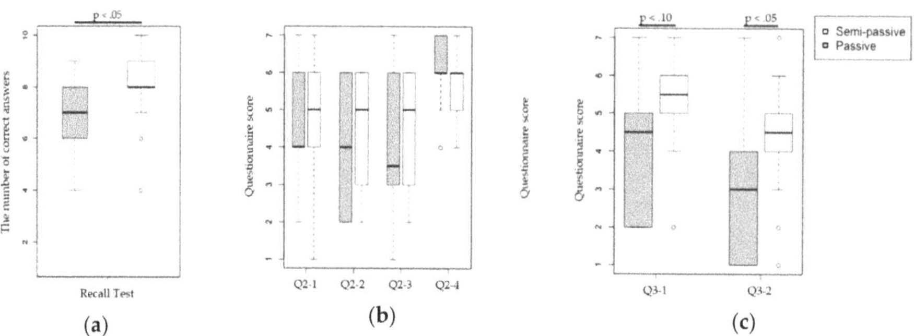

Figure 6. The result of (**a**) the recall test, and the (**b**) strength of subjective message, and (**c**) empathic concern questionnaires.

4. Discussion

During the analysis of the recall test, we found that the participants remembered more words of the conversation in the semi-passive condition than in the passive one, which supports the first hypothesis. In the experiment, the participants attended a 10-min conversation where the androids asked them something roughly once every minute in the semi-passive condition. The questions in the recall test were chosen so that the answer words appeared between these interaction blocks of asking and answering in the semi-passive condition. In other words, the questions were independent of what the androids asked. Therefore, the interaction block was considered to encourage participants to remember not only the words that appeared in the block, but also in the entire conversation.

Regarding participants' evaluation of the strength of message, we did not find any improvement in the semi-passive condition compared to the passive one, which does not support the second hypothesis. In other words, the semi-passive form of the conversation did not effectively contribute to how much the participants were convinced by the subjective message from the android. On the other hand, higher empathic concern toward the androids was observed in the semi-passive condition than in the passive one, supporting the third hypothesis. In other words, the participants were moved to follow the misery situation of the androids. These two potentially inconsistent results may imply that the subjective message from the androids in the semi-passive conversation was successfully conveyed not through being felt that they were strongly requested but rather through spontaneously evoking the subject's empathic concern.

It is worth considering how the current result is limited by the fact that we tested only with one scenario including a subjective message from the androids. The remaining question is whether the improvement in recall performance is also maintained in a scenario without such a message, which should be beneficial for the general purpose of information conveyance. Therefore, a further experiment should use content with only objective information. On the other hand, even if it is limited to cases with subjective aspects, the results of this study should still be good news for information media because the development of affective applications have been considered [21]. However, the current result might be limited to the android robot, which is very human-like. Therefore, it is also worth investigating how much the information media should be human-like so that the effect of the semi-passive form of conversation can be utilized.

5. Conclusions

In this paper, we proposed efficient media to convey information based on what we call semi-passive conversation. To test the effect, a conversation system using two androids was implemented, which consisted of a scenario-based conversation system having a function of sometimes briefly listening to the user's agreement or disagreement to the conversation. Regarding objective information, the result of the recall test showed that the participants who attended the semi-passive conversation remembered more words in the conversation than those who attended the passive conversation which involved the same content except for the interaction blocks that checked the participants' agreement or disagreement. Regarding subjective information, the participants showed stronger agreement for the action of the androids, which was interpreted as indicating that spontaneously empathic concern was evoked more in the semi-passive conversation than in the passive one. Further experiments changing the subjective quality of the contents and human-likeness of the robots are important for future work.

Author Contributions: Conceptualization, T.N., Y.Y., K.O. and H.I.; methodology, T.N., Y.Y. and K.O.; software, T.N.; validation, T.N.; formal analysis, T.N.; investigation, T.N.; resources, Y.Y., K.O. and H.I.; data curation, T.N.; writing—original draft preparation, T.N.; writing—review and editing, Y.Y. and K.O.; visualization, T.N.; supervision, Y.Y. and K.O.; project administration, H.I.; funding acquisition, H.I.

Funding: This work was supported by the Japan Science and Technology Agency (JST) Exploratory Research for Advanced Technology: ERATO, grant number JPMJER1401, Japan.

Conflicts of Interest: The authors declare no conflict of interest.

References

1. Softbank, P. Available online: http://www.softbank.jp/en/robot/ (accessed on 15 June 2019).
2. Shiomi, M.; Kanda, T.; Ishiguro, H.; Hagita, N. Interactive Humanoid Robots for a Science Museum. In Proceedings of the 1st ACM SIGCHI/SIGART Conference on Human-Robot Interaction—HRI '06, Salt Lake City, UT, USA, 2–3 March 2006; ACM Press: New York, NY, USA, 2006; Volume 22, p. 305. [CrossRef]
3. Kanda, T.; Shiomi, M.; Miyashita, Z.; Ishiguro, H.; Hagita, N. An Affective Guide Robot in a Shopping Mall. In Proceedings of the 4th ACM/IEEE International Conference on Human Robot Interaction—HRI '09, La Jolla, CA, USA, 9–13 March 2009; ACM Press: New York, NY, USA, 2009; p. 173. [CrossRef]
4. Noma, M.; Saiwaki, N.; Itakura, S.; Ishiguro, H. Composition and Evaluation of the Humanlike Motions of an Android. In Proceedings of the 2006 6th IEEE-RAS International Conference on Humanoid Robots, Genoa, Italy, 4–6 December 2006; pp. 163–168. [CrossRef]
5. Sakamoto, D.; Kanda, T.; Ono, T.; Ishiguro, H.; Hagita, N. Android as a Telecommunication Medium with a Human-like Presence. In Proceedings of the ACM/IEEE International Conference on Human-Robot Interaction—HRI '07, Arlington, VA, USA, 10–12 March 2007; ACM Press: New York, NY, USA, 2007; p. 193. [CrossRef]
6. Koyama, N.; Tanaka, K.; Ogawa, K.; Ishiguro, H. Emotional or Social? In Proceedings of the 5th International Conference on Human Agent Interaction—HAI '17, Bielefeld, Germany, 17–20 October 2017; ACM Press: New York, NY, USA, 2007; pp. 203–211. [CrossRef]
7. Watanabe, M.; Ogawa, K.; Ishiguro, H. Can Androids Be Salespeople in the Real World? In Proceedings of the 33rd Annual ACM Conference Extended Abstracts on Human Factors in Computing Systems—CHI EA '15, Seoul, Korea, 18–23 April 2015; ACM Press: New York, NY, USA, 2007; Volume 2, pp. 781–788. [CrossRef]
8. Zue, V.; Seneff, S.; Glass, J.R.; Polifroni, J.; Pao, C.; Hazen, T.J.; Hetherington, L. JUPITER: A Telephone-Based Conversational Interface for Weather Information. *IEEE Trans. Speech Audio Process.* **2000**, *8*, 85–96. [CrossRef]
9. Komatani, K.; Ueno, S.; Kawahara, T.; Okuno, H.G. Flexible Guidance Generation Using User Model in Spoken Dialogue Systems. In Proceedings of the 41st Annual Meeting on Association for Computational Linguistics—ACL '03, Sapporo, Japan, 7–12 July 2003; Association for Computational Linguistics: Morristown, NJ, USA, 2003; Volume 1, pp. 256–263. [CrossRef]
10. Misu, T.; Kawahara, T. Speech-Based Interactive Information Guidance System Using Question-Answering Technique. In Proceedings of the 2007 IEEE International Conference on Acoustics, Speech and Signal Processing—ICASSP '07, Honolulu, HI, USA, 15–20 April 2007; pp. IV-145–IV-148. [CrossRef]

11. Shiomi, M.; Sakamoto, D.; Takayuki, K.; Ishi, C.T.; Ishiguro, H.; Hagita, N. A Semi-Autonomous Communication Robot. In Proceedings of the 3rd International Conference on Human Robot Interaction—HRI '08, Amsterdam, The Netherlands, 12–15 March 2008; ACM Press: New York, NY, USA, 2008; p. 303. [CrossRef]
12. Komatsu, T.; Yamada, S. Adaptation Gap Hypothesis: How Differences between Users' Expected and Perceived Agent Functions Affect Their Subjective Impression. *J. Syst. Cybern. Inform.* **2011**, *9*, 67–74.
13. Arimoto, T.; Yoshikawa, Y.; Ishiguro, H. Multiple-Robot Conversational Patterns for Concealing Incoherent Responses. *Int. J. Soc. Robot.* **2018**, *10*, 583–593. [CrossRef]
14. Hayashi, K.; Sakamoto, D.; Kanda, T.; Shiomi, M.; Koizumi, S.; Ishiguro, H.; Ogasawara, T.; Hagita, N. Humanoid Robots as a Passive-Social Medium. In Proceedings of the ACM/IEEE International Conference on Human-Robot Interaction—HRI '07, Arlington, VA, USA, 9–11 March 2007; ACM Press: New York, NY, USA, 2007; p. 137. [CrossRef]
15. Kanda, T.; Ishiguro, H.; Ono, T.; Imai, M.; Mase, K. Multi-Robot Cooperation for Human-Robot Communication. In Proceedings of the 11th IEEE International Workshop on Robot and Human Interactive Communication, Berlin, Germany, 27 September 2002; pp. 271–276. [CrossRef]
16. Ishi, C.T.; Liu, C.; Ishiguro, H.; Hagita, N. Evaluation of Formant-Based Lip Motion Generation in Tele-Operated Humanoid Robots. In Proceedings of the 2012 IEEE/RSJ International Conference on Intelligent Robots and Systems, Vilamoura, Portugal, 7–12 October 2012; pp. 2377–2382. [CrossRef]
17. Slothlib - Revision 77. Available online: http://svn.sourceforge.jp/svnroot/slothlib/ (accessed on 20 August 2019).
18. Juman—Kurohashi—Kawahara Lab. Available online: http://nlp.ist.i.kyoto-u.ac.jp/index.php?JUMAN (accessed on 20 August 2019).
19. Pedregosa, F.; Varoquaux, G.; Gramfort, A.; Michel, V.; Thirion, B.; Grisel, O.; Blondel, M.; Müller, A.; Nothman, J.; Louppe, G.; et al. Scikit—learn: Machine learning in Python. *J. Mach. Learn. Res.* **2011**, *12*, 2825–2830.
20. Gensim: Topic Modelling for Humans. Available online: https://radimrehurek.com/gensim/ (accessed on 20 August 2019).
21. Boccanfuso, L.; Barney, E.; Foster, C.; Ahn, Y.A.; Chawarska, K.; Scassellati, B.; Shic, F. Emotional Robot to Examine Different Play Patterns and Affective Responses of Children with and without ASD. In Proceedings of the 2016 11th ACM/IEEE International Conference on Human-Robot Interaction (HRI), Christchurch, New Zealand, 7–10 March 2016; Computer Science Yale University: New Haven, CT, USA, 2016; pp. 19–26. [CrossRef]

Article

Twin-Robot Dialogue System with Robustness against Speech Recognition Failure in Human-Robot Dialogue with Elderly People

Takamasa Iio [1,2,*], Yuichiro Yoshikawa [3,4], Mariko Chiba [5], Taichi Asami [5], Yoshinori Isoda [5] and Hiroshi Ishiguro [3,4]

1 Faculty of Engineering, Information and Systems, University of Tsukuba, Tsukuba 305-8573, Japan
2 JST PRESTO, Kawaguchi-shi 332-0012, Japan
3 Graduate School of Engineering Science, Osaka University, Toyonaka 560-0043, Japan; yoshikawa@irl.sys.es.osaka-u.ac.jp (Y.Y.); ishiguro@irl.sys.es.osaka-u.ac.jp (H.I.)
4 JST ERATO, Kawaguchi-shi 332-0012, Japan
5 NTT DOCOMO, Inc., Tokyo 100-6150, Japan; mariko.chiba.eh@nttdocomo.com (M.C.); taichi.asami.mr@nttdocomo.com (T.A.); isoday@nttdocomo.com (Y.I.)
* Correspondence: iio@iit.tsukuba.ac.jp; Tel.: +81-29-853-5115

Received: 23 December 2019; Accepted: 17 February 2020; Published: 23 February 2020

Featured Application: Dialogue robot system for elderly people who have less opportunities to talk to other people.

Abstract: As agents, social robots are expected to increase opportunities for dialogue with the elderly. However, it is difficult to sustain a dialogue with an elderly user because speech recognition frequently fails during the dialogue. Here, to overcome this problem, regardless of speech recognition failure, we developed a question–answer–response dialogue model. In this model, a robot took initiative in the dialogue by asking the user various questions. Moreover, to improve user experience during dialogue, we extended the model such that two robots could participate in the dialogue. Implementing these features, we conducted a field trial in a nursing home to evaluate the twin-robot dialogue system. The average word error rate of speech recognition was 0.778. Despite the frequently high number of errors, participants talked for 14 min in a dialogue with two robots and felt slightly strange during the dialogue. Although we found no significant difference between a dialogue with one robot and that with two robots, the effect size of the difference in the dialogue time with one robot and that with two robots was medium (Cohen's d = −0.519). The results suggested that the presence of two robots might likely encourage elderly people to sustain the talk. Our results will contribute to the design of social robots to engage in dialogues with the elderly.

Keywords: dialogue system; elderly people; two robots

1. Introduction

For elderly people, it is important to have opportunities to talk with someone daily. The act of talking with someone is a fundamental action for building social connectedness and reducing the feeling of social isolation. Social disconnectedness and perceived social isolation are likely to cause health risks [1–3], such as dementia [4,5], depression [6], and early death [7]. Therefore, it is important to increase opportunities for elderly people to have dialogues with other people.

However, increasing such opportunities are not as easy as they may seem. To see why, we considered a case in Japan, which is one of the countries with the highest rate of people aged 65 or over. Ninety percent of the elderly people who live with their families have dialogues every day in some ways including telephone calls and e-mails. However, only 54.3% of those who live alone do so [8]. To increase opportunities for those living alone to have a dialogue, it was desirable for their family members, friends, neighbors, and hired caregivers to support them. Nonetheless, it was not easy for them to continue spending much time with the elderly people. In other words, there are limitations in human resources to support elderly people.

Social robots, including computer agents, are expected to increase opportunities for the elderly to engage in dialogues. Various applications of social robots for elderly people, such as schedule management, cognitive games, physical exercise suggestions, and information-provision [9,10], have been proposed so far. Although these applications can be useful in maintaining the health of elderly users, the aim of the applications is not to sustain a dialogue. Studies of communication for elderly people, for example Erber [11], Caris-Verhallen et al. [12], and Grainger [13], are premised on the idea that dialogue is important for elderly people. Furthermore, based on this idea, interventions that attempt to motivate residents of nursing care homes to interact with each other have been proposed; for example, staff training to raise awareness and to encourage caregivers to stimulate residents to interact [14–16]. Like these studies, under the assumption that much of the beneficial features of human social interaction is carried by dialogue, it does make sense to investigate whether these beneficial features of dialogue can also be realized by a non-human social agent. The first step towards producing and investigating such an application is to develop a system that can sustain a dialogue with elderly people for some time.

However, it is quite hard for robots to sustain a dialogue with the elderly. Kopp et al. [17] pointed out that "elderly users often have selectively impaired abilities, e.g., for auditory perception, articulation, adapting to a recommended interaction style, adhering to a clean turn-taking structure, or comprehending content of high information density [18,19]". In particular, the difficulty of speech recognition in a dialogue with elderly people [18] is a critical issue in sustaining a coherent dialogue. In commonly used chat-bot systems, speech recognition failures basically because of nonsense responses and results in dialogue breakdown. It is unclear how a robot could sustain a coherent dialogue for a while under the situation where speech recognition would frequently fail.

Our goal is to develop a robot dialogue system that can sustain a coherent dialogue with elderly people for some time and also provide good user experience with the dialogues. To achieve this goal, we propose a question–answer–response dialogue model in which a robot takes the initiative in the dialogue by asking a user, various questions. Moreover, we propose an approach to extend the model such that two robots can participate in the dialogue. To evaluate how these features influence user's dialogue time and user's dialogue experience, we implement a dialogue system with two robots called the twin-robot dialogue system and conduct a field trial using the twin-robot dialogue system in a nursing home. We report the details of the trial and the results and finally, we discuss the implication of the results.

2. Related Work

Over the years, several categories of robot technologies have been proposed to support elderly people [9,10,20]. However, there are only two basic categories. One is robot technologies that physically support humans. These technologies include smart wheelchairs [21,22], prosthetic hands, and exoskeletons [23,24]. With these technologies, robots are considered as tools for extending the human body rather than as independent agents. The other is robot technologies that socially support humans. This category is further divided into two subcategories. One is robot technologies that support daily life tasks, such as medication and task schedule management [17,25,26], monitoring (fall detection) [27,28], household tasks [29,30], and shopping support [31]. The other subcategory is robot technologies that support health maintenance and psychological well-being improvement, such as pet-type robots (Paro) [32], conversational robots (including computer agents), AIBO [33,34], and NeCoRo [35]. Note that this categorization is formal. In fact, robot systems that belong to both categories have also been proposed. For example, there have been wheelchairs that talk to elderly people who ride them [22] and robots that chat while shopping [31].

In this study, we review past studies of robot technologies for supporting the elderly through conversations. A mobile robotic assistant, Pearl [25], provided a reminder of the elderly daily activities, such as to visit the toilet every three hours and also to take medication. Regarding Pearl, a caregiver for an elderly client had input his/her daily activities in advance, and Pearl reminded the person based on the schedule. MonAMI Reminder [26] was also a schedule management assistant that allowed users to register their own schedule. Due to the difficulty in speech recognition, the input to the agent was provided with a digital pen and paper. Reminders were output, by voice, through embodied agents on a device. These robots and the agent have not been evaluated from the viewpoint of quality of a conversation. Ryan was a conversation robot for elderly people with dementia and depression [36]. Ryan has been developed by DreamFace Technologies, LLC. This robot has a head projection system that displays an animated avatar onto a mask and can show an emotive and expressive face. The robot has a touch-screen interface on its torso, which can be used as music player, narrated photo album, and video player. Furthermore, the robot can play cognitive games with a user using the screen and remind the user of daily activities with simple chats. Abodollahi et al. [36] installed this robot in a room for elderly participants and asked them to live together for 4 to 6 weeks. As a result, the average one-turn conversation between a participant and the robot was 198 times per day, which is relatively large. However, this paper does not show specific results, such as examples of dialogues, duration of a dialogue, and accuracy of speech recognition. Therefore, the quality of a dialogue was quite unclear.

A virtual assistant, Billy ('Billie'), to accompany and guide the elderly throughout the day was developed [17,37]. Billy basically performs schedule management like the MonAMI Reminder. However, with Billy, all inputs and outputs can be done through spoken-dialogue and natural confirmation signals like nodding or non-lexical cues. Furthermore, Billy can provide suggestions for leisure activities depending on the user. In their study, they took into account that elderly people seldom speak clearly, they cannot understand high-density dialogue, and they cannot perform turn-taking well. In view of this, they developed a robust and reliable interaction design called social cooperative behavior for the schedule management system. Although field trials started, the specific results have not yet been reported. Their system was sophisticated in terms of schedule management. However, with regards to achieving a conversation for a long while, it appeared to have low conversation support.

For robots specialized in assisting physical or cognitive activities, Ifbot could provide some activity programs such as Japanese language quizzes, singing songs, mouth exercise, and arithmetic. Although they used speech recognition in the programs, they said; "Almost all participants may have been dissatisfied with the robot's speaking and voice recognition functions." [38]. Matilda is a human-like (in appearance and attributes; e.g., voice, expressions, gestures, emotions) assistive communication robot (service and companion) in nursing homes in Australia [39]. Although users' impression of Matilda was assessed through a field trial in the nursing homes (Australia), the dialogue between elderly participants and the robot was not mentioned. Minami et al. [40] developed a dialogue robot system that chats with elderly people watching TV. This robot could provide responses by extracting social media comments related to TV programs. In addition to this function, the robot improves users' dialogue experience such as backchannel and repetition. However, the study did not evaluate the system with elderly people, and it did not consider dialogue breakdown due to speech recognition failures. Otaki et al. [41] developed a robot system that supported the co-imagination method for elderly people. With this robot system, elderly people talk to each other with a specified theme while looking at photos taken in their early lives. This method is designed to train the cognitive functions, which especially decline with aging, at an early stage of dementia [42]. The robot acts as a moderator of the conversation. Because the robot was remote-controlled by an operator, it is unclear how successful robot moderation between elderly participants is under the situation of frequent speech recognition failures. Sakakibara et al. [43] proposed a system that dynamically generated dialogue scenarios for counseling patients with dementia. In the study, personalized conversations were generated using the history obtained in conversation and linked open data. However, they did not care about the situation of speech recognition failures and the system had not yet been evaluated. Jokinen proposed constructive dialogue models [44] and an architecture based on the model [45] for socially intelligent robots. A robot with this architecture can be aware of potentially interesting topics and of the user's attention, interest, and understanding through multimodal signals. Dialogue content is managed by topic tracking and anticipating possible continuations, calculated by coherence measures using the semantic distance between possible topics. With this architecture, Jokinen et al. [45] developed an application, using a robot assistant that instructed the human user on various task procedures related to elder care support services. However, the architecture is not applied to engage in dialogues with the elderly and is not concerned with speech recognition failures.

Through the above review, past studies of social assistive robots for elderly people appear to have paid little attention to speech recognition failure. They have not evaluated for how long dialogues with elderly participants had been continued. In contrast to the past studies, this paper focuses on developing and evaluating a robot dialogue system that can sustain a dialogue even in a situation where speech recognition frequently fails.

3. Question–Answer–Response Dialogue Model

We developed a question–answer–response dialogue model in which a robot takes the initiative in a dialogue by asking a user various questions in order to sustain the dialogue regardless of speech recognition failures. This model has two features:

- The model regards a dialogue as a transition process which comprises four states.
- Every time a state is transitioned to the next state, a system selects a suitable utterance (among a set of utterances defined in the next state) for the robot.

Here, we explain the details of the model.

3.1. States

The four states in a dialogue are as follows: question, answer, backchannel, and comment. Figure 1 shows part of a dialogue constructed from the four states. The question state means the robot asks the user a question. The answer state means the user answers the robot. The backchannel state means the robot provides a backchannel, such as "uh-huh", "I see", or "really?". The comment state means that the robot comments on the question or the answer from the user.

Figure 1. Dialogue constructed from the four states.

There are three reasons for adopting such a structure:

- The structure makes a dialogue robust against speech recognition failures. Although speech recognition may fail, a robot can sustain a coherent dialogue by providing an ambiguous response; independent of user answer. Consider the dialogue in Figure 1. The user might answer "yes", "no", or "so-so" instead of "I don't remember". However, the response of the robot sounds coherent in any of the three ways. It means that the robot can sustain a coherent dialogue even when the speech recognition result is doubtful.

- The structure makes it easy to control the direction of a dialogue. A dialogue with the elderly is basically non-task-oriented (i.e., chat). However, the dialogue often has some type of implicit purposes. For example, it would be expected to stimulate users' cognitive function abilities or to get users to pay attention to their daily lifestyle. To achieve such a purpose, a robot needs to mainly control the direction of the dialogue such that it can talk with the user about certain topics. Because the structure compels the robot to take the initiative in a dialogue, the robot would control the direction of the dialogue more easily than if it had been a free-style chat.
- The structure makes it easier to create a scenario of a dialogue. In the present natural language processing technology, it is still difficult to generate robot responses suitable for user utterances automatically. In particular, it is quite difficult to generate robot responses that facilitate dialogue with implicit purposes. In view of this, we had to create a scenario manually, consulting experts of such dialogues (i.e., caregivers). In general, creating a scenario from scratch is difficult. A scenario writer needs to consider many factors; for example, what kind of story the robot tells, how the story unfolds, whether each topic of the story remains coherent, when and how the robot asks, and for how long the robot should take the initiative. In contrast to the difficulty in considering such factors, using the proposed structure, a scenario writer needs to only create pairs of questions asked by a robot and responses to answers the user might give. This is relatively easier than creating scenarios from scratch.

3.2. Transition Rules

The state transition diagram is shown in Figure 2. Table 1 shows the transition rules for each state. The system transforms from one state to another when it receives some events. In this study, we defined three events: an event in which a robot completed an utterance (Robot Utterance Completion: RUC), an event in which the system recognized user utterance (User Utterance Recognition: UUR), and an event in which user utterance has not been recognized for a certain period of time (Timeout: TO).

The reason for separating the backchannel state from the comment state is to deal with a situation where a user remained mute. In general, when a user answered something, the robot should provide a backchannel like "I see" or "uh-huh" to show that the robot is listening to the user. However, when a user does not answer, the robot should not provide any backchannel because the backchannel would make the user feel strange. At that time, the system skips the backchannel state and transitions to the comment state as shown in the state transition diagram (Figure 2).

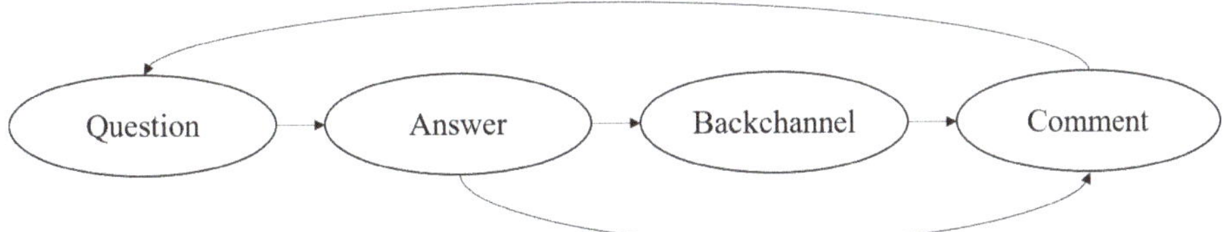

Figure 2. State transition diagram Of a dialogue with an elderly person.

Table 1. State transition table of the state transition diagram. The events are abbreviated as follows: RUC: Robot utterance completion, UUR: User utterance recognition, and TO: User utterance timeout.

State	RUC	UUR	TO
Question	Answer	-	-
Answer	-	Backchannel	Comment
Backchannel	Comment	-	-
Comment	Question	-	-

3.3. Utterance Selection in Each State

Every state (except for the answer state) has a set of objects for the robot utterances and an algorithm for selecting an object from the set.

The question state has a set of objects shown in the Listing 1. When the system transitions to the question state for the first time, it selects an object among the set randomly. After the second time, it selects an object with the same topic as the previous question among the set, except for objects that have already been selected. The reason for maintaining the same topic is that it would be unnatural to change topics with every question. However, to prevent the user from getting bored, the topic is changed every four times.

Listing 1: Objects of the question state.

```
[
  {
    Question: "Have you ever been abroad?",
    Topic: "travel"
  },
  {
    Question: "Where do you want to travel to?",
    Topic: "travel"
  },
  {
    Question: "What was your favorite food in childhood?",
    Topic: "childhood"
  },
  ...
]
```

The answer state has no set. The system selects no objects because it waits for the user to answer.

The backchannel state has a set of objects shown in the Listing 2. The question key is used to associate a backchannel with a question. The keyword, 'key', is a list of answers expected of the user in the answer state. The backchannel key is a sentence with a backchannel. The algorithm for selecting a backchannel is as follows: First, the system selects objects having the same question key as the previous question and it searches whether any of keywords of each object is in the speech recognition results. If a keyword is in the speech recognition results, the object having that keyword is selected. In contrast, if there is no keyword in the speech recognition results, the system selected the object without keywords (i.e., default backchannel). For example, as shown in Figure 1, when the user answered "I don't remember." to the question; "Have you ever been abroad?", the second object of the Listing 2 is selected.

Listing 2: Objects in the backchannel state.

```
[
  {
    Question: "Have you ever been abroad?",
    Keyword: ["yes", "yeah"]
    Backchannel: "Cool!"
  },
  {
    Question: "Have you ever been abroad?",
    Backchannel: "I see."
  },
  {
    Question: "Where do you want to visit?",
    Keyword: ["kyoto", "tokyo", "osaka"]
    Backchannel: "Yeah."
  },
  ...
]
```

The comment state has the same set of objects as in the backchannel state (Listing 3). The selection algorithm is also the same as that of the backchannel state.

Listing 3: Objects in the comment state.

```
[
  {
    Question: "Have you ever been abroad?",
    Keyword: ["yes", "yeah"]
    Comment: "You are a global person. I also want to go someday."
  },
  {
    Question: "Have you ever been abroad?",
    Comment: "I would like to go abroad someday."
  },
  {
    Question: "Where do you want to visit?",
    Keyword: ["kyoto", "tokyo", "osaka"]
    Comment: "That's an exciting city. I want to go there too."
  },
  ...
]
```

4. Participation of Multiple Robots

Although the question–answer–response dialogue model is expected to be robust against speech recognition failures, the dialogue generated by the model might be tedious for users. This is because the dialogue system is a one-way model; the robot asks a question, the user answers it, and the robot responds to the answer. If this continues for a while, the user might feel bored and stop the dialogue early.

We let two robots participate in the dialogue to decrease the tediousness of the model. There have been reports of advantages of using multiple robots in a dialogue. For example, when multiple robots participate in a conversation, a user seems to become insensitive to unnaturalness about consistency in a dialogue [46,47]. In addition, the user tends to feel that he or she can talk easily with multiple robots [48,49] and the user is likely to experience eye contact with the robots [50]. Karatas et al. [51] developed a multi-agent system that interacts with a driver in a car and showed that using multiple agents reduces cognitive loads of the driver compared to using a single agent. Sakamoto et al. [52] conducted a field trial in which two robots provide information to passersby at the station and found that passersby were more likely to stop when two robots were talking than when a single robot was talking. Iio et al. [47] demonstrated that visitors in an exhibition hall tended to have a longer dialogue with multiple robots than a single one.

In this study, we defined the participation framework of a dialogue in which two robots and one user participated, according to Goffman [53]. The participation framework contains a speaker, an addressee, and a side-participant. The turn-taking system was implemented based on Sacks et al. [54]. The rules of turn-taking of each state in our system are described below.

In the question state, the rule of selecting a speaker differs between the first time and the second time or after. In the first time, either of the robots is selected as a speaker randomly. Another robot becomes a side-participant. After the first time, the speaker of the previous question is selected as a speaker. However, when the topic changes from the previous question, the side-participant of the previous question is selected as a speaker.

This approach is based on the concept of common ground [55] among the participants. When a speaker asks a question on a certain topic, the addressee and the side-participant would have common ground that the speaker appears to be interested in the topic. Such common ground would allow the speaker to continue asking questions on the same topic. Therefore, it is reasonable for the speaker of the previous question to continue questions. However, when the topic of a question changes, it is not easy to interpret the sudden topic shift on the common ground. Arimoto et al. [46] reported that the unnaturalness of the sudden topic shift would be alleviated by changing the speaker. Therefore, it is reasonable for the side-participant of the previous question to become a speaker when the topic changes from the previous question.

In the answer state, the user becomes a speaker. The speaker of the previous question would be regarded as an addressee from the viewpoint of the concept of adjacency pair [56].

In the backchannel state, the speaker of the previous question is selected as a speaker again. This is grounded on the concept of the sequence-closing third [57]. Since the backchannel state is regarded as post-expansion of the question–answer adjacency pair, it is reasonable that the speaker of the previous question, which were addressed in the previous answer state, become a speaker.

In the comment state, a speaker depends on a speech recognition result of the previous answer state. Although the speaker of the previous question is basically selected as a speaker, the side-participant of the previous question is selected when the result has a kind of negative expression (including "No", "Nothing", "Never", etc.) or timeout. When the side-participant is selected, the side-participant speaks to the speaker; in other words, the speaker is selected as an addressee. In this manner, the speaker can easily continue asking a question in the next question state. A user's answer with negative expressions might indicate low interest in the topic. Here, if the side-participant expresses interest in the topic by commenting on the previous question, it appears to be reasonable for the speaker to continue asking a question on the same topic from the viewpoint of common ground [55] because at least one participant shows the interest in the topic.

5. System

We developed a twin-robot dialogue system including the two features: the question–answer–response dialogue model and the participation of two robots in a dialogue. The hardware components of the system are shown in Figure 3 and the system architecture is shown in Figure 4.

A microphone array collects sounds. The sounds are integrated though noise a reduction process by a microphone array. The integrated sound is then sent to the automatic speech recognition module, which recognizes the user utterance. We used a cloud speech recognition service provided by NTT

Docomo. The service receives a voice and returns the voice recognition results, which are sent to the utterance selection module. According to the selection rules (see Section 3.3), the utterance selection module selects an utterance from the database. The selected utterance is sent to the robot controllers. The voice recognition results are also sent to the nodding generation module as a signal of user speech. The nodding generation module sends a nodding motion to the robot controllers. Nodding is a motion for expressing that the robot is listening to the user in the nodding generation module. This motion is always executed in the answer state whenever the system received a speech recognition result. The robot controllers interpret the utterance with motions and execute them. After the execution is completed, the completion signal is sent to the utterance selection module. The utterance selection module selects a next utterance. As such, the system repeats selecting and executing an utterance according to a speech recognition result and its own behavior execution.

A social-conversational robot developed by VSTONE, CommU, was used as the dialogue partner in our system. This robot is desktop sized at 304 mm high, 180 mm wide, and 131 mm deep, weighing 938 g. CommU has three degrees of freedom (df) for its waist, 3 df for its neck, and 2 df for each eyes. The robot has two LEDs in its cheeks. The robot controller was a software server, which received a command, such as "speak" or "nod". The robot controller controlled the robot according to a received ommand.

Figure 3. Hardware components of the system and the connections between them.

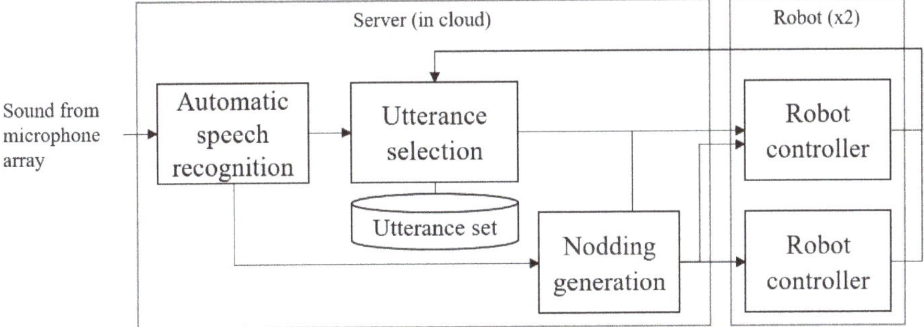

Figure 4. The system architecture.

6. Field Trial

6.1. Purpose

We conducted a field trial in a nursing home. There were two purposes for this field trial. The first was to investigate whether the twin-robot dialogue system can sustain a coherent dialogue with elderly people for a certain time and the second was to evaluate whether the system can provide good user experience of a dialogue.

6.2. Participants

Thirty elderly residents in a nursing home participated in the trial: 26 females and 4 males. They were native Japanese speakers. Their average age was 86.3 years (SD = 7.5). According to caregivers, 13 participants had no dementia, 4 had mild dementia, and the other 13 had advanced dementia.

The participants were recruited by caregivers of the nursing home. Before the trial, we explained the purpose and the procedure to the caregivers and asked them to recruit candidates who would like to participate in the trial. We sent an instruction document of the trial to their families and asked the families to fill out a consent form. The consent form had been approved by the ethical committee of Osaka University. The candidates whose families agreed that they should participate in the trial became the participants.

Furthermore, two caregivers participated in the trial to observe participants' behavior.

6.3. Scenarios

To achieve the purpose, it was desirable to design an experiment that could clarify two basic research questions: Whether or not the system has the question–answer–response dialogue model, and whether a single robot or two robots engage in a dialogue. However, without the question–answer–response dialogue model, it was obvious that the elderly could not continue a dialogue with the system. The reasons are as follows: the chatbot model commonly used in non-task-oriented chat systems generates responses based on the results of speech recognition. When speech recognition fails, the chatbot model generates a response that does not match the context of the dialogue. Since speech recognition frequently fails during a dialogue with elderly people, the system with the chatbot model would give unrelated responses in the dialogue in most cases. Therefore, we designed scenarios depending on whether only a single robot or two robots participate in a dialogue, which are as follows:

1. **One-robot scenario**. One robot participated in a dialogue. The robot performed tasks according to the question–answer–response dialogue model (see Section 3).
2. **Two-robot scenario**. Two robots participated in a dialogue. The robots performed tasks according to the question–answer–response dialogue model (see Section 3). They take turns according to the rules described in Section 4.

The field trial was a between-participant design. The participants were assigned to each scenario in such a way as to balance the dementia level of participants in of each scenario as shown in Table 2.

Table 2. Distribution of participants' cognitive capacities.

Cognitive Capacities	Scenario		Total
	One-Robot	Two-Robot	
Noting or mild	6	6	12
Severe	5	7	12
Total	11	13	24

6.4. Procedure

The procedure was as follows: A caregiver escorted a participant to a place of trial (Figure 5). The caregiver had the participant sit down on a chair in front of the robot. If the participant was using a wheelchair, the caregiver put the participant with the wheelchair in front of the robot. After escorting, the caregiver moved to a position behind the participant. Thus, the caregiver was invisible to the participant during the trial. Then, a controller greeted the participant and explained the task. The instruction was as follows: "This robot starts to talk to you in a little while. Please talk with it." After the instruction, the controller started the system and the robot started a dialogue. As the participants were native Japanese speakers, the field trial was conducted in the Japanese language. The procedure that a caregiver takes an elderly person to the robot, encourages him or her to talk with the robot, and watches him or her from behind would be reasonable at least in the phase of introducing the robot system.

The dialogue continued following the flowchart of Figure 6. The robot said the introduction first. Next, the robot started a dialogue. In every 5 min during the dialogue, the robot asked the participant whether to continue the dialogue or not. When the participant gave a positive answer, the robot continued the dialogue. Otherwise, a negative answer ended the dialogue.

Here, we should note an inappropriate case caused by speech recognition errors. If the robot recognized that the participant answered positively even though the participant actually answered negatively, the robot would have continued the dialogue. Because such a situation must be avoided, an experimenter force-quit a program for a dialogue as soon as possible.

Considering the burdens of a participant, we limited the dialogue time to 15 min even if the participant would like to continue. The dialogue was recorded by video cameras.

The caregiver had observed the dialogue and filled out in a questionnaire about the participant behavior. When the dialogues ended or were force-quit, the caregiver took the participant to a place away from the robot. Then, an interviewer asked the participant a simple question. After that, the caregiver took the participant back to his or her room, and then escorted the next participant to the place of trial.

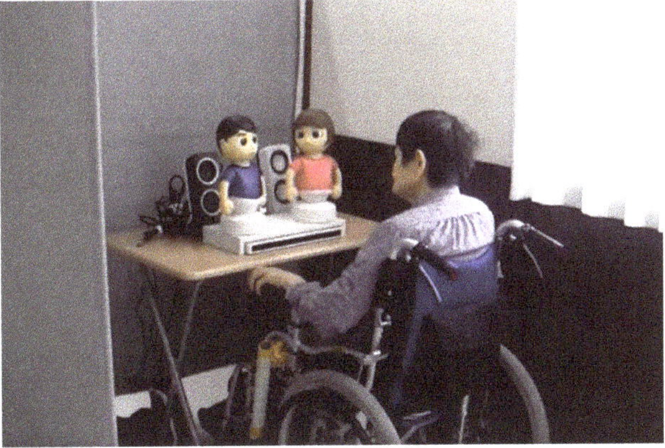

Figure 5. Twin-Robot dialogue system that talks with an elderly person.

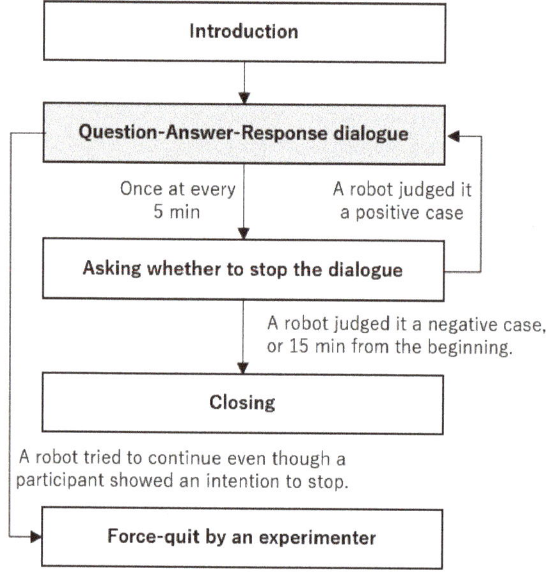

Figure 6. Flowchart of a dialogue in the trial.

6.5. Dialogue Contents

The number of questions we created was 55, which enabled a dialogue to run for approximately 27 min because it required approximately 30 s in the cycle for the robot to ask a question, receive an answer from the user, and respond to the answer. The details are shown in Table 3. The questions, backchannels, and comments were created by an expert in robot speech creation and elderly care.

Table 3. A part of the questions we prepared for the trial.

Topic	Number	Example
Childhood	16	"Where did you usually play?" "What toy did you want?" "Did you like to go to school?"
Travel	20	"Do you like travel?" "Where have you traveled so far?" "What was the best food you have had in travel?"
Health	19	"Have you had severe illness so far?" "Do you like walking?" "Do you have food you eat for health?"

6.6. Measurements

We measured the following values: word error rate (WER), dialogue time, user utterance time, and subjective impressions of the participant and caregiver.

6.6.1. Word Error Rate

WER is a typical metric of the accuracy of speech recognition [58]. In this experiment, the WER is for errors that occur when the robot recognizes the participant's speech. The WER was calculated as follows:

$$WER = (S + D + I)/N \tag{1}$$

where S, D, I, and N are the number of substitutions, deletions, insertions, and words in the reference, respectively. To compute the WER, we transcribed all participants utterances in the dialogue. The WER was used to confirm the difficulty of speech recognition in a dialogue with elderly people.

6.6.2. Dialogue Time

Dialogue time is the time from when the robot starts a dialogue until the robot ends the dialogue or a participant leaves the seat. Dialogue time was used to evaluate how long the twin-robot dialogue system can sustain a dialogue with elderly people.

6.6.3. Participant Utterance Time

Participant utterance time is the time that a participant spent speaking during a dialogue. By watching the video of the dialogue, we recorded the time when the participant started speaking and when he or she ended speaking. Participant utterance time was used to investigate whether participants participated positively during a dialogue.

6.6.4. Participant Subjective Impression

We asked a participant the following question: "Did you feel something strange in that dialogue with the robot?" The question was asked to elicit their viewpoint of the naturalness of the dialogue. Indeed, we needed to use formal psychological measures to obtain accurate results; however, this was quite challenging to accomplish for the elderly with dementia.

6.6.5. Caregiver Subjective Impression

We asked the caregivers the following question: "Did the participant talk with the robot more positively, comparing to when he or she talked with staffs." The question used a 7-point Likert scale, in which one means 'strongly disagree', and seven means 'strongly agree'. The question was asked in order to find out if the participant had been in the same or different state as usual.

7. Results

We obtained the videos and the questionnaire results of 24 participants and analyzed the data. Although there had been 30 participants altogether, one participant could not continue a dialogue because he was not able to hear the robot's voice at all, while the other five participants had halted the dialogue owing to technical problems (e.g., network trouble, program bugs). The two-robot scenario had 13 participants, and one-robot scenario had 11 participants.

We used the Mann–Whitney U test to compare data between the scenarios, and the alpha-level set at 0.05. We used a computer software 'jamovi' [59] for this test.

7.1. Word Error Rate

Figure 7 shows the averages of the WERs. The total average of the WERs was 0.778 (SD = 0.144). The average of the WERs of the one-robot scenario was 0.789 (SD = 0.123), and that of the two-robot scenario was 0.768 (SD = 0.163). There was no significant difference between the scenarios (U = 62.0, p = 0.608, Cohen's d = 0.148).

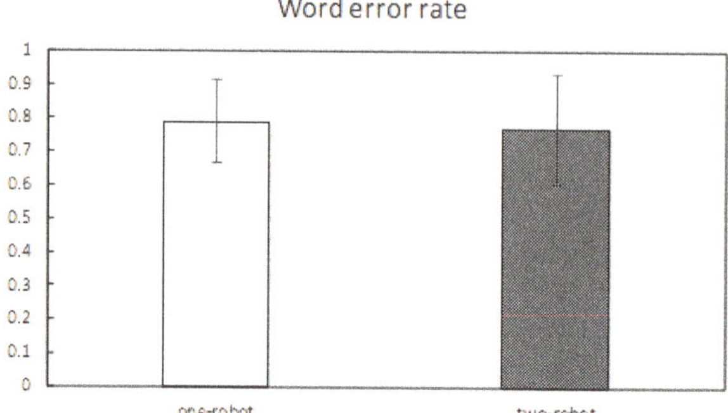

Figure 7. The averages of the word error rate (WER) in each scenario.

7.2. Dialogue Time

Figure 8 shows the averages of the dialogue time. The total average of the dialogue time was 12 min 51 s (SD = 4 min 52 s). The average of the dialogue time of one-robot scenario was 11 min 30 s (SD = 5 min 18 s), and that of two-robot scenario was 14 min (SD = 4 min 20 s). Here, milliseconds were round off. There was no significant difference between the scenarios (U = 48.0, $p = 0.188$, Cohen's d = −0.519).

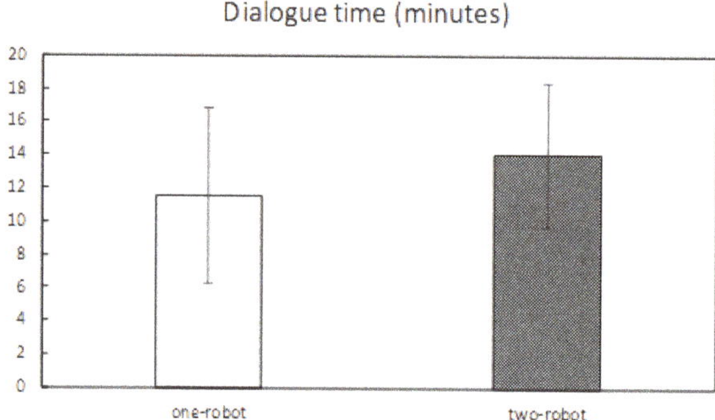

Figure 8. Averages of the dialogue time in each scenario.

7.3. Participant Utterance Time

Figure 9 shows the averages of the participant utterance time. The total average of the participant utterance time was 3 min 31 s (SD = 3 min 14 s). The average of the participant utterance time of one-robot scenario was 3 min 5 s (SD = 3 min 21 s), and that of two-robot scenario was 3 min 53 s (SD = 3 min 13 s). There was no significant difference between the scenarios (U = 61.0, $p = 0.569$, Cohen's d = −0.242).

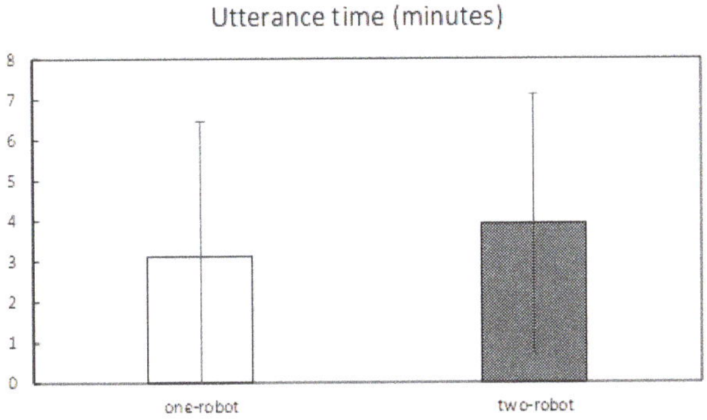

Figure 9. Averages of the user utterance time in each scenario.

7.4. Participant Subjective Impression

Figure 10 shows the results of the question to the participants, which was whether the participants have felt something strange in a dialogue. The numbers of the participants who answered "Yes", "No", and nothing were 3 (13%), 17 (71%), and 4 (17%), respectively in total. Those numbers were 1 (9%), 8 (73%), and 2 (18%) in one-robot scenario, and were 2 (15%), 9 (69%), and 2 (15%) in two-robot scenario, respectively.

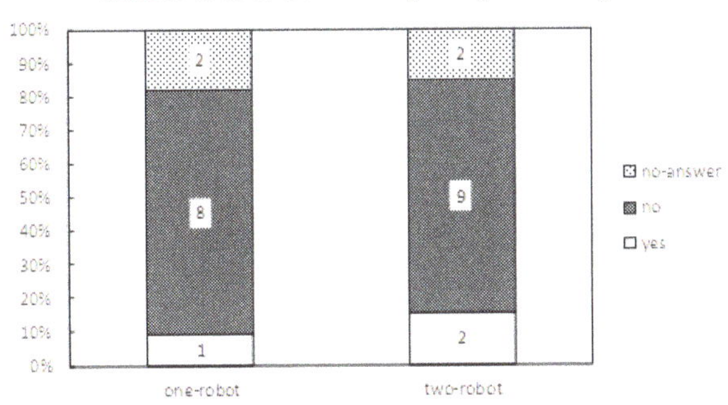

Figure 10. Results of the question to the participants in each scenario.

7.5. Caregiver Subjective Impression

Figure 11 shows the averages of the scores of the question to the caregivers, which is whether the participants had talked with the robot more positively than usual. The total average of the scores was 4.92 (SD = 1.89). The average of the scores of the one-robot scenario was 4.55 (SD = 1.86), and that of two-robot scenario was 5.23 (SD = 1.92). There was no significant difference between the scenarios (U = 61.5, p = 0.568, Cohen's d = −0.186).

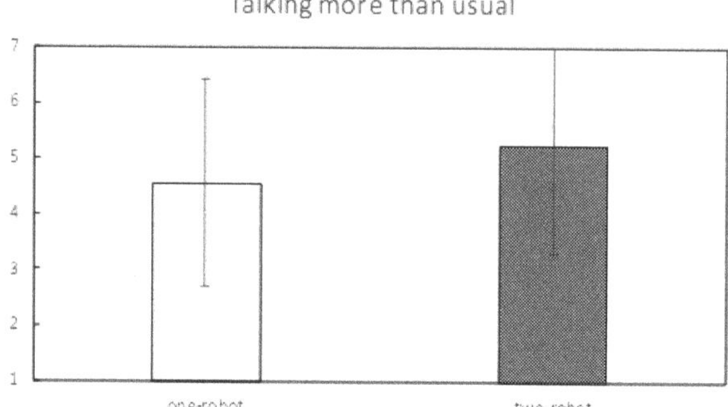

Figure 11. Averages of the scores of the question to the caregivers.

8. Discussion and Conclusions

8.1. Interpretation of the Results

The total average of the WERs was 0.778. This means that approximately 78% of the words in the utterances of the participants were mis-recognized. In general, it would be too difficult to continue a dialogue with this speech recognition accuracy. Despite the difficult situation, the system continued the dialogues for 12 min 51 s on average. This suggests that the twin-robot dialogue system could sustain a dialogue for a certain time regardless of speech recognition failures.

The average of the participant utterance time was 3 min 31 s, which was approximately 27% of the average dialogue time (cf. the average of the robot utterance times was approximately 5 min 51 s). In other words, the ratio of the participant utterance times to the robot utterance time was approximately 3:5. Because the gap between the utterance time of the participants and the robot was not so much, the participants can be considered to have positively participated in dialogue with the twin-robot dialogue system.

Regarding subjective impressions, 71% of participants answered that there was nothing strange in the dialogues with the robot. We believe that the system could have provided a dialogue without breakdown for many participants. In addition, the caregivers answered that they felt that the participants had been speaking more positively than usual. Because such positive participation might have involved a novelty effect that none of the participants has spoken to a robot before or an experimenter effect that the participants received special attention in the context of this experiment, we cannot justify whether the system was able to encourage some participants to participate more actively. To clarify the effect of the system on the positive participation, a long-term study is required.

In contrast, there was no significant difference between the one-robot scenario and two-robot scenario in each measurement. Therefore, it is still unclear if the use of two robots is effective in improving the user experience of dialogue. Nevertheless, regarding dialogue time, the effect size was medium (Cohen's $d = -0.519$). The results suggested that the presence of two robots might likely encourage elderly people to sustain the talk.

8.2. Effects on the Elderly with Dementia

The caregivers who observed the trial remarked, during an interview after the trial, that participants with dementia appeared to have really enjoyed the dialogue. To comprehend their opinion, we grouped the results of the following question posed to the caregivers, "Did the participant talk with the robot more positively, comparing to when he or she talk with staffs?" at the level of dementia (Figure 12). Although two-way ANOVA showed no interaction between groups and scenarios, the graph appears to suggest the trend that participants with severe dementia spoke more actively than usual, in accordance with the observation caregivers made during the trial and confirmed during the interview after the trial.

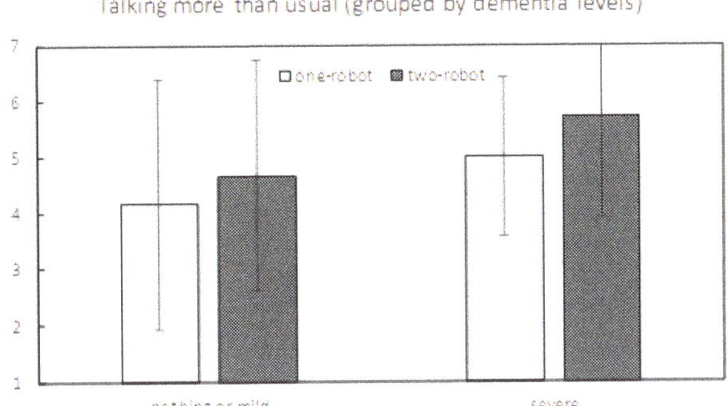

Figure 12. Averages of the scores of the question to the caregivers, grouped by dementia levels of the participants.

Furthermore, we grouped the participant utterance times precisely as we did for the caregiver impressions (Figure 13). This graph shows that some of the participants were talking for more than 5 min (i.e., Participants 2, 15, 18, 19, and 20). Especially, there were three participants with severe dementia in the two-robot scenario. Although we cannot conclude with certainty based on such small data, this result appears to provide a new hypothesis that some elderly people with severe dementia may actively speak when placed in a two-robot scenario.

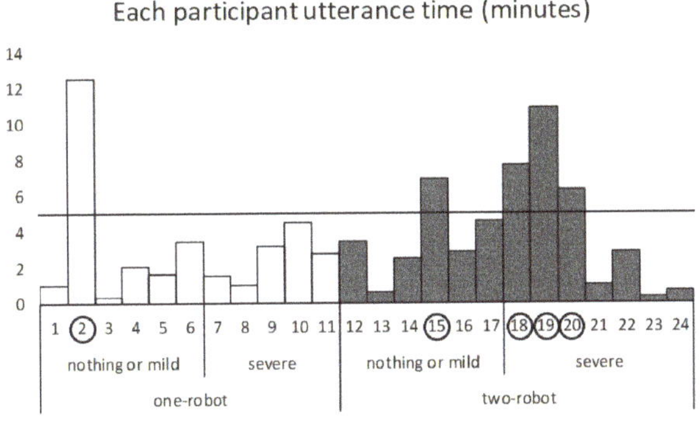

Figure 13. Utterance time of each participant grouped by dementia levels.

Further research are required to investigate whether using multiple robots is better for elderly people with severe dementia.

8.3. Influence of Topics on Participants' Utterance Time

In order to investigate whether the topic of the questions influence the verbal participation time of the participants, we calculated the mean of the utterance time of the participants in each topic. The results were as follows: In the one-robot scenario, the utterance time of the topic of travel, health, and childhood were 11.9 s (SD = 21.3), 6.8 s (SD = 5.1), and 14.7 s (SD = 27.4). In the two-robot scenario, the utterance time of the topic of travel, health, and childhood were 9.3 s (SD = 10.5), 9.2 s (SD = 11.5), and 12.0 s (SD = 12.7). We analyzed the results using two-way mixed ANOVA, which has scenario factor and topic factor. The results showed that there was no main effect in the scenario factor ($F(1,20) = 0.01$, $p = 0.920$), no main effect in the topic factor ($F(2,40) = 2.094$, $p = 0.136$), and no interaction between the two factors ($F(2,40) = 0.933$, $p = 0.402$). Therefore, it is unlikely that differences in topics had a systematic effect on utterance time.

Because the variance of the utterance time is very large, the influence of the topic of the question on the utterance time appeared to be considerably dependent on the individual. As an interesting example, we found that a topic stimulated participants' memory and the participants began to talk about their life for a long time. Specifically, participant 2 spent about 2 min talking about her initial visit to the nursing home when the robot asked, "Have you ever had a honeymoon?". Moreover, in response to the question, "What class did you like in elementary school?" she had talked about her childhood struggles for about 3 min. Participant 18 spent about 3 min talking about their experiences during World War II when asked "Have you lived around here since you were a child?". Their utterance times were quite long considering that usual answers of other participants were only one or two sentences. More surprisingly, even the caregivers, who have been interacting with the participants in their daily lives, did not know the stories of the participants until that time. These examples are interesting from the aspect of robots being potentially able to elicit a much deeper story from the elderly if the robot chooses topics adjusted to the individual.

8.4. Pros and Cons of Our System

We found several pros and cons of the twin-robot dialogue system through the field trial.

- Leading a dialogue by a robot

  Pros. Participants who have no topic to discuss might have easily taken part in a dialogue. In general, it is quite challenging for people to initiate a dialogue unless they have topics they would want to talk about. We found that many participants had no topic to discuss with the robot. By the robot initiating a dialogue, those participants could have participated in the dialogue without worrying about initiating it.
  Cons. Dialogue initiation by the robot may have frustrated some participants in case they had something they would have preferred to talk about with the robot.

- Patterning a dialogue

  Pros. Participants who are not good at communicating smoothly might have easily followed a dialogue because the user could have predicted the flow of the dialogue. This aspect should be important in dialogues for elderly people with declining cognitive ability.
  Cons. Participants who have no communication problems might have felt bored earlier during a dialogue if the dialogue was monotonous.

- Choosing robot responses by using keyword match of user answers

  Pros. This method was clearly robust against speech recognition failures. In our question–answer–response dialogue model, if the speech recognition result contains words of the keyword attribute, the backchannel (comment) associated with the keyword is selected. Otherwise, the backchannel (comment) associated with no keyword (i.e., the default backchannel (comment)) is selected. Therefore, when the speech recognition result is a broken sentence, the default backchannel (comment) is selected in most cases. Because the default backchannel (comment) is a sentence that is coherent for any answer, the dialogue was usually coherent even if the speech recognition fails.
  Cons. There are two situations for a dialogue to break down. First, there is the case where a user asks a robot a question while the robot is in the "answer mode". Here, the sentence associated with the default attribute is selected unless it was time for the user to be posing a question to the robot. Because the sentence of the default attribute is not meaningful to the

question, the dialogue would become unnatural. Second, there is the case where a keyword is matched owing to speech recognition failures, although this will rarely happen. For example, let us consider the following situation: a robot asks "Which countries do you want to travel, France or England?". Although a user answers France, the speech recognition result could be England. At this point, the dialogue would be strange because the robot would choose "England" as the response. To avoid this, we need more sophisticated algorithms.

8.5. Application of the System

The ability to talk with elderly people is becoming increasingly important for social robots because dialogues play an important role in building human–robot relationships. Social robots instruct elderly people to take medicines, exercise, undergo cognitive training, and suggest lifestyle improvements in order to sustain physical and mental wellbeing. Instructions in such situations may not work well if the relationship with elderly people—in other words, a sense of trust, security, and familiarity—is not built. Conversely, if the relationship between elderly people and social robots is well formed, the instruction will be more effective. Therefore, not only robots as companions but also a various other robots will need to have a certain level of dialogue with humans. The proposed twin-robot dialogue system would be useful from the viewpoint that they can sustain dialogues up to 12 min 51 s.

Author Contributions: Conceptualization, H.I. and Y.I.; methodology, Y.Y. and T.I.; software, M.C., T.A., and T.I.; validation, Y.Y. and T.A.; formal analysis, T.I. and M.C.; investigation, T.I., Y.Y., and M.C.; resources, Y.Y. and T.A.; data curation, T.I. and M.C.; writing—original draft preparation, T.I.; writing—review and editing, T.I., Y.Y., and T.A.; visualization, T.I.; supervision, H.I. and Y.I.; project administration, H.I. and Y.I.; funding acquisition, H.I. and Y.I. All authors have read and agreed to the published version of the manuscript.

Funding: This research was funded by NTT DOCOMO, INC., JST, PRESTO Grant Number JPMJPR1851, JST, ERATO Grant JPMJER1401 and JSPS KAKENHI Grant Number 19H05691.

Acknowledgments: We appreciate participants and caregivers who had participated in the field trial, and the facility manager that had admitted us to do the field trial.

Conflicts of Interest: Mariko Chiba, Taichi Asami and Yoshinori Isoda are working at NTT DOCOMO, INC., which is one of the funders of this study.

References

1. House, J.S.; Landis, K.R.; Umberson, D. Social relationships and health. *Science* **1988**, *241*, 540–545. [CrossRef] [PubMed]
2. Cornwell, E.Y.; Waite, L.J. Social disconnectedness, perceived isolation, and health among older adults. *J. Health Soc. Behav.* **2009**, *50*, 31–48. [CrossRef] [PubMed]
3. Holt-Lunstad, J.; Smith, T.B.; Baker, M.; Harris, T.; Stephenson, D. Loneliness and social isolation as risk factors for mortality: A meta-analytic review. *Perspect. Psychol. Sci.* **2015**, *10*, 227–237. [CrossRef] [PubMed]
4. Kotwal, A.A.; Kim, J.; Waite, L.; Dale, W. Social function and cognitive status: Results from a US nationally representative survey of older adults. *J. Gen. Intern. Med.* **2016**, *31*, 854–862. [CrossRef]
5. Poey, J.L.; Burr, J.A.; Roberts, J.S. Social connectedness, perceived isolation, and dementia: Does the social environment moderate the relationship between genetic risk and cognitive well-being? *Gerontologist* **2017**, *57*, 1031–1040. [CrossRef]
6. Heikkinen, R.L.; Kauppinen, M. Depressive symptoms in late life: A 10-year follow-up. *Arch. Gerontol. Geriatr.* **2004**, *38*, 239–250. [CrossRef]
7. Berkman, L.F.; Syme, S.L. Social networks, host resistance, and mortality: A nine-year follow-up study of Alameda County residents. *Am. J. Epidemiol.* **1979**, *109*, 186–204. [CrossRef]
8. Japan Cabinet Office: Annual Report on the Ageing Society: 2018 (Summary), <Viewpoint 2> Establishing a New Paradigm of Health through Science and Technology in Ageing Society. Available online: https://www8.cao.go.jp/kourei/english/annualreport/2018/2018pdf_e.html (accessed on 20 February 2020).
9. Broadbent, E.; Stafford, R.; MacDonald, B. Acceptance of healthcare robots for the older population: Review and future directions. *Int. J. Soc. Robot.* **2009**, *1*, 319. [CrossRef]

10. Kachouie, R.; Sedighadeli, S.; Khosla, R.; Chu, M.T. Socially assistive robots in elderly care: A mixed-method systematic literature review. *Int. J. Hum.-Comput. Interact.* **2014**, *30*, 369–393. [CrossRef]
11. Erber, N.P. Conversation as therapy for older adults in residential care: The case for intervention. *Int. J. Lang. Commun. Disord.* **1994**, *29*, 269–278. [CrossRef]
12. Caris-Verhallen, W.M.; Kerkstra, A.; Bensing, J.M. The role of communications in nursing care for elderly people: A review of the literature. *J. Adv. Nurs.* **1997**, *25*, 915–933. [CrossRef] [PubMed]
13. Grainger, K. Communication and the institutionalized elderly. In *Handbook of Communication and Aging Research*; Routledge: Abingdon, UK, 2004; pp. 479–497.
14. Allen, C.I.; Turner, P.S. The effect of an intervention programme on interactions on a continuing care ward for older people. *J. Adv. Nurs.* **1991**, *16*, 1172–1177. [CrossRef] [PubMed]
15. Allen-Burge, R.; Burgio, L.D.; Bourgeois, M.S.; Sims, R.; Nunnikhoven, J. Increasing communication among nursing home residents. *J. Clin. Geropsychol.* **2001**, *7*, 213–230. [CrossRef]
16. Nieuwenhuis, R. Breaking the speech barrier. *Nurs. Times* **1989**, *85*, 34. [PubMed]
17. Kopp, S.; Brandt, M.; Buschmeier, H.; Cyra, K.; Freigang, F.; Krämer, N.; Straßmann, C. Conversational Assistants for Elderly Users–The Importance of Socially Cooperative Dialogue. In Proceedings of the AAMAS Workshop on Intelligent Conversation Agents in Home and Geriatric Care Applications, Stockholm, Sweden, 15 July 2018.
18. Young, V.; Mihailidis, A. Difficulties in automatic speech recognition of dysarthric speakers and implications for speech-based applications used by the elderly: A literature review. *Assist. Technol.* **2010**, *22*, 99–112. [CrossRef]
19. Yaghoubzadeh, R.; Kramer, M.; Pitsch, K.; Kopp, S. Virtual agents as daily assistants for elderly or cognitively impaired people. In Proceedings of the International Workshop on Intelligent Virtual Agents, Edinburgh, UK, 29–31 August 2013; pp. 79–91.
20. Feil-Seifer, D.; Mataric, M.J. Defining socially assistive robotics. In Proceedings of the 9th International Conference on Rehabilitation Robotics—ICORR 2005, Chicago, IL, USA, 28 June–1 July2005; pp. 465–468.
21. Gomi, T.; Griffith, A. Developing intelligent wheelchairs for the handicapped. In *Assistive Technology and Artificial Intelligence*; Springer: Berlin/Heidelberg, Germany, 1998; pp. 150–178.
22. Shiomi, M.; Iio, T.; Kamei, K.; Sharma, C.; Hagita, N. Effectiveness of social behaviors for autonomous wheelchair robot to support elderly people in Japan. *PLoS ONE* **2015**, *10*, e0128031. [CrossRef] [PubMed]
23. Kazerooni, H.; Steger, R.; Huang, L. Hybrid control of the Berkeley lower extremity exoskeleton (BLEEX). *Int. J. Robot. Res.* **2006**, *25*, 561–573. [CrossRef]
24. Kiguchi, K.; Rahman, M.H.; Sasaki, M.; Teramoto, K. Development of a 3DOF mobile exoskeleton robot for human upper-limb motion assist. *Robot. Auton. Syst.* **2008**, *56*, 678–691. [CrossRef]
25. Pollack, M.E.; Brown, L.; Colbry, D.; Orosz, C.; Peintner, B.; Ramakrishnan, S.; Thrun, S. Pearl: A mobile robotic assistant for the elderly. In Proceedings of the AAAI Workshop on Automation as Eldercare, Edmonton, AB, Canada, 28–29 July 2002; Volume 2002, pp. 85–91.
26. Beskow, J.; Edlund, J.; Granström, B.; Gustafson, J.; Skantze, G.; Tobiasson, H. The MonAMI Reminder: A spoken dialogue system for face-to-face interaction. In Proceedings of the Tenth Annual Conference of the International Speech Communication Association, Brighton, UK, 6–10 September 2009.
27. Noury, N.; Rumeau, P.; Bourke, A.K.; ÓLaighin, G.; Lundy, J.E. A proposal for the classification and evaluation of fall detectors. *IRBM* **2008**, *29*, 340–349. [CrossRef]
28. Iio, T.; Shiomi, M.; Kamei, K.; Sharma, C.; Hagita, N. Social acceptance by senior citizens and caregivers of a fall detection system using range sensors in a nursing home. *Adv. Robot.* **2016**, *30*, 190–205. [CrossRef]
29. Dario, P.; Guglielmelli, E.; Laschi, C.; Teti, G. MOVAID: A personal robot in everyday life of disabled and elderly people. *Technol. Disabil.* **1999**, *10*, 77–93. [CrossRef]
30. Graf, B.; Hans, M.; Schraft, R.D. Mobile robot assistants. *IEEE Robot. Autom. Mag.* **2004**, *11*, 67–77. [CrossRef]
31. Iwamura, Y.; Shiomi, M.; Kanda, T.; Ishiguro, H.; Hagita, N. Do elderly people prefer a conversational humanoid as a shopping assistant partner in supermarkets? In Proceedings of the 6th International Conference on Human-Robot Interaction, Lausanne, Switzerland, 6–9 March 2011; pp. 449–456.
32. Shibata, T.; Wada, K. Robot therapy: A new approach for mental healthcare of the elderly–a mini-review. *Gerontology* **2011**, *57*, 378–386. [CrossRef] [PubMed]

33. Kanamori, M.; Suzuki, M.; Oshiro, H.; Tanaka, M.; Inoguchi, T.; Takasugi, H.; Yokoyama, T. Pilot study on improvement of quality of life among elderly using a pet-type robot. In Proceedings of the 2003 IEEE International Symposium on Computational Intelligence in Robotics and Automation. Computational Intelligence in Robotics and Automation for the New Millennium (Cat. No. 03EX694), Kobe, Japan, 16–20 July 2003; Volume 1, pp. 107–112.
34. Tamura, T.; Yonemitsu, S.; Itoh, A.; Oikawa, D.; Kawakami, A.; Higashi, Y.; Fujimoto, T.; Nakajima, K. Is an entertainment robot useful in the care of elderly people with severe dementia? *J. Gerontol. Ser. A Biol. Sci. Med. Sci.* **2004**, *59*, M83–M85. [CrossRef] [PubMed]
35. Libin, A.; Cohen-Mansfield, J. Therapeutic robocat for nursing home residents with dementia: Preliminary inquiry. *Am. J. Alzheimer's Dis. Other Dement.* **2004**, *19*, 111–116. [CrossRef] [PubMed]
36. Abdollahi, H.; Mollahosseini, A.; Lane, J.T.; Mahoor, M.H. A pilot study on using an intelligent life-like robot as a companion for elderly individuals with dementia and depression. In Proceedings of the 2017 IEEE-RAS 17th International Conference on Humanoid Robotics (Humanoids), Birmingham, UK, 15–17 November 2017; pp. 541–546.
37. Yaghoubzadeh, R.; Buschmeier, H.; Kopp, S. Socially cooperative behavior for artificial companions for elderly and cognitively impaired people. In Proceedings of the 1st International Symposium on Companion-Technology, Ulm, Germany, 23–25 September 2015.
38. Kanoh, M.; Oida, Y.; Nomura, Y.; Araki, A.; Konagaya, Y.; Ihara, K.; Kimura, K. Examination of practicability of communication robot-assisted activity program for elderly people. *J. Robot. Mechatron.* **2011**, *23*, 3. [CrossRef]
39. Khosla, R.; Chu, M.T.; Kachouie, R.; Yamada, K.; Yamaguchi, T. Embodying care in Matilda: An affective communication robot for the elderly in Australia. In Proceedings of the 2nd ACM SIGHIT International Health Informatics Symposium, Miami, FL, USA, 28–30 January 2012; pp. 295–304.
40. Minami, H.; Kawanami, H.; Kanbara, M.; Hagita, N. Chat robot coupling machine responses and social media comments for continuous conversation. In Proceedings of the 2016 IEEE International Conference on Multimedia and Expo Workshops (ICMEW), Seattle, WA, USA, 11–15 July 2016; pp. 1–6.
41. Otaki, H.; Otake, M. Interactive Robotic System Assisting Image Based Dialogue for the Purpose of Cognitive Training of Older Adults. In Proceedings of the 2017 AAAI Spring Symposium Series, Stanford, CA, USA, 27–29 March 2017.
42. Otake, M.; Kato, M.; Takagi, T.; Asama, H. The coimagination method and its evaluation via the conversation interactivity measuring method. In *Early Detection and Rehabilitation Technologies for Dementia: Neuroscience and Biomedical Applications*; IGI Global: Hershey, PA, USA, 2011; pp. 356–364.
43. Sakakibara, S.; Saiki, S.; Nakamura, M.; Yasuda, K. Generating personalized dialogue towards daily counseling system for home dementia care. In Proceedings of the International Conference on Digital Human Modeling and Applications in Health, Safety, Ergonomics and Risk Management, Vancouver, BC, Canada, 9–14 July 2017; pp. 161–172.
44. Jokinen, K. Dialogue models for socially intelligent robots. In Proceedings of the International Conference on Social Robotics, Qingdao, China, 28–30 November 2018; pp. 127–138.
45. Jokinen, K.; Nishimura, S.; Watanabe, K.; Nishimura, T. Human-robot dialogues for explaining activities. In Proceedings of the 9th International Workshop on Spoken Dialogue System Technology, Singapore, 18–20 April 2019; pp. 239–251.
46. Arimoto, T.; Yoshikawa, Y.; Ishiguro, H. Multiple-robot conversational patterns for concealing incoherent responses. *Int. J. Soc. Robot.* **2018**, *10*, 583–593. [CrossRef]
47. Iio, T.; Yoshikawa, Y.; Ishiguro, H. Retaining Human-Robots Conversation: Comparing Single Robot to Multiple Robots in a Real Event. *J. Adv. Comput. Intell. Intell. Inform.* **2017**, *21*, 675–685. [CrossRef]
48. Todo, Y.; Nishimura, R.; Yamamoto, K.; Nakagawa, S. Development and evaluation of spoken dialog systems with one or two agents through two domains. In Proceedings of the International Conference on Text, Speech and Dialogue, Pilsen, Czech Republic, 1–5 September 2013; pp. 185–192.
49. Shibahara, Y.; Yamamoto, K.; Nakagawa, S. Effect of sympathetic relation and unsympathetic relation in multi-agent spoken dialogue system. In Proceedings of the 2016 International Conference On Advanced Informatics: Concepts, Theory and Application (ICAICTA), George Town, Malaysia, 16–19 August 2016; pp. 1–6.

50. Arimoto, T.; Yoshikawa, Y.; Ishiguro, H. Nodding responses by collective proxy robots for enhancing social telepresence. In Proceedings of the Second International Conference on Human-Agent Interaction, Tsukuba, Japan, 28–30 October 2014; pp. 97–102.
51. Karatas, N.; Yoshikawa, S.; Okada, M. Namida: Sociable driving agents with multiparty conversation. In Proceedings of the Fourth International Conference on Human Agent Interaction, Singapore, 4–7 October 2016; pp. 35–42.
52. Sakamoto, D.; Hayashi, K.; Kanda, T.; Shiomi, M.; Koizumi, S.; Ishiguro, H.; Ogasawara, t.; Hagita, N. Humanoid robots as a broadcasting communication medium in open public spaces. *Int. J. Soc. Robot.* **2009**, *1*, 157–169. [CrossRef]
53. Goffman, E. *Forms of Talk*; University of Pennsylvania Press: Philadelphia, PA, USA, 1981.
54. Sacks, H.; Schegloff, E.A.; Jefferson, G. A simplest systematics for the organization of turn taking for conversation. In *Studies in the Organization of Conversational Interaction*; Academic Press: Cambridge, MA, USA, 1978; pp. 7–55.
55. Clark, H.H.; Carlson, T.B. Hearers and speech acts. *Language* **1982**, *58*, 332–373. [CrossRef]
56. Schegloff, E.A.; Sacks, H. Opening up closings. *Semiotica* **1973**, *8*, 289–327. [CrossRef]
57. Schegloff, E.A. *Sequence Organization in Interaction: A Primer in Conversation Analysis I (Vol. 1)*; Cambridge University Press: Cambridge, UK, 2007.
58. McCowan, I.A.; Moore, D.; Dines, J.; Gatica-Perez, D.; Flynn, M.; Wellner, P.; Bourlard, H. *On the Use of Information Retrieval Measures for Speech Recognition Evaluation (No. REP_WORK)*; IDIAP: Martigny, Switzerland, 2004.
59. The Jamovi Project. Jamovi (Version 0.9) [Computer Software]. 2019. Available online: https://www.jamovi.org (accessed on 20 February 2020).

MDPI
St. Alban-Anlage 66
4052 Basel
Switzerland
Tel. +41 61 683 77 34
Fax +41 61 302 89 18
www.mdpi.com

Applied Sciences Editorial Office
E-mail: applsci@mdpi.com
www.mdpi.com/journal/applsci

www.ingramcontent.com/pod-product-compliance
Lightning Source LLC
LaVergne TN
LVHW071608170726
843515LV00008B/2345
* 9 7 8 3 0 3 6 5 2 8 4 6 5 *